Springer Transactions in Civil and Environmental Engineerir

Editor-in-Chief

Springer Transactions in Civil and Environmental Engineering (STICEE) publishes the latest developments in Civil and Environmental Engineering. The intent is to cover all the main branches of Civil and Environmental Engineering, both theoretical and applied, including, but not limited to: Structural Mechanics, Steel Structures, Concrete Structures, Reinforced Cement Concrete, Civil Engineering Materials, Soil Mechanics, Ground Improvement, Geotechnical Engineering, Foundation Engineering, Earthquake Engineering, Structural Health and Monitoring, Water Resources Engineering, Engineering Hydrology, Solid Waste Engineering, Environmental Engineering, Wastewater Management, Transportation Engineering, Sustainable Civil Infrastructure, Fluid Mechanics, Pavement Engineering, Soil Dynamics, Rock Mechanics, Timber Engineering, Hazardous Waste Disposal Instrumentation and Monitoring, Construction Management, Civil Engineering Construction, Surveying and GIS Strength of Materials (Mechanics of Materials), Environmental Geotechnics, Concrete Engineering, Timber Structures.

Within the scopes of the series are monographs, professional books, graduate and undergraduate textbooks, edited volumes and handbooks devoted to the above subject areas.

More information about this series at http://www.springer.com/series/13593

T. G. Sitharam · Ravi Jakka ·
Sreevalsa Kolathayar
Editors

Latest Developments in Geotechnical Earthquake Engineering and Soil Dynamics

Editors
T. G. Sitharam
Indian Institute of Technology Guwahati
Guwahati, Assam, India

Ravi Jakka
Department of Earthquake Engineering
Indian Institute of Technology Roorkee
Roorkee, Uttarakhand, India

Sreevalsa Kolathayar
Department of Civil Engineering
National Institute of Technology Karnataka
Mangalore, Karnataka, India

ISSN 2363-7633 ISSN 2363-7641 (electronic)
Springer Transactions in Civil and Environmental Engineering
ISBN 978-981-16-1470-5 ISBN 978-981-16-1468-2 (eBook)
https://doi.org/10.1007/978-981-16-1468-2

This Springer imprint is published by the registered company Springer Nature Singapore Pte Ltd.
The registered company address is: 152 Beach Road, #21-01/04 Gateway East, Singapore 189721, Singapore

Preface

This book volume contains the state-of-the-art contributions from invited speakers of the 7th International Conference on Recent Advances in Geotechnical Earthquake Engineering and Soil Dynamics, 2021 (7ICRAGEE).

We thank all the staff of Springer for their full support and cooperation at all the stages of the publication of this book. We do hope that this book will be beneficial to students, researchers and professionals working in the field of Geotechnical Earthquake Engineering and Soil Dynamics. The comments and suggestions from the readers and users of this book are most welcome.

Guwahati, India
Roorkee, India
Suratkal, India

T. G. Sitharam
Ravi Jakka
Sreevalsa Kolathayar

Acknowledgements

We (editors) thank all the invited speakers of 7th International Conference on Recent Advances in Geotechnical Earthquake Engineering and Soil Dynamics, 2021 (7ICRAGEE), who have contributed articles to this book. We could bring this volume out in time only due to the invited authors' timely contribution and cooperation.The editors also thank and acknowledge the service of the anonymous reviewers for their valuable time and efforts.

Contents

1 **Single-Frequency Method for Computing Seismic Earth Pressures** 1
Scott J. Brandenberg, Maria Giovanna Durante, and Jonathan P. Stewart

2 **Three-Dimensional Centrifuge and Numerical Modeling of Underground Structures Subjected to Normal Faulting** 11
Charles Wang Wai Ng, Qipeng Cai, and Sina Baghbanrezvan

3 **Liquefaction Mitigation Measures: A Historical Review** 41
Ikuo Towhata

4 **Liquefaction-Induced Pile Downdrag from Full-Scale Testing** 87
Kyle Rollins and Cameron Lusvardi

5 **Cyclic Resistance and Large Deformation Characteristics of Sands Under Sloping Ground Conditions: Insights from Large-Strain Torsional Simple Shear Tests** 101
Gabriele Chiaro

6 **High-Speed Trains with Different Tracks on Layered Ground and Measures to Increase Critical Speed** 133
Amir M. Kaynia

7 **Numerical Simulation of Coir Geotextile Reinforced Soil Under Cyclic Loading** 151
Jayan S. Vinod, Abdullah Al-Rawabdeh, Ana Heitor, and Beena K. Sarojiniamma

8 **Assessing the Effect of Aging on Soil Liquefaction Resistance** 163
Ronald D. Andrus and Barnabas Bwambale

9 **Uncertainties in Small-Strain Damping Ratio Evaluation and Their Influence on Seismic Ground Response Analyses** 175
Sebastiano Foti, Mauro Aimar, and Andrea Ciancimino

10 Large Deformation Analysis of Coseismic Landslide Using Material Point Method 215
Gang Wang, Kewei Feng, and Duruo Huang

11 The State of Art on Equivalent State Theory for Silty Sands 225
Md Mizanur Rahman

12 Forensic Evaluation of Long-Distance Flow in Gently Sloped Ground During the 2018 Sulawesi Earthquake, Indonesia 247
Hemanta Hazarika, Divyesh Rohit, Takashi Kiyota, Mitsu Okamura, Siavash Manafi Khajeh Pasha, and Sukiman Nurdin

13 Empirical Predictions of Fourier Amplitude and Phase Spectra Including Local Site Effects for Simulation of Design Accelerograms in Western Himalayan Region 281
Ishwer Datt Gupta

14 Regional–Local Hybrid Seismic Hazard and Disaster Modeling of the Five Tectonic Province Ensemble Consisting of Westcentral Himalaya to Northeast India 307
Sankar Kumar Nath, Chitralekha Ghatak, Arnab Sengupta, Arpita Biswas, Jyothula Madan, and Anand Srivastava

15 Geosynthetics in Retaining Walls Subjected to Seismic Shaking . . . 359
G. Madhavi Latha, A. Murali Krishna , G. S. Manju, and P. Santhana Kumar

16 Studies on Modeling of Dynamic Compaction in a Geocentrifuge 373
B. V. S. Viswanadham and Saptarshi Kundu

17 A State of Art: Seismic Soil–Structure Interaction for Nuclear Power Plants 393
B. K. Maheshwari and Mohd. Firoj

18 Seismic Stability of Slopes Reinforced with Micropiles—A Numerical Study 411
Priyanka Ghosh, Surya Kumar Pandey, and S. Rajesh

19 Deformation Modulus Characteristics of Cyclically Loaded Granular Earth Bed for High-Speed Trains 423
Satyendra Mittal and Anoop Bhardwaj

20 Disturbance in Soil Structure Due to Post-cyclic Recompression 433
Ashish Juneja and A. K Mohammed Aslam

21 Application of Soft Computing in Geotechnical Earthquake Engineering 443
Pijush Samui

22 Resilient Behavior of Stabilized Reclaimed Bases 455
Sireesh Saride and Maheshbabu Jallu

23 Computing Seismic Displacements of Cantilever Retaining Wall Using Double Wedge Model 475
Prajakta R. Jadhav and Amit Prashant

24 Importance of Site-Specific Observations at Various Stages of Seismic Microzonation Practices 489
Abhishek Kumar

25 Influence of Bio- and Nano-materials on Dynamic Characterization of Soils 499
K. Rangaswamy, Geethu Thomas, and S. Smitha

26 Dynamic Characterization of Lunar Soil Simulant (LSS-ISAC-1) for Moonquake Analysis 513
Kasinathan Muthukkumaran, T. Prabu, and I. Venugopal

27 Dynamic Response of Monopile Supported Offshore Wind Turbine in Liquefied Soil 525
Sumanta Haldar and Sangeet Kumar Patra

28 Nonlinear Ground Response Analysis: A Case Study of Amingaon, North Guwahati, Assam 539
Arindam Dey, Shiv Shankar Kumar, and A. Murali Krishna

Editors and Contributors

About the Editors

Prof. T. G. Sitharam is currently the Director of Indian Institute of Technology Guwahati (IIT), India. He is a KSIIDC Chair Professor in the area of Energy and Mechanical Sciences and Senior Professor at the Department of Civil Engineering, Indian Institute of Science, Bengaluru (IISc). He was the founder Chairman of the Center for Infrastructure, Sustainable Transport and Urban Planning (CiSTUP) at IISc, and is presently the Chairman of the AICTE South Western Zonal Committee, Regional office at Bengaluru and Vice President of the Indian Society for Earthquake Technology (ISET). Professor Sitharam is the founder President of the International Association for Coastal Reservoir Research (IACRR). He has been a Visiting Professor at Yamaguchi University, Japan; University of Waterloo, Canada; University of Dolhousie, Halifax, Canada; and ISM Dhanbad, Jharkhand, and was a Research Scientist at the Center for Earth Sciences and Engineering, University of Texas at Austin, Texas, USA until 1994.

Prof. Ravi Jakka is working as Associate Professor in the Department of Earthquake Engineering, Indian Institute of Technology, Roorkee. He is also currently serving as Secretary, Indian Society of Earthquake Technology (ISET). He has graduated in Civil Engineering from Andhra University Engineering College in the year 2001. He has obtained masters and doctorate degrees from IIT Delhi in the years 2003 and 2007 respectively. His areas of interest are Dynamic Site Characterization, Soil Liquefaction, Seismic Slope Stability of Dams, Landslides, Foundations & Seismic Hazard Assessment. He has published over 100 articles in reputed international journals and conferences. He has supervised over 35 Masters Dissertations and 6 Ph.D. thesis, while he is currently guiding 10 Ph.D. thesis. He has received prestigious DAAD and National Doctoral fellowships. He has obtained University Gold Medal from Andhra University. He also received 'Young Geotechnical Engineer Best Paper Award' from Indian Geotechnical Society. He was instrumental in the development of Earthquake Early Warning System for northern India, a prestigious national project. He is also the Organizing Secretary to 7th International Conference on Recent Advances in Geotechnical Earthquake Engineering.

Prof. Sreevalsa Kolathayar pursued M.Tech. from Indian Institute of Technology (IIT) Kanpur, Ph.D. from Indian Institute of Science (IISc) and served as International Research Staff at UPC BarcelonaTech Spain. He is presently Assistant Professor in the Department of Civil Engineering, National Institute of Technology, Karnataka. Dr. Kolathayar has authored three books and over 65 research papers. His broad research areas are geotechnical earthquake engineering, geosynthetics & geonaturals, and water geotechnics. He is currently the Secretary of the Indian chapter of International Association for Coastal Reservoir Research (IACRR), and Executive Committee Member of Indian Society of Earthquake Technology. In 2017, The New Indian Express honored Dr. Kolathayar with 40 under 40—South India's Most Inspiring Young Teachers Award. He is the recipient of ISET DK Paul Research Award from Indian Society of Earthquake Technology, IIT Roorkee. He received "IEI Young Engineers Award" by The Institution of Engineers (India), in recognition of his contributions in the field of Civil Engineering.

Contributors

Mauro Aimar Politecnico di Torino, Turin, Italy

Abdullah Al-Rawabdeh School of Civil Mining and Environmental Engineering, University of Wollongong, Wollongong, NSW, Australia

Ronald D. Andrus Glenn Department of Civil Engineering, Clemson University, Clemson, SC, USA

S. Baghbanrezvan Hong Kong University of Science and Technology Clear Water Bay, Kowloon, Hong Kong

Anoop Bhardwaj Indian Institute of Technology (IIT), Roorkee, India

Arpita Biswas Department of Geology and Geophysics, Indian Institute of Technology Kharagpur, Kharagpur, India

Scott J. Brandenberg Department of Civil and Environmental Engineering, University of California, Los Angeles, CA, USA

Barnabas Bwambale ECS Southeast, LLP, Fayetteville, NC, USA

Q. P. Cai Hua Qiao University, Xiamen, China

Gabriele Chiaro University of Canterbury, Christchurch, New Zealand

Andrea Ciancimino Politecnico di Torino, Turin, Italy

Arindam Dey Indian Institute of Technology Guwahati, Guwahati, Assam, India

Maria Giovanna Durante Department of Civil, Architectural, and Environmental Engineering, University of Texas, Austin, TX, USA

Kewei Feng Hong Kong University of Science and Technology, Hong Kong SAR, China

Mohd. Firoj Department of Earthquake Engineering, IIT Roorkee, Roorkee, India

Sebastiano Foti Politecnico di Torino, Turin, Italy

Chitralekha Ghatak Department of Geology and Geophysics, Indian Institute of Technology Kharagpur, Kharagpur, India

Priyanka Ghosh Department of Civil Engineering, IIT Kanpur, Kanpur, India

Ishwer Datt Gupta Row House 04, Suncity, Anandnagar, Pune, India

Sumanta Haldar Indian Institute of Technology Bhubaneswar, Jatni, Bhubaneswar, India

Hemanta Hazarika Kyushu University, Fukuoka, Japan

Ana Heitor School of Civil Engineering, University of Leeds, Leeds, UK

Duruo Huang Tsinghua University, Beijing, China

Prajakta R. Jadhav Department of Civil Engineering, Indian Institute of Technology Gandhinagar, Ahmedabad, India

Maheshbabu Jallu Department of Civil Engineering, Indian Institute of Technology, Hyderabad, TS, India

Ashish Juneja Indian Institute of Technology Bombay, Powai, Mumbai, India

Amir M. Kaynia Norwegian Geotechnical Institute (NGI), Oslo, Norway; Norwegian University of Science and Technology (NTNU), Trondheim, Norway

Takashi Kiyota Tokyo University, Tokyo, Japan

Abhishek Kumar Indian Institute of Technology, Guwahati, India

Shiv Shankar Kumar National Institute of Patna, Patna, Bihar, India

Saptarshi Kundu Department of Civil Engineering, Indian Institute of Technology Bombay, Mumbai, India

Cameron Lusvardi Brigham Young University, Provo, UT, USA

Jyothula Madan Department of Geology and Geophysics, Indian Institute of Technology Kharagpur, Kharagpur, India

G. Madhavi Latha Indian Institute of Science, Bengaluru, India

B. K. Maheshwari Department of Earthquake Engineering, IIT Roorkee, Roorkee, India

G. S. Manju Indian Institute of Science, Bengaluru, India

Satyendra Mittal Indian Institute of Technology (IIT), Roorkee, India

A. K Mohammed Aslam Yonsei University, Seoul, South Korea

A. Murali Krishna Indian Institute of Science, Bengaluru, India;
Indian Institute of Technology Tirupati, Tirupati, Andhra Pradesh, India

Kasinathan Muthukkumaran Department of Civil Engineering, National Institute of Technology, Tiruchirapalli, Tamilnadu, India

Sankar Kumar Nath Department of Geology and Geophysics, Indian Institute of Technology Kharagpur, Kharagpur, India

C. W. W. Ng Hong Kong University of Science and Technology Clear Water Bay, Kowloon, Hong Kong

Sukiman Nurdin Tadulako University, Palu, Indonesia

Mitsu Okamura Ehime University, Ehime, Japan

Surya Kumar Pandey Department of Civil Engineering, IIT Kanpur, Kanpur, India

Siavash Manafi Khajeh Pasha IMAGEi Consultant, Tokyo, Japan

Sangeet Kumar Patra Indian Institute of Technology Bhubaneswar, Jatni, Bhubaneswar, India

T. Prabu Department of Civil Engineering, National Institute of Technology, Tiruchirapalli, Tamilnadu, India

Amit Prashant Civil Engineering, Indian Institute of Technology Gandhinagar, Ahmedabad, India

Md Mizanur Rahman University of South Australia, Adelaide, Australia

S. Rajesh Department of Civil Engineering, IIT Kanpur, Kanpur, India

K. Rangaswamy Department of Civil Engineering, NIT Calicut, Kozhikode, Kerala, India

Divyesh Rohit Kyushu University, Fukuoka, Japan

Kyle Rollins Brigham Young University, Provo, UT, USA

Pijush Samui Department of Civil Engineering, NIT Patna, Patna, Bihar, India

P. Santhana Kumar Indian Institute of Science, Bengaluru, India

Sireesh Saride Department of Civil Engineering, Indian Institute of Technology, Hyderabad, TS, India

Beena K. Sarojiniamma Division of Civil Engineering, School of Engineering, Cochin University of Science and Technology, Kochi, Kerala, India

Arnab Sengupta Department of Geology and Geophysics, Indian Institute of Technology Kharagpur, Kharagpur, India

S. Smitha Department of Civil Engineering, NIT Calicut, Kozhikode, Kerala, India

Anand Srivastava Department of Geology and Geophysics, Indian Institute of Technology Kharagpur, Kharagpur, India

Jonathan P. Stewart Department of Civil and Environmental Engineering, University of California, Los Angeles, CA, USA

Geethu Thomas Department of Civil Engineering, NIT Calicut, Kozhikode, Kerala, India

Ikuo Towhata Kanto Gakuin University, Yokohama, Kanagawa, Japan

I. Venugopal C&MG LEOS, U R Rao Satellite Centre, Indian Space Research Organization, Bengaluru, India

Jayan S. Vinod School of Civil Mining and Environmental Engineering, University of Wollongong, Wollongong, NSW, Australia

B. V. S. Viswanadham Department of Civil Engineering, Indian Institute of Technology Bombay, Mumbai, India

Gang Wang Hong Kong University of Science and Technology, Hong Kong SAR, China

Chapter 1
Single-Frequency Method for Computing Seismic Earth Pressures

Scott J. Brandenberg, Maria Giovanna Durante, and Jonathan P. Stewart

1.1 Introduction

1.1.1 Mononobe–Okabe Method

Seismic earth pressures on earth retention structures are often computed using the 'Mononobe–Okabe' (or M–O) method, which was originally formulated by Okabe (1926) and experimentally verified by Mononobe and Matsuo (1929). The M–O method assumes that horizontal, k_h, and vertical, k_v, body forces act on an active Coulomb-type wedge in frictional soil, which in turn results in an incremental change of the lateral earth pressure coefficient, K_E, over the static active earth pressure coefficient, K_A, as indicated in Fig. 1.1. Although the M–O method is the standard of practice and has been incorporated into numerous design documents (e.g., NCHPR 2008; BSSC 2020), it exhibits two possible numerical instabilities when k_h is large. Given that $\eta = \tan^{-1}[k_h/(1-k_v)]$, the first numerical instability is that the M–O solution is complex-valued when $\eta > \phi - \beta$ due to the $\sin(\phi - \eta - \beta)$ term. The second numerical instability is that the M–O solution is infinite when $\eta = \pi/2 - \delta - \theta$ due to the $\cos(\delta + \theta + \eta)$ term. These conditions are encountered frequently in high seismicity regions, where shaking intensity is anticipated to be strong.

S. J. Brandenberg (✉) · J. P. Stewart
Department of Civil and Environmental Engineering, University of California, Los Angeles, CA 90095-1593, USA
e-mail: sjbrandenberg@ucla.edu

J. P. Stewart
e-mail: jstewart@seas.ucla.edu

M. G. Durante
Department of Civil, Architectural, and Environmental Engineering, University of Texas, Austin, TX 78712-0273, USA
e-mail: mgdurante@utexas.edu

T. G. Sitharam et al. (eds.), *Latest Developments in Geotechnical Earthquake Engineering and Soil Dynamics*, Springer Transactions in Civil and Environmental Engineering, https://doi.org/10.1007/978-981-16-1468-2_1

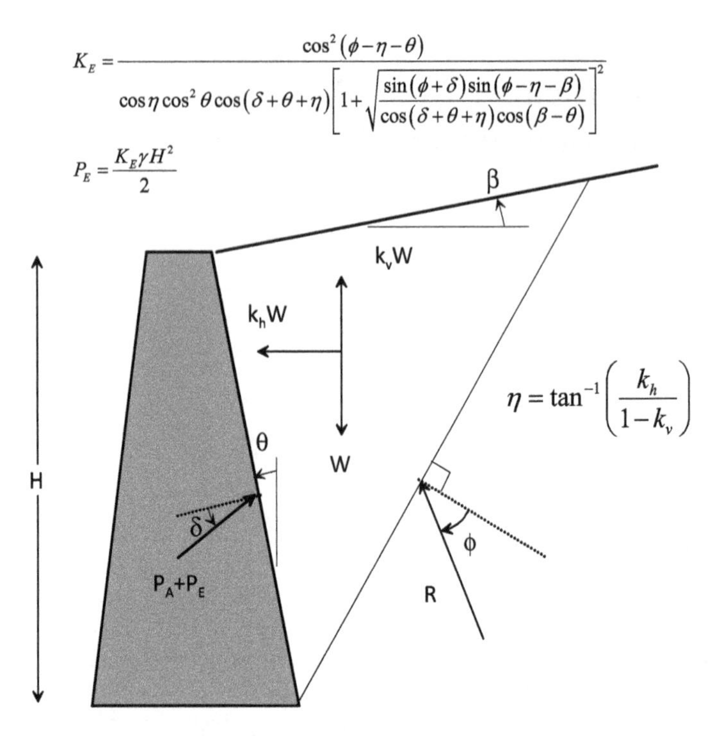

Fig. 1.1 Schematic of Mononobe–Okabe earth pressure solution

Seed and Whitman (1970) observed that the M–O solution for the seismic increment of active force is approximately linear with respect to k_h, with $K_E = 3/4k_h$ for values of k_h up to 0.4 g, which is a remarkable linear proportionality for a purely plastic solution. Because this equation is simple and stable, it is often used in lieu of the M–O solution, even when $k_h > 0.4$, which lies beyond the range intended by Seed and Whitman.

1.1.2 Elastodynamic Continuum Solutions

Elastodynamic continuum solutions (e.g., Wood 1973; Veletsos and Younan 1994a, b; Younan and Veletsos 2000; Vrettos et al. 2016) solve the equations of motion to derive earth pressure solutions for walls retaining elastic soil. To facilitate analytical solutions, these elastodynamic methods assume that the retained soil rests atop a rigid base, where the input ground motion is applied. These solutions tend to produce large earth pressures at the resonant frequencies of the retained soil because of the large soil displacements (relative to the base) that occur at those frequencies.

However, most retained soil deposits rest atop a compliant base rather than a rigid base, and the resonant frequencies associated with high earth pressures predicted by most elastodynamic solutions are frequently unrealistic.

1.1.3 *Elastodynamic Winkler Solution*

Brandenberg et al. (2015) developed an elastodynamic Winkler solution in which soil–structure interaction is modeled using Winkler stiffness intensity terms, and the ground motion at the surface of the retained soil is assumed to be known, as illustrated in Fig. 1.2. Using the surface motion, this method solves the problems arising from resonances associated with the rigid base condition. The Winkler solution also permits incorporation of wall flexibility and vertically inhomogeneous shear wave velocity profiles. Solutions were shown to provide reasonable agreement with measurements from a set of centrifuge modeling experiments and with continuum numerical simulations.

The method was formulated using a frequency-domain solution and a single-frequency solution. In the frequency-domain solution, the seismic excitation is input as a surface displacement time series. The Fourier transform of the motion is computed, and earth pressures are computed for each frequency component. A time series of the earth pressure resultant and bending moment at the base of the wall are then obtained using an inverse Fourier transform. In the single-frequency solution, the seismic excitation is modeled as a single displacement amplitude and associated frequency. Brandenberg et al. provided some basic preliminary guidance

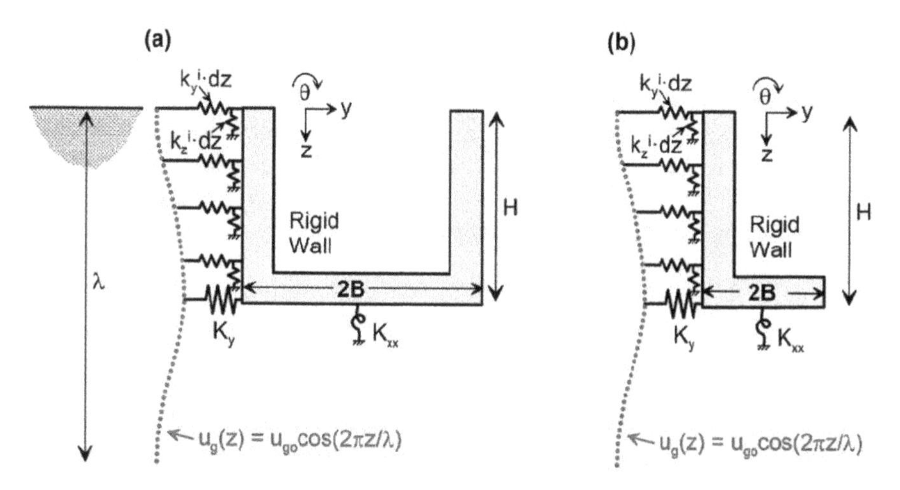

Fig. 1.2 Schematic of elastodynamic Winkler solution by Brandenberg et al. (Brandenberg et al. 2015) for **a** a U-shaped wall and **b** a free-standing retaining wall

on selecting the representative frequency and amplitude, but did not perform a detailed study. This paper provides a detailed study of the appropriate selection of surface displacement amplitude and frequency.

1.2 Equivalent Single-Frequency Solution Parameters

Expressions for equivalent surface motion amplitude and frequency for the single-frequency solution are derived here by performing frequency-domain solutions for a set of ground motions and selecting single-frequency parameters that match the maximum seismic earth pressure resultant obtained from the frequency-domain solution. The steps are as follows: (1) define the transfer function for earth pressure resultant for the frequency-domain solution for a cantilevered rigid wall retaining uniform elastic soil, (2) select a suite of earthquake ground motions, and compute the earth pressure resultant time series, P_{E}, using the frequency-domain solution, and (3) select input parameters for the single-frequency solution such that the earth pressure resultant, $P_{\mathrm{E\_SF}}$, matches that from the frequency-domain solution, $P_{\mathrm{E\_FD}}$.

1.2.1 Transfer Functions for Frequency-Domain Solution

The expression for the seismic earth pressure resultant for a rigid cantilevered wall retaining uniform elastic soil excited by vertically propagating shear waves is obtained by integrating earth pressure over the height of the wall, as indicated in Eq. 1.1, where H is the wall height, k_y^i is the Winkler stiffness intensity, u_{g0} is the surface displacement amplitude, $k = \omega/V_{\mathrm{S}}$ is the wave number, ω is angular frequency, and V_{S} is shear wave velocity. Note that $u_{g0}(\cos kz—\cos kH)$ defines the relative displacement between the wall and the free-field soil column. The solution for the time series of P_{E} is obtained by taking the Fourier transform of the surface displacement time series, solving Eq. 1.1 for each frequency component and subsequently taking the inverse Fourier transform to obtain the P_{E} time series.

$$P_{\mathrm{E}} = \int_0^H k_y^i u_{g0}(\cos kz - \cos kH)\mathrm{d}z = k_y^i u_{g0}\left(\frac{\sin kH}{k} - H\cos kH\right) \tag{1.1}$$

The expression for Winkler stiffness intensity developed by Kloukinas et al. (2012) is given by Eq. 1.2, where ν is the soil Poisson ratio, G is soil shear modulus, and $\lambda = 2\pi/k$ is wavelength.

$$k_y^i = \frac{\pi}{\sqrt{(1-\nu)(2-\nu)}}\frac{G}{H}\sqrt{1-\left(\frac{4H}{\lambda}\right)^2} \tag{1.2}$$

1.2.2 Earthquake Ground Motion Selection

A set of earthquake ground motions were selected from the PEER NGA West 2 ground motion database (Ancheta et al. 2014). We desired to select a set of ground motions that exhibited a distribution of mean period that matched the motions in the NGA West 2 database, where mean period, T_m, is defined by Rathje et al. (2004) in Eq. 1.3, where i is an index counter defining the number of frequencies in a Fourier transform, C is the amplitude of the Fourier coefficients, and f is frequency.

$$T_m = \frac{\sum_i C_i^2(1/f_i)}{\sum_i C_i^2}, \quad \begin{array}{l} 0.25\,\text{Hz} \le f_i \le 20\,\text{Hz} \\ \Delta f \le 0.05\,\text{Hz} \end{array} \tag{1.3}$$

Figure 1.3 illustrates the cumulative distribution of T_m for the entire NGA West 2 database (dashed line) and for the subset of 100 motions selected for this study (dots). The NGA West 2 database defines T_m for the RotD50 component of the ground motion (black dots in Fig. 1.3). We utilized the two horizontal components of the selected ground motions here, represented by blue and red dots in Fig. 1.3.

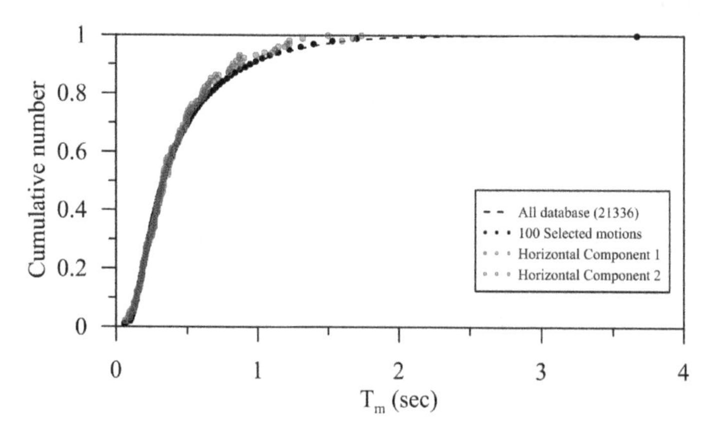

Fig. 1.3 Cumulative distribution functions for T_m from NGA West 2 database and 100 selected ground motions

1.2.3 Selection of Single-Frequency Parameters

After computing the peak values of P_E for the selected ground motions, we computed values using a single-frequency solution in which the input frequency was $1/T_m$ and the ground motion amplitude was $u_{g0} = PGV/\omega$, where PGV is the peak horizontal velocity. This procedure followed the preliminary recommendation by Brandenberg et al. (2015), where PGV/ω was a crude approximation for the appropriate surface displacement to use as an input based on the assumption that it corresponds to the most energetic portion of the ground motion. Figure 1.4 shows results of the residuals defined as $\ln(P_{E\_FD}) - \ln(P_{E\_SF})$. Residuals are near zero for $V_S T_m/H$ larger than about 10, but inconsistencies arise at lower values because the transfer function defined by Eq. 1.1 exhibits nodes where the normalized force amplitude becomes zero. When the value of $V_S T_m/H$ falls near one of these nodes in the single-frequency solution, a very low value of earth pressure is predicted. This is unrealistic because other frequencies contribute substantially to the problem, and selection of a single-frequency cannot capture the broadband nature of a real ground motion.

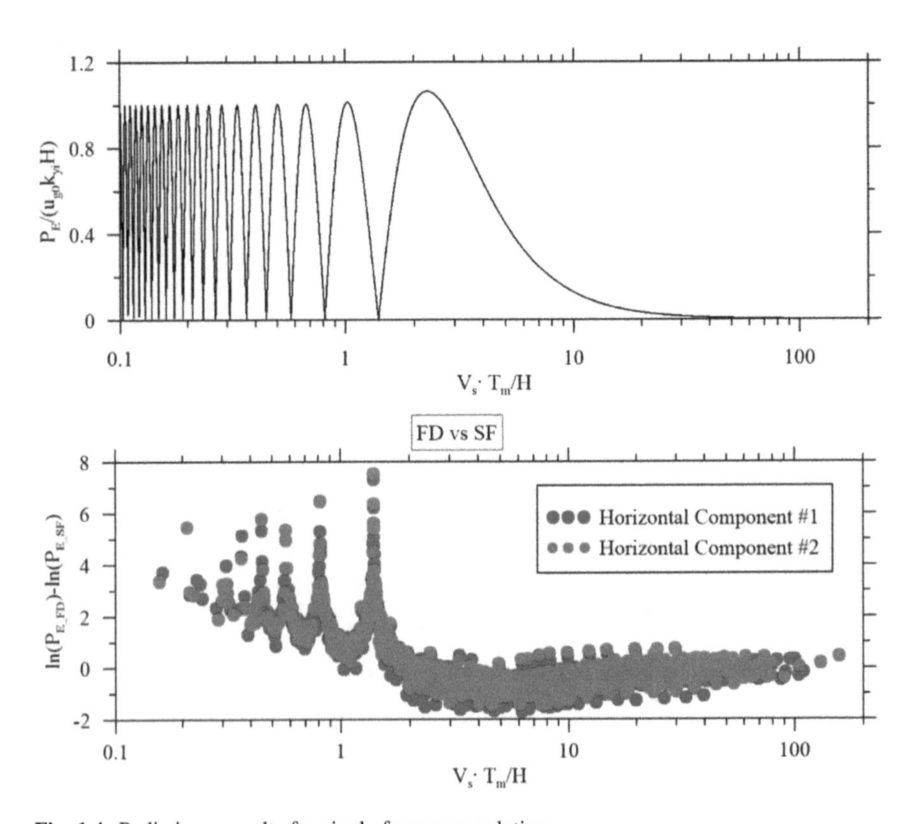

Fig. 1.4 Preliminary results for single-frequency solution

To overcome this problem, we define a simplified transfer function expression for P_E using the single-frequency solution in Eq. 1.4. The expression applies for the case of a wall on a rigid base. The intent of this expression is to recognize that the nodes arise from seismic waves with short wavelengths relative to the wall height, resulting in a complex pattern of compressive and extensional stress changes at the soil–wall interface. The solution in this high-frequency region becomes very sensitive to wavelength, and we believe it is prudent to select an envelope in this region that captures the peaks of the transfer function. This admittedly introduces some conservatism to the single-frequency solution compared with the frequency-domain solution.

$$\frac{P_{E\_SF}}{k_y^i H u_{g0}} = \min\left(1, \frac{\sin kH}{kH} - \cos kH\right) \tag{1.4}$$

Residuals utilizing Eq. 1.4 for the single-frequency solution are shown in Fig. 1.5. Utilizing Eq. 1.4 smoothed out the large spikes in the residuals apparent in Fig. 1.4 and reduced the value of the residuals in this high-frequency region. However, a trend persists that must be removed to obtain unbiased results.

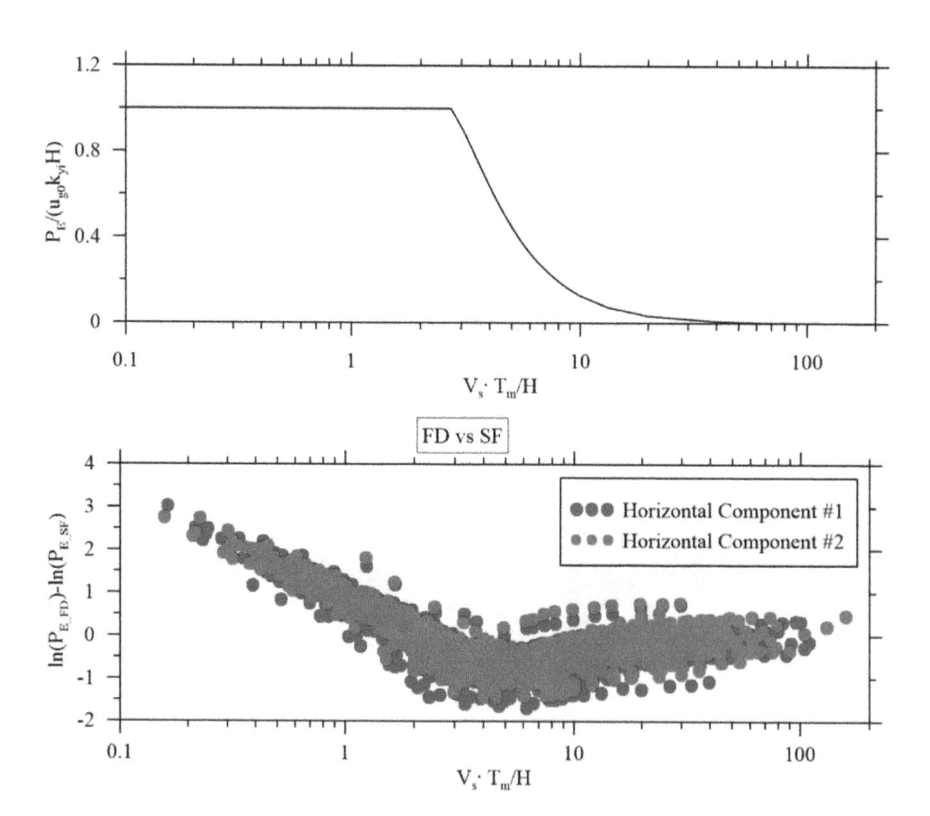

Fig. 1.5 Residuals using Eq. 1.6 for the single-frequency solution

The trend in Fig. 1.5 is removed using a combination of two approaches. First, we cap the frequency modifier on the Winkler stiffness intensity as indicated in Eq. 1.5. The frequency modifier term accounts for the effects of wave propagation in the retained soil, which reduces Winkler stiffness intensity. By adjusting the frequency modifier term, we are able to increase the Winkler stiffness intensity, thereby increasing the value of P_{E_SF}, thereby reducing residuals. This adjustment also prevents imaginary values of ξ_{freq} that arise at high frequency, thereby assuring real-valued outputs for P_{E_SF}.

$$k_y^i = \frac{\pi}{\sqrt{(1-\nu)(2-\nu)}}\frac{G}{H}\xi_{freq}, \xi_{freq} = \max\left(\sqrt{1-\left(\frac{4H}{T_m V_S}\right)^2}, 0.7\right) \tag{1.5}$$

Second, we utilize the relationship between u_{g0} and PGV provided in Eq. 1.6. This expression empirically removes the bias that remains after utilizing Eq. 1.5.

$$u_{g0} = \frac{PGV}{\omega}\xi_{ugo,SF}, \xi_{ugo,SF} = \begin{cases} 0.65 \text{ for } \frac{V_S T_m}{H} < 2.5 \\ 0.332 \log\frac{V_S T_m}{H} + 0.518 \text{ for } 2.5 \le \frac{V_S T_m}{H} \le 20 \\ 0.95 \text{ for } \frac{V_S T_m}{H} > 20 \end{cases} \tag{1.6}$$

Residuals utilizing Eqs. 1.4, 1.5 and 1.6 to compute the P_{E_SF} are shown in Fig. 1.6. The mean value of the residuals is −0.043, indicating a small conservative bias in the proposed single-frequency solution. The standard deviation of the residuals is 0.28. This standard deviation is not a quantification of error relative to measured values, but rather is an indication of the difference between the frequency-domain solution and the single-frequency solution.

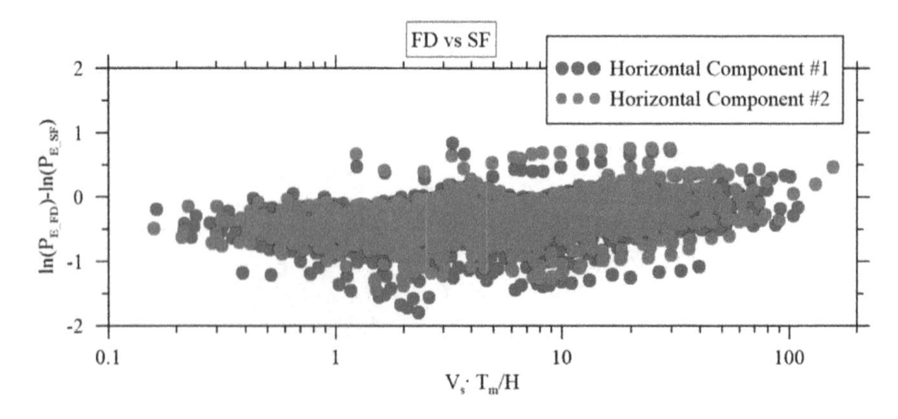

Fig. 1.6 Residuals using Eqs. 1.6, 1.7 and 1.8 for the single-frequency solution

1.3 Conclusions

This paper presented a calibration of a single-frequency solution to match frequency-domain solutions for seismic earth pressures acting on rigid cantilevered walls supporting uniform elastic soil. The results were obtained by selecting ground motions with a range of T_m that is characteristic of the NGA West 2 ground motion database, running frequency-domain solutions to obtain the peak seismic earth pressure resultant, selecting $1/T_m$ as the representative frequency in the single-frequency calculation and adjusting the values of u_{g0} and k_y^i to provide a close match between the frequency-domain and single-frequency solutions. The resulting procedure exhibits a slight conservative bias in that the predicted seismic earth pressure resultants are slightly larger than the frequency-domain solutions.

The various calibration equations utilized to produce agreement between the single-frequency and frequency-domain solutions do introduce simplifying assumptions that users should be aware of. The frequency-domain solution illustrated in Fig. 1.4 exhibits 'nodes' at frequencies where the positive pressure increments are perfectly balanced by negative pressure increments, resulting in zero net seismic force. This aspect of behavior occurs when wavelength is small relative to wall height, and earth pressures vary significantly due to small changes in frequency in this range. This behavior is not well-suited to a single-frequency solution, so the simplifying assumption was made to use the upper bound earth pressures in this region. This is likely the reason why the single-frequency solution produces a slight conservative bias and also likely contributes to dispersion in the errors introduced by the single-frequency solution relative to the frequency-domain solution.

Several additional factors not addressed herein also influence mobilization of seismic earth pressures. Those factors include (i) wall flexibility, (ii) vertical inhomogeneity of soil shear wave velocity, (iii) soil plasticity, (iv) plasticity in the wall structural elements, (v) formation of gaps at the soil–wall interface and (vi) development of excess pore pressure in the retained soil. These factors lie beyond the scope of this paper, though (i) and (ii) have been addressed in Brandenberg et al. (2017), Durante et al. (2018).

References

Ancheta TD, Darragh RB, Stewart JP, Seyhan E, Silva WJ, Chiou BS-J, Wooddell KE, Graves RW, Kottke AR, Boore DM, Kishida T, Donahue JL (2014) NGA-West2 database. Earthq Spectra 30:989–1005

Brandenberg SJ, Mylonakis G, Stewart JP (2015) Kinematic framework for evaluating seismic earth pressures on retaining walls. J Geotech Geoenviron Eng 141(7):04015031

Brandenberg SJ, Mylonakis G, Stewart JP (2017) Approximate solution for seismic earth pressures on rigid walls retaining inhomogeneous elastic soil. Soil Dyn Earthquake Eng 97:468–477

BSSC (2020) Seismic lateral earth pressures. In: NEHRP recommended provisions for seismic regulations for new buildings and other structures, part 3: resource papers. Building Seismic Safety Council, Federal Emergency Management Agency, Washington D.C. (in press)

Durante M, Brandenberg SJ, Stewart JP, Mylonakis G (2018) Winkler stiffness intensity for flexible walls retaining inhomogeneous soil. Geotech Earthquake Eng Soil Dyn V Numer Model Soil Struct Interact 473–482

Kloukinas P, Langoussis M, Mylonakis G (2012) Simple wave solution for seismic earth pressures on non-yielding walls. J Geotech Geoenviron Eng 138(12):1514–1519

Okabe S (1926) General theory of earth pressure and seismic stability of retaining wall and dam. J Jpn Soc Civil Eng 12(4):34–41

Mononobe N, Matsuo M (1929) On the determination of earth pressures during earthquakes. Proc World Engrg Congr 9:179–187

NCHRP (National Cooperative Highway Research Program) (2008) Seismic analysis and design of retaining walls, buried structures, slopes, and embankments. Rep. 611 (Anderson DG, Martin GR, Lam IP, Wang JN (eds)). National Academies, Washington D.C.

Rathje RM, Faraj F, Russell S, Bray JD (2004) Empirical relationships for frequency content parameters of earthquake ground motions. Earthq Spectra 20(1):119–144

Seed HB, Whitman RV (1970) Design of earth retaining structures for dynamic loads. In: Proceedings of ASCE specialty conference on lateral stresses in the ground and design of earth retaining structures, vol 1. Cornell University, Ithaca, NY, pp 103–147

Veletsos AS, Younan AH (1994a) Dynamic soil pressures on rigid retaining walls. Earthquake Eng Struct Dyn 23(3):275–301

Veletsos AS, Younan AH (1994b) Dynamic modeling and response of soil-wall systems. J Geotech Eng 120(12):2155–2179

Vrettos C, Beskos DE, Triantafyllidis T (2016) Seismic pressures on rigid cantilever walls retaining elastic continuously non-homogeneous soil: an exact solution. Soil Dyn Earthquake Eng 82:142–153

Wood JH (1973) Earthquake induced soil pressure on structures, Report No. EERL 73-05. California Institute of Technology, Pasadena, CA

Younan AH, Veletsos AS (2000) Dynamic response of flexible retaining walls. Earthquake Eng Struct Dyn 29:1815–1844

Chapter 2
Three-Dimensional Centrifuge and Numerical Modeling of Underground Structures Subjected to Normal Faulting

Charles Wang Wai Ng, Qipeng Cai, and Sina Baghbanrezvan

2.1 Introduction

Earthquake-induced fault rupture propagation in overlying soil has a significant impact on underground structures adjacent to the fault. Extensive damage to pile foundation and tunnel has been recorded in recent catastrophic earthquakes such as the 1999 Chi-chi and Kocaeli earthquakes and the 2008 Wenchuan earthquake (Wang et al. 2001; Dong et al. 2003; Anastasopoulos and Gazetas 2007; Faccioli et al. 2008; Li 2008; Wang et al. 2009). Seismic codes worldwide advise against construction in the vicinity of an active fault (EAK 2000; EC8 2002; GB50011-2010 2010), but it remains a challenge to determine a reasonable distance to the free-field rupture outcrop for the safe design of a pile foundation. Meanwhile, the safe distance to the free-field rupture outcrop is generally not applicable to tunnel design in a seismically active zone because faults have a large zone of influence on tunnels. Hence, it is critical to identify the potential damage zone to protect the tunnel.

For a soil layer overlying a bedrock fault, the dip angle of the fault rupture in the soil generally varies with distance to the ground surface (Bray et al. 1994). The fault rupture path is strongly influenced by a variety of factors such as the properties of the overlying soil, the type of bedrock fault, the magnitude of bedrock fault movement and even an underground structure (Cole and Lade 1984; Loukidis 2009). When there is an underground structure, the fault rupture path may be substantially modified, and the rupture may still propagate from the bedrock to the ground surface. The interaction between fault rupture in soil and a shallow foundation has recently attracted considerable attention (Anastasopoulos et al. 2007,

C. W. W. Ng (✉) · S. Baghbanrezvan
Hong Kong University of Science and Technology Clear Water Bay, Kowloon, Hong Kong
e-mail: cecwwng@ust.hk

Q. Cai
Hua Qiao University, No. 668, Jimei Avenue, Xiamen 361021, China

T. G. Sitharam et al. (eds.), *Latest Developments in Geotechnical Earthquake Engineering and Soil Dynamics*, Springer Transactions in Civil and Environmental Engineering, https://doi.org/10.1007/978-981-16-1468-2_2

2008, 2009; Ahmed and Bransby 2009; Ashtiani et al. 2015). A heavily loaded shallow foundation may divert the rupture completely away from the structure (Anastasopoulos et al. 2007). The development of such interaction mechanisms may be even more evident for pile foundations due to the load transfer to a deeper soil layer. Because of the complex interaction mechanism between a pile foundation and fault rupture, determination of the safe distance to the free-field rupture outcrop for piles remains a major challenge (Anastasopoulos et al. 2013; Cai and Ng 2016).

Field case studies following the 2008 Wenchuan earthquake in China revealed that fault movement caused more severe damage to tunnels than did the seismic waves (Li 2008; Wang et al. 2009). The tunnel lining had been sheared off and had even collapsed due to permanent ground deformation. Several model tests have been conducted to investigate the interaction between a tunnel and fault rupture propagation in soil (Baziar et al. 2014, 2019; Kiani et al. 2016; Cai et al. 2019). These studies have illustrated the importance of the proper estimation of forces in a tunnel lining induced by faulting as well as identifying a potential damage zone for the design of tunnels.

This keynote paper consists of two major parts. In the first part, three-dimensional centrifuge model test results of a single pile and a pile group (1×3) subjected to normal fault propagation in dry Toyoura sand are reported. The single pile was located on the footwall side of the bedrock fault line. For the pile group, three piles were connected with an elevated pile cap in which one of the piles located on the footwall side with the same distance as the single pile. A numerical parametric study using a strain-softening Mohr–Coulomb model in FLAC3D was conducted to investigate the effects of the distance from the pile foundation to the free-field fault rupture outcrop on the responses of the pile foundation. In addition, the distance to the free-field fault rupture outcrop was examined for both the single pile and the pile group. In the second part of the paper, the influence of normal faulting on a tunnel in sand is investigated through a series of three-dimensional centrifuge model tests and numerical back-analyses. This keynote paper summarizes and reinterprets published data from Cai and Ng (2016), Cai et al. (2017, 2019).

2.2 Problem Definition

Figure 2.1a shows a pile foundation in soil deposits subjected to underlying normal faulting. The horizontal distance from the furthest edge of the pile cap with a width of B (or single pile with a diameter of d) to the free-field fault rupture outcrop is defined by S. The horizontal distance of the soil at the ground surface from the free-field fault rupture outcrop is defined by X. The vertical component of the magnitude and the dip angle of the bedrock fault are h and α, respectively. Figure 2.1b shows a tunnel which is also subjected to underlying normal faulting.

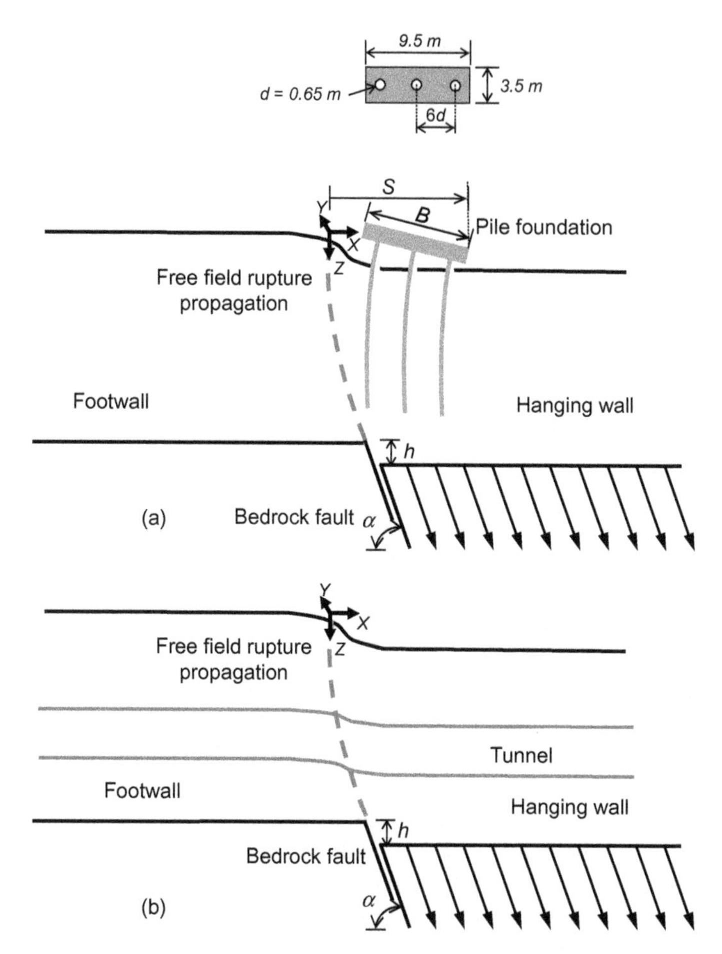

Fig. 2.1 Problem definition: interaction between fault rupture propagation and **a** pile foundation, **b** tunnel

2.3 Three-Dimensional Centrifuge and Numerical Modeling of Pile-Faulting and Tunnel-Faulting Interaction

2.3.1 Experimental Program and Setup

The centrifuge model tests reported in this keynote were conducted at the Hong Kong University of Science and Technology (HKUST). The 400 *g*-t geotechnical centrifuge at the HKUST is equipped with advanced simulation capabilities including the world's first in-flight biaxial (2D) shaker (Ng et al. 2004), an advanced four-axis robotic manipulator and a state-of-the-art data acquisition and

Table 2.1 Scaling laws relevant to centrifuge modeling

Quantity	Scaling law (model/prototype)
Length	*1/N*
Displacement	*1/N*
Stress	1
Strain	1
Density	1
Force	$1/N^2$
Bending moment	$1/N^3$
Axial rigidity (*EA*)	$1/N^2$
Flexural stiffness (*EI*)	$1/N^4$

control system (Ng et al. 2001; Ng et al. 2002; Ng 2014). This beam centrifuge with a diameter of 8.4 m is equipped with two swinging platforms, one for static tests and one for dynamic tests. All the tests reported in this paper were performed with an effective centrifugal acceleration of 50 g. The scaling laws relevant to this study are summarized in Table 2.1 (Taylor 1995).

Figure 2.2a shows the schematic diagram of the centrifuge model setup with a single pile in test SP. The dimensions of the model container were 350 × 1170 400 mm (width × length × height) in model scale. The shaft length and the distance to the bedrock fault line of the model pile were 300 mm and 91 mm, respectively. Thus, the single pile is located on the footwall. To form the 3 × 1 pile group (for test PG), two more piles were installed on the hanging wall side of the single pile (refer to Fig. 2.1a for details).

Figure 2.2b shows a schematic elevation view of the model container with a tunnel in test *U*. The sand layer was 500 mm thick and the tunnel crown was located 100 mm below the model surface. For the other test (test *M*), the tunnel crown was located 200 mm below the model surface. The hydraulic cylinder and the hanging wall base were installed at the bottom of the strongbox to simulate normal faulting. Driven by a hydraulic cylinder, the hanging wall block could move along a target direction at an angle of 70° with respect to the horizontal (Ng et al. 2012; Cai et al. 2013; Cai and Ng 2014). More details of the hydraulic cylinder and the unconstrained boundary container are given later in the paper.

2.3.2 *Model Pile and Model Tunnel*

Figure 2.3a shows a schematic elevation view of the single pile. The model piles were made from hollow square aluminum tubes with a width of 10.0 mm and a thickness of 0.7 mm. The pile shaft was coated with a layer of epoxy 1.5 mm in thickness to protect the attached strain gauge and to provide a uniform pile–soil interface (Ng et al. 2013, 2014, 2015, 2017; Ng and Lu 2014). The final width of the square section of the model pile was 13 m (0.65 m in prototype when tested at

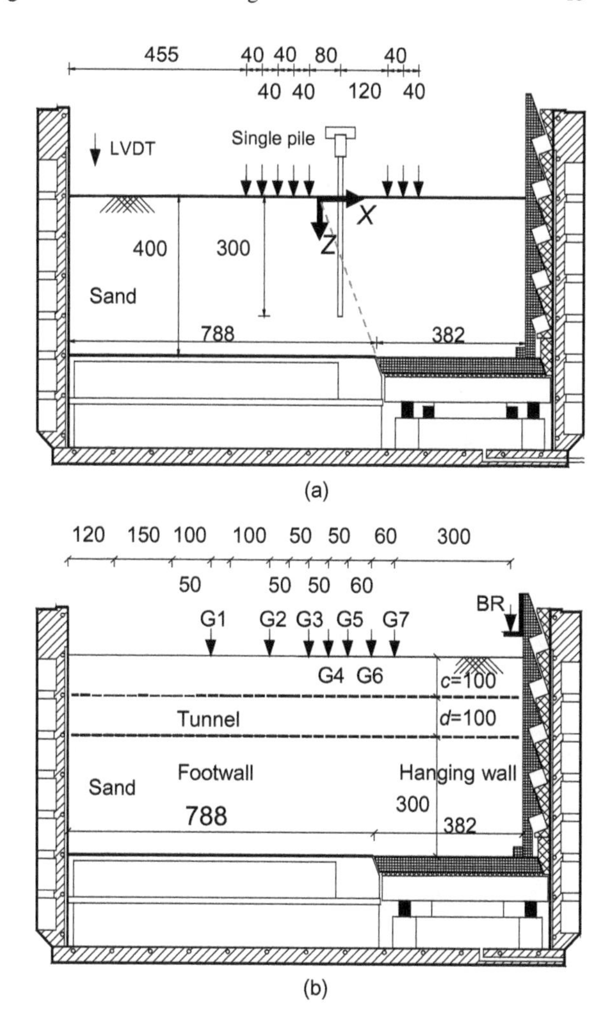

Fig. 2.2 Cross-section of the experimental apparatus for: **a** the single pile test; and **b** the tunnel test

50 g). For the 3 × 1 pile group, three single piles were firmly fixed to a relatively rigid pile cap with a center-to-center spacing of 78 mm (3.90 m in prototype) (see Fig. 2.3b). The pile cap was elevated by 155 mm (7.75 m in prototype) and so the embedded depth of each pile was 300 mm (15.00 m in prototype). Given Young's modulus of aluminum alloy (=70 GPa) and the epoxy coating (=2 GPa), the bending stiffness of the model pile was 75.3 kN m^2 in model scale (4.71 × 108 kN m^2 in prototype).

Figure 2.3c shows a schematic elevation view of the model tunnel. The model tunnel was made from an aluminum alloy tube. The outer diameter (D) and the lining thickness were 100 and 3 mm, respectively, equivalent to 5,000 and 150 mm in prototype scale when tested at 50 g. The model tunnel was 1,150 mm long, equivalent to 57.5 m in prototype. The scaling law for the flexural stiffness of the whole model tunnel is $1/N^4$. By assuming Young's modulus (E_c) of 33 GPa (ACI

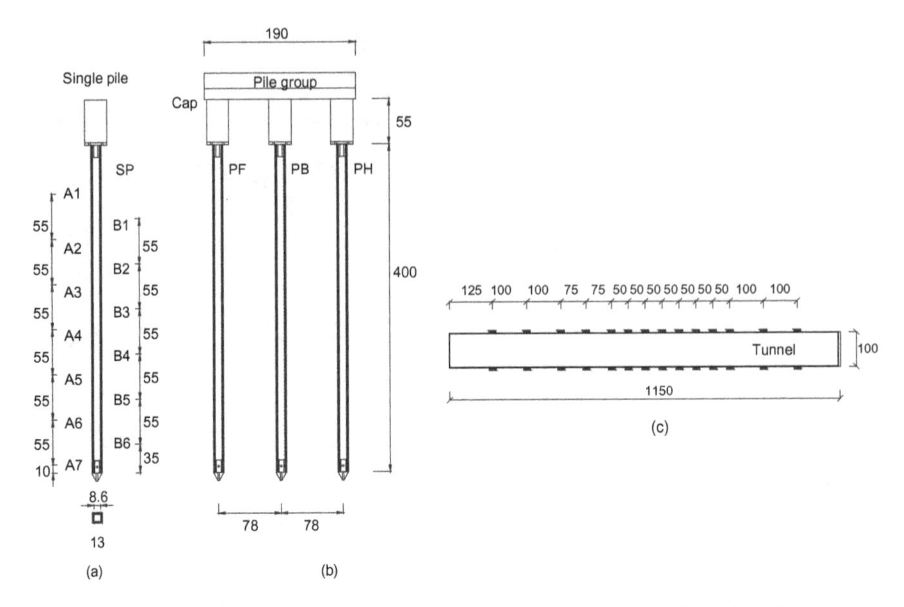

Fig. 2.3 **a** Model single pile; **b** model pile group; and **c** model tunnel (dimensions in mm)

2011), the tunnel lining thickness is equivalent to 360 mm in prototype scale in the longitudinal direction of the tunnel. The two ends of the tunnel were not connected to the model box and no additional fixity was imposed.

2.3.3 Model Preparation

For simplicity, dry Toyoura sand was used in all the tests reported. Toyoura sand is a uniform fine sand having a mean grain size (D_{50}) of 0.17 mm, a maximum void ratio of 0.977, a minimum void ratio of 0.597, a specific gravity of 2.65 and an angle of friction at the critical state ϕ'_{cv} of 31° (Ishihara 1993).

Figure 2.4 shows the prepared model package for the centrifuge tests. The pluvial deposition method was adopted to prepare all the centrifuge models. Dry Toyoura sand was rained onto the base of the model box from a height of 500 mm above the sand bed to give a medium-dense sand layer (with a relative density of 62–65%). For tests on the pile foundation (test SP and test PG), once the surface of the sand bed had reached the level where the pile toe should be, the pile was temporarily fixed in position before the sand deposition process was resumed. For tests on the tunnel (test *U* and test *M*), the model tunnel was laid flat on the sand bed once the surface of the sand bed had reached the level where the invert of the tunnel should be.

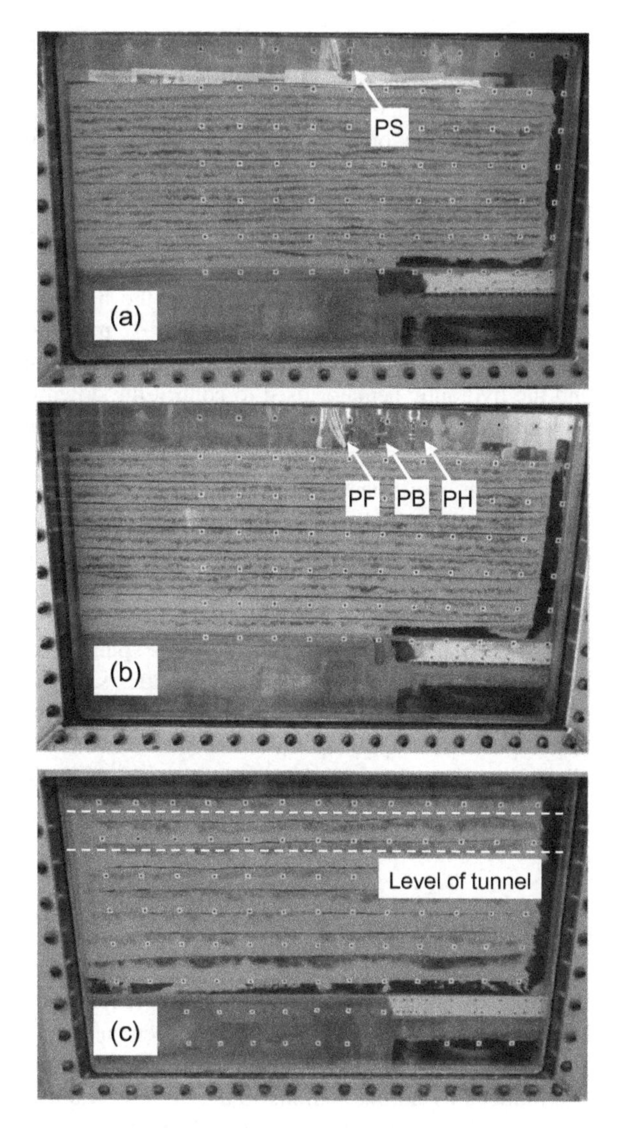

Fig. 2.4 Model package before the centrifuge test on **a** the single pile (test SP); **b** the pile group (test PG); and **c** the tunnel (test *U*)

2.3.4 *Instrumentation and Centrifuge Model Test Procedure*

After each model preparation, the model package was transferred to the centrifuge platform. Typical instrumentation layouts for the tests on the single pile (test SP) and the tunnel (test *U*) are shown in Fig. 2.2a, b, respectively. Linear variable differential transformers (LVDTs) were mounted on the model surface and at the top of the boundary wall on the hanging wall side to measure settlements of the ground surface and the bedrock hanging wall. Four digital cameras were installed to record soil deformation during faulting in-flight. The digital images were then analyzed using the Geo-PIV program developed by White et al. (2003).

A vertical load of 0.5 kN (1,250 kN in prototype when tested at 50 g) was applied to the pile top in the single pile test and each individual pile in the elevated pile group. The working load was chosen based on the ultimate capacity of the prototype pile, which was 0.81 kN (2,036 kN in prototype) estimated as a summation of shaft resistance and end bearing capacity. The centrifuge was then spun up to 50 g and reached a steady state. Subsequently, normal fault movement was simulated by a downward movement of the hanging wall block in five steps of $h = 8$ mm, 16 mm, 26 mm, 36 mm and 42 mm (i.e., $h = 0.4$ m, 0.8 m, 1.3 m, 1.8 m and 2.1 m in prototype) Fig. 2.5 shows a cross section of the model container and the hydraulic actuator used to control the downward fault movement together with the unconstrained boundary. The effects of using a constrained and unconstrained boundary were reported by Cai et al. (2015).

The fault movement was controlled by draining off the oil in the hydraulic cylinder resulting in the downward movement of the platen. The vertical settlement of the platen was controlled by four platen guides around the cylinder. After spinning down the centrifuge, post-experiment observations were made and the failure patterns on the model surfaces were recorded.

For the tunnel tests, the centrifuge was spun up to an acceleration of 50 g and sufficient time was allowed for the transducers to stabilize. Subsequently, normal faulting was simulated in-flight with a fault movement of $h = 16$ mm (i.e., 0.8 m in prototype). After applying a normal fault movement, a sufficient time lapse was allowed until readings of transducers became stable. Ground surface settlement and the induced axial strains were measured during the normal faulting.

2.3.5 Numerical Back-Analysis of Centrifuge Tests

A series of three-dimensional finite-difference analyses were performed using FLAC3D to examine the interaction between an underground structure and faulting.

Figure 2.6a shows a typical numerical mesh for analyzing the failure of a single pile induced by normal faulting in the SP test. The mesh and pile configuration had the same dimensions in the numerical run as in the centrifuge test. The boundary conditions adopted in the finite-difference analysis were roller-support on the four vertical sides and pin-support at the base of the mesh. The soil was modeled using four-node tetrahedral elements. The pile was modeled by a ‘pile’ element in FLAC3D (Itasca 2000).

Dry Toyoura sand with a density of 1533 kg/m^3 was described using an elastoplastic constitutive model with a strain-softening Mohr–Coulomb failure criterion. The peak friction angle (ϕ_{peak}), critical-state friction angle (ϕ_c), dilation angle (ψ), elasticity modulus (E) and Poisson ratio (υ) of the sand were taken to be 38°, 31°, 10°, 36 MPa and 0.2, respectively (Cai et al. 2019). The post-peak

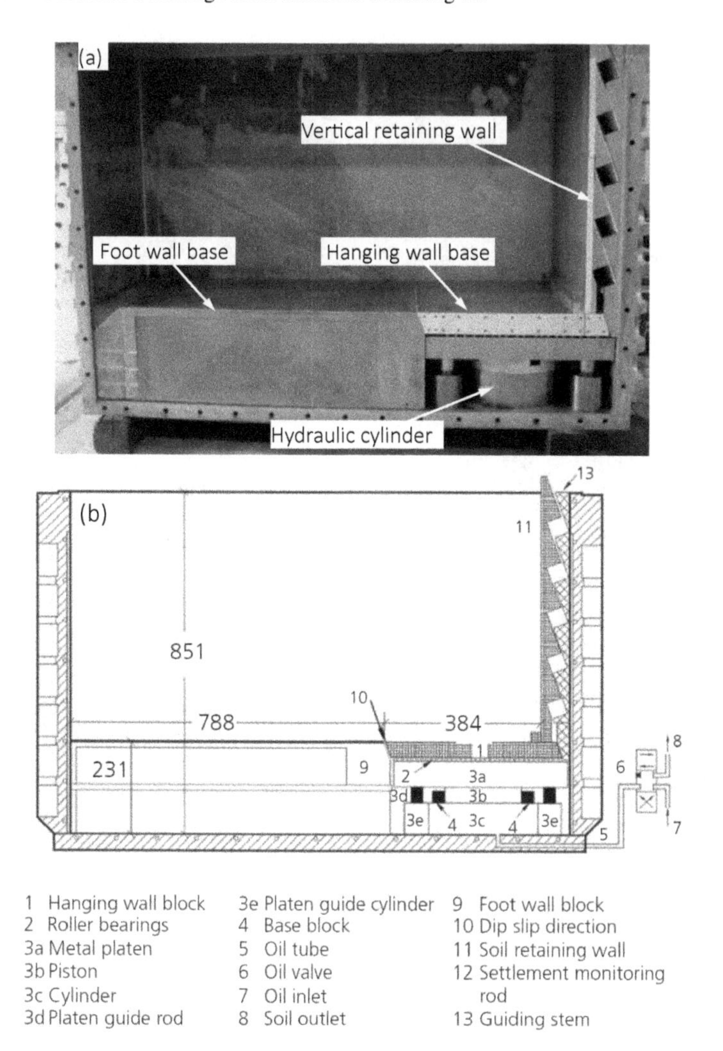

Fig. 2.5 Cross section of the container for centrifuge tests: **a** hydraulic cylinder for simulation of normal fault movement and **b** unconstrained boundaries (all dimensions in mm)

strain-softening behavior of the soil was modeled by a linear reduction in the friction angle with the accumulated plastic strain. This modeling technique has been deemed suitable for simulating fault rupture propagation (Anastasopoulos et al. 2007; Ng et al. 2012; Cai and Ng 2014; Cai et al. 2019).

The model pile was modeled as a linear elastic material with Young's modulus of 70 GPa and Poisson ratio of 0.2. The pile–soil interface was modeled as a spring-slider system (Itasca 2000). Stiffness in the normal direction (k_n) and that in the shear direction (k_s) were both estimated by

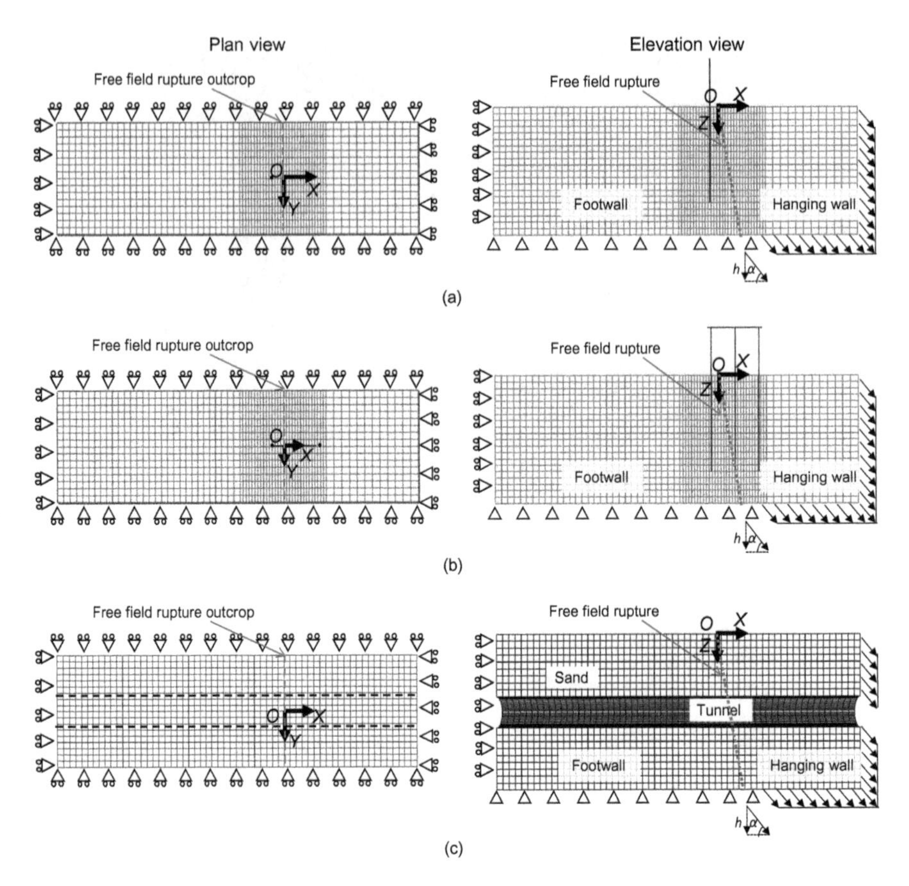

Fig. 2.6 Meshes and boundary conditions for numerical analysis of **a** the single pile; **b** the pile group; and **c** the tunnel

$$\max\left[\frac{(K + 3G/4)}{\Delta z_{\min}}\right] \tag{2.1}$$

where K is the bulk modulus; G is the shear modulus; and $\Delta z_{\min}$ is the smallest dimension of an adjoining zone in the normal direction. Since $\Delta z_{\min}$ = 2 mm in this study, k_n and k_s were both taken to be 1.75×10^7 kPa/m. The friction angle of the pile–soil interface (ϕ_s) was taken to be the critical-state friction angle of the surrounding soil ($\phi_c = 31°$).

In each numerical analysis, prior to the simulation of bedrock fault movement, model dimensions were used but the gravity was increased to 50 g to obtain the same initial in situ stresses as those in the centrifuge. Inclined downward displacements were then applied at the grid points along the vertical side and the base of the soil mesh on the hanging wall side to simulate normal faulting. The grid point displacements of the soil were the same as those of the bedrock fault in centrifuge tests.

Figure 2.6a, b shows the numerical mesh of single pile and pile group subjected to normal faulting, respectively. The single pile and pile group configurations in the numerical analysis were identical to that in the centrifuge tests. To simplify the analysis, the pile cap was also simulated by a 'pile' element. Each pile top was fixed to the pile cap and a total vertical loading of 1.5 kN (3750 kN in prototype) was applied to the pile cap.

Figure 2.6c shows the numerical mesh for analyzing the response of the tunnel subjected to normal faulting (test *M*). The tunnel was modeled by 'liner' elements. The mesh of the tunnel lining had the same distribution as the adjacent grids of the soil. To investigate the stiffening effects provided by the tunnel, three different tunnel stiffness were adopted in numerical simulation, in which Young's modulus of the tunnel (E_T) was reduced to 0.65 E_T and 0.17 E_T (Cai et al. 2019). The friction angle between the tunnel lining and the surrounding soil was taken to be 20° (i.e., two-thirds of the critical-state friction angle of soil). Since model dimensions were used, an enhanced gravity of 50 g was also applied in numerical simulations. This was followed by applying an inclined downward displacement at the grid points along the vertical side and the base of the soil mesh on the hanging wall side to simulate normal faulting. No displacement was imposed on the two ends of the tunnel to simulate the free boundary condition adopted in the centrifuge test. Numerical back-analysis of the centrifuge test *U* was also conducted.

2.3.6 *Parametric Study of Pile-Fault-Distance and Tunnel Depth*

To investigate the effects of pile-fault-distance on the responses of a single pile and a pile group, a series of numerical analyses were conducted by varying the horizontal distance (*S*) (see Fig. 2.1) from the pile or pile group edge to the free-field fault outcrop, as summarized in Table 2.2. The performance of the single pile and pile group was further evaluated in terms of pile displacements, redistribution of axial forces and the induced bending moments.

Table 2.2 Summary of the numerical analysis of pile-fault interaction

Series	Description	Distance of pile face or pile cap edge to free-field fault outcrop '*S*' (m) (refer to Fig. 2.1)
1	Single pile	−21.2, −16.2, −12.2, −6.2, −2.5, −0.5, 3.5, 9.5, 15.5, 21.5, 28.5
2	Pile group	−15.5, −15.0, −14.5, −14.0, −13.5, −13.0, −12.5, −12.0, −10.5, −8.5, −6.5, −4.5, −2.5, −0.5, 1.5, 3.5, 5.5, 7.5, 9.5, 11.5, 13.5, 15.5, 17.5, 19.5, 21.5, 23.5, 25.5, 27.5, 29.5, 31.5

Table 2.3 Summary of the numerical analysis plan of tunnel-fault interaction

Series	Description	Young's modulus of tunnel (GPa)	Tunnel depth d_T (m)
1	Back analyses of centrifuge tests	70	7.5, 12.5
2	Effects of tunnel stiffness	11.6, 45.5	7.5, 12.5
Tunnel length: 57.5 m			

Regarding the tunnel, two different embedded tunnel depths were compared to investigate the effects of tunnel depth on tunnel responses to faulting. A summary of the numerical analysis of tunnel-faulting interaction is given in Table 2.3.

2.4 Interpretation of Three-Dimensional Centrifuge Tests and Numerical Simulations

The experimental and numerical results were analyzed to reveal the interaction between fault rupture and underground structures. Data presented here are converted into prototype scale, unless stated otherwise.

2.4.1 *Ground Surface Settlements Adjacent to the Single Pile and Pile Group*

Figure 2.7a compares the measured and computed ground surface settlements in the single pile test. Both the measured and computed results show that the ground surface on the footwall side remained stationary while that on the hanging wall side settled with the bedrock fault movement. Computed results further illustrate that the differential settlement mainly occurred at $-1.5\text{ m} < X < 3.5\text{ m}$, where X is the horizontal distance from the free-field rupture outcrop (see Figs. 2.1 and 2.6). For the smallest vertical fault movement of $h = 0.4$ m simulated, the differential settlement reveals an error function-type profile as suggested by Cai and Ng (2013):

$$d_Z = \left[1 + \mathrm{erf}\left(\frac{X}{2.70} + 0.15\right)\right] \tag{2.2}$$

where d_Z is the ground surface settlement.

For the larger vertical movements (i.e., h larger than 0.4 m), a localized differential settlement and scarp were developed at the ground surface on the hanging wall side of the pile and the measured settlement profile could be fitted well with the error function qualitatively, although the settlement profile underestimated by the error function quantitatively (Cai and Ng 2016).

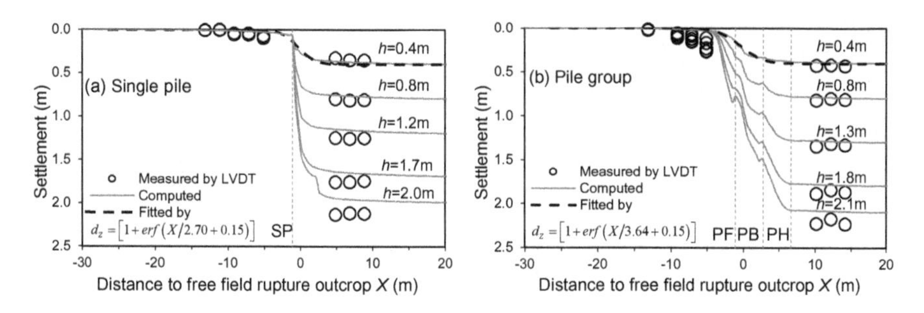

Fig. 2.7 Development of ground surface settlement induced by faulting from **a** the single pile test; **b** the pile group test

Figure 2.7b shows the ground surface settlement with the presence of a pile group. Compared with the differential settlement zone for the single pile, that for the pile group is extended by 2.5 times (i.e., −4 m < X < 8.5 m). For the vertical fault movement of h = 0.4 m simulated, the differential settlement may be captured by an error function-type profile as follows:

$$d_Z = \left[1 + \mathrm{erf}\left(\frac{X}{3.64} + 0.15\right)\right] \tag{2.3}$$

No scarp was formed at the ground surface. In addition, a relatively small ground settlement was observed around pile PF, suggesting a relative displacement between the pile and the surrounding soil. For h larger than 0.4 m, the ground surface differential settlement further developed without formation of a scarp around the pile group (Cai and Ng 2016).

2.4.2 *Normal Fault Propagation in Sand and Fault-Pile Interaction*

Figure 2.8 compares the results of experiments and numerical simulations after faulting for a single pile (h = 2 m) and a pile group (h = 2.1 m). A fault rupture was found to have extended from the bedrock fault to the ground surface with a dip angle of 80° in the single pile test (see Fig. 2.8a). A scarp emerged on the hanging wall side of the pile. Figure 2.8b shows the computed plastic shear strain for the same test. The plastic shear strain was mainly induced near the fault rupture suggesting the fault rupture was of a shearing type. In addition, both the centrifuge and numerical results suggest that the fault rupture refracted (i.e., became steeper) at the soil–bedrock interface. The same phenomenon has also been observed in centrifuge tests conducted on free-field sand beds (Anastasopoulos et al. 2007) and was probably related to the more dilatant behavior of sand (Bray et al. 1994; Anastasopoulos et al. 2007).

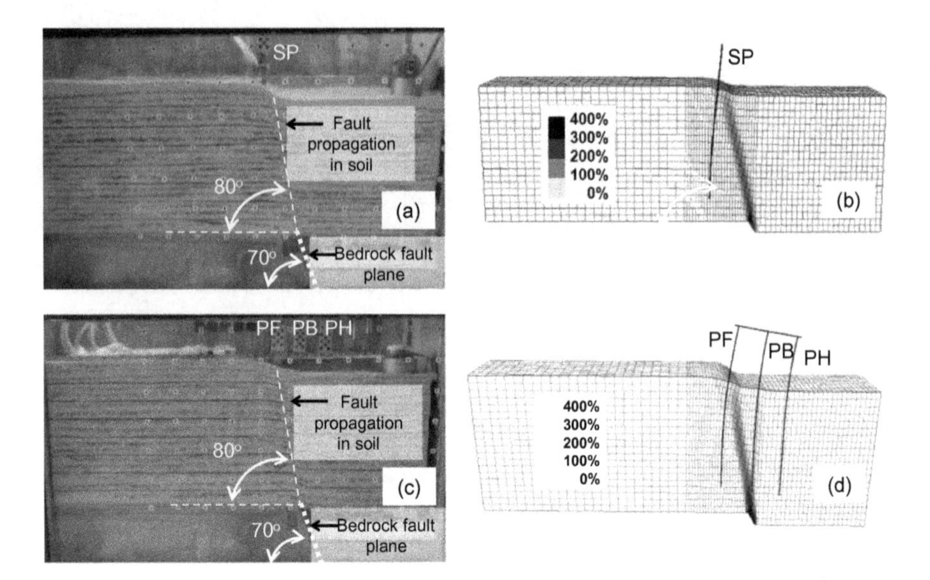

Fig. 2.8 Interaction between fault propagation and **a** the single pile from the centrifuge test; **b** the single pile from the numerical analysis; **c** the pile group from the centrifuge test; and **d** the pile group from the numerical analysis

Figure 2.8c shows the soil deformations after faulting for the pile group. The location of the induced fault rupture was similar to that observed for the single pile case. The fault rupture developed between the two piles PF and PB. As shown in Fig. 2.8d, the numerical back-analysis reveals that the fault rupture passed through the tip of pile PB and emerged at the ground surface on the footwall at the location of pile PF. Compared with the single pile case, the presence of two more piles (i.e., PB and PH) in the three-pile group did not alter the location and angle of the fault rupture, except the differential settlement zone was extended (as shown in Fig. 2.7b). In their numerical analyses of pile-fault interaction, Anastasopoulos et al. (2013) observed that the pile group altered the path of fault rupture. They found that the rupture passed through the tip of the piles and emerged at the ground surface when the pile group was positioned further away on the hanging wall side. While the main rupture zone was not affected, a secondary rupture was generated between the piles resulting in the modification of the fault rupture path. Generally, the observations from the centrifuge tests and numerical back-analyses were consistent, lending confidence to the other reported results.

2.4.3 Pile Top Displacement and Tilting

Figure 2.9a compares the measured and computed displacements at the single pile tip during five steps of faulting simulation. The measured and computed results are

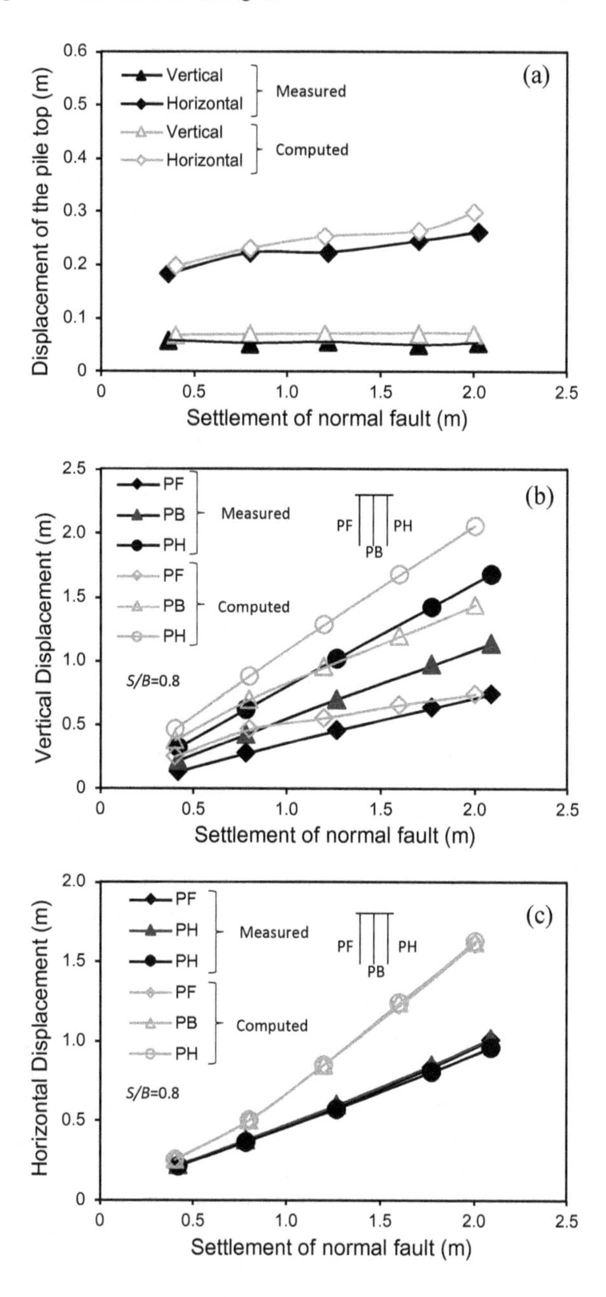

Fig. 2.9 Measured and computed **a** displacement of the single pile; **b** vertical displacement of the pile group; and **c** horizontal displacement of the pile group

fairly consistent with each other and revealed that the settlement of the single pile was induced during the first step of faulting (h = 0.4 m) with no further increase in the subsequent steps (i.e., h = 0.8, 1.3, 1.8 and 2.0 m). Most of the pile horizontal displacement was also observed during the first step of faulting but with a slight increase during the subsequent steps.

As illustrated in Fig. 2.9b, the measured vertical displacements of the three piles (PF, PB and PH) in the pile group varied linearly with the settlements of the hanging wall during the five steps of faulting simulation. The computed results show a consistent trend but with some overestimation. The linearly increasing difference among the measured settlements of the three piles reveals that the pile group tilted toward the hanging wall after each step of faulting. The same phenomenon was also observed in the numerical analyses of a 2 × 4 pile group conducted by Anastasopoulos et al. (2013). The PB and PH piles were pulled downwards and sideways due to the settlement of the hanging wall, whereas pile PF resisted on the footwall side. Consequently, the piles on the hanging wall side (PB and PH) experienced large bending moments at their heads.

Based on technical code for building pile foundations (JGJ 94-2008), the ultimate bending moment capacity of the 0.65 m diameter pile with the maximum design reinforcement ratio of 0.75% is 550 kN m. When h = 2.1 m, the computed maximum bending moments of piles for PF, PB and PH are 2,385 kN m, 2,859 kN m and 2,310 kN m, respectively. This means that the induced bending moments are at least four times larger than the ultimate design capacity. Regarding axial force induced in each pile, the computed maximum tensile force induced in pile PB is 1,212 kN, which exceeds the ultimate axial tension capacity of 771 kN and the pile will be damaged in a tensile-failure mode. Anastasopoulos and Gazetas (2007) reported tensile failure in piles due to normal faulting during the Kocaeli 1999 earthquake. The piles crossing the fault rupture failed under tension and the piles adjacent to the surface rupture showed tensile cracking. Hence, the induced tensile stress must be considered in any design analysis.

Figure 2.9c compares the measured and computed horizontal displacements of the three piles during faulting. The measured horizontal displacements increased linearly with the magnitude of faulting simulated and were almost identical suggesting a rigid body movement of the pile group in the horizontal direction. Similar results were also obtained from the numerical simulation. This implies that the three piles were well connected by the pile cap and displaced toward the hanging wall horizontally after faulting. The observed differences in the vertical displacements of the three piles (refer to Fig. 2.9b) must have been resisted by the bending moment of the pile cap. The induced maximum sagging bending moment in the pile cap was 6913 kN m, which exceeds 42% of its bending moment capacity. Consequently, an increase in the thickness and reinforcement ratio of the pile cap must be considered in the structural design (Anastasopoulos et al. 2013).

Due to displacements of the PB and PH piles, the pile cap was pulled and tilted towards the hanging wall after faulting. Based on the computed results, the pile PF resisted the faulting actions through an increase in axial compression of 1,057 kN (52% of axial compression capacity of 2,038 kN). Moreover, the computed horizontal displacements of the three piles were identical, and they also increased with the magnitude of faulting simulated. However, the differences in the computed and measured values increased with the magnitude of faulting, probably because the interface between each pile and its surrounding soil was not properly simulated numerically.

2.4.4 *Influence of Pile Location on Pile Responses: Numerical Parametric Study*

Based on the numerical parametric analyses, Fig. 2.10a illustrates the effects of the distance from the farthest face of single pile to the free-field fault outcrop (*S*) on pile displacement, when it was subjected to bedrock fault movement of $h = 0.4$ m. The *S* is normalized by the pile diameter. Based on the single pile response, three characteristic zones (I, II, III) may be identified. When the single pile was located on the footwall side and far from the bedrock fault (i.e., $S/D < -10$), no significant displacement or tilting was expected and hence the pile remained stationary after faulting. This is consistent with the results of centrifuge tests on single piles by Yao and Takemura (2020) who found that the piles out of the fault zone were practically unaffected by fault propagation. It may be defined as a safe zone (*I*) (i.e., $S/D < -10$) for any building to be constructed.

As the location of the single pile approached the bedrock fault line from $S/D = -10$ m onward, the horizontal displacement and tilting of the pile increased and reached their peak values when the pile reached $S/D = -0.8$ and then decreased to their respective, almost steady values at $S/D = 5$. These results agree well with Yao and Takemura's (2020) observation that the maximum horizontal displacement of the single pile would occur when the pile is located near the bedrock fault line where the normal fault rupture propagates through the single pile. On the other hand, the vertical displacement of the pile increased from $S/D = -10$ onward, almost linearly with S/D to reach a nearly steady value at $S/D = 5$. Thus, the normalized distance of $-10 < S/D < 5$ may be defined as the transition zone (II) in which special design considerations are likely to be needed.

When the single pile was located at a distance $S/D \geq 5$ on the hanging wall side, the pile displaced by almost the same amount as the hanging wall (i.e., $h = 0.4$ m), revealing a vertical rigid body movement together with its surrounding soil. On the other hand, the pile displaced horizontally by about 50% of the settlement but without any pile tilting. This implies that the pile simply translated

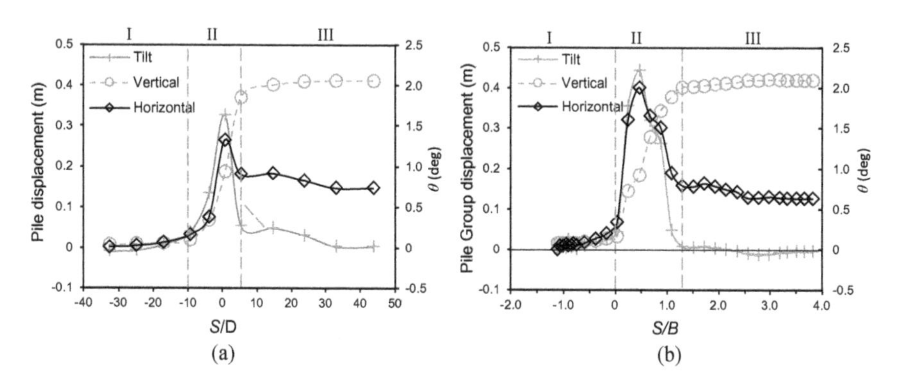

Fig. 2.10 Influence of pile location on the response of **a** single pile and **b** pile group

along a locus inclining at about 63.4° to the horizontal, as opposed to the dip angle of 80° observed in Fig. 2.8a. This zone may be defined as the translational zone (III) when $S/D \geq 5$. In this zone, the fault rupture could be accommodated by the rigid body movement of the structure (Oettle and Bray 2013).

Figure 2.10b shows the computed response of the 3 × 1 pile group to the faulting with a magnitude of 0.4 m for different distances (S) from the furthest edge of the pile group cap to the free-field fault rupture outcrop. Each S in the figure is normalized by the width B (i.e., 9.5 m in prototype) of the pile group cap (as defined in Fig. 2.1). Very similar responses to the single pile were computed, except the magnitudes were slightly different. Thus, the three characteristic zones can also be identified. Similar to the single pile case, $S/B < 0$ can be defined as the safe zone (I), and $0 < S/B < 1.3$ as the transition zone (II). The translational zone (III) occurs where $S/B \geq 1.3$. The peak vertical displacement of 0.4 m and the corresponding horizontal displacement of 0.15 resulted in a translational locus of about 69° to the horizontal, which was slightly steeper than that in the single pile case. Similar characteristic zones were also observed by Anastasopoulos et al. (2013) through numerical simulation of a 2 by 4 pile group subjected to normal fault propagation.

To further investigate the response and load redistribution of individual piles in the elevated pile group at different locations, Fig. 2.11 shows the computed axial forces (above the ground surface) of the three piles (PF, PB and PH) when $h = 0.4$ m. The same three characteristic zones can be easily identified as those based on pile group displacements shown in Fig. 2.10b. In zone I, the three piles share almost the same amount of applied working load of 1250 kN each. This zone may be called a safe zone (I) as shown in Fig. 2.10b.

In zone II ($0 < S/B < 1.3$), the middle pile PB behaves very differently to the other two piles PF and PH. The compressive force in PB increases to a maximum of 2853 kN which is 40% larger than the ultimate axial compressive capacity of

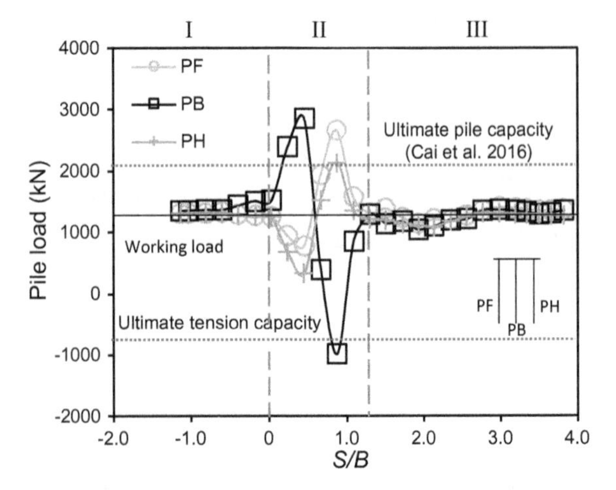

Fig. 2.11 Redistribution of axial forces within the pile group at different locations

2,038 kN. Obviously, this should be considered in any design analysis. On the contrary, the compressive axial forces in PF and PH decrease to their minimum values of 671 kN and 319 kN, respectively. Interestingly, a neutral point can be identified, i.e., $S/B = 0.6$, at which the three piles are all subjected to the same working load before faulting. When the pile cap is located at S/B larger than 0.6, the axial compressive force induced in pile PB decreases and reaches its largest tension of 986 kN at $S/B = 0.9$. This induced tension is 128% of the ultimate tension capacity (i.e., 771 kN) and the pile could be damaged in a tensile-failure mode. Hence, the induced tensile stress must be considered in any design analysis. In contrast, the axial compressive forces in PF and PH increase to their maximum values of 2,668 kN and 2,139 kN, respectively, to maintain the vertical force equilibrium. The axial force in both piles exceeds the ultimate axial compressive capacity and reach to 131 and 105% of its ultimate axial compressive capacity of 2,038 kN, respectively. As the pile group moves away from the fault line, the tensile force in PB and compressive forces in PF and PH all return to their initial working values at $S/B = 1.3$. Based on the computed results, one may define $0 < S/B < 1.3$ as the transition zone (II) in which the axial force is significantly redistributed among the three piles. Within this zone, it is fairly clear that a minimum of 131% of the ultimate axial compressive capacity should be considered for all three piles. On the other hand, a minimum tensile capacity of about 986 kN should be allowed for pile PB. When the pile group is located at $S/B \geq 1.3$, no redistribution of the axial forces among the three piles can be found. This is consistent with the translational zone (III) identified in Fig. 2.10b.

Figure 2.12 shows the effects of pile group location on the response of the pile group subjected to different levels of normal fault movement from numerical simulations. When the pile group is located on the footwall side and far from the bedrock fault (i.e., zone I with $S/B < 0$), no significant displacement and tilting are observed and hence the pile cap remained stationary after faulting. In this zone, the induced tilting of the pile cap is within the allowable tilting limit of 0.11° (i.e., 0.2%) suggested by Eurocode 7 (CEN 2001) for buildings.

In zone II ($0 < S/B < 1.3$), the horizontal displacement progressively increases while the vertical displacement and tilting reach their peak values following a subsequent decrease as the pile group approaches at $S/B = 1.3$. The rotation (or tilting) of the pile cap increases with the normal fault settlement within this zone. Irrespective of the level of normal fault settlement in zone II, the tilting of the pile cap exceeds the allowable tilting limit of buildings suggested by Eurocode 7 (CEN 2001). In zone III, the vertical and horizontal displacements and hence the tilting of the pile group reach their ultimate values. Similar to zone I, the induced tilting of the pile cap by the five levels of normal fault movements is within the allowable tilting limit suggested by Eurocode 7 (CEN 2001). This suggests that the tilting remains constant (i.e., close to zero) in zones I and III and hence the range of the three identified zones for pile group location is not affected by the magnitude of the normal fault movement. Similar behavior was observed by Anastasopoulos et al.

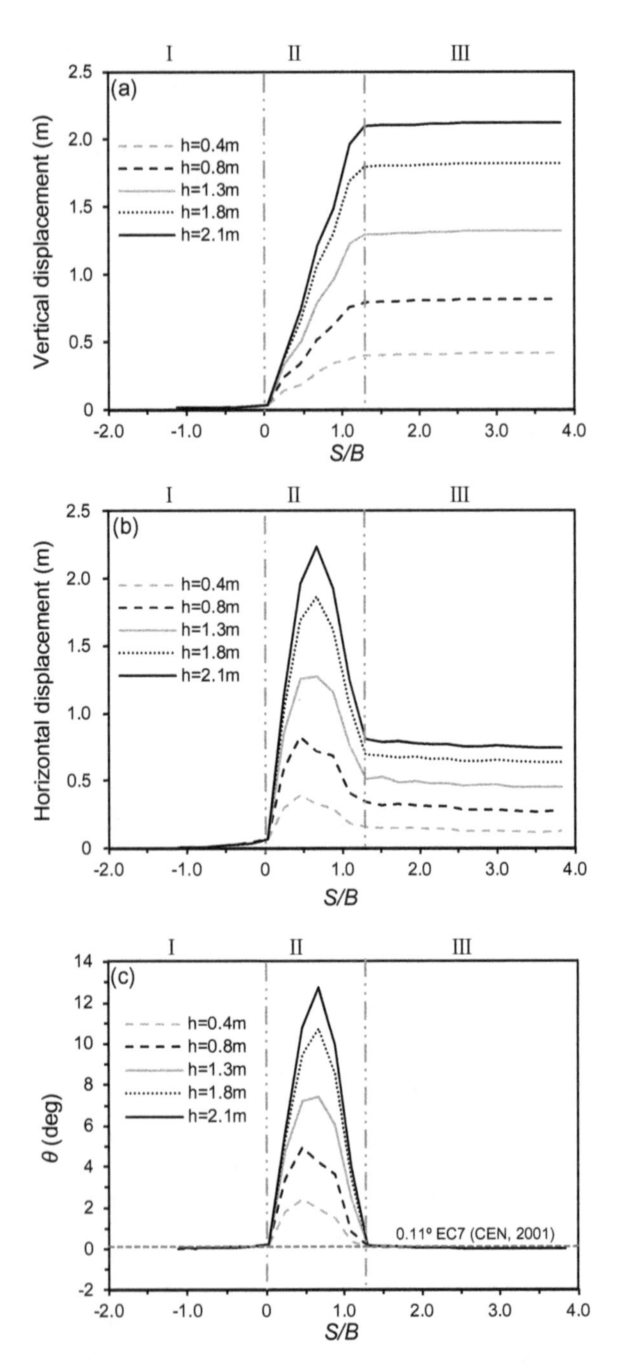

Fig. 2.12 Effects of normalized distance from pile group to fault rupture on the computed: **a** vertical displacement; **b** horizontal displacement; and **c** tilting of the pile group

(2013) for a 2 × 4 pile group suggesting the effectiveness of the chosen distance normalization (i.e., *S*/*B*) for identifying the three characteristic zones.

Figure 2.13 shows the effects of pile group location on the computed maximum and minimum bending moments induced in piles PF, PB and PH subjected to the five levels of normal fault movements. The induced bending moments in the figure are normalized by the ultimate bending capacity of the piles. The ultimate moment capacity of the pile with a diameter of 0.65 m and the maximum design reinforcement ratio of 0.75% (based on Technical code for building pile foundations (JGJ 94-2008)) is 550 kN m. In zone I ($S/B < 0$), the piles remain at the footwall sustaining (relatively) limited distortion since the propagating fault rupture has not passed through the piles (see Fig. 2.12c). Hence, the bending moments of the piles remain almost zero. As the fault rupture hits the piles in zone II, the pile group acts as a retaining system of the soil mass behind the piles resulting in an increase in their tilting (see Figs. 2.8 and 2.12) and bending moments. The tilting and the associated bending moments reach their peak values in zone II. For a normal fault settlement of 0.4 m, the bending moments of piles PF, PH and PB reach 152%, 130% and 160% of their bending capacity, respectively. Furthermore, the induced maximum sagging bending moment in the pile cap reach to 4401 kN m, which is 90% of its bending capacity. The maximum induced moment in the piles increase with settlements of the normal fault. For the normal fault settlement of 2.1 m, the bending moments of piles PF, PH, and PB reach 430%, 550% and 580% of their bending capacity, respectively. Similar observations were reported by Anastasopoulos et al. (2013) suggesting that special design considerations such as increases in the reinforcement and pile diameter are likely to be needed for the piles in this zone. The induced maximum sagging bending moment in the pile cap reach to 6913 kN m, which is 140% of its bending capacity. Consequently, an increase in the thickness and reinforcement ratio is required during the structural design of the pile cap (Anastasopoulos et al. 2013).

As the fault rupture passes beyond the pile cap in zone III ($S/B > 1.3$), the pile group follows the movements of the hanging wall in its downward and outward translational directions, and hence, the induced bending moments are smaller than the bending capacity of the piles. Based on the results revealed in Figs. 2.10, 2.11, 2.12, 2.13, it is reasonable to define the three characteristic zones, i.e., safe, transition and translational, for design analyses.

2.4.5 *Ground Surface Settlement Along the Longitudinal and Transverse Tunnel Directions*

Figure 2.14 compares measured and computed ground surface settlement profiles along the tunnel axis with a fault magnitude of $h = 0.8$ m. The computed vertical displacements of the tunnel crown are also given in the figure for comparison. As illustrated in Fig. 2.14a, the measured ground surface settlement above the tunnel

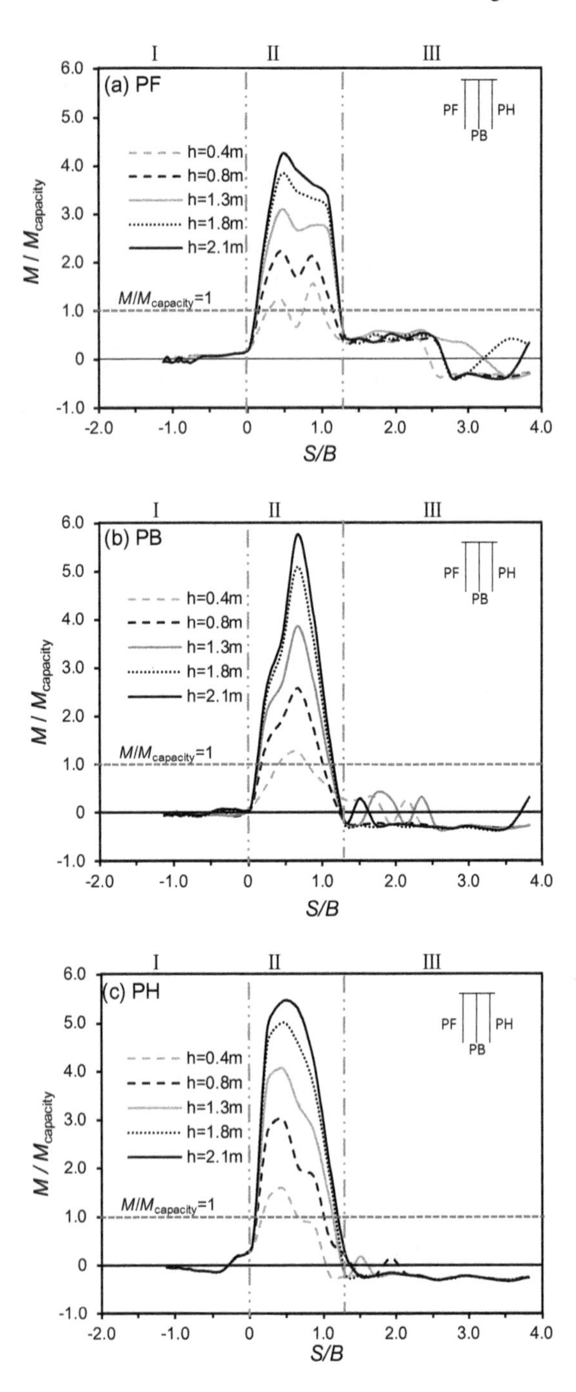

Fig. 2.13 Effects of distance from pile group to fault rupture on the computed bending moment of pile **a** PF; **b** PB; and **c** PH

with a tunnel depth (d_T) of 7.5 m increases gradually from the footwall side to the hanging wall side, resulting in an average ground surface gradient of 2.1%, which is consistent with the computed results of about 2.0%. Similarly, the average computed tunnel crown settlement gradient is 2.0% which far exceeds the recommended serviceability limit of 0.1% for a tunnel by the Land Transport Authority (LTA 2000) in Singapore, as expected. It is also found that computed settlements at the tunnel crown for $X > -17.5$ m far exceed the recommended serviceability limit of 15 mm by the Land Transport Authority (LTA 2000), as expected. Figure 2.14b compares measured and computed ground surface settlement profiles above the tunnel when it is located at a depth of 12.5 m after the same fault movement. There is no major difference between the measured and computed results when the tunnel is located at either 7.5 m or 12.5 m. The measured ground settlement profile is consistent with the computed ground surface profile and with tunnel crown settlements with an average slope of 2.2%, except when X is greater than 0 from the free-field fault rupture outcrop. Similar to the tunnel located at a depth of 7.5 m, the average computed slope of the tunnel crown settlement far exceeds the limit of 0.1% (LTA 2000). The increasing discrepancy between measured and computed ground surface profiles when $X > 0$ may be attributed to the continuum assumption made in the numerical simulations, opposite to the particulate behavior of model sand used in the experiments, especially near the fault line. Furthermore, the settlements at the tunnel crown for $X > -19$ m exceed the recommended serviceability limit of 15 mm by (LTA 2000), irrespective of the tunnel depth. All these results suggest that special design consideration should be given for tunnel located nearby a fault zone.

Figure 2.15 compares measured and computed ground surface settlements perpendicular to the longitudinal tunnel axis when the fault magnitude is $h = 0.8$ m. To investigate the stiffening effects provided by the tunnel, three different magnitudes of tunnel stiffness (i.e., $E_T I = 4.71 \times 10^8$ kN m^2, $0.65 E_T I = 3.06 \times 10^8$ kN m^2 and $0.17 E_T I = 0.80 \times 10^8$ kN m^2) are considered in the numerical analyses. Comparing the computed results of three different tunnel stiffness, it is found that there is no major difference in the computed soil settlement profiles. When the tunnel is located at 7.5 m deep (Fig. 2.15a), the measured induced

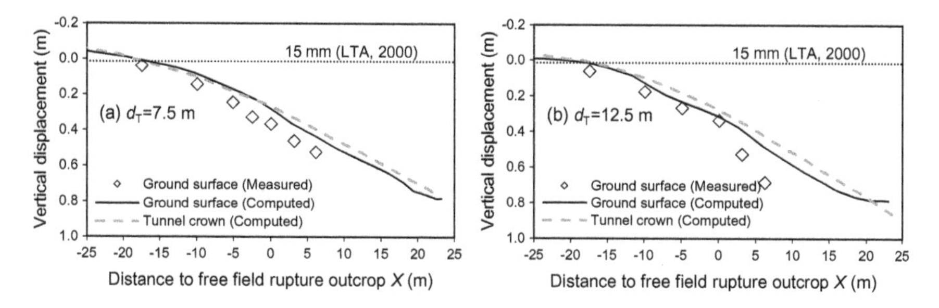

Fig. 2.14 Development of ground surface settlement induced by faulting with the tunnel axis at **a** $d_T = 7.5$ m; and **b** $d_T = 12.5$ m

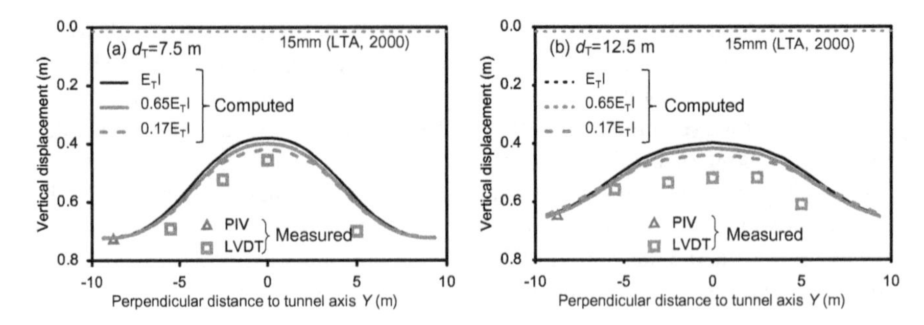

Fig. 2.15 Development of ground surface settlements induced by faulting ($h = 0.8$ m) perpendicular to the longitudinal tunnel axis with tunnel depth located at **a** $d_T = 7.5$ m; and **b** $d_T = 12.5$ m

settlement of the ground surface above the tunnel axis (at $Y = 0$ m) is 0.45 m, which reveals that the soil above the tunnel experiences a settlement far exceeding the allowable limit of 15 mm recommended by LTA (2000) in Singapore. The ground surface settlement displays an inverted *U* shape profile, where the minimum settlement occurs at the location of the tunnel. As expected, the shielding effects of the tunnel are the highest near the tunnel axis, and it decreases with an increase of the distance from its longitudinal axis. The measured ground surface settlement reaches the maximum value of 0.72 m at $Y = -8.75$ and 8.75 m. Although the measured surface settlements are underestimated slightly by the numerical analyses, the computed results are generally in good agreement with the measurements.

Figure 2.15b compares measured and computed ground surface settlement profiles induced by faulting perpendicular to the longitudinal tunnel axis when the tunnel is located at a depth of 12.5 m. The measured soil settlement profile similar to that of the tunnel located at 7.5 m is observed, except that the increase in the burial depth of the tunnel results in a slight increase in ground surface settlement (i.e., the settlement increases to 0.50 m at $Y = 0$ m). Similar to the tunnel at a depth of 7.5 m, the ground surface settlement exceeds the allowable limit of 15 mm recommended by LTA (2000). The ground surface settlement profile shows a similar inverted *U* shape profile to the tunnel depth at a depth of 7.5 m, except the settlement profile spreads wider with an increase in the tunnel depth. Consequently, the maximum ground surface settlement is smaller for the deeper tunnel, and it reaches 0.65 m at $Y = -8.75$ and 8.75 m. The observed wider settlement profile of the deeper tunnel near the bedrock is consistent with the results of Baziar et al. (2014) who conducted a series of centrifuge tests on a tunnel subjected to reverse faulting. They found that an increase in the tunnel depth can cause the rupture to be modified near the tunnel and propagate in a wider zone of soil layer above the tunnel.

2.4.6 Propagation of Normal Fault and Fault-Tunnel Interaction

Figure 2.16a shows the computed plastic shear strain contours when the tunnel is located at $d_T = 7.5$ m. While the hanging wall block moved at an angle of 70° with respect to the horizontal, the fault rupture in the soil underneath the tunnel extends upward with a dip angle of 80°, similar to the sand bed strengthened by the presence of the single pile and the pile group (refer to Fig. 2.8). An increase in the dip angle of fault rupture from 70° in bedrock to 80° in soil is probably related to the dilatant behavior of sand. The same phenomenon has also been observed in centrifuge tests conducted on free-field sand beds (Anastasopoulos et al. 2007). For the soil above the tunnel, no fault rupture can be observed. This suggests that the soil above the tunnel is shielded from the shearing deformation arising from faulting. When the tunnel is located at $d_T = 12.5$ m, a similar propagation of fault rupture is computed as shown in Fig. 2.16b. The fault rupture is shielded from propagating upward after reaching the tunnel. The shielding effects of the embedded tunnel have also been observed by Baziar et al. (2014) in centrifuge tests in which the tunnel axis was parallel to the bedrock fault plane. In their study, a scarp emerged at the ground surface since the fault rupture might have bypassed the tunnel and the extent of shielding depended on the horizontal tunnel distance from the fault rupture. However, the shielding effect of the tunnel is more apparent when the tunnel is perpendicular to the bedrock fault plane and no scarp could emerge at the ground surface due to the stiffening effects of the tunnel.

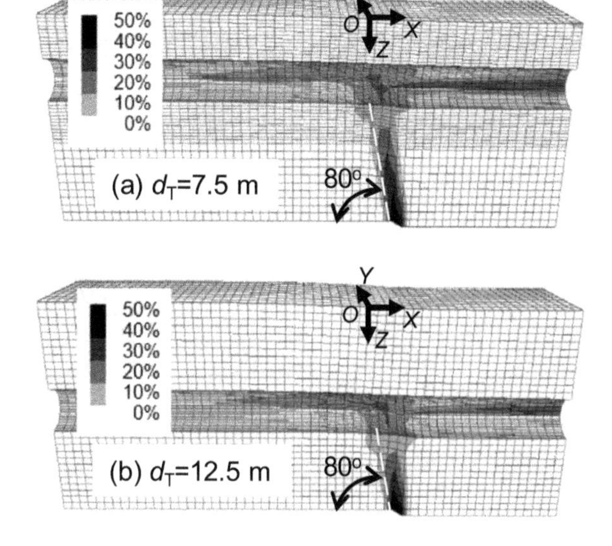

Fig. 2.16 Computed plastic shear strain contours due to fault propagation when the tunnel is located at **a** $d_T = 7.5$ m; and **b** $d_T = 12.5$ m

Figure 2.17 compares the measured and computed additional longitudinal tensile strains induced at the tunnel crown after faulting with a magnitude of $h = 0.8$ m. The measured values and computed results are in excellent agreement. The distributions of the longitudinal strains at the tunnel crown show that the tunnel was subjected to a hogging moment after normal faulting. Tensile strains were induced at the tunnel crown and the peak tensile strain was located near the middle of the model tunnel (i.e., $X = -6.5$ m) in both centrifuge tests. The measured peak tensile strains were 1,697 $\mu\varepsilon$ and 2,566 $\mu\varepsilon$ when the tunnel was located at the depths of 7.5 m and 12.5 m, respectively. The measured tensile strain decreased to zero toward the two free ends of each tunnel.

By comparing the results of strains induced in the tunnels located at the depths of 7.5 and 12.5 m, it is clear that the maximum induced tensile strain (i.e., 2,566 $\mu\varepsilon$) at $d_T = 12.5$ m is about 50% larger than that (i.e., 1,697 $\mu\varepsilon$) at $d_T = 7.5$ m. This is because the extent of the hogging deformation of the tunnel increased with burial depth. The part of the tunnel exceeding the limiting cracking tensile strain of 150 $\mu\varepsilon$ for unreinforced concrete (ACI 2001) is located at $-29.5 \text{ m} < X < 16.5$ m and $-35.0 \text{ m} < X < 13.0$ m when the tunnel is 7.5 m and 12.5 m deep, respectively. In other words, the length of excessive tensile zone is 46 m and 48 m for d_T equals to 7.5 m and 12.5 m, respectively. This implies that sufficient reinforcements must be provided to limit tensile cracking to an acceptable level in a fault zone. Based on centrifuge tests, similar results were reported by Baziar et al. (2014) suggesting that a deeper tunnel near the bedrock fault requires larger flexibility to prevent cracking due to fault movement. An alternative design method is to set flexible joints to accommodate deformation due to rupturing for the tunnels crossing an active fault (Kiani et al. 2016).

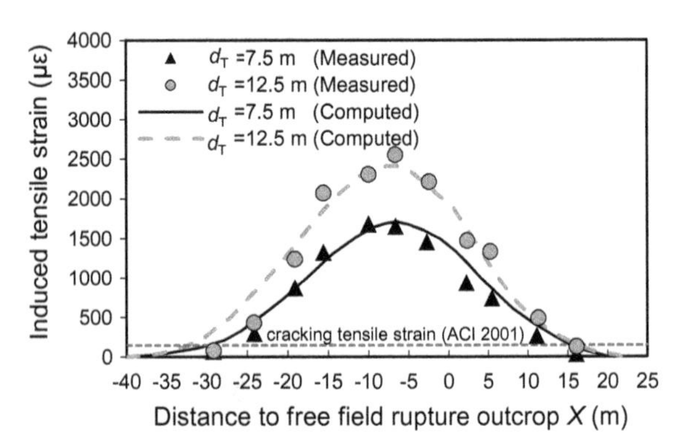

Fig. 2.17 Longitudinal tensile strains induced at the tunnel crown (h = 0.8 m)

2.5 Summary and Conclusion

Based on a series of centrifuge model tests and numerical analyses on pile foundations and tunnel subjected to normal faulting, the following conclusions may be drawn.

For a single pile located on the footwall, both measured and computed vertical displacements of the pile are consistent, and they are independent of the four levels of faulting movement simulated. On the other hand, the horizontal displacement of the pile increases almost linearly with the magnitude of the faulting induced. Three characteristic zones may be identified: (1) safe zone—I (i.e., $S/D \leq -10$), where no significant displacement or tilting is expected and hence the pile remains stationary after faulting; (2) transition zone—II (i.e., $-10 < S/D < 5$) where the pile is displaced and tilted toward the hanging wall and hence special design considerations are needed; and (3) translational zone—III (i.e., $S/D \geq 5$), where the pile is displaced translationally but without tilting.

Similarly, three characteristic zones, i.e., safe zone I ($S/B \leq 0$), transition zone II ($0 < S/B < 1.3$) and translational zone—III ($S/B \geq 1.3$), can also be identified for pile groups. The range of the three identified zones is not affected by the magnitude of the normal fault movements simulated.

For the pile group located at $S/B = 0.8$ in the transition zone—II ($0 < S/B < 1.3$), the piles on the hanging wall side (PB and PH) are pulled downward and sideways due to the settlement of the hanging wall, whereas pile PF resisted on the footwall side. With a fault movement of $h = 2.1$ m, for an example, the computed maximum bending moments of piles PF, PB and PH are 2,385 kN m, 2,859 kN m and 2,310 kN m, respectively, as compared with the ultimate moment capacity of 550 kN m. The computed maximum tensile force induced in pile PB is 1,212 kN which exceeds the ultimate tension capacity of 771 kN by 57% and the pile will be damaged in a tensile-failure mode. Hence, the induced bending moment and tensile stress must be considered for designs.

In the transition zone, the middle pile PB behaves very differently to the other two piles PF and PH. For example, when $h = 0.4$ m, the compressive axial force in PB increases with S/B and reaches the maximum of 2,853 kN which exceeds 40% of its ultimate axial compressive capacity. However, the axial force induced in pile PB decreases when the pile group is located at $S/B = 0.9$ and the largest induced tension in pile PB reaches 986 kN, which is 28% larger than its tension capacity. This implies that the performance of the middle pile is mostly affected by the location of pile group.

The tilting of the pile cap and the associated bending moments increase with the magnitude of normal fault settlement within the transition zone II ($0 < S/B < 1.3$), whereas the tilting remains constant (i.e., close to zero) in zones I and III, irrespective of the magnitude of fault settlement. In zone II, the tilting of the pile cap far exceeds the allowable tilting limit of buildings suggested by Eurocode 7 (CEN 2001). The tilting and the associated bending moments reach their peak values in this zone. As expected, the maximum induced moment in the piles increases with the magnitude of fault settlements. During the normal fault settlement of 2.1 m, for example, the bending moments of piles PF, PH, and PB reach 430%, 550% and 580% of their bending moment capacity, respectively. The maximum sagging bending moment induced in the pile cap reaches 6,913 kN m, which is 142% of its bending moment capacity. Consequently, an increase in thickness and reinforcement in pile cap should be considered. In the translational zone III ($S/B \geq 1.3$), the pile group follows the movements of the hanging wall in its downward and outward translational directions and hence the induced bending moments are smaller than the bending moment capacity of the piles.

When there is a tunnel constructed in sand stratum, the ground above the longitudinal axis of the tunnel is shielded from the shearing deformation due to faulting. As expected, the gradient of induced tunnel crown settlement far exceeds the recommended serviceability limit of 0.1% for a tunnel by the Land Transport Authority (LTA 2000) for both tunnels at depths of 7.5 and 12.5 m. Perpendicularly to the longitudinal tunnel axis (i.e., transverse direction), the ground surface settlement reveals an inverted *U* shape profile. This suggests that the shielding effects of the tunnel are higher near the longitudinal tunnel axis and it decreases with an increase of the distance from the tunnel axis. Due to the 0.8 m faulting, for example, the tunnel is subjected to a hogging bending moment and longitudinal tensile strain is induced at the tunnel crown. When the tunnel depth d_T increases from 7.5 to 12.5 m, the maximum induced tensile strains increase by about 50%. The length of the tunnel exceeding the limiting cracking tensile strain of 150 $\mu\varepsilon$ for unreinforced concrete (ACI 2001) also increases slightly with the tunnel depth. For the cases investigated, the length of the excessive tensile zone is 46 m and 48 m for d_T equals to 7.5 m and 12.5 m, respectively. This implies that sufficient reinforcement should be provided.

Acknowledgements The authors would like to acknowledge the financial support provided by the National Natural Science Foundation of China (grant nos. 51778249 and 51608170), the Natural Science Foundation of Fujian Province of China (project no. 2018J01072), the Promotion Program for Young and Middle-aged Teachers in Science and Technology Research of Huaqiao University (project no. ZQN-PY216), and Huaqiao University (project no. 16BS509).

References

ACI (American Concrete Institute) (2001) Control of cracking in concrete structures. ACI 224R-01. Farmington Hills, MI

ACI (American Concrete Institute) (2011) Building code requirements for structural concrete and commentary. ACI 318M-11. Farmington Hills, MI

Ahmed W, Bransby MF (2009) Interaction of shallow foundations with reverse faults. J Geotech Geoenviron Eng ASCE 135(7):914–924

Anastasopoulos I, Gazetas G (2007) Foundation-structure systems over a rupturing normal fault: part I. Observations after the Kocaeli 1999 earthquake. Bull Earthq Eng 5(3):253–275

Anastasopoulos I, Gazetas G, Bransby MF, Davies MCR, El Nahas A (2007) Fault rupture propagation through sand: finite-element analysis and validation through centrifuge experiments. J Geotech Geoenviron Eng ASCE 133(8):943–958

Anastasopoulos I, Callerio A, Bransby MF, Davies MCR, El Nahas A, Faccioli E, Gazetas G, Masella A, Paolucci R, Pecker A, Rossignol E (2008) Numerical analyses of fault-foundation interaction. Bull Earthq Eng 6(4):645–675

Anastasopoulos I, Gazetas G, Bransby MF, Davies MCR, El Nahas A (2009) Normal fault rupture interaction with strip foundations. J Geotech Geoenviron Eng ASCE 135(3):359–370

Anastasopoulos I, Kourkoulis R, Gazetas G, Tsatsis A (2013) Interaction of piled foundation with rupturing normal fault. Geotechnique 63(12):1042–1059

Ashtiani M, Ghalandarzadeh A, Towhata I (2015) Centrifuge modeling of shallow embedded foundations subjected to reverse fault rupture. Can Geotech J 77(2):558–566

Baziar MH, Nabizadeh A, Lee CJ, Hung WY (2014) Centrifuge modeling of interaction between reverse faulting and tunnel. Soil Dyn Earthq Eng 65:151–164

Baziar MH, Nabizadeh A, Khalafian N, Lee CJ, Hung WY (2019) Evaluation of reverse faulting effects on the mechanical response of tunnel lining using centrifuge tests and numerical analysis. Géotechnique (in press)

Bray JD, Seed RB, Cluff LS, Seed HB (1994) Earthquake fault rupture propagation through soil. J Geotech Eng 120(3):543–561

Cai QP, Ng CWW (2013) Analytical approach for estimating ground deformation profile induced by normal faulting in undrained clay. Can Geotech J 50(4):413–422

Cai QP, Ng CWW (2014) Effects of the tip depth of a pre-existing fracture on surface fault ruptures in cemented clay. Comput Geotech 56:181–190

Cai QP, Ng CWW (2016) Centrifuge modeling of pile-sand interaction induced by normal faulting. J Geotech Geoenviron Eng ASCE 142(10):04016046

Cai QP, Ng CWW, Luo GY, Hu P (2013) Influences of pre-existing fracture on ground deformation induced by normal faulting in mixed ground conditions. J Cent S Univ 20(2):501–509

Cai Q, Ng CWW, Hu P (2015) Boundary effects on ground surface rupture induced by normal faulting. Géotechnique Lett 5(3):161–166

Cai QP, Ng CWW, Chen XX, Guo LQ (2017) Failure mechanism and setback distance of single pile subjected to normal faulting. Chin J Geotech Eng 39(4):720–726 (in Chinese)

Cai QP, Peng JM, Ng CWW, Shi JW, Chen XX (2019) Centrifuge and numerical modelling of tunnel intersected by normal fault rupture in sand. Comput Geotech 111:137–146

CEN (2001) Eurocode 7, part 1: geotechnical design: general rules, final draft prEN 1997-1. European Committee for Standardization (CEN), Brussels, Belgium

Cole DA Jr, Lade PV (1984) Influence zones in alluvium over dip-slip faults. J Geotech Eng 110(5):599–615

Dong JJ, Wang CD, Lee CT, Liao JJ, Pan YW (2003) The influence of surface ruptures on building damage in the 1999 Chi-Chi earthquake: a case study in Fengyuan city. Eng Geol 71(1–2):157–179

EAK: Greek Seismic Code, Organization of Seismic Planning and Protection, Athens (2000) (in Greek)

EC8 (2002) Eurocode 8: design of structures for earthquake resistance. European Committee for Standardization (CEN)

Faccioli E, Anastasopoulos I, Gazetas G, Callerio A, Paolucci R (2008) Fault rupture-foundation interaction: selected case histories. Bull Earthq Eng 6(4):557–583

GB50011-2010 (2010) Code for seismic design of buildings. China Building Industry Press, Beijing (in Chinese)

Ishihara K (1993) Liquefaction and flow failure during earthquakes. Geotechnique 43(3):351–415

Itasca FLAC (2000) Fast lagrangian analysis of continua. Itasca Consulting Group Inc., Minneapolis

JGJ 94-2008 (2008) China Academy of Building Research (CABR), Technical code for building pile foundations, Ministry of Construction, China.

Kiani M, Akhlaghi T, Ghalandarzadeh A (2016) Experimental modeling of segmental shallow tunnels in alluvial affected by normal faults. Tunnel Undergr Space Technol 51:108–119

Land Transport Authority (LTA) (2000) Code of practice for railway protection. Development and Building Control Department, Singapore

Li TB (2008) Failure characteristics and influence factor analysis of mountain tunnels in epicenter zones of great Wenchuan earthquake. J Eng Geol 16(3):742–750 (in Chinese)
Loukidis D, Bouckovalas GD, Papadimitriou AG (2009) Analysis of fault rupture propagation through uniform soil cover. Soil Dyn Earthq Eng 29(11–12):1389–1404
Ng CWW (2014) The 6th ZENG Guo-Xi lecture: the state-of-the-art centrifuge modelling of geotechnical problems at HKUST. J Zhejiang Univ Sci A (Appl Phys Eng) 15(1):1–21
Ng CWW, Lu H (2014) Effects of the construction sequence of twin tunnels at different depths on an existing pile. Can Geotech J 51(2):173–183
Ng CWW, Van Laak PA, Tang WH, Li XS, Zhang LM (2001) The Hong Kong geotechnical centrifuge. In: Lee CF, Lau CK, Ng CWW, Kwong AKL, Pang PLR, Yin JH, Yue ZQ (eds) 3rd international conference soft soil engineering. Swets & Zeitlinger, Hong Kong, pp 225–230
Ng CWW, Van Laak PA, Zhang LM, Tang WH, Zong GH, Wang ZL, Xu GM, Liu SH (2002) Development of a four-axis robotic manipulator for centrifuge modeling at HKUST. In: Proceedings international conference physical modelling in geotechnics, St. John's Newfoundland, Canada, pp 71–76
Ng CWW, Li XS, Van Laak PA, Hou DYJ (2004) Centrifuge modeling of loose fill embankment subjected to uni-axial and bi-axial earthquakes. Soil Dyn Earthq Eng 24(4):305–318
Ng CWW, Cai QP, Hu P (2012) Centrifuge and numerical modeling of fault rupture propagation in clay with and without a pre-existing fracture. J Geotech Geoenviron Eng ASCE 138(12): 1492–1502
Ng CWW, Lu H, Peng SY (2013) Three-dimensional centrifuge modelling of the effects of twin tunnelling on an existing pile. Tunn Undergr Space Technol 35:189–199
Ng CWW, Soomro MA, Hong Y (2014) Three-dimensional centrifuge modelling of pile group responses to side-by-side twin tunnelling. Tunn Undergr Space Technol 43:350–361
Ng CWW, Hong Y, Soomro MA (2015) Effects of piggyback twin tunnelling on a pile group: three-dimensional centrifuge and numerical modelling. Géotechnique 65(1):38–51
Ng CWW, Wei J, Poulos HG, Liu HL (2017) Effects of multi-propped excavation on an adjacent floating pile. J Geotech Geoenviron Eng ASCE 143(7):04017021
Oettle NK, Bray JD (2013) Geotechnical mitigation strategies for earthquake surface fault rupture. J Geotech Geoenviron Eng ASCE 139(11):1864–1874
Taylor RN (1995) Geotechnical centrifuge technology. Blackie Academic and Professional, London
Wang WL, Wang TT, Su JJ, Lin CH, Seng CR, Huang TH (2001) Assessment of damage in mountain tunnels due to the Taiwan Chi-Chi earthquake. Tunn Undergr Space Technol 16(3): 133–150
Wang ZZ, Gao B, Jiang Y, Yuan S (2009) Investigation and assessment on mountain tunnels and geotechnical damage after the Wenchuan earthquake. Sci China Ser E Technol Sci 52(2): 546–558
White DJ, Take WA, Bolton MD (2003) Soil deformation measurement using particle image velocimetry (PIV) and photogrammetry. Geotechnique 53(7):619–631
Yao C, Takemura J (2020) Centrifuge modeling of single piles in sand subjected to dip-slip faulting. J Geotech Geoenviron Eng 146(3):04020001

Chapter 3
Liquefaction Mitigation Measures: A Historical Review

Ikuo Towhata

3.1 Introduction

Seismic liquefaction during strong earthquakes has been a very important issue of geotechnical earthquake engineering for more than half a century. Many technological efforts have been made, and nowadays, a number of mitigation measures are available. Nonetheless, recent earthquakes in Japan and New Zealand induced substantial liquefaction damage and demonstrated that still the problems are not fully solved. It is thus important to consider why the past efforts could not fully solve the problem and what are missing. From this viewpoint, the present paper attempts to review the past efforts and indicate their limitations, thereby making clear the future direction of technology.

3.2 Overview of Liquefaction-Induced Damage

To properly discuss the mitigation of liquefaction problem, it is essential to understand the damaging mechanism. As is well known, liquefaction converts stable sandy (cohesionless) subsoil to extremely soft medium in which the effective stress is very low and pore water pressure is high. Consequently, the bearing capacity of subsoil is lost and the surface structures suffer serious loss of bearing capacity. Figure 3.1 illustrates one of the most famous examples of liquefaction damage that occurred in Niigata in 1964 during the Niigata earthquake of *M*w (moment magnitude) = 7.6. This site used to be a part of the channel of the Shinano River that was 400-m wide. In 1922, the main stream was diverged in the upstream section and the flood flow in Niigata City decreased. This allowed the local community to fill one part of the channel in 1920s with readily available beach sand and

I. Towhata (✉)
Kanto Gakuin University, Yokohama, Kanagawa, Japan

T. G. Sitharam et al. (eds.), *Latest Developments in Geotechnical Earthquake Engineering and Soil Dynamics*, Springer Transactions in Civil and Environmental Engineering, https://doi.org/10.1007/978-981-16-1468-2_3

to convert the channel to a residential area. The sand filling was conducted by just throwing sand into water. Hence, the consequent ground condition was loose sandy water-saturated and therefore prone to liquefaction. 40 years later when this site was still very young in a geological sense, liquefaction did happen. The acceleration records in the basement of one of the tilted buildings show PGA (maximum acceleration) equal to 159 and 155 cm/s^2 in EW and NS directions, respectively.

Light underground structure floats when the surrounding subsoil liquefies. Floating occurs when the net unit weight of the structure is lighter than that of the surrounding water-saturated liquefied sand which is approximately 18–20 kN/m^3. Figure 3.2 shows one of the examples of such floating that occurred in Yokohama City during the 2011 Tohoku earthquake (Mw = 9.0). The relatively new appearance of this floated underground car park suggests that backfill subsoil was young. PGA at the nearby K-Net Yokohama site (KNGH10) was 165 and 138 cm/s^2 in EW and NS directions, respectively.

Fig. 3.1 Tilting of apartment building in Kawagishi-cho site of Niigata (by reconnaissance team, department of Civil Engg, Univ. Tokyo)

Fig. 3.2 Floating of underground parking lot in Yokohama City in 2011

Fig. 3.3 Liquefaction-induced lateral displacement of quay wall in Kobe Harbor

Liquefaction induces displacement not only in the aforementioned vertical directions but also in the horizontal direction. Being called lateral flow, the horizontal displacement is driven by gravity and hence is oriented from higher to lower places. Figure 3.3 shows such displacement of a quay wall in Port Island of Kobe Harbor in 1995. The lateral displacement was at maximum 7 m in this harbor and the port operation was suspended for two years. This long term of suspension profoundly affected the harbor business and local economy. This is an important evidence that business continuity planning (BCP) is important in geotechnical problems.

Kobe Harbor had not been prepared for strong earthquakes. When Port Island was constructed by discharging crushed weathered granite into the sea, the most important geotechnical problem was consolidation settlement in the soft marine clay. Earthquake did not attract much concern because the previous strong earthquake in the region was the one in 1596 with magnitude of around 7.0. Afterward, no significant event had happened and the local community had forgotten the memory. Empiricism is thus not reliable.

Figure 3.4 presents the vertical variation of PGA in a liquefied area of Port Island. The top 16 m is the liquefaction-prone man-made island. The PGA at the surface is weaker than those in the natural unliquefied layers. It is thus important that liquefaction reduces the shear rigidity and strength of soil, making the liquefied sandy layer a natural base isolator that cannot transmit strong shear wave (S-wave) to the surface. Accordingly, the surface acceleration becomes weak. Although the reduced acceleration may allow to assume smaller seismic design load, it is not good news because permanent displacement of structures (tilting, floating and lateral displacement) affects their serviceability of structures. One may be able to expect the liquefaction-induced base isolation when the surface crust is thick and stable. In such a case, the distance between the shallow foundation and the liquefiable deep layer is long, and presumably, the lateral size of liquefiable

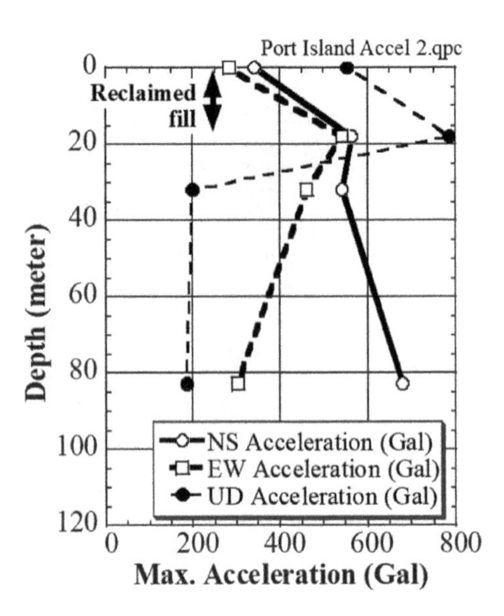

Fig. 3.4 Vertical variation of PGA in Port Island of Kobe Harbor in 1995 (Development Bureau, City of Kobe)

layer has to be big enough to affect the surface structure. Although much is known about the size of local liquefiable sandy pocket, it seems difficult to expect such a condition.

The irony in Fig. 3.4 is that the unfavorable liquefaction in subsoil reduces the acceleration and seismic force in superstructures, while prevention of liquefaction reduces this 'benefit.' One may apt to prevent liquefaction to the optimum extent, thereby keeping the subsoil stable and maintaining the surface acceleration still low. This is an attractive but difficult task because mechanical behavior of loose sand near liquefaction is full of uncertainty. To date, engineers have been trying to avoid onset of liquefaction and make superstructures resistant or resilient against strong shaking by such measures as aseismic design, base isolation and seismic control.

3.3 Causative Mechanism of Liquefaction and Principles of Its Mitigation

The mechanism of onset of liquefaction is illustrated in Fig. 3.5. The situation starts with a loose granular structure of sand particles. When strong earthquake occurs, the particles are subject to oscillated shear stress and move back and forth (Fig. 3.5a). This cyclic motion of particles makes them fall into big voids among grains (Fig. 3.5b; negative dilatancy). If sand is dry, this negative dilatancy occurs immediately and the overall volume of sandy ground is quickly compacted. In reality, the sand is saturated with incompressible water and the volume contraction is required to push the water out of the ground surface. This drainage process takes tens of minutes that is much longer than the duration of seismic shaking. In the

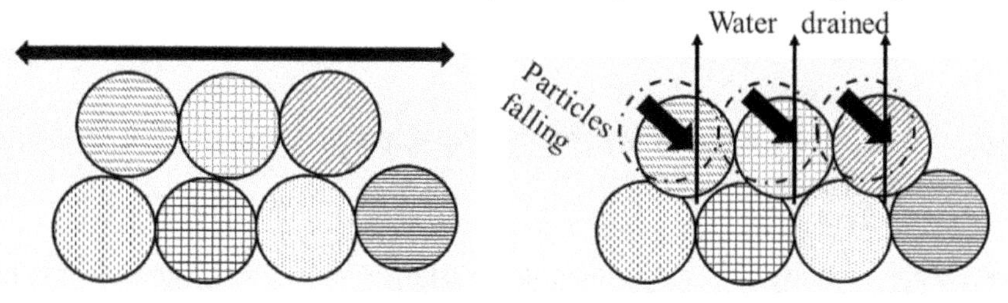

Fig. 3.5 Process toward onset of liquefaction in loose water-saturated packing of sand particles

meantime, particles cannot fall down into voids that are occupied by water. Therefore, during shaking, sand particles are suspended (floating) in pore water. In this phase, the particle-to-particle contact disappears. In other words, the effective stress disappears and shear strength of sand becomes null. Thus, sandy ground behaves very similar to mud water. In brief, the procedure toward liquefaction-induced ground deformation is characterized by

- Cyclic motion of sand grains that have no bonding with one another,
- Tendency to fall into voids among loosely packed particles,
- Incompressibility of pore water occupying the voids, and
- Long time for pore water to flow out of the ground surface.

This situation is found in

- Young man-made islands,
- Abandoned river channel and ponds, filled with sand naturally or artificially and
- Backfill of lifeline facilities.

The story in Fig. 3.5 clearly suggests how to prevent liquefaction; those that happen in Fig. 3.5 should be prevented or altered. Table 3.1 indicates the relation between what happens during liquefaction and how to prevent them. It deserves attention that there is a different viewpoint about liquefaction damage mitigation. It is said therein that liquefaction damage is tolerated and that we have to be prepared for quick restoration. This idea is less costly and meaningful for inexpensive structures because seismic liquefaction does not occur frequently and inexpensive

Table 3.1 Principles of mitigation of liquefaction disaster

Process toward liquefaction	Principles of mitigation
a Loose packing of sand particles	→ Compaction/densification
b Falling of particles into big voids	→ Bonding/grouting of grains
c Significant cyclic motion of particles	→ Mechanical constraint against ground motion
d Long time needed for drainage	→ Installation of water channel; gravel drains
e Incompressibility of pore water	→ Dewatering or injection of gas bubbles

infrastructures may not be able to afford such costly mitigation measures as suggested in Table 3.1.

3.4 Mitigation Measures for Newly Constructed Structures

Installation of mitigation measures prior to construction has a clear and straightforward procedure. The best kind of mitigation is chosen by considering the in situ condition as well as financial restriction. This section classifies the available measures into two groups. The first group aims to prevent the onset of liquefaction, whereas the second intends to allow some extent of liquefaction and damage (displacement) for the purpose of cost saving. Note that 'prevention of liquefaction' in this paper means that the triggering of liquefaction is prevented and that the calculated factor of safety against liquefaction (F_L) is greater than unity under the specified design earthquake load.

As known well, the factor of safety is calculated by the following formula in practice;

$$F_L \equiv \frac{\text{CRR}(\text{SPT}-N, \text{Grain size, Plasticity etc.})}{\text{CSR}(\text{PGA, Depth})} \tag{3.1}$$

in which 'CRR' stands for the soil's resistance against liquefaction and is a function of SPT-N or other sounding data, grain size and, if possible, plasticity index and other physical properties. 'CSR' in Eq. 3.1 designates the seismic load and is a function of PGA at the ground surface together with the depth from the surface. Strictly speaking, PGA at the surface is influenced by onset or mitigation of liquefaction (Fig. 3.4), and design cannot specify PGA. Nevertheless, practice specifies PGA in the range of 0.15–0.35 g as an input parameter (g = gravity acceleration). Note that this use of PGA can be replaced by that of the amplitude of cyclic shear stress together with the effective vertical stress prior to earthquakes, if the cyclic stress can be assessed by dynamic response analysis.

3.4.1 Prevention of Liquefaction for New Structures

This section supposes a situation in which mitigation measure is installed in an open space, prior to construction of structures. In an open space, the installation is subject to minimum constraint. In this respect, Mitchell (2008) classified existing technologies into densification, drainage, void filling (grouting), underground columns and walls, and remove/replace. This section touches upon densification and drainage that are often practiced for new structures, while others will be discussed in later sections.

Compaction/Densification As shown in Table 3.1, compaction is an effective idea to prevent liquefaction. SPT-*N* value of compacted sand is made greater and CRR as well as F_L value in Eq. 3.1 increases. The effect of compaction lasts permanently without maintenance efforts. This is an important advantage of compaction.

Sandy ground is compacted either by underground vibration or surface tamping. The former is able to compact down to deeper depth (typically GL-20 m or deeper), while the latter can compact within a few meters from the surface but with lower cost.

The former pushes additional sand or gravel into subsoil and achieves compaction. Its earliest example was vibroflotation by which horizontal shaking was induced by a device at the bottom of a machine and the surface depression was filled with additionally supplied sand/gravel (D'Appolonia 1954). This technology was applied to liquefaction mitigation (Mogami and Watanabe 1957) and successfully protected oil tank foundation during the 1964 Niigata earthquake (Watanabe 1965); see Fig. 3.6. After this achievement, compaction became trusted for mitigation of liquefaction.

Stone columns can achieve compaction effect as well if they are installed by vibration or other measures without pushing the existing soil out of the surface. If the employed stone has permeability higher than the surrounding soil, additional effect of accelerated drainage and dissipation of pore water pressure (Table 3.1d) can be expected. Moreover, similar compaction effect can be achieved by driving timber piles into subsoil (Mitchell and Wentz 1991; Numata et al. 2012).

Sand compaction pile is a newer method for densification of sandy ground. A vibrator is placed at the top of a device and a casing pipe with sand goes down with vibration. After reaching the specified depth, the bottom of the casing is opened and the casing moves back and forth in the vertical direction, thus enlarging the diameter of the column made of the newly pushed sand. Both original and new body of sand are thus made dense. To facilitate the migration of sand out of the casing, pressurized air is provided into the casing. This air injection reduces the degree of saturation in the ground and further improves the mitigative effect (Table 3.1e).

(a) Without compaction

(b) With vibroflotation

Fig. 3.6 Performance of oil tank foundation on liquefaction-prone sand in Niigata in 1964 (Watanabe 1965)

The problem of vibratory compaction was its noise and ground vibration both of which disturbed the environment. Possible damage to fragile structures in the vicinity was a problem as well. The revised machine for sand compaction pile employs static rotation of the casing around the vertical axis and does not cause such problems. Figure 3.7 illustrates the procedure of the static construction of a sand compaction pile. Figure 3.8 shows the extent of noise near the static compaction. The author did not feel noise or ground vibration while watching the procedure. Figure 3.9 depicts the situation after the 2011 Tohoku earthquake where compaction was going on and liquefaction occurred only in the uncompacted part.

Another short coming of compaction is the induced lateral displacement of subsoil. This is a natural consequence of compaction, or installation of additional material into ground. Therefore, compaction in the vicinity of existing structures has to be avoided. The minimum distance from existing structures is said to be 5 m but the author has seen a case at about 1 m.

Compaction by hitting the ground surface is obviously less costly than underground compaction. However, its effect may be limited to shallow depth. Dynamic consolidation is one of the technologies of this type (Ménard and Broise 1975) in which a heavy weight of 10 tons falls freely from the height of 20 m or so, and this falling energy is spent on making a big depression at the surface. This depression is filled with sand and compaction is repeated until desired density is attained. Hansbo (1978) discussed the application of this technology to compaction of sandy ground. Because of its lower cost, dynamic consolidation has been applied to mitigation of liquefaction (Tanaka and Sasaki 1989). The author found a verification of mitigation by this technology at Lukang in Taiwan after the 1999 Chi-chi earthquake ($Mw = 7.7$). At the time of the earthquake, dynamic consolidation was going on as mitigation of future liquefaction (Hwang et al. 2003) and only the unimproved area suffered liquefaction (Fig. 3.10). Figure 3.11 illustrates the variation of soil properties before and after dynamic consolidation. Shen et al. (2018) conducted more elaborate studies on this case. Thus, improvement by dynamic consolidation is evident.

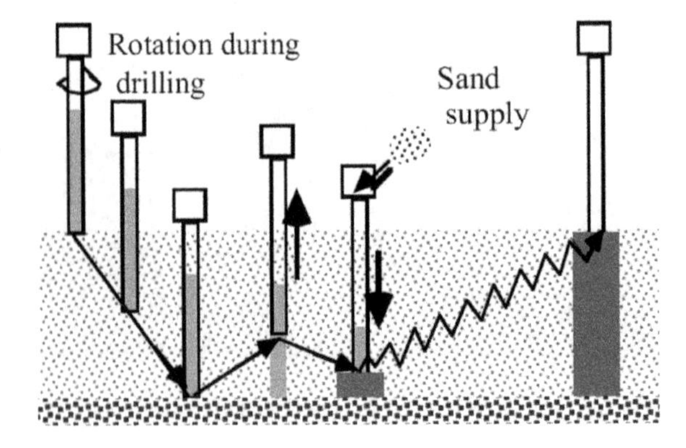

Fig. 3.7 Procedure of static compaction (after Fudo-Tetra Company)

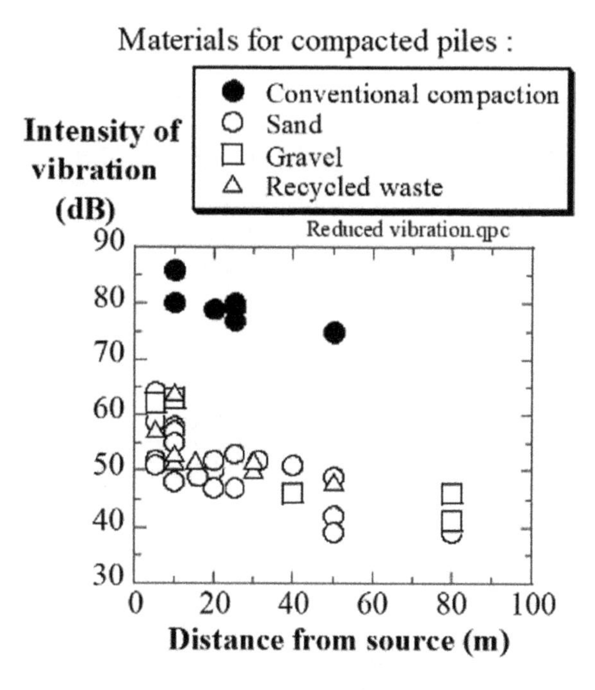

Fig. 3.8 Noise measured near the static compaction (Fudo-Tetra Company)

Fig. 3.9 Mitigative effect of static compaction demonstrated by the 2011 Tohoku earthquake

It is possible in practice to employ smaller tampers to compact only the top few meters, leaving the lower subsoil still prone to liquefaction. This topic will be addressed later in Fig. 3.21.

Grouting Mixing of sand with cement prevents free movement of sand particles during earthquakes, thus preventing the falling of particles into voids (Table 3.1b).

(a) Improved area without liquefaction

(b) Liquefaction and sand ejecta in unimproved area

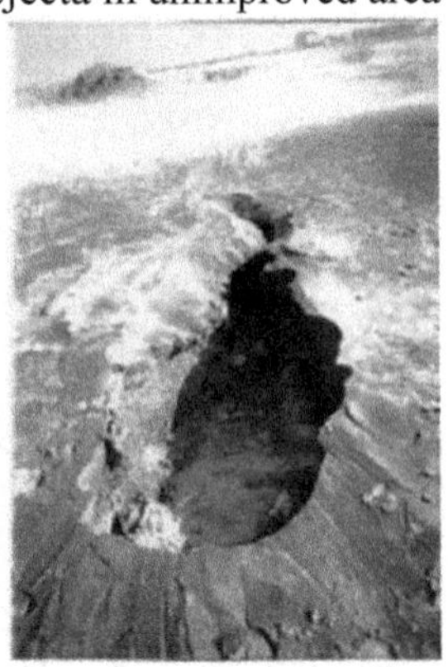

Fig. 3.10 Mitigative effect of dynamic consolidation in Lukang of Taiwan demonstrated during the 1999 Chi-chi earthquake (Kaiyo Kogyo Company)

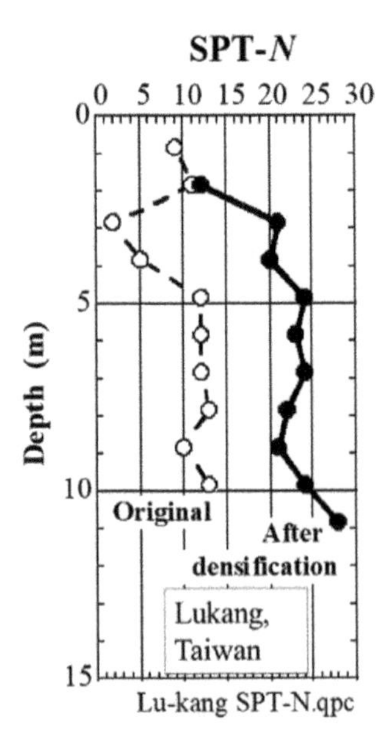

Fig. 3.11 SPT-N values before and after dynamic compaction

The most frequently used bonding material is cement and is mixed with soil either by a mechanical mixing machine (Fig. 3.12) or by a jet grouting machine. The former machine is more reliable but bigger and heavier than the latter.

The good point of grouting is its high reliability without need for future maintenance, if installed successfully, as well as its quiet construction without noise and ground vibration. Moreover, no lateral displacement of ground occurs because soil

Fig. 3.12 Ongoing grouting by mechanical mixing

is not compacted but only mixed with cement. Thus, ground improvement in busy urban area is possible. On the contrary, in case of jet grouting, treatment of waste muddy water is costly due to environmental regulation.

The grouting technology started with overall improvement; the improvement ratio $Ar = 100\%$. This was however soon found to be costly, typically 10 times more expensive than sand compaction of the same volume of soil.

Efforts for cost reduction resulted in underground walls of grid configuration that constrain cyclic shear deformation of subsoil during shaking (Table 3.1c and Fig. 3.13). This technology had been installed in the foundation of a hotel building in Kobe Harbor before the 1995 earthquake. Since the hotel building successfully survived the disaster among extremely damaged harbor structures, this technology came to be trusted (Uchida and Konishi 2015).

The major issue in design of grid-type wall is the decision on the interval of walls. Basically, the narrower the interval is, the more effective is its mitigation of liquefaction. In early 1990s, it was thought that the interval should be 80% or less of the thickness of liquefaction-prone sand (Fig. 3.14). Although this idea came

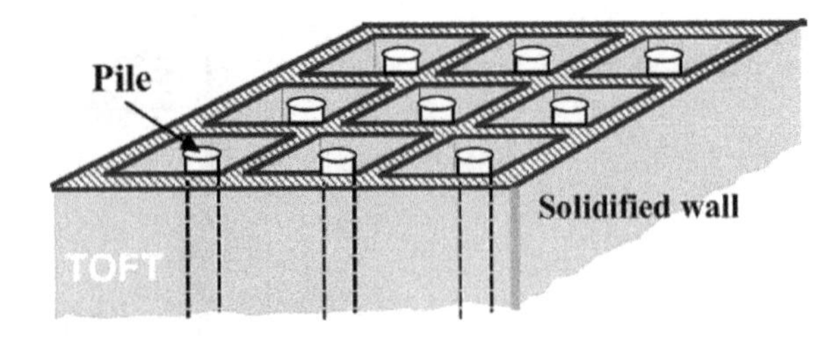

Fig. 3.13 Constraint of shear deformation of soil by grid-type underground walls

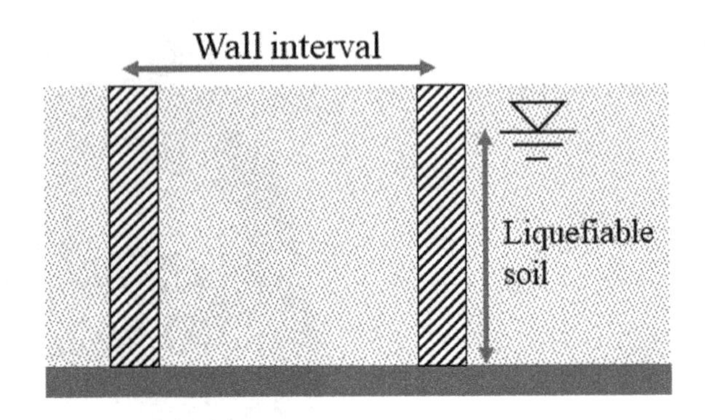

Fig. 3.14 Schematic illustration on relationship between thickness of liquefaction-prone soil and wall interval

from observation in many shaking model tests, it was prone to misunderstanding that the thicker the liquefaction-prone soil is (the stronger the design earthquake is), the wider interval is allowed. This misunderstanding was totally wrong, and nowadays, the interval is decided by running dynamic FE analyses in which the induced shear stress amplitude is evaluated and put in the calculation of the factor of safety, F_L (Eq. 3.1). A simplified method (Taya et al. 2008; Uchida et al. 2016) may be useful in the preliminary stage of foundation design, although their applicability is limited to a certain extent.

Gravel drains for quick dissipation of excess pore water pressure The essence of liquefaction-induced damage is the unacceptably large deformation (Figs. 3.1–3.3) that is induced when the effective stress is very low. To terminate this critical stress condition, measures to quickly dissipate the high excess pore water pressure are executed by installing gravel columns. Because the permeability of gravel (grain size is of the order of mm) is 10 times or more higher than that of sand, ground water changes its flow from the long vertical path to the short horizontal one (Fig. 3.15). Accordingly, the consolidation time needed to dissipate the high excess pore water pressure is shortened from tens of minute to seconds. Thus, liquefied subsoil does not have time to deform significantly.

The mitigative effects of gravel drain were verified in Kushiro Harbor during the 1993 Kushiro-oki earthquake of $M\text{w} = 7.6$. While the untreated subsoil liquefied profoundly, the quays with gravel drain installation did not suffer from liquefaction (Iai et al. 1996). After the disaster, it was confirmed that the remaining body of gravel drain did not have a problem of clogging (Fig. 3.16). Prior to this

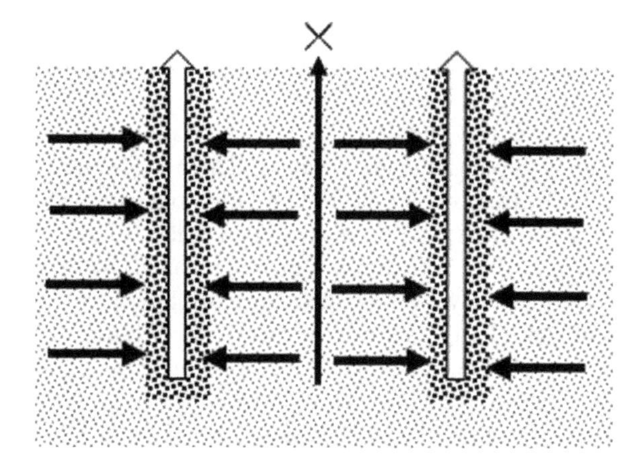

Fig. 3.15 Change of flow path by installed gravel drains

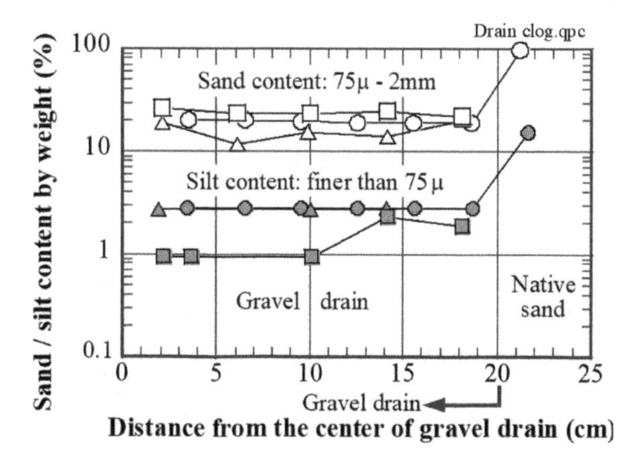

Fig. 3.16 Grain size in gravel drains and surrounding sand after liquefaction during the 1993 Kushiro-oki earthquake: excavation of gravel drains after earthquake (Iai et al. 1996)

earthquake, a series of full-scale shaking model tests had been conducted at the Port and Harbor Research Institute (Iai 1988), and it had been found that the pore pressure rise was held significantly lower within the gravel drain than in the surrounding liquefaction-prone sand. To facilitate design of gravel drains, Ito et al. (1991) carried out laboratory tests wherein water flow and possible grain migration were reproduced. Accordingly, it was proposed that the grain size ratio of D (Gravel drain; 15% passing)/D (Sand; 85% passing) <9.2 keeps the sand migration within 2 cm from the gravel-sand boundary. Note that this ratio (=9.2) is greater than what is often employed for filters because gravel drains works for a short duration of earthquake shaking while filters have much longer service time. It is also interesting in their study that shaking during the tests did not affect the extent of clogging.

Prefabricated vertical drain (PVD) has a very similar function as gravel drains. Sassa et al. (2017) addressed a successful case history of its use during the 2011 Tohoku earthquake. The spacing of drains therein was 45–65 cm. This short interval comes from the limited water flow capacity of PVDs with small cross section. With regard to this shortcoming, Rollins et al. (2004) proposed the use of bigger diameter of 75–150 mm, thereby making the interval equal to 1–2 m. This reduced the construction cost.

Noteworthy, however, is that dissipation of pore water pressure is associated with a certain amount of consolidation settlement. When the author visited the Kushiro Harbor after the 1993 Kushiro-oki earthquake of magnitude = 7.5, the site of gravel drains had developed more or less 15 cm of settlement (as illustrated later in Fig. 3.19), which was quickly repaired by overlaying pavement, while the important quay wall avoided large displacement. Thus, the gravel drains quickly terminate the critical state of very low effective stress in subsoil and reduce the unfavorable residual deformation of structures.

Figure 3.17 illustrates the installation of gravel drains in the vicinity of fragile wooden houses. The advantage of gravel drain lies in very limited environmental disturbance during installation (the casing has spirals on the outside and soil is moved vertically toward the surface; not in the lateral direction) and lack of noise/vibration.

The mitigative effect depends on the spacing among drains. Current design practice depends on seepage analysis (Seed and Booker 1977) by which variation of pore water pressure with time is calculated while considering the generation of pore pressure by seismic shaking. Because laboratory tests indicate that sand becomes

Fig. 3.17 Installation of gravel drains

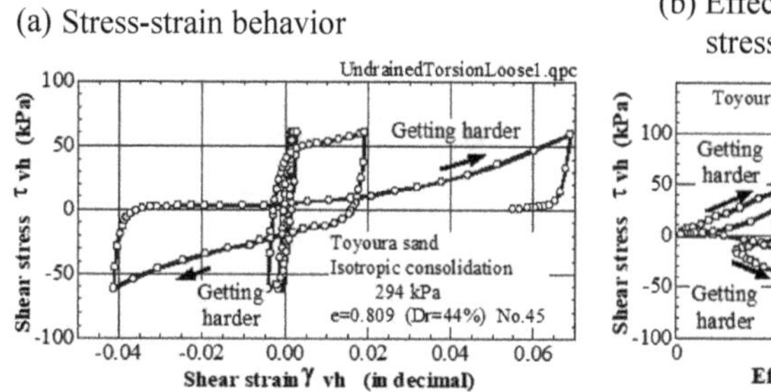

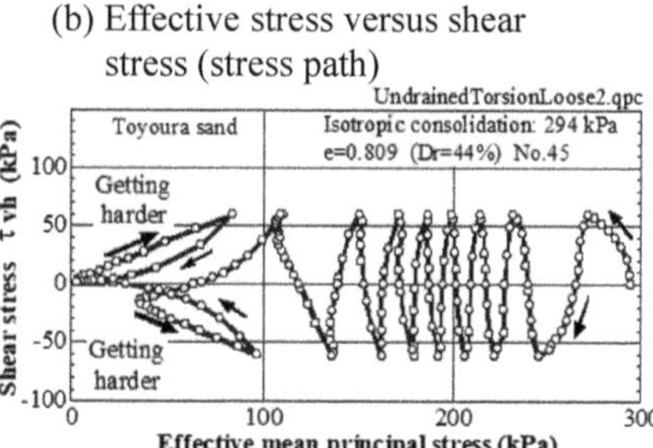

Fig. 3.18 Typical behavior of loose sand undergoing cyclic undrained test in torsional shear device (Towhata 1982)

substantially softer and strain increases when the effective stress in the original ground becomes less than 50% of its initial value (Fig. 3.18), the design of gravel drains aims to keep the effective stress more than 50% of the initial value. This aim becomes substantially difficult when the design earthquake (PGA in Eq. 3.1) becomes stronger (recent trends) and the rate of pore pressure generation becomes faster. Note that Onoue et al. (1987) studied the effects of the permeability in the gravel drain shaft that is hypothesized to be infinite in Seed and Booker's original theory.

Noteworthy is that gravel drain allows pore water to come out of the ground. Hence, the ground surface depresses (Fig. 3.19). This deformation can be easily restored by overlaying asphalt etc. and is not considered to be damage.

Nowadays, Japanese practice does not make much use of gravel drains because of the increased level of design earthquake after the 1995 Kobe earthquake; see Fig. 3.20. The author misses the above-mentioned good points of the gravel drain and supposes that the negative situation today is overconservative. Therefore, the following points have to be studied so that the gravel drain technology may revive.

- Although excess pore water pressure may exceed 50% level in the liquefaction-prone subsoil, the effective stress inside the drains is held high. This implies that drain columns maintain certain rigidity and prevent development of large deformation in the surrounding subsoil.
- Since the permeability of gravel drain is held high, the high excess pore water pressure in the liquefaction-prone soil dissipates quickly. There is no time for the subsoil to develop large deformation.
- Gravel drains designed against an old weaker design earthquake successfully prevented disaster during the stronger shaking caused by the 2011 Tohoku earthquake (Sasaki et al. 2012).
- The future of gravel drain technology should be studied from the viewpoint of seismic performance design in which induced deformation is assessed and judged whether it is acceptable or not.

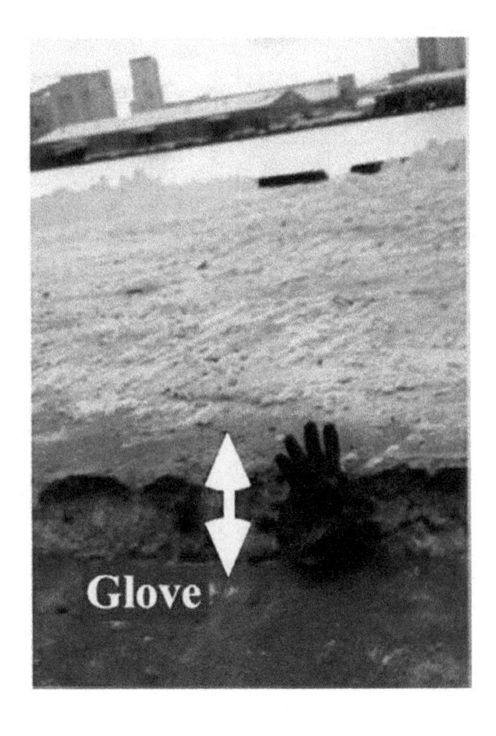

Fig. 3.19 Subsidence of ground surface after earthquake due to drainage of ground water through gravel drains

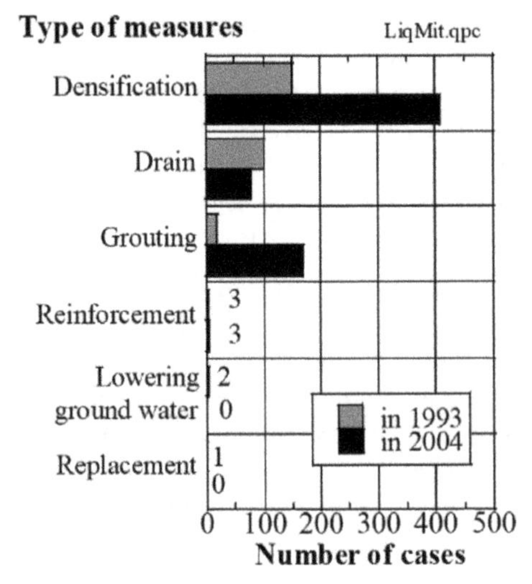

Fig. 3.20 Market share of liquefaction mitigation measures before and after the 1995 Kobe earthquake

3.4.2 Allowing for Limited Extent of Liquefaction for New Structures

Not all kinds of structure require full prevention of subsoil liquefaction. From the viewpoint of cost versus benefit, inexpensive structure may accept a certain extent of deformation induced by liquefaction. This is particularly the case for personal houses and lifelines for which funding for seismic safety is very limited.

It has been known for a long time that liquefaction in deeper soil is less hazardous to surface structures than that in shallow soil (Tatsuoka et al. 1980). Asada (1998) reported his past reconnaissance study on damage of houses that was caused by subsurface liquefaction during the 1983 Nihonkai Chubu earthquake of $M\mathrm{w} = 7.7-7.9$. He inferred that damage extent is reduced if surface unliquefied crust is sufficiently thick. This means that thick surface crust is able to support the weight of a light structure such as personal house. Also, no underground structure moves if embedded in this surface crust. Based on this knowledge, Ishihara (1985) proposed a simple design chart. Figure 3.21 plots H_1 (thickness of unliquefied surface crust) and H_2 (thickness of underlying liquefiable soil) for safe and unsafe situations under specified PGA. It is inferred that liquefaction at depth does not affect the ground surface if H_1 is big enough. A thick surface crust can develop sufficient bearing capacity and resistance against punching failure under the weight

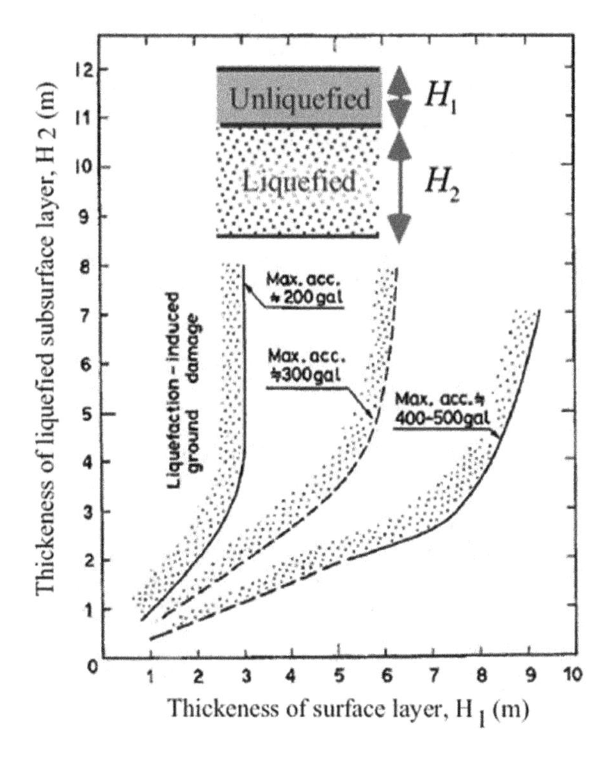

Fig. 3.21 Chart for safety judgment based on thickness of surface unliquefied crust and thickness of underlying liquefiable layer (Ishihara 1985)

of light structures (two-storeyed house, for example). In Fig. 3.21, the significance of H_1 is thus clear.

A question arises on the significance of the thickness of H_2. The current practice for judgment of liquefaction ($F_L < 1$ or not) relies on SPT-N (Eq. 3.1) and minor difference in N value drastically affects this Yes-or-No judgment and the values of these layer thicknesses. In this regard, a revision was proposed in which H_1 is kept unchanged because of its clear significance and H_2 was replaced by P_L that is less affected by N value's error. This P_L is a weighted average of $(1 - F_L)$ over the top 20 m of soil (Tatsuoka et al. 1980). The new proposal is depicted in Fig. 3.22.

The aim of Fig. 3.22 is safety evaluation of subsoil in residential land. The owners of such land are individuals who do not or cannot afford expensive ground improvement works. Different from important infrastructures that are required to be safe during strong earthquakes for the sake of resilient community, houses do not have to be very stable during strong but rare seismic events. In this regard, the calculation of F_L values in Fig. 3.22 considers the aging effect that increases the liquefaction resistance of aged sand (Towhata et al. 2016b). With this provision, the assessed F_L value is greater than conventional liquefaction risk assessment and private owners can avoid the expenditure for possibly unnecessary ground improvement.

Lifeline is prone to liquefaction (Figs. 3.23 and 3.24) and induced ground deformation (Fig. 3.25). Sewage pipeline is prone to liquefaction damage because it is often embedded at susceptible depth and pipe connection is easy to be separated upon ground displacement. Liquefied sand (often backfill sand) comes through this separation into pipe and makes liquid flow difficult. Its cleaning is a tiresome job. Also, possible change of the pipeline slope hinders liquid flow under gravity. Except very important trunk lines, lifelines are of very limited construction budget per unit length. Moreover, the land where lifelines are buried is hardly owned by

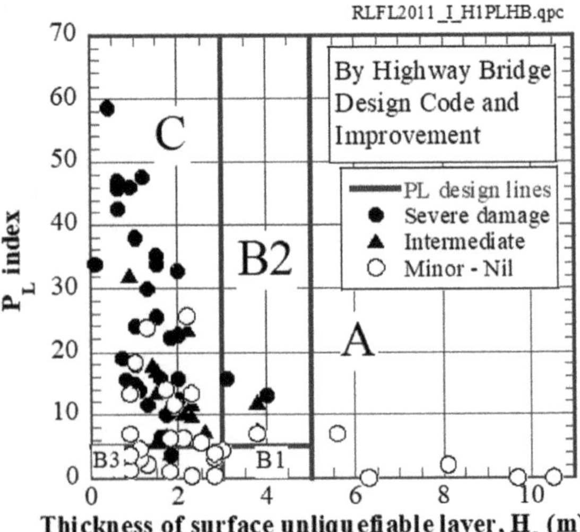

Fig. 3.22 Revised chart for safety of houses by means of H_1 and P_L; A: liquefaction unlikely, B: low probability and C: high probability of liquefaction (Towhata et al. 2016a)

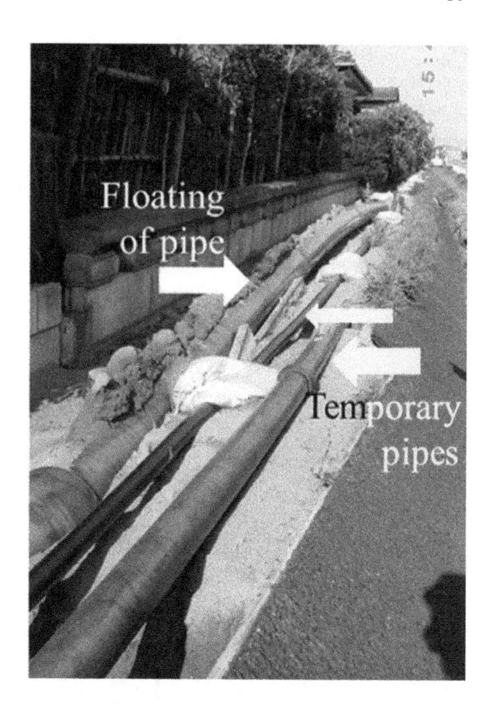

Fig. 3.23 Floating of sewage pipeline as a consequence of liquefaction (Itako, Japan, during the 2011 Tohoku earthquake)

Fig. 3.24 Liquefied sand coming into sewage pipeline from disconnection after 2011 Tohoku earthquake (Urayasu City Government)

the lifeline operator. Therefore, it is difficult for the operator to improve ground along the lifeline route. To date, what can be done is the use of flexible pipe material (steel in place of iron), flexible pipe joint or liquefaction-resistant backfill (sand mixed with 2% cement). Note that it is difficult to excavate cement-mixed backfill sand later for maintenance or replacement of pipes.

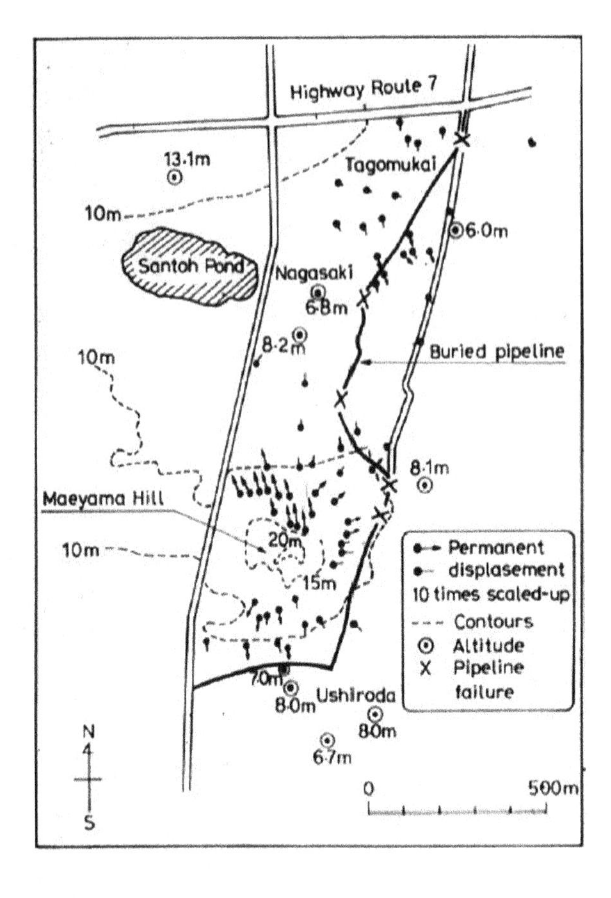

Fig. 3.25 Bending, separation and compressional failure of gas pipeline due to lateral flow of liquefied sandy slope in Noshiro City during the 1983 Nihonkai Chubu earthquake (data after Hamada et al. 1986)

3.5 Mitigation Measures Under Existing Structures

In the early phase of economic development of nations, the public concern addresses planning and construction of new infrastructures, whereas safety and environmental protection are of the secondary value. It is unfortunate but true that the value of human life is not very high. After some stage of development is achieved, concern increases about safety against natural disasters, inclusive of subsoil liquefaction. Then the problem is how to protect the built environment (existing structures) from liquefaction underneath. The ground surface has been occupied by structures, and there is no more space for big machines to improve soil. Possible ground displacement upon compaction affects existing structures as well. Solution of such a problem is difficult and 'costly.' This section discusses the issue of existing infrastructures for which public sectors can afford necessary and reasonable cost. Issues of liquefaction-prone residential land with existing houses will be discussed in Sect. 3.6.

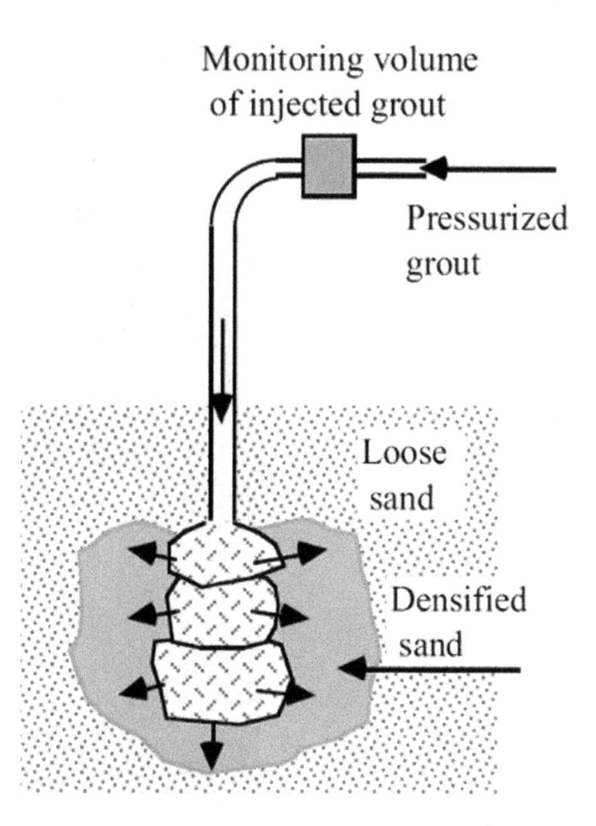

Fig. 3.26 Schematic sketch of compaction grouting

3.5.1 *Prevention of Liquefaction Under Existing Structures*

Compaction grouting Compaction grouting (Fig. 3.26) pushes grout into subsoil. The volume of this pushed-in material causes volumetric contraction of the original sand. By repeating the procedure at different depths, a column is formed to support the superstructure. Because the machine is small, the entire procedure can be performed in the basement floor of an existing building (Fig. 3.27). Mechanics of sandy soil suggests that static compression of cohesionless soil is not easy but the practice of compaction grouting pulsates the pressure of grout, thus enabling significant volume contraction of the surrounding sand. This situation is similar to sand compaction pile in which the casing moves back and forth in the vertical direction, thus pulsating the in situ stress and facilitating compaction (Fig. 3.7).

Injection of colloidal silica Injection of colloidal silica follows different philosophy. Liquid of very low viscosity seeps into sandy ground. Because the size of the resolved silica particle is smaller than micron, the liquid can flow into fine sand and silty sand that are highly prone to liquefaction (Yonekura and Kaga 1992). The pressure needed is not so high as for compaction grouting. Therefore, the needed pumping machine is small. In a few days after injection, the liquid gets solidified

Fig. 3.27 Ongoing compaction grouting in Yokohama; the floor of basement was removed

(Fig. 3.28) and prevents liquefaction. Injection can also be performed through an obliquely drilled hole (Fig. 3.29). Hence, improvement under existing structure is possible. Similar technology has been studied by Gallagher et al. (2006, 2007) as well. The chemical composition of the employed liquid is not open to the public.

When the target soil is composed of coarser grains, microcement suspension (particle size being of order of a few microns) can seep into ground and improve the soil at lower cost (Shimoda et al. 1979; Wang et al. 2017).

Durability of colloidal silica grout is an important issue of concern. Sasaki et al. (2019) reported unconfined compression strength of soil that was improved long ago by colloidal silica. Soil samples were collected from the site and tested in the laboratory. Figure 3.30 illustrates the variation of the strength with time for both liquid-type grout (ASF) and suspension of fine particles (HBS). It is shown therein that no decay happened for more than ten years. Further, because the gel time (time for solidification) is short, no health problem has been known about colloidal silica grouting.

Fig. 3.28 Mass of sand solidified by injection of colloidal silica

Fig. 3.29 Injection of colloidal silica through inclined drilled hole

(a) Temporal change of unconfined compression strength

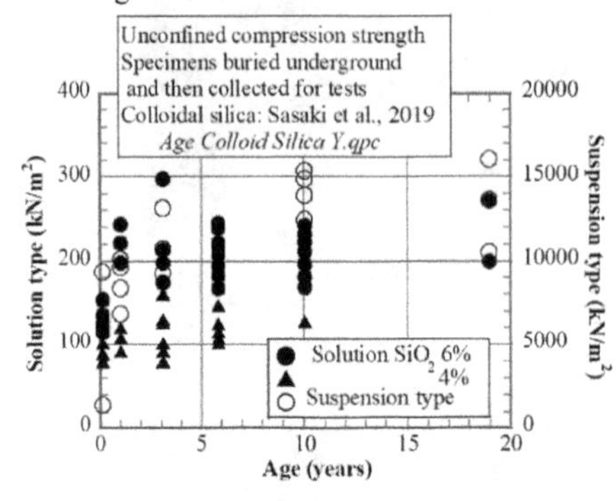

(b) Change of liquefaction resistance with age from 3 to 10 years

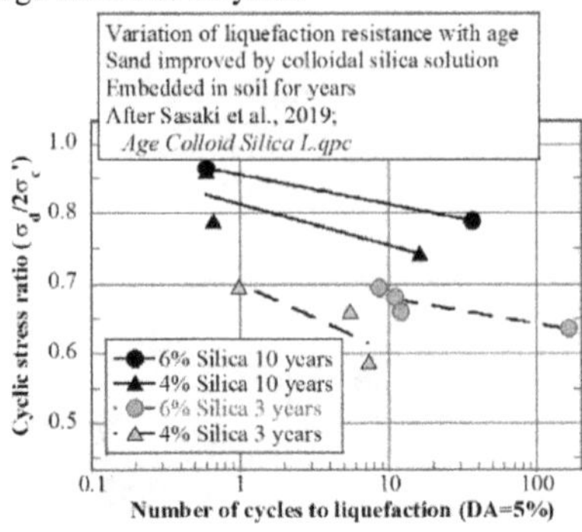

Fig. 3.30 Durability of soil improved by colloidal silica (Sasaki et al. 2019)

Dewatering Liquefaction resistance can be drastically improved by reducing the degree of saturation down to, typically, 90% or less (Table 3.1e). Among several possible technologies for this aim, lowering of ground water level has been in practice. Concerns about safety of oil tanks against liquefaction disaster increased in 1980s. The Japanese Government installed a new safety regulation to increase the spacing among tanks. Because this regulation was applied only to new construction, existing tanks were able to continue operation without retrofitting, which was good for the oil refinery industries. However, this situation changed afterward and older tanks were requested to install safety measures. Most of the available ground improvement technologies were found unacceptable by industries because temporary removal of existing tanks during ground improvement and reinstallation later were considered new construction by the law and stricter rule had to be applied, i.e., greater spacing among tanks and accordingly a smaller number of tanks in the refinery. Since the reduced number of tanks meant less business, one of the refineries in Kawasaki near Tokyo decided to lower the ground water level while maintaining all the refinery facilities untouched (Kawasaki City, year unknown; Tsue 2013). Figure 3.31 illustrates the idea in which the refinery site is surrounded by impervious underground wall and water inside the wall was pumped up. The impervious layer at the bottom together with the surrounding wall reduced the water supply from outside. No consolidation problem has occurred due to favorable ground conditions.

Structural measure In case that soil improvement is not feasible, liquefaction mitigation resorts to structural measures. The first choice is pile foundation. End bearing piles can maintain stability of high-rise buildings even during liquefaction. Noteworthy is the reduction of lateral resistance of liquefied subsoil around pile. One may suppose that the effective stress is null in liquefied soil and that no lateral resistance can be expected from such soil. However, laboratory shear tests (Fig. 3.18) indicate that the stress–strain curve is of banana shape and attains certain

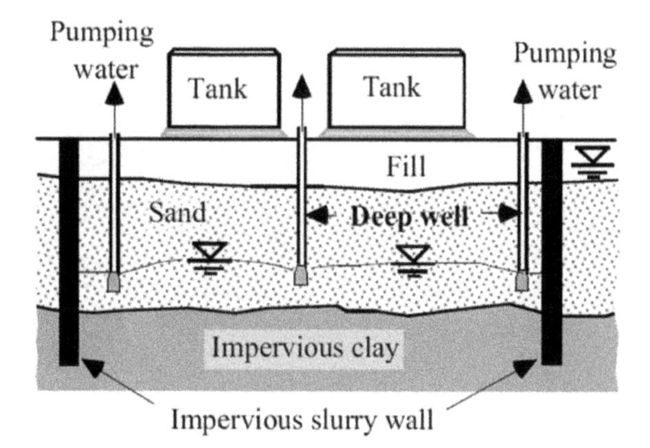

Fig. 3.31 Example of oil refinery with ground water lowering for liquefaction mitigation

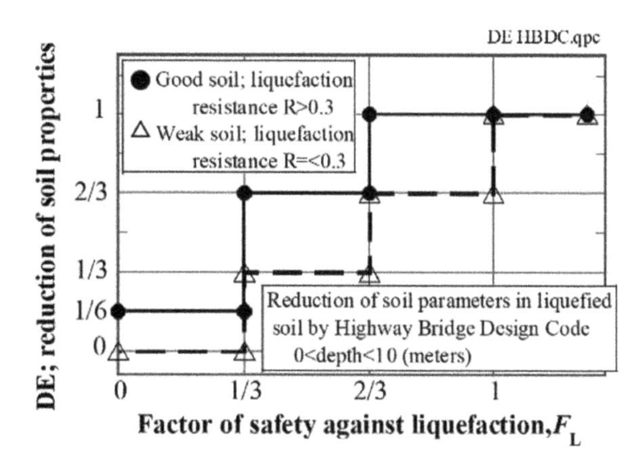

Fig. 3.32 Example of reduction of soil properties by extent of liquefaction (after Highway Bridge Design Code)

shear resistance after deformation. Thus, the lateral stiffness and ultimate resistance of pile should be reduced to a reasonable extent (not to zero) in accordance with the extent of liquefaction. In other words, different F_Ls correspond to different extents of reduction. This provision is available in the Highway Bridge Design Code of Japan (Fig. 3.32) and its experimental back ground was provided by Iwasaki (1981). Note that his data interpretation is full of engineering judgment. Another issue is the differential subsidence between pile-supported buildings and surrounding liquefied ground (Fig. 3.33). This situation induces breakage of lifelines connected to a building and difficulty in operation of emergency vehicles going out of a building.

Rocking foundation of a building is an inexpensive base isolation (Fig. 3.34) in which the impact between the base slab of a building and the foundation ground dissipates energy and the earthquake response is reduced. Because of its simplicity, rocking foundation has been investigated by many people experimentally and analytically (Huckelbridge and Clough 1977; Priestley et al. 1978; Muto and

Fig. 3.33 Differential settlement between pile-supported building and surrounding liquefied ground (in Urayasu 2011)

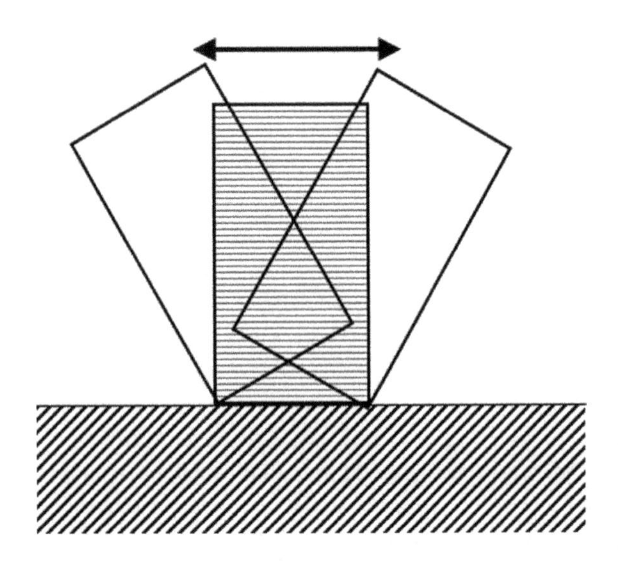

Fig. 3.34 Concept of rocking foundation

Kobayashi 1979; Toki et al. 1980; Hayashi 1996; Mergos and Kawashima 2005; Kim et al. 2013, among many others).

Those studies assumed that the base ground is rigid or elastic. It seems that practice of building construction did not pay much attention to this hypothesis. Many modern buildings in Christchurch, New Zealand, that had rocking foundation suffered tilting or permanent deformation of foundation ground, partly, because of liquefaction and plastic deformation of subsoil (Fig. 3.35). Although those buildings did not suffer structural damage as intended by the principle of rocking foundation, the induced tilting was not acceptable from the serviceability viewpoint and the affected buildings had to be demolished. Thus, communication between building engineers and geotechnical engineers was not sufficient.

Fig. 3.35 Inclination of building with shallow foundation resting on liquefied subsoil (in Christchurch 2011)

3.5.2 Allowing for Limited Extent of Liquefaction Under Existing Structures

Ground improvement under existing structures is not easy or costly, even if possible. It is therefore an important choice to allow limited liquefaction, if damage is allowable, or to accept liquefaction and be prepared for quick restoration. This is often the case when the concerned structure is inexpensive such as lifelines, river levees and houses. Although efforts have been made to make the liquefaction extent limited, achievements so far are not enough. Due to this reason, this section discusses achievements for river levees only and those for other kinds of structure will be touched upon in Sects. 3.6 and 3.7.

Figure 3.36 illustrates the liquefaction damage of Yodo River near the river mouth during the 1995 Kobe earthquake. Because the subsidence exceeded 3 m, the high water level of the river (at high tide of the sea) was almost overtopping the levee. Because the area behind the levee had suffered substantial consolidation settlement in the twentieth century and the ground surface had already been lower than the sea level, the possible overtopping of the unlimited volume of sea water could have totally damaged the area. This event demonstrated the importance of seismic-resistant design of river levees.

In contrast, the conventional philosophy of earthquake problems of river levee addressed quick restoration, while no seismic resistance was installed. This philosophy was based on the empirical knowledge that earthquakes and floods unlikely occur at the same time. Moreover, the construction budget for levees has not been sufficient. Quick restoration of damage (within two weeks before heavy rain and flood occur) was considered reasonable and economical. This idea was supported by the empirical fact that liquefaction-induced subsidence of levee never exceeds 75% of its original height (Fig. 3.37). It is very likely that the remaining height of a levee (minimum 25% of height) is still high enough to prevent overtopping of two-week high water level. If so, efforts are concentrated on preparation of quick

Fig. 3.36 Liquefaction-induced subsidence of Yodo River levee

restoration by accumulating construction materials and making emergency contracts with local construction industries. If the remaining height is not enough, levees should be reinforced against liquefaction. This requires a probabilistic way of thinking. If necessary, the foundation subsoil is improved against liquefaction by compaction, grouting or gravel drains (Sasaki et al. 2012). Note that this philosophy failed in the case of Fig. 3.36 near the sea where high tide (equivalent flood) occurs twice a day.

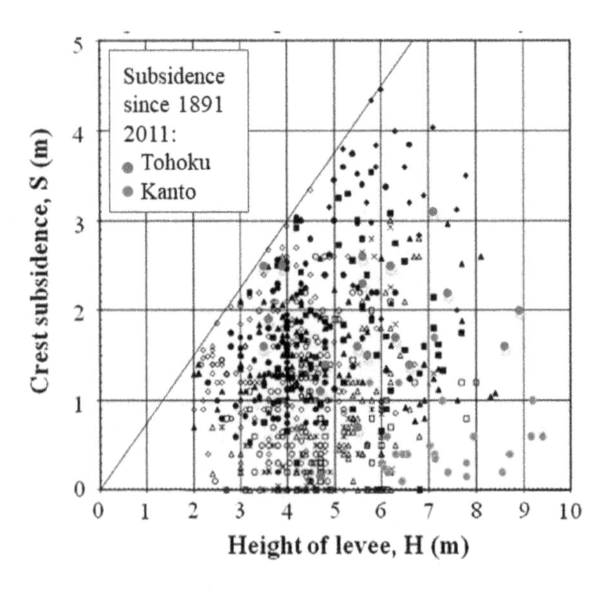

Fig. 3.37 Subsidence of river levees caused by past earthquakes (data by MLIT)

Fig. 3.38 Mechanism of levee subsidence by subsoil liquefaction and its mitigation by sheet pile walls (Mizutani and Towhata 2001)

Figure 3.38a illustrates the results of a 1-G shaking model test in which a model embankment rested on liquefaction-prone sand. After the onset of liquefaction, the subsoil spread laterally and induced subsidence in the vertical direction. To mitigate this mechanism, sheet piles were installed under the toes of the embankment. Figure 3.38b shows that the subsidence was reduced.

Two issues are noteworthy in sheet pile mitigation. First, sheet pile wall on the landside of a levee is not a good idea because ground water flow during high flooding is stopped by an impervious wall and water remains under the levee, thus increasing the risk of levee breaching. Hence, a wall made by sand compaction is more appropriate. Second, sheet pile walls cannot perfectly prevent the subsidence of the levee top. As suggested by Fig. 3.38b, subsidence under the top and uplift under the slopes can occur irrespective of the wall confinement.

It has been supposed that liquefaction is a problem of subsoil underlying levee. However, the recent problem of levee is liquefaction inside the levee. Figure 3.39 is an example of damage of a levee resting on clayey subsoil. Note that there is no evidence of liquefaction such as sand ejecta or ground fissures on the ground surface. Despite that, the levee slope moved down substantially. Studies after the 2011 Tohoku earthquake revealed that liquefaction occurred within many levees and induced distortion (Sasaki et al. 2012). This finding was a drastic change from the conventional idea that river levee is situated above the ground water level and that liquefaction therein is unlikely. In reality, (rain) water remains in the levee, and also, the levee body subsides into clayey subsoil and gets submerged as a consequence of consolidation settlement (Fig. 3.40).

Mitigation of this kind of liquefaction in levee is very difficult with available budget. Problems are

- Levee has been constructed by stages over historical time and information on its internal structure is hardly known.
- Much is not known about extent of subsidence into clayey subsoil.
- Because levee is a long linear structure on heterogeneous subsoil, ground investigation technology cannot capture the soil condition in detail.

Fig. 3.39 Deformation of levee resting on vulnerable soil induced by internal liquefaction (Naruse River during the 2011 Tohoku earthquake)

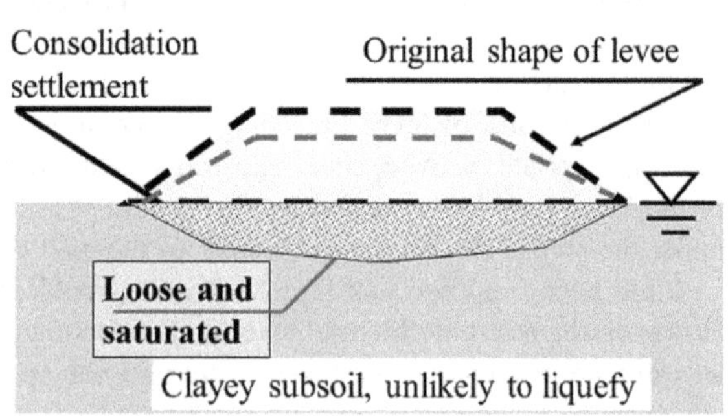

Fig. 3.40 Schematic illustration of liquefaction inside levee

- The major mission of levee is prevention of flood disaster. Levee should not breach during high water level. Seepage through or under levee should not be promoted by liquefaction-resistant measure.

3.6 Mitigation Measures Under Existing Houses

The 2011 Tohoku earthquake affected many residential areas located on earth fills in hilly areas and on man-made islands in coastal and lake areas. Because the damage was so vast, the national government launched a special program by which the restoration of damaged residential areas can receive financial supports. Note that such a financial support to improve the value of personal properties (real estate) is an exceptional measure because there is a strict rule not to spend people's tax money on benefit of private sectors. Because of its special nature, the program required the following conditions to be satisfied by local counterparts:

- Subsoil liquefaction occurs uniformly whether the area is public or private.
- Therefore, both public space (streets and lifelines) and private space (residential land) are reinforced against future earthquake/liquefaction together.
- Cost of ground improvement in the public space (streets) is paid by the national government, while that for the private land should be shared 50–50% by the public sector and the residents.
- Local residents make unanimous agreement to carry out the program.

It is naturally believed that individual resident cannot organize ground improvement project under house, that the unanimous agreement should be attained to ask local government to organize the project on behalf of individuals, and that the local government bears responsibility to individuals. These issues were understood to mean that no mistake is allowed in the procedure. No technical challenge was allowed. Only those technologies that had been validated through practice were eligible. Accordingly, ground water lowering (Fig. 3.31) and underground grid wall (Fig. 3.13) were candidates.

3.6.1 Ground Water Lowering in Residential Land

An earlier example of this technology is found in Tsukiji District of Amagasaki City to the west of Osaka. This district was constructed by land reclamation in the sea in the first half of the seventeenth century which was 350 years before the 1995 Kobe earthquake. Old houses in this site was completely destroyed by liquefaction and strong shaking in 1995 and the reconstruction of the community was a big issue. Further, this area had suffered consolidation settlement problem (ground surface being lower than sea level) for decades due to pumping of ground water. After the 1995 earthquake, it was decided to solve two problems simultaneously. The procedure for this consisted of total removal of houses in this district, filling 1.5 m soil at the surface as a compensation for ground subsidence, installation of drainage pipes and reconstruction of houses (CRTDED 2007). The pipes keep the water level low enough to attain reasonable thickness ($H_1 = 3$ m) of surface unliquefiable crust (Fig. 3.21). The ground improvement in this project was simple because all the existing houses had been demolished and the ground surface was open to works. The collected water flowed down through pipes under gravity and was collected for pumping/drainage into sea by electric pumps at the final point. The entire project was completed in 2007 that was 12 years after the disaster. For details, refer to Suwa and Fukuda (2014). Oshima et al. (2020) conducted ground investigation in Tsukiji to show that the seismic load 'CSR' in Eq. 3.1 decreased after water level lowering and earth filling but the resistance 'CRR' did not change because subsoil was not touched. Consequently, the factor of safety 'F_L' increased.

Soil mechanics states that ground water pumping often triggers consolidation settlement in underlying clayey layer. However, the author could not find any evidence of subsidence in Tsukiji (Fig. 3.41). This is probably because the extent of water lowering is small and, more importantly, the ground water level around 1960, when industrial water pumping was severest, was lower than that after 1995, thus keeping the clayey subsoil still overconsolidated in the twenty first century.

After 2011, several municipalities planned to carry out ground water lowering under the governmental financial support. Figure 3.42 illustrates the site in Itako City that used to be a lake and was converted by earth filling to a firm land and then to a residential land. Drainage pipes were installed and water flows out under gravity. Figure 3.43 is the case in Kuki City where water is coming out under gravity. In these two examples, drainage was driven by gravity and energy was free. In these cities, the extent of lowering water level was limited and consolidation settlement in underlying clayey soils was not significant.

Several other municipalities could not conduct ground improvement project. In many of them, local people refused personal expenditure for installation and maintenance of water lowering facilities. Only exception is Chiba City, east of Tokyo, where drainage pipes are installed and maintained without people's financial burden. According to Yasuda (2016a, 2018), the major issues are possibility of sufficient drainage and ground water lowering by installing pipes at reasonable cost. On this issue, field pumping tests have been carried out to verify that pipe installation only under streets, not under existing houses, is sufficient. Thereby, water supplied by precipitation and ground water flow (both in the horizontal direction and upward direction from underlying aquifer) should be drained out.

Embankment for residential land in hilly terrains has a similar problem. Constructed by filling soil in small valleys, residential embankment receives a substantial ground water coming into former valleys and becomes heavier and prone to liquefaction. Figure 3.44 shows sand ejecta in such circumstances in Sendai City, Japan, after the 2011 Tohoku earthquake. To mitigate the risk at a

Fig. 3.41 Tsukiji district of Amagasaki in 2012

Fig. 3.42 Improved residential land in Itako

Fig. 3.43 Drainage of water from Kuki City (photo by Prof. J. Koseki)

reasonable cost, while maintaining existing houses at the surface, ground water lowering by lateral drain pipes is recommended (Yasuda 2016b). Because of the sloping ground, drainage occurs under gravity action and maintenance cost is low.

Injection of air bubble is a smaller version of dewatering and has an advantage that it can be executed in a smaller and convenient scale. In practice, however, desaturation by injection of air bubbles is subject to two technical problems. The first is the uniformity of air bubble distribution. Probably, this issue is solved to a certain extent by injecting bubbles at many points in subsoil. The second is the long-term durability of air bubbles in the improved site. Oxygen gas dissolves into water easily, and nitrogen gas does so as well but more slowly. Further, all the bubbles migrate with the flow of ground water. Okamura et al. (2009) carried out in situ air bubble injection to verify this technology. The distribution of air bubbles was monitored by electric method, and it was shown that there is nonuniformity to a certain extent. However, it was found further that uniformity was improved by employing higher injection pressure. Another important finding was that frozen

Fig. 3.44 Sand ejecta in residential embankment in hilly terrain near Sendai after the 2011 Tohoku earthquake (sand still remaining on road surface 17 days after the earthquake)

samples, that were collected one month later, exhibited that the degree of saturation was still reasonably low. However, for practice, the life of desaturation has to be longer. To facilitate maintenance, Fujii et al. (2019) proposed to monitor the *P*-wave velocity that should be less than 1000 m/s for a reasonable extent of desaturation.

3.6.2 Underground Grid Wall

Urayasu and Chiba cities (Fig. 3.45) are close to each other, both facing the Tokyo Bay and having experienced liquefaction disasters in their reclaimed lands. The difference is the thickness of underlying clayey soil. Chiba does not have a big river in its vicinity and the thickness of clay is limited, while Urayasu is near big river mouths where the thickness of soft clay is 40 m. Since the beginning of land reclamation in 1960s, Urayasu had been suffering from settlement problems until the end of the century (Chiba Prefectural Government 2011). Thus, there was a fear when method of seismic ground improvement was being discussed that ground water lowering might reactivate the consolidation settlement. Although there was an opinion that consolidation in the underlying clayey strata is unlikely because of the ample water supply from the deeper artesian pressure aquifer, the local government decided not to take risk of ground subsidence.

Being located at a river mouth, the subsoil in Urayasu is heterogeneous, consisting of former water channels and sand bars. Moreover, the land reclamation procedure (dredging of sea bed soil and earth filling) makes substantially different

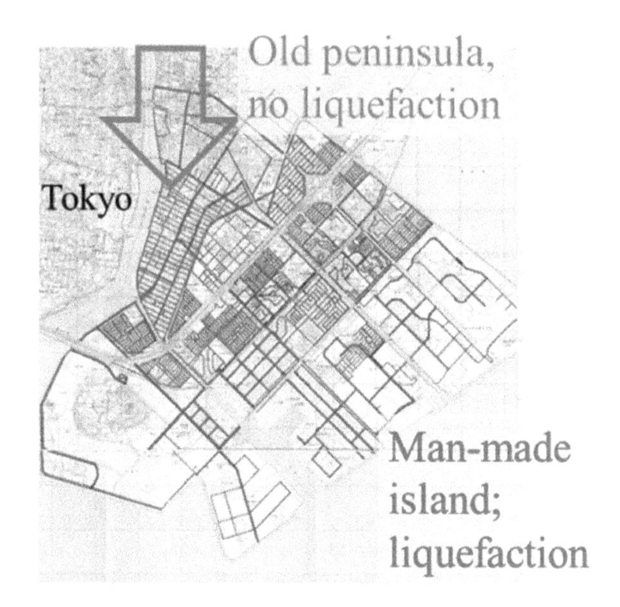

Fig. 3.45 Location of Urayasu City South East of Tokyo

Fig. 3.46 Ongoing implementation of under ground wall by small jet grouting machine

soil sediments, clayey and sandy, from place to place. Underground grid walls were decided to be constructed in such subsoil by bringing in small jet grouting machines in narrow spaces among houses (Fig. 3.46). The depth of walls was set at the typical depth of liquefaction-prone sandy soil (young alluvium under man-made island). In some places, the sandy soil was found deeper but a recent study (Takahashi et al. 2012) showed that slightly insufficient depth of wall does not affect the overall function of the grid wall.

The project stopped in April of 2017 during the preliminary construction when jet grouting was going on in clayey part of subsoil. Pressurized grout could not go uniformly around the device and flowed in wrong directions. Although liquefaction mitigation is not necessary in clay, the overall stability of grid wall (Fig. 3.47)

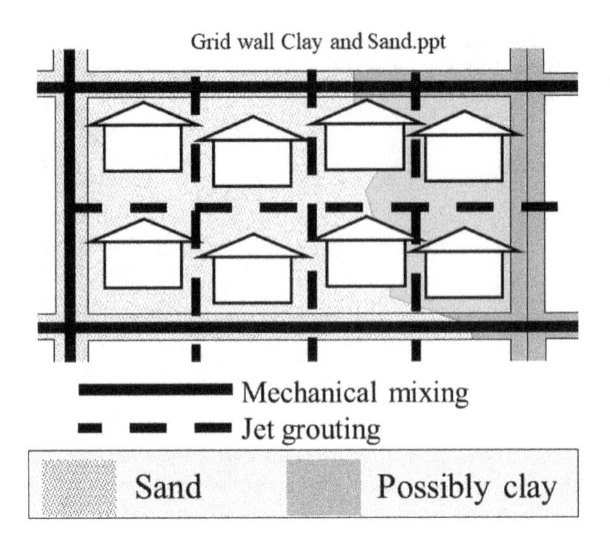

Fig. 3.47 Rectangular configuration of underground grid wall in residential area

during earthquakes required full implementation of walls. Lack of wall in clayey part would affect the stability during earthquakes. Moreover, many plastic drains that were embedded in clay during island construction got caught by the jet grouting machines and hindered the grout ejection. Technically, these problems were solved by raising the jet pressure and increasing the number of machine operation (twice or even three times operation of grouting procedure). However, the elongated construction period was not appreciated by local community and the unanimous agreement on the continuation of the project was declined. Finally, only one part of the city with 33 families accepted the project (Fig. 3.46).

For individual houses, reinforcement of foundation and shallow soil by compaction or micropiles is commonly practiced. Rigid concrete slab with steel reinforcement makes jacking-up easy after possible tilting.

3.7 Emerging Topics

The previous sections addressed difficulties associating existing structures. This problem has not been well solved yet. A typical problem of this kind is that of existing buried lifelines. As shown in Fig. 3.23, the basic mechanism of damage is the floating of pipes induced by liquefaction of loose water-saturated sandy backfill. Cement mixing of the backfill soil is a good solution but is possible only when the pipe is replaced and the backfill is completely excavated along the pipe. This situation does not occur frequently.

The authors proposed to install mitigation measures only at a small portion of the backfill, requiring a limited amount of excavation (Otsubo et al. 2016). One of the proposed ideas is illustrated in Fig. 3.48. A small part of the backfill is excavated

Fig. 3.48 Prevention of pipe floating by embedded support

and a columnar support is installed between the pipe and the surface pavement. When liquefaction occurs in the backfill, the column prevents the pipe from floating, while supported by the rigid surface pavement. Because the backfill soil is not mixed with cement or other solidification agent, future excavation for maintenance is not difficult.

As for grouting, efforts have been made to reduce the cost for grouting from the original 100% grouting to grid-type walls (Fig. 3.13) whose improvement ratio (*AR*) is more or less 30%. Further reduction may be possible by replacing walls by columns, although the constraining effect decreases and pore pressure likely develops more. Figure 3.49 illustrates the idea of 1-G model test on level ground (Bahmanpour et al. 2019). Pore pressure development was monitored during very strong shaking (maximum 500 cm/s^2, 10 Hz and 150 cycles) that addressed very rare seismic event as occurred in 2011 with *M*w = 9.0. It was shown that pore pressure development was delayed by the existence of columns (Fig. 3.50). Interestingly, Martin et al. (2004) reported a case where vertical columns of 2 and 7% improvement ratio prevented liquefaction during the 1999 Kocaeli earthquake of *M*w = 7.6 in Turkey. Takahashi et al. (2016) carried out centrifuge model tests to reveal that columns can reduce lateral flow of liquefied slope (Fig. 3.51). It was an important finding that an irregular configuration of columns constrains the soil flow more effectively than columns of regular (square) configuration, while the total number of the columns was held unchanged. Such effect is called pile pinning. Boulanger et al. (2007) conducted centrifugal model tests and discussed methodologies to quantitatively predict the reduced ground displacement. Turner and Brandenberg (2015) performed numerical analyses to show the size of strong interaction between pile and flowing subsoil that increases with the size of the flowing ground.

Fig. 3.49 1-G model test on effects of rigid columns on mitigative effect of liquefaction (Bahmanpour et al. 2019)

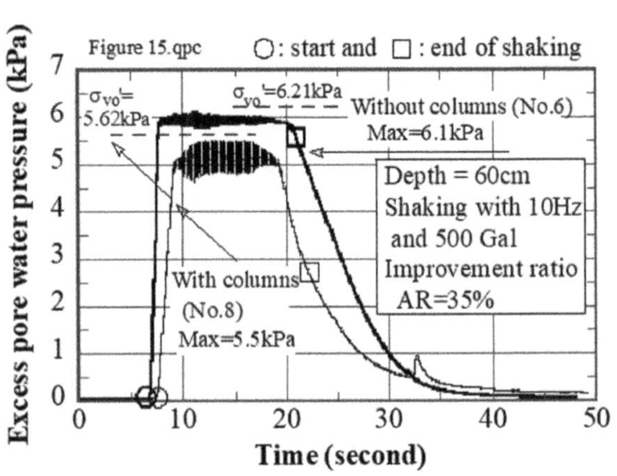

Fig. 3.50 Reduction of pore pressure rise by installed vertical columns (Bahmanpour et al. 2019)

Efforts toward less expensive ground improvement are always important. Blasting is one of the candidates. Efforts so far could not achieve SPT-N = 20 that is required to prevent liquefaction against very rare earthquake (PGA = typically 350 gal) (Towhata 2008). Figure 3.52 illustrates one of the past efforts.

Biocementation expects that microbials produce bonding materials from injected nutrients. Because needed cost is low, many studies have been carried out on it. The improvement in the mechanical properties of soil has been proven in laboratory specimens by DeJong et al. (2006), Terajima et al. (2009), Kawasaki (2015) and many others. Montoya et al. (2013) improved a bigger model ground for centrifugal shaking tests to validate the microbial improvement of liquefaction resistance of sandy ground.

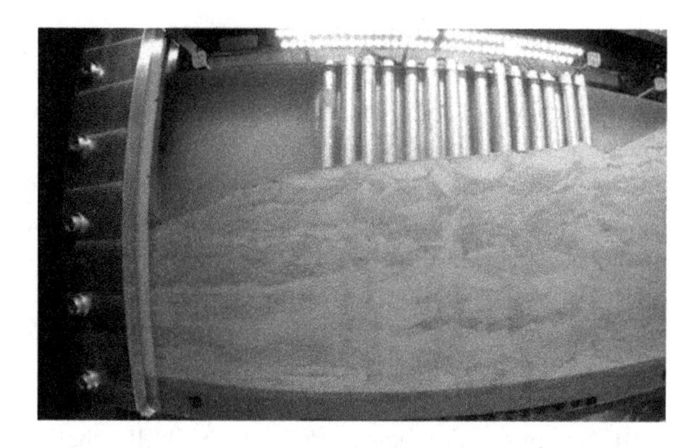

Fig. 3.51 Mitigation of lateral flow of liquefied slope by installed vertical columns (photograph was taken by the author) (Towhata et al. 2015)

Fig. 3.52 Ongoing blasting for compaction of sandy ground by Sato Kogyo Company (Towhata 2008)

The author supposes that one of the remaining problems is the natural enemy that exists in situ and may kill useful microbials. Laboratory tests are benefited by non-existence of such an enemy, and many studies have successfully improved soils. In this regard, Fukue (2015) states that he was able to improve in situ soil by microbial production of $CaCO_3$ bonding. Another problem is difficulty in making controlled and uniform ground improvement by microbials. In business, engineers have to make a contract to achieve the desired extent of ground improvement uniformly in the project area within an estimated cost and time. Solving these issues requires more efforts.

Lowering of ground water was discussed already to show that it can effectively reduce the degree of saturation and increase the liquefaction resistance of soil. Another method of desaturation is injection or creation of gas bubbles in soil. Injection of air into soil makes pore water compressible (Table 3.1e) and has been studied by Ishihara et al. (2003), Okamura et al. (2006, 2011), Okamura and Noguchi (2009), Yegian et al. (2007) and others. Durability of the improvement relies on the persistence of lowered degree of saturation (typically 90% or less). The first reason for disappearance of air bubble in soil is resolution of O_2 gas into pore water, while N_2 gas is less soluble. The model tests by Yegian et al. (2007) imply that the extent of gas resolution is not very significant. What is more significant is the migration of bubbles with ground water flow (advection and diffusion). Therefore, the degree of saturation has to be maintained periodically by injecting gas. Note that gas can be generated by electrolysis or microbial activity as well. Further, it seems difficult to achieve uniform desaturation by air injection or other methods. This point needs attention in practice.

Recycling of industrial wastes is commonly practiced in geotechnical engineering. An example of this type in liquefaction perspective is the use of tire chips (Fig. 3.53) that are mixed with soil for earth filling. Figure 3.54 compares results of shaking table tests in which an embedded pipe floats due to liquefaction of backfill sand. No compaction was made of the backfill, whether with or without tire chips. It is shown in this figure that floating was reduced by mixed tire chips, whereas pore pressure generation was reduced only to a certain extent.

The author supposes that the tire chip effect is twofold. First, chips constrain deformation of sand as soil reinforcement. This effect cannot be expected when chips are of block shape. Then, the second mechanism is the compressibility of rubber that can reduce the pore pressure generation (Table 3.1e). Dynamic deformation characteristics of sand-tire chip mixture was studied by Hazarika (2013).

There are still mysterious aspects in mitigation of liquefaction. Figure 3.55 illustrates the distribution of liquefaction in Port Island of Kobe during the 1995 Kobe earthquake. While liquefaction occurred profoundly in the peripheral area of the Island (Fig. 3.3), the central part did not liquefy. This is not because of the

Fig. 3.53 Size and shape of example tire chips (Nguyen 2007)

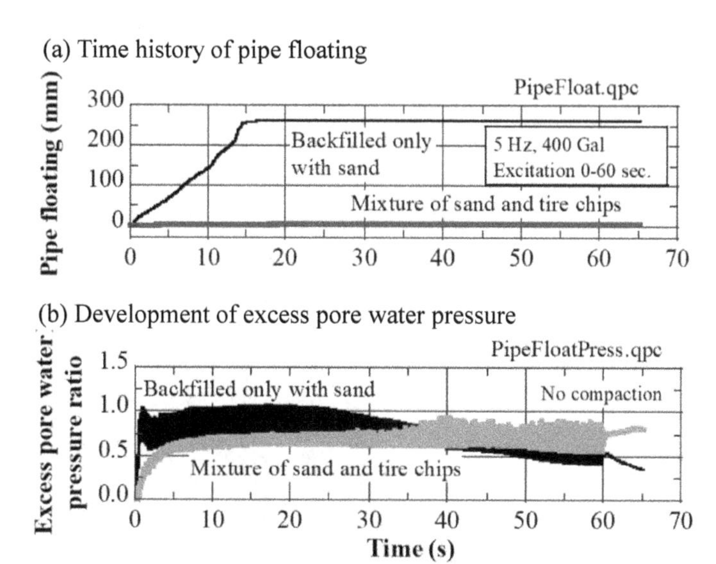

Fig. 3.54 1-G shaking model test on floating of pipe embedded in sand with and without tire chip mixture (shaking at 400 Gal at 5 Hz from 0 to 60 s) (Nguyen 2007)

liquefaction mitigation except in a small part where vibro-compaction was practiced. As stated before, profound efforts were made during construction of this island to reduce the consolidation settlement. For this purpose, sand drains and preloading were implemented. It is supposed nowadays that construction of sand drains compacted to some extent not only the underlying clay layer but also the surface sandy soil, that preloading brought the sandy soil to a state of overconsolidation by which liquefaction resistance is increased (Ishihara and Takatsu 1979), and that operation of many construction machines caused vibration and compacted unintentionally the vulnerable sand. It is recalled here that ground improvement of river levees in Miyagi of Japan successfully prevented liquefaction during the very strong 2011 Tohoku earthquake, although those measures were designed against a weaker design earthquake (Sasaki et al. 2012). It seems that current mitigation design is conservative to some extent.

In addition to technological topics so far referred to, non-technical topics need attention. Cost and financial burden are always important. In particular, people who were offered the governmental project of ground improvement were extremely nervous about financial burden. In an extreme case, one local community declined the proposal because of annual maintenance charge of only about US 80 dollars per year. Similar situation is frequently found in local governments.

The principle of life cycle cost (Towhata et al. 2009) is a probabilistic way of thinking that small additional expenditure on disaster mitigation can reduce the damage cost during rare but heavy future disaster and that the total cost, consisting

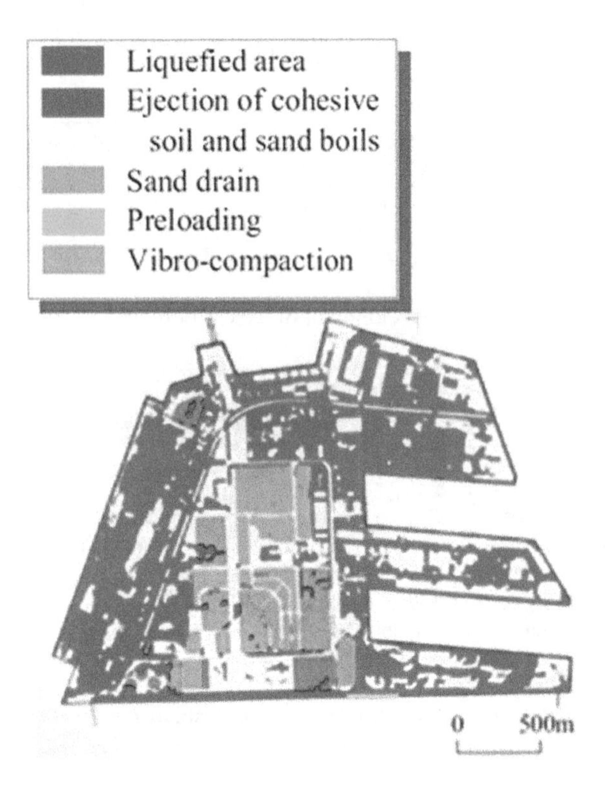

Fig. 3.55 Distribution of liquefaction in Port Island of Kobe during the 1995 Kobe earthquake (JGS 1996)

of those for construction, maintenance and damage, is reduced. In other words, cost for damage mitigation is something like insurance. Financial framework for 'insurance' needs more attention.

3.8 Conclusion

The present paper reviewed the historical development of liquefaction mitigation measures since 1950s, while, certainly, some technology has its origin well before that period. The liquefaction mitigation started as ground improvement to prevent liquefaction for new construction and later developed into more difficult types for prevention or reduction of liquefaction effect for existing structures. As was repeated in the main text, existing structure is a very difficult target and the current technology is still expensive and/or time consuming. More efforts are needed to improve this situation.

References

Asada A (1998) Simplified assessment of liquefaction-induced damage of houses & its mitigation, Personal report (in Japanese)

Bahmanpour A, Towhata I, Sakr M, Mahmoud M, Yamamoto Y, Yamada S (2019) The effect of underground columns on the mitigation of liquefaction in shaking table model experiments. Soil Dyn Earthquake Eng 116(1):15–30

Boulanger RW, Chang D, Brandenberg SJ, Armstrong RJ, Kutter BL (2007) Seismic design of pile foundations for liquefaction effects. Earthquake Geotechnical Engineering, Springer, Dordrecht, pp 277–302

Chiba Prefectural Government (2011) Report on current situation of consolidation and ground settlement in Chiba, publ. Environment and Life Department (in Japanese)

CRTDED Committee for Recovery of Tsukiji District from Earthquake Disaster (2007) Recovery of Tsukiji, Amagasaki, from liquefaction disaster, record of recovery from earthquake disaster (in Japanese)

D'Appolonia E (1954) Loose sands—their compaction by vibroflotation. In: Symposium on dynamic testing of soils, vol 156. ASTM STP, pp 138–162

DeJong JT, Fritzges MB, Nüsslein K (2006) Microbially induced cementation to control sand response to undrained shear. J Geotech Geoenviron Eng 132(1):1381–1392

Fujii N, Asada H, Yabe H, Yamaura M (2019) Technological progress of air injection desaturation method. Monthly J Jpn Geotech Soc 67(1):32–33 (in Japanese)

Fukue M (2015) An experimental study on the stabilization of coral gravel beach using bio-cement —cementation of coral gravels due to the precipitation of carbonate. J Jpn Soc Exp Mech 15 (3):231–238 (in Japanese)

Gallagher PM, Pamuk A, Abdoun T (2006) Stabilization of liquefiable soils using colloidal silica grout. J Mater Civil Eng 19(1):33–40

Gallagher PM, Conlee CM, Rollins KM (2007) Field testing of passive site stabilization. J Geotech Geoenviron Eng ASCE 133(2):186–196

Hamada M, Yasuda S, Isoyama R, Emoto K (1986) Generation of permanent ground displacements induced by soil liquefaction. Proc JSCE (376/III-f6):211–220 (in Japanese)

Hansbo S (1978) Dynamic consolidation of soil by a falling weight. Ground Eng 11(5):27–31

Hayashi Y (1996) Damage reduction effect due to basemat uplift of buildings. J Struct Constr Eng Architectural Inst Jpn 485:53–62 (in Japanese)

Hazarika H (2013) Paradigm shift in earthquake induced geohazards mitigation—emergence of nondilatant geomaterials. In: Indian geotechnical conference, Roorkee

Huckelbridge AA, Clough RW (1977) Earthquake simulation tests of a nine story steel frame with columns allowed to uplift, EERC report, University of California Berkeley, 77/23

Hwang JH, Yang CW, Chen CH (2003) Investigation on soil liquefaction during the Chi-chi earthquake. Soils Found 43(6):107–123

Iai S (1988) Large scale model tests and analyses of gravel drains. Report of the Port and Harbour Research Institute, vol 27, pp 25–150

Iai S, Tanaka Y, Ando H, Tanabe T, Sumi K (1996) On possibility of clogging in gravel drains that prevented liquefaction during the 1993 Kushiro Oki earthquake. In: 51st annual convention of JSCE, vol 3A, pp 244–245 (in Japanese)

Ishihara K (1985) Stability of natural deposits during earthquakes, theme lecture. In: 11th international conference on soil mechanics and foundation engineering, San Francisco, vol 1, pp 321–376

Ishihara K, Takatsu H (1979) Effects of overconsolidation and Ko conditions on the liquefaction characteristics of sands. Soils Found 19(4):59–68

Ishihara M, Okamura M, Oshita T (2003) Desaturating sand deposit by air injection for reducing liquefaction potential. In: 2003 Pacific conference on earthquake engineering, Paper No. 89

Ito K, Okita Y, Matsuzawa H (1991) Study on clogging criterion for gravel drain. Proc JSCE 439:53–62 (in Japanese)

Iwasaki T (1981) Dynamic soil-structure interaction with emphasis on geotechnical engineering aspects. Monthly Mag JGS 29(9):7–10 (in Japanese)

JGS; Japanese Geotechnical Society (1996) Report on geotechnical aspects of Hanshin Awaji earthquake disasters. Reference 1:232

Kawasaki S (2015) Present status of ground improvement technologies using microbial functions. J Min Mater Process Inst Jpn 131:155–163 (in Japanese)

Kawasaki City: Mitigation of liquefaction of oil refinery in Kawasaki. http://www.city.kawasaki.jp/840/cmsfiles/contents/0000051/51556/I_honbun1(33-70).pdf. Retrieved 27 Mar 2020 (in Japanese) (year unknown)

Kim DS, Lee SH, Kim DK, Ha JG (2013) Prospective of ground motion evaluation in Korean seismic code: site classification, response spectrum and SFSI. In: 5th international geotechnical symposium, Incheon, pp 87–95

Martin JR, Olgun CG, Mitchell JK, Durgunoglu HT (2004) High-modulus columns for liquefaction mitigation. J Geotech Geoenviron Eng 130(6):561–571

Ménard L, Broise Y (1975) Theoretical and practical aspect of dynamic consolidation. Géotechnique 25(1):3–18

Mergos PE, Kawashima K (2005) Rocking isolation of a typical bridge pier on spread foundation. J Earthquake Eng 9(2):395–414

Mitchell JK (2008) Mitigation of liquefaction potential of silty sands, from research to practice in geotechnical engineering, geotechnical special publication, vol 180. ASCE, Reston, VA, pp 433–451

Mitchell JK, Wentz FJ (1991) Performance of improved ground during the Loma Prieta earthquake, EERC report, University of California Berkeley, pp 91–12

Mizutani T, Towhata I (2001) Model tests on mitigation of liquefaction-induced subsidence of dike by using embedded sheet-pile walls. In: Proceedings of 4th international conference recent advances in geotechnical earthquake engineering and soil dynamics, San Diego, Paper Number 5.24

Mogami T, Watanabe T (1957) Field experiment on vibroflotation method part 2. Monthly Mag JSSMFE 5(5):23–27 (in Japanese)

Montoya BM, DeJong JT, Boulanger RW (2013) Dynamic response of liquefiable sand improved by microbial-induced calcite precipitation. Géotechnique 63(4):302–312

Muto K, Kobayashi T (1979) Nonlinear rocking analysis of nuclear reactor buildings—simultaneous horizontal and vertical earthquake inputs. Trans Architectural Inst Jpn 276:69–77 (in Japanese)

Nguyen AC (2007) Mitigation of liquefaction-induced damages to buried pipes by backfilling with tire chips. Master thesis, University of Tokyo

Numata A, Motoyama H, Kubo H, Oshida M (2012) Log piling method to mitigate liquefaction damage. In: 15th world conference on earthquake engineering, Lisbon

Okamura M, Noguchi K (2009) Liquefaction resistance of unsaturated non-plastic silt. Soils Found 49(2):221–229

Okamura M, Ishihara M, Tamura K (2006) Degree of saturation and liquefaction resistance of sand improved with sand compaction pile. J Geotech Geoenviron Eng ASCE 132(2):258–264

Okamura M, Takebayashi M, Nishida K, Fujii N, Jinguji M, Imasato T, Yasuhara H, Nakagawa E (2009) In-situ test on desaturation by air injection and its monitoring. Proc JSCE C 65(3):756–766 (in Japanese)

Okamura M, Takebayashi M, Nishida K, Fujii N, Jinguji M, Imasato T, Yasuhara H, Nakagawa E (2011) In-situ desaturation test by air injection and its evaluation through field monitoring and multiphase flow simulation. J Geotech Geoenviron Eng ASCE 137(7):643–652

Onoue A, Mori N, Takano J (1987) In-situ experiment and analysis on well resistance of gravel drains. Soils Found 27(2):42–60

Oshima A, Yasuda K, Yamada S, Suwa S, Takahashi S, Fukai A (2020) Site investigation results and verification of liquefaction countermeasure effect by groundwater lowering method at Tsukiji in Amagasaki City. J Soc Mater Sci Jpn 69(1):97–104 (in Japanese)

Otsubo M, Towhata I, Hayashida T, Shimura M, Uchimura T, Liu B, Taeseri D, Cauvin B, Rattez H (2016) Shaking table tests on mitigation of liquefaction vulnerability for existing embedded lifelines. Soils Found 56(3):348–364

Priestley MJN, Evison RJ, Carr AJ (1978) Seismic response of structures free to rock on their foundations. Bull NZ Natl Soc Earthquake Eng 11(3):141–150

Rollins KM, Goughnour RR, Anderson JKS, Wade SF (2004) Liquefaction hazard mitigation by prefabricated vertical drains. In: 5th international conference on case histories in geotechnical engineering, New York, vol 4

Sasaki Y, Towhata I, Miyamoto K, Shirato M, Narita A, Sasaki T, Sako S (2012) Reconnaissance report on damage in and around river levees caused by the 2011 off the Pacific coast of Tohoku Earthquake. Soils Found 52(5):1016–1032

Sasaki T, Yonekura R, Shimada S (2019) Long term durability test of improved ground by chemical grouting method in site. In: 54th national convention of JGS, Saitama, pp 481–482 (in Japanese)

Sassa S, Yamazaki H, Hayashi K, Yoshioka Y (2017) Prevention of propagation of liquefaction, sand boils and surface deformations by drain method. Proc JSCE B3 73(2):I_276-I_281 (in Japanese)

Seed HB, Booker JR (1977) Stabilization of potentially liquefiable sand deposits using gravel drains. J Geotech Eng ASCE 103(GT7):757–768

Shen M, Martin JR, Ku CS, Lu YC (2018) A case study of the effect of dynamic compaction on liquefaction of reclaimed ground. Eng Geol 240:48–61

Shimoda M, Hayakawa H, Hosoda H (1979) The properties of ultra fine grouting material and the application work. J Res Onoda Cem Company 31(2):31–53 (in Japanese)

Suwa S, Fukuda M (2014) Case study on countermeasure to liquefaction by dewatering method. J Soc Mater Sci Jpn 63(1):21–27 (in Japanese)

Takahashi H, Morikawa Y, Tsukuni S, Yoshida M, Fudaka H (2012) Study on reduction of wall depth for mitigation of liquefaction by means of solidified grid underground wall. Report of the Port and Airport Research Institute 51 (in Japanese)

Takahashi H, Takahashi N, Morikawa Y, Towhata I, Takano D (2016) Efficacy of pile-type improvement against lateral flow of liquefied ground. Géotechnique 66(8):617–626

Tanaka S, Sasaki T (1989) Sandy ground improvement for liquefaction at Noshiro Thermal Power station. Monthly Mag JSSMFE 37(3):86–90 (in Japanese)

Tatsuoka F, Iwasaki T, Tokida K, Yasuda S, Hirose M, Imai T, Kon-no M (1980) Standard penetration tests and soil liquefaction potential evaluation. Soils Found 20(4):95–111

Taya Y, Uchida A, Yoshizawa M, Onimaru S, Yamashita K, Tsukuni S (2008) Simple method for determining lattice intervals in grid-form ground improvement. Jpn Geotech J 3(3):203–212 (in Japanese)

Terajima R, Shimada S, Oyama T, Kawasaki S (2009) Fundamental study of siliceous biogrout for eco-friendly soil improvement. Proc JSCE C 65(1):120–130 (in Japanese)

Toki K, Sato T, Miura F (1980) Separation and sliding between soil and structure during strong ground motion. Proc JSCE 302:31–41 (in Japanese)

Towhata I (1982) Effects of stress axes rotation on deformation of sand undergoing cyclic shear. Ph.D. thesis, University of Tokyo (in Japanese)

Towhata I (2008) Geotechnical earthquake engineering. Springer, p 633

Towhata I, Yoshida I, Ishihara Y, Suzuki S, Sato M, Ueda T (2009) On design of expressway embankment in seismically active area with emphasis on life cycle cost. Soils Found 49 (6):871–882

Towhata I, Morikawa Y, Takahashi H, Takahashi N, Sugawa T (2015) Centrifuge model tests on mitigation against liquefied-soil lateral flow by using cement treated soil columns. In: 16th European conference on soil mechanics and geotechnical engineering, Edinburgh, Paper No. 1199

Towhata I, Yasuda S, Yoshida K, Motohashi A, Sato S, Arai M (2016a) Qualification of residential land from the viewpoint of liquefaction vulnerability. Soil Dyn Earthquake Eng J 91:260–271

Towhata I, Taguchi Y, Hayashida T, Goto S, Shintaku Y, Hamada Y, Aoyama S (2016b) Liquefaction perspective of soil ageing. Géotechnique 67(6):467–478

Tsue M (2013) Liquefaction mitigation practice for oil refinery plant. In: Proceedings of 43rd autumn convention of Japan Petroleum Institute, Session 2D12, pp 207–208 (in Japanese)

Turner BJ, Brandenberg SJ (2015) Pile pinning and interaction of adjacent foundations during lateral spreading. J Deep Found Inst 9(2):92–102

Uchida A, Konishi K (2015) Evolution of grid−form deep mixing walls as liquefaction countermeasure. Monthly Mag JGS 63(8):12–15 (in Japanese)

Uchida A, Taya Y, Honda T, Tsukuni S, Konishi K (2016) Applicability of simple method for determining space of grid in soil cement mixing walls. Jpn Geotech J 11(3):259–267 (in Japanese)

Wang W, Hashimoto K, Tsukamoto Y, Hyodo T, Kajiwara S, Udagawa K (2017) Laboratory tests using permeation grouting of ultra microfine cement with high water-cement ratio for soil liquefaction countermeasure. In: 52nd national convention of JGS, Nagoya, pp 1675–1676 (in Japanese)

Watanabe T (1965) Verified improvement effects of vibroflotation upon Niigata earthquake. Monthly Mag JSSMFE 13(2):27–33 (in Japanese)

Yasuda S (2016a) Liquefaction mitigation in urban developed areas. Bull JAEE 28:18–23 (in Japanese)

Yasuda S (2016b) Reconstruction of residential land after Tohoku earthquake. Found Eng Equipment 44(2):22–27 (in Japanese)

Yasuda S (2018) Problems of ground improvement raised in recent earthquakes and their subsequent response. J Soc Mater Sci Jpn 67(1):1–9 (in Japanese)

Yegian M, Esellerbayat E, Alshawabkeh A, Ali S (2007) Induced-partial saturation for liquefaction mitigation experimental investigation. J Geotech Geoenviron Eng ASCE 133(4):372–380

Yonekura R, Kaga M (1992) Current chemical grout engineering in Japan, grouting soil improvement and geosynthetics, geotechnical special publication 30, vol 1. ASCE, pp 725–736

Chapter 4
Liquefaction-Induced Pile Downdrag from Full-Scale Testing

Kyle Rollins and Cameron Lusvardi

4.1 Introduction

Frequently, deep foundations extend through potentially liquefiable loose to medium dense sand layers and bear on more competent layers at depth as shown in Fig. 4.1a. Prior to liquefaction, the applied pile head load, P, is transferred to the soil through positive side friction, Q_s, and the load in the pile decreases as shown in Fig. 4.1b. The load at the base of the pile is carried by end-bearing resistance, Q_b, which requires some settlement to develop as illustrated by the toe resistance versus settlement (Q_b–z) curve in Fig. 4.1c. When liquefaction occurs, skin friction in the liquefied layers is expected to decrease to near zero, and many design procedures use this value to evaluate the consequences of friction loss and pile settlement as shown in Fig. 4.1b. The reduction in positive skin friction in the liquefied layers leads to an increase in load at the toe of the pile, and mobilization of the increased end-bearing resistance leads to additional pile settlement as shown in Fig. 4.1c, d. As the earthquake-induced pore pressures dissipate in the liquefiable layer and settlement occurs, negative skin friction develops at the pile–soil interface in the clay layer above the liquefied layer, increasing the load in the pile (see Fig. 4.1b). In addition, the skin friction at the pile–soil interface in the liquefied layer is likely to increase as the excess pore pressure decreases. Therefore, the negative skin friction that ultimately develops in the liquefied layers will likely be higher than zero and will induce even greater load in the pile. The increased negative skin friction in the liquefied and non-liquefied layers further increases the required end-bearing resistance (see Fig. 4.1c) and leads to additional pile settlement as shown in Fig. 4.1d.

K. Rollins (✉)
Brigham Young University, 430 EB, Provo, UT 84604, USA
e-mail: rollinsk@byu.edu

C. Lusvardi
Reaveley Engineers and Association, 675 E 500 S #400, Salt Lake City, UT 84102, USA

T. G. Sitharam et al. (eds.), *Latest Developments in Geotechnical Earthquake Engineering and Soil Dynamics*, Springer Transactions in Civil and Environmental Engineering, https://doi.org/10.1007/978-981-16-1468-2_4

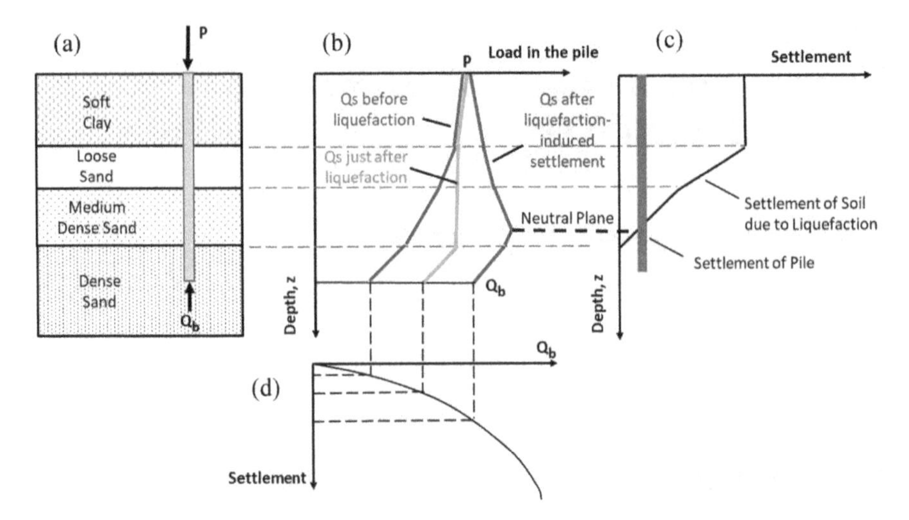

Fig. 4.1 Plot showing **a** soil profile around pile, **b** load in the pile versus depth, **c** settlement versus depth and **d** end-bearing resistance (Q_b) versus settlement before and after liquefaction-induced settlement

The neutral plane, shown in Fig. 4.1, represents the point where pile settlement and soil settlement are equal. Negative friction develops above the neutral plane, and positive friction develops below the neutral plane. As a result, the maximum load in the pile occurs at the neutral plane. The location of the neutral plane is normally obtained by trial and error so that the pile head load, P, plus the negative friction above the neutral plane is equal to the positive skin friction and end-bearing resistance, Q_b, below the neutral plane.

In the absence of test results, some investigators have used theoretical concepts to predict the behavior of piles when subjected to liquefaction-induced drag loads. Boulanger and Brandenberg (2004) defined negative skin friction in the liquefied zone in terms of the effective stress during reconsolidation, but concluded that the negative skin friction could be assumed to be zero with little error in the computed pile force or settlement. Fellenius and Siegel (2008) applied the unified design of piles approach that was developed for downdrag in clays, to the problem of downdrag in liquefied sand, once again assuming that negative skin friction in the liquefied zone would be zero. They also conclude that liquefaction above the neutral plane would not increase the load in the pile based on the concept that negative friction would already be present prior to liquefaction.

To understand better the development of negative skin friction on piles in liquefied sand and the resulting pile response, the author and his collaborators have conducted a number of full-scale field tests involving blast-induced liquefaction as summarized in Table 4.1. These tests provide an increasing set of relatively consistent results that will be summarized in this paper. Blast-induced liquefaction was first used to investigate the lateral resistance of piles in liquefied sands (Weaver

Table 4.1 Summary of blast-induced liquefaction downdrag tests

Site location	Pile type	Soil profile	References
Vancouver, Canada	Driven steel pipe: 32.4 cm diameter, 21 m long	6 m of cohesive soil over loose clean sand (D_r = 40%)	Rollins and Strand (2006), Rollins et al. (2018)
Christchurch, New Zealand	Three augercast piles: 61 cm diameter; 8.5, 12, and 14 m long	1.5 m of cohesive soil over medium dense silty sand (D_r = 60%)	Rollins et al. (2018), Rollins and Hollenbaugh (2015)
Mirabello, Italy	Bored micropiles: 25 cm diameter, 15 m long	6 m of cohesive soil over 3 m of sandy silt and 18 m of silty sand	Amoroso et al. (2018)
Turrell, Arkansas, USA	Three driven piles *H* pile: (H14 × 117), 28 m long Pipe pile: 46 cm diameter, 24 m long PSC pile: 46 cm square, 22.5 m long three bored piles 1.22 m diameter, 27.6 m long 1.82 m diameter, 21.3 m long 1.22 m diameter, 28 m long	9 m of cohesive soil over silty sand and sandy silt	Kevan et al. (2019), Ishimwe et al. (2018)

et al. 2005; Rollins et al. 2005) and has become widely used to investigate a number of ground improvement strategies (Wentz et al. 2015; Ashford et al. 2004; Gallagher et al. 2007).

4.2 Driven Pile Downdrag Testing in Vancouver, Canada

Rollins and Strand (2006) conducted a full-scale load test using a 32.4 cm diameter steel pipe pile driven to a depth of 21 m in Vancouver, Canada. As shown in Fig. 4.2, the soil profile, with a water table at 3.5 m, consisted of non-liquefiable soils to a depth of about 5 m underlain by loose liquefiable sand ($D_r \approx 40\%$) to a depth of 15 m. The loose sand was underlain by sand with a relative density of 50–60%.

Reacting against the load frame, the hydraulic jacks initially applied a load of 536 kN which was about 50% of the ultimate pile resistance using the Davisson criteria. Based on pore pressure transducer measurements, detonation of the sequence of explosive charges produced liquefaction from about 5.5–13 m. Reconsolidation of the liquefied sand produced 27 cm of settlement or about 3% volumetric strain, similar to what would be expected for liquefaction produced by an earthquake.

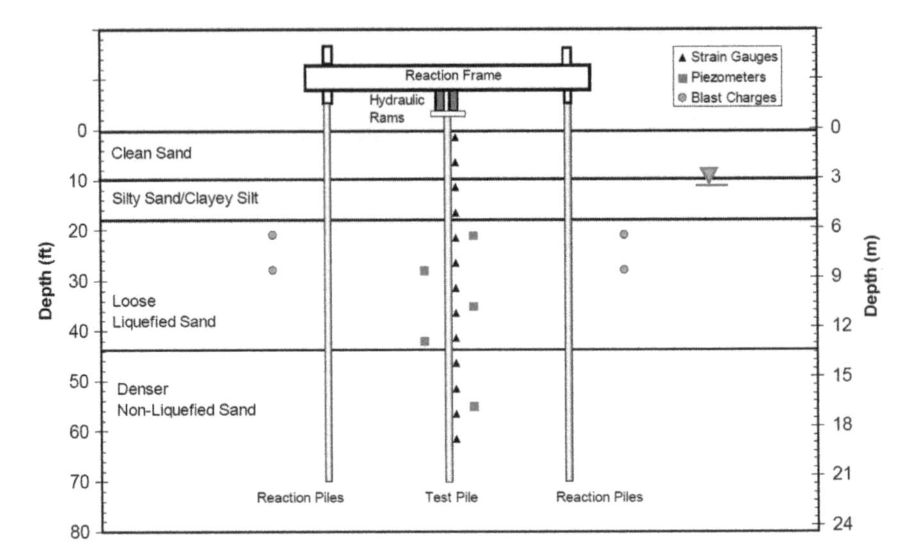

Fig. 4.2 Schematic drawing showing soil profile at the Vancouver, Canada, test site along with test pile, reaction piles, pore pressure transducers, strain gauges and blast charge locations

Figure 4.3 provides a summary of the load in the pile versus depth before liquefaction, immediately after liquefaction and at the completion of pore pressure dissipation. Prior to blasting, pile head load was transferred to the surrounding soil primarily by side friction. At the onset of blasting, the test pile settled slightly so that the load applied by the hydraulic jacks dropped by 156 kN at the top of the pile. When this 156 kN load was re-applied, this load was resisted by positive skin friction from the pile head downward in the upper section of the pile consistent with observations by Bozozuk (1981) for a pile in clay subjected to downdrag and then reloaded. It should be noted that the total measured skin friction from the ground surface to a depth of 6 m immediately prior to blasting was approximately 166 kN. Therefore, the redevelopment of positive skin friction due to this applied load appears to be reasonable. The load of 536 kN was maintained throughout the remainder of the test by adding hydraulic fluid to the jack as the pile began to settle and relieved the load. This apparently maintained the positive friction in the upper 6 m of the pile. This result indicates that it would be desirable to apply dead load to the top of the pile in future tests to avoid the complication of re-application of pile head load.

Following liquefaction, load transfer within the liquefied zone dropped to near zero, and the load originally carried by positive skin friction in this zone was transferred to the lower end of the pile where liquefaction had not developed. As a result, at the base of the liquefied zone, the load in the pile increased by 130 kN after blast-induced liquefaction, and the pile settled about 4.5 mm as a result of the mobilization of skin friction and end-bearing in the underlying sand layer.

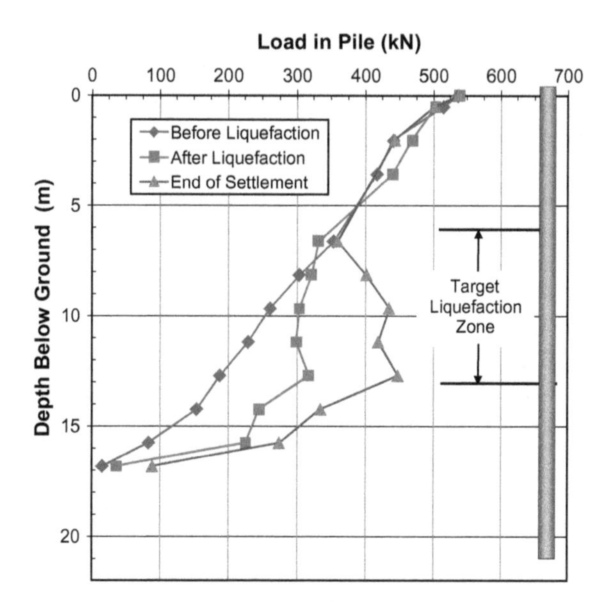

Fig. 4.3 Pile load versus depth curves before blasting, immediately after blasting and after settlement of the liquefied layer at Vancouver, Canada (Rollins and Strand 2006)

Once excess pore pressure had dissipated and settlement had stopped, the load versus depth curve in the previously liquefied zone developed a negative slope as shown in Fig. 4.3. The negative slope indicates that negative skin friction had developed in this zone and was applying drag load to the pile. As the pore pressures dissipated and effective stresses increased, the skin friction at the pile interface also increased and produced a drag load of about 100 kN. The drag load produced during reconsolidation was approximately one-half of the positive skin friction force prior to liquefaction. This load was once again transferred to the sand below the liquefied zone with a resulting additional pile settlement of about 2.5 mm or a total pile settlement of 7 mm.

4.3 Augercast Pile Downdrag Testing in Christchurch, New Zealand

Rollins et al. (2018) report results of blast liquefaction tests on three 60 cm diameter continuous flight auger piles in Christchurch, New Zealand. The three test piles were installed in a triangular arrangement at 2 m center-to-center spacing to depths of 8.5, 12 and 14 m, respectively. In these tests, the soil profile consisted of a 1.5 m thick layer of sandy silt underlain by poorly graded medium dense clean sand to a depth of 10.5 m. This layer was in turn underlain by interbedded layers of medium dense to dense clean sand.

Two blast-induced liquefaction downdrag tests were performed on the piles to evaluate their performance with and without applied static load. In the first blast test, there was no load applied to the piles. The detonation of a sequence of small explosive charges liquefied a layer of sand from the water table at 1.5 m to a depth of about 13 m. Ground settlement was approximately 4 cm immediately around the group but higher beyond it. Because the ground settled more than the piles (1–2 cm), negative skin friction developed in each case.

Plots of the load in each pile as a function of depth interpreted from the strain gauge readings are provided in Fig. 4.4 for the conditions 60 min after blasting when liquefaction-induced settlement was completed. Because no pile head load is applied, any load in the piles is induced by negative skin friction or drag load above the neutral plane.

Clearly, the negative skin friction is not zero at the end of consolidation in Fig. 4.4. The neutral plane is visible in each of the plots as the point where the load in the pile begins to decrease. Because the neutral plane in each case is located

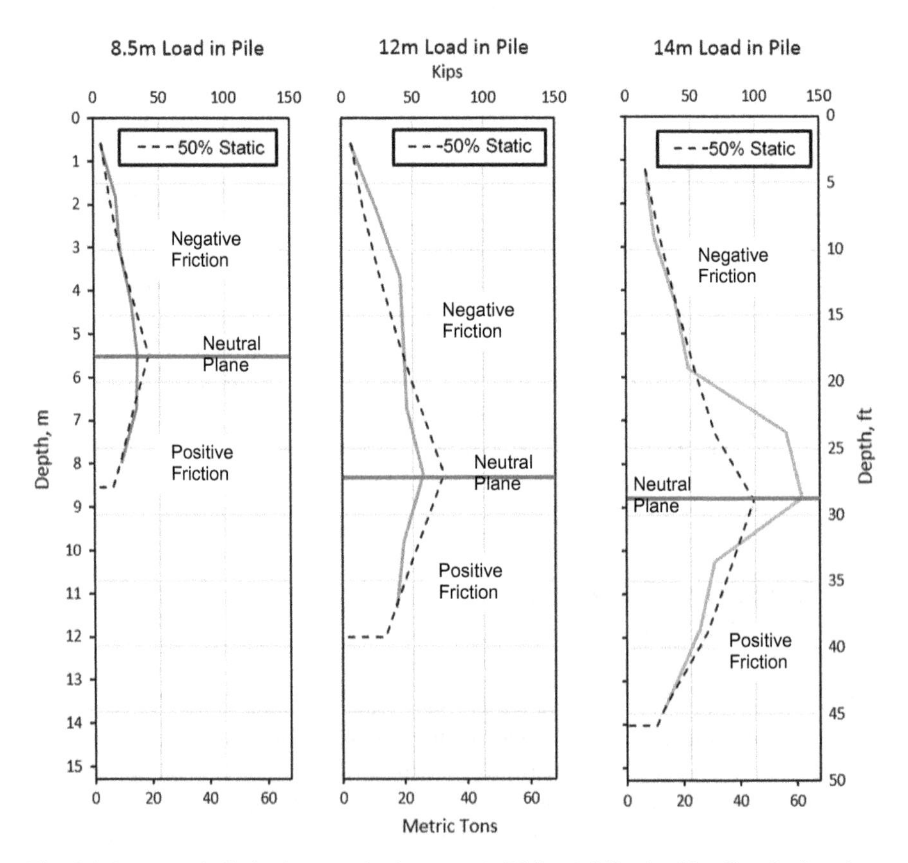

Fig. 4.4 Interpreted pile load versus depth curves (solid lines) following blast liquefaction along with predicted curves (dashed lines) assuming skin friction equal to 50% of measured average positive skin friction from the static load test. The neutral plane is shown in each plot with a horizontal line separating negative skin friction above from positive skin friction below (Rollins et al. 2018)

within the liquefied layer, rather than at the bottom of the liquefied layer as suggested by some design procedures, positive skin friction below the neutral plane also develops within the liquefied zone as reconsolidation occurs. The depth to the neutral plane increased as the length of the pile increased suggesting that the pile settlement decreased as the pile length increased.

About one month after the initial blast-induced liquefaction tests, static load tests were performed on each pile using dead weights (Rollins and Hollenbaugh 2015). Figure 4.4 also presents dashed lines showing the load in the pile assuming 50% of the average positive skin friction found in the static load test along the pile length where liquefaction occurred. Because the neutral plane is located within the liquefied zone, both negative and positive skin frictions are reduced by 50% in the computations. Agreement with the measured curves is generally very good and confirms the reduced skin friction value obtained from the test in Vancouver.

Following the static load tests, a total of 300 tons of dead load was distributed among the test piles prior to a subsequent blast-induced liquefaction downdrag test. The load carried by each test pile was measured by a load cell on the top of each pile. In this test, liquefaction developed from a depth of 3–7 m below the ground surface. Because of the load on the piles, they settled more than the surrounding ground and positive skin friction developed even within the liquefied layers. Skin friction within the non-liquefied layers was roughly the same as that measured before liquefaction, while skin friction in the liquefied zones immediately after reconsolidations was about 40% of the pre-liquefaction skin friction.

4.4 Micropile Downdrag Testing in Mirabello, Italy

Amoroso et al. (2018) describe results from a blast-induced downdrag test conducted on a 25-cm diameter micropile at a test site in Mirabello, Italy, where liquefaction was observed in the M_w 6.1 Emilia Romagna earthquake in 2012. As shown in Fig. 4.5, the soil profile consists of about 6 m of non-liquefiable cohesive soil underlain by a 2-m thick sandy silt layer and a 10-m thick sand layer. The test pile extended to a depth of 17 m but was not loaded.

The blasting sequence liquefied a layer from about 6–12 m, resulting in about 15 cm of settlement at the ground surface although the test pile only settled about 15 mm. A Sondex settlement profilometer was used to record settlement with depth as shown in Fig. 4.5. As excess pore pressure dissipated and the sand reconsolidated, the liquefied layers from 6 to 13 m settled about 11.5 cm ($\approx$2% volumetric strain). The cohesive surface layer largely settled as a block on top of the liquefied layer and settlement below 12 m was relatively minor (less than 1.5 cm). The volumetric strain in the liquefied zone produced by blasting is consistent with that expected in an earthquake based on predictive equations developed by Ishihara and Yoshimine (1992) and Zhang et al. (2002).

During reconsolidation of the liquefied soil, negative skin friction developed from the ground surface to the neutral plane at a depth of about 12 m where pile

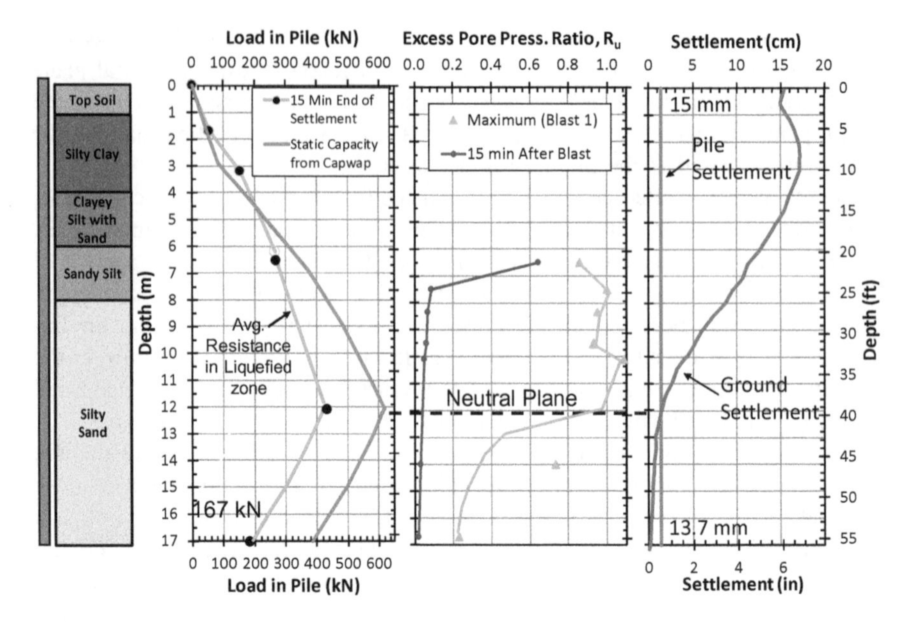

Fig. 4.5 Plots showing soil profile, load in the pile and excess pore pressure ratio, along with soil and pile settlement following blast-induced liquefaction

settlement and soil settlement were equal as shown in Fig. 4.5. The negative skin friction in the non-liquefied layers was similar to the positive skin friction prior to liquefaction; however, the average negative friction in the liquefied layers was only about 50% of the positive skin friction based on CAPWAP measurements without liquefaction. End-bearing resistance was mobilized at the toe of the pile as a result of negative friction that produced a settlement of 13.7 mm at the toe, equal to about 4% of the pile diameter, even without any load at the top of the pile.

4.5 Driven and Bored Pile Downdrag Testing in Turrell, Arkansas, USA

Kevan et al. (2019) and Ishimwe et al. (2018) report results from blast-induced downdrag tests conducted on three driven piles and three drilled shafts at a test site near the Mississippi River in Turrell, Arkansas. The driven piles consisted of an *H* pile (HP14 × 117), a 46-cm diameter pipe pile and a 46-cm square pre-stressed concrete pile. The drilled shafts consisted of two 1.22-m diameter shafts and one 1.82-m diameter shaft. The test piles were loaded using a steel pile cap and steel beams.

The soil profile consisted of a 9-m thick layer of non-liquefiable clay underlain by liquefiable silty sand and sand layers that were in turn underlain by a dense sand layer as shown in Fig. 4.6. Three separate blast tests were performed involving one bored

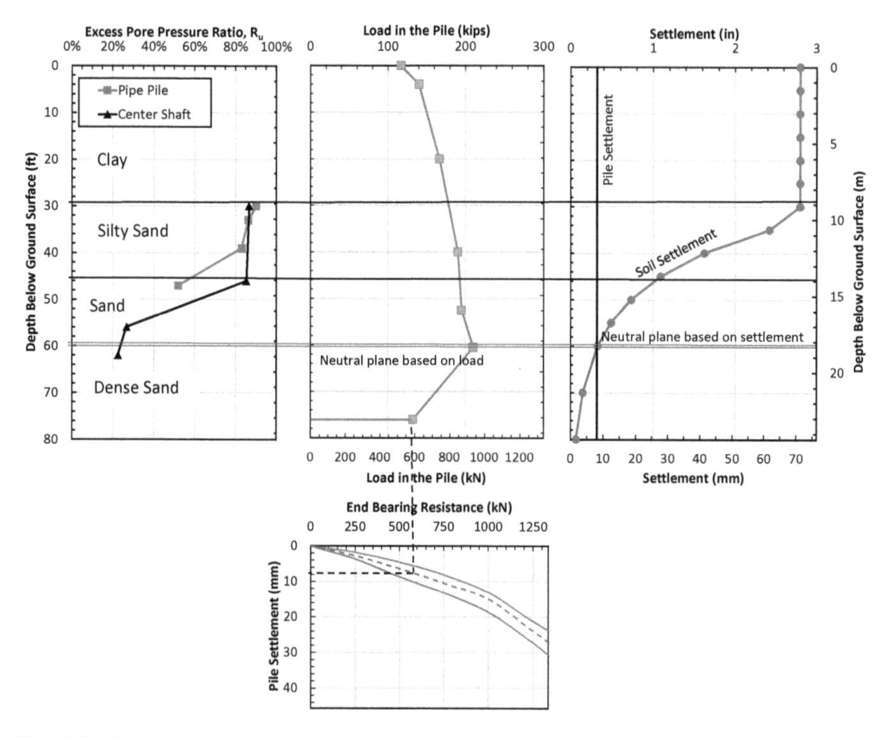

Fig. 4.6 Plots showing excess pore pressure ratio, pile and soil settlement, load in the pile versus depth and the end-bearing resistance versus pile settlement curve for the steel pipe pile (Kevan et al. 2019)

pile (drilled shaft) and one driven pile with results for the pipe pile provided in Fig. 4.6. Liquefaction was typically induced within the 5-m thick silty sand layer and elevated pressure extended into the underlying sand. Liquefaction produced ground surface settlements of about 7.5–10 cm or a volumetric strain of about 1.0–1.5%.

The end-bearing resistance versus settlement curve was developed from generic shapes recommended by O'Neill and Reese (1999). For the pipe pile test shown in Fig. 4.6, the neutral plane, where the pile settlement equals the soil settlement (8 mm), was located at a depth of about 18.3 m, and this point also corresponded with the maximum load in the pile. The average skin friction in the liquefied soil layers following reconsolidation was calculated as 38% of the static resistance prior to liquefaction. In contrast, the skin friction in the non-liquefied clay was within 4% of the pre-blast resistance in the clay layer, but about 20% higher in the sand layer below the liquefied zone. We observed similar results for the other driven and bored test piles.

Although significant negative skin friction was developed along the length of the deep foundations and liquefaction-induced settlement was substantial, measured settlement of the deep foundations was normally within acceptable levels for the pile head loads involved. This is indicative of the relatively high end-bearing resistance in these piles. The measured pile settlement was generally consistent with the neutral plane concept obtained by balancing applied pile head force and negative friction with the positive friction and displacement-compatible end-bearing resistance after reducing skin friction in the liquefied layers.

4.6 Procedure for Determining the Neutral Plane for Piles in Liquefied Sand

Based on the results from the full-scale tests previously discussed in this paper, the following step-by-step procedure is recommended for determining the location of the neutral plane.

- Compute the liquefaction-induced settlement versus depth profile.
  Using liquefaction settlement analysis procedures, such as those suggested by Ishihara and Yoshimine (1992) and Zhang et al. (2002), compute the settlement with depth within the zone of liquefaction as shown in Fig. 4.1c. Assume that non-liquefied layers above the liquefied zone will simply move downward as a block.
- Estimate the location of the neutral plane, and compute the load in the pile versus depth.
  The negative friction above the neutral plane progressively increases the load in the pile with depth relative to the applied load at the ground surface. Above the neutral plane, assume that negative friction will be equal to the positive friction in the non-liquefied layers above the liquefied layer. Furthermore, assume that negative friction within the liquefied layers will be equal to 50% of the positive friction in the layer prior to liquefaction. Below the neutral plane, assume that positive friction in the liquefied layers is 50% of the positive friction prior to liquefaction with positive friction in the non-liquefied layers remaining the same (see Fig. 4.1b).
- Compute the settlement at the toe of the pile.
  The settlement at the toe (S_{toe}) can be computed using the equation

$$S_{toe} = S_{NP} - \Sigma(PL/AE) \tag{4.1}$$

where S_{NP} is the soil settlement at the neutral plane, and $\Sigma(PL/AE)$ is the change in settlement in the pile due to elastic compression between the neutral plane and the toe of the pile. P is the average load in the pile within a depth interval L, A is the cross-sectional area of the pile, and E is the elastic modulus of the pile. For very stiff piles, the toe settlement will be equal to the settlement at the neutral plane.

- Use the Q_b–z curve to determine if the mobilized Q_b for computed toe settlement is equal to required, Q_b.

Using a Q_b–z curve, such as that recommended by O'Neill and Reese (1999), compute the Q_b value mobilized for the toe displacement computed in the previous step.

- Compare the Q_b mobilized at the toe of the pile with that required for equilibrium between the downward and upward forces on the pile

Compare the Q_b value in the previous step with the Q_b value required for vertical force equilibrium using the equation

$$Q_b = P - Q_{S-N} - Q_{S-P} \tag{4.2}$$

where P is the pile head load, $Q_{S\text{-}N}$ is the negative skin friction from the ground surface to the neutral plane, and $Q_{S\text{-}P}$ is the positive skin friction below the neutral plane to the toe of the pile.

- Revise the location of the neutral plane until the mobilized Q_b based on toe settlement agrees with the Q_b computed with Eq. 4.2.

Once convergence is obtained, the neutral plane can then be used to compute the maximum load in the pile and the pile head (S_{head}) settlement using the equation

$$S_{head} = S_{NP} + \Sigma(PL/AE) \tag{4.3}$$

where Σ(PL/AE) is the sum of the elastic compression between the neutral plane and the pile head with variables as defined previously. For very stiff piles, the pile head settlement will be the same as the settlement at the neutral point.

4.7 Conclusions

Based on the results from the four separate full-scale blast liquefaction test sites, the following conclusions have been made:

1. In liquefied soils, negative and positive skin frictions after liquefaction and reconsolidation are typically 40–55% of the positive skin friction before liquefaction.
2. In non-liquefied soils above a liquefied layer, negative friction was approximately equal to the positive friction prior to liquefaction.
3. Despite the variation in pile type (driven piles, large bored piles, augercast piles and micropiles), the side friction following liquefaction was generally consistent, suggesting that this may be a typical result.
4. The depth to the neutral plane increased (and pile settlement decreased) as pile length increased.
5. In general, the neutral plane was not located at the base of the liquefied layer.
6. Measured pile settlement was generally consistent with the neutral plane concept after balancing applied pile head force plus negative friction force with the

positive friction force and the displacement-compatible end-bearing resistance, Q_b. Pile settlement is highly dependent on the stiffness of the Q_b−z curve not simply the magnitude of liquefaction-induced soil settlement.

Acknowledgements Funding to summarize case histories involving liquefaction-induced downdrag on piles was provided by the National Cooperative Highway Report Program (NCHRP). Funding for the Vancouver testing was provided by the TRB Ideas Deserving Exploratory Analysis (IDEAS) program of the Transportation Research Board. Funding for the Christchurch, New Zealand testing was primarily provided by a grant from the US National Science Foundation (Grant CMMI-1408892) with supplemental funding from the Pacific Earthquake Engineering Research (PEER) Center (Research Agreement Number: 1110-NCTRKR), the Federal Highway Administration and the Utah Department of Transportation Research Division. Funding for the Mirabello, Italy testing was mainly funded by the FIRB-Abruzzo project ('Indagini ad alta risoluzione per la stima della pericolosità e del rischio sismico nelle aree colpite dal terremoto del 6 aprile 2009', http://progettoabruzzo.rm.ingv.it/it). Finally, funding for the Turrell, Arkansas testing was provided primarily by the Arkansas Highway and Transportation Department with additional support from the National Science Foundation (Grant CMMI-1650576). This financial support is gratefully acknowledged; however, the conclusions and opinion expressed in this paper do not necessarily represent those of the sponsors.

References

Amoroso S, Rollins KM, Lusvardi C, Monaco P, Milana G (2018) Blast-induced liquefaction results at the silty-sand site of Mirabello, Emilia Romagna region, Italy. Geotech Earthquake Eng Soil Dyn V ASCE 10

Ashford SA, Rollins KM, Lane JD (2004) Blast-induced liquefaction for full-scale foundation testing. J Geotech Geoenviron Eng 130(8):798–806

Boulanger RW, Brandenberg SJ (2004) Neutral plane solution for liquefaction-induced down-drag on vertical piles. Geotech Eng Transp Projects 470–478

Bozozuk M (1981) Bearing capacity of pile preloaded by downdrag. In: 10th international conference on soil mechanics and foundation engineering, pp 631–636

Fellenius BH, Siegel TC (2008) Pile drag load and downdrag in a liquefaction event. J Geotech Geoenviron Eng 134(9):1412–1416

Gallagher PM, Conlee CT, Rollins KM (2007) Full-scale field testing of colloidal silica grouting for mitigation of liquefaction risk. J Geotech Geoenviron Eng 133(2):186–196

Ishihara K, Yoshimine M (1992) Evaluation of settlements in sand deposits following liquefaction during earthquakes. Soils Found 32(1):173–188

Ishimwe E, Coffman RA, Rollins KM (2018) Analysis of post-liquefaction axial capacities of driven pile and drilled shaft foundations. IFCEE, pp 272–283

Kevan L, Rollins KM, Coffmann R, Ishimwe E (2019) Full-scale blast liquefaction testing in arkansas USA to evaluate pile downdrag and neutral plane concepts. Springer, Earthquake Geotechnical Engineering for Protection and Development of Environment and Constructions, pp 648–655

O'Neill M, Reese L (1999) Drilled shafts construction procedures and design methods, vol 99. FHWA Publication No, FHWA IF, p 025

Rollins K, Strand S (2006) Downdrag forces due to liquefaction surrounding a pile. In: Proceedings of 8th US national conference on earthquake engineering

Rollins K, Hollenbaugh J (2015) Liquefaction induced negative skin friction from blast-induced liquefaction tests with auger-cast piles. In: Proceedings of 6th international conference on earthquake geotechnical engineering, New Zealand Geotechnical Society, Christchurch, New Zealand

Rollins KM, Gerber TM, Lane JD, Ashford SA (2005) Lateral resistance of a full-scale pile group in liquefied sand. J Geotech Geoenviron Eng 131(1):115–125

Rollins KM, Strand SR, Hollenbaugh JE (2018) Liquefaction induced downdrag and dragload from full-scale tests. In: Developments in earthquake geotechnics. Springer, pp 89–109

Weaver TJ, Ashford SA, Rollins KM (2005) Response of 0.6 m cast-in-steel-shell pile in liquefied soil under lateral loading. J Geotech Geoenviron Eng 131(1):94–102

Wentz F, van Ballegooy S, Rollins K, Ashford S, Olsen M (2015) Large scale testing of shallow ground improvements using blast-induced liquefaction. In: 6th international conference on earthquake geotechnical engineering. New Zealand Geotechnical Society

Zhang G, Robertson P, Brachman RW (2002) Estimating liquefaction-induced ground settlements from cpt for level ground. Can Geotech J 39(5):1168–1180

Chapter 5
Cyclic Resistance and Large Deformation Characteristics of Sands Under Sloping Ground Conditions: Insights from Large-Strain Torsional Simple Shear Tests

Gabriele Chiaro

5.1 Introduction

Soil liquefaction is a phenomenon that typically occurs in saturated loose sandy soil deposits during earthquakes. Its effects are most evident in sloping ground, where the substantial liquefaction-induced loss of soil shear strength and stiffness results typically in lateral ground displacement ranging from a few centimeters to several hundreds of meters. Figure 5.1 shows two examples of large deformation often observed in the field following major earthquakes. While the consequences of liquefaction (e.g., damage to piles and access of bridges, roads, flood-prevention systems and riverbanks, embankment dams, landslides, etc.) have been well documented during past and recent earthquakes (Hamada et al. 1994; Cubrinovski et al. 2010, 2011; Kiyota et al. 2011; Chiaro et al. 2015a, 2017a, 2018), there is still lack of knowledge as to the mechanics for large shear deformation (triggering and driving forces, large shear strain development characteristics and shear strength recovery process) in liquefied sandy soils. This limits our ability to identify susceptible soil deposits in advance and prevent potential catastrophic failures from occurring.

In any seismic event, the development of large ground deformation represents a major hazard to many engineering structures and buried lifeline facilities. Therefore, when evaluating liquefaction, it is important to assess whether or not a given soil in its in situ density and stress state has the potential for large ground deformation, including flow-type failure (Verdugo and Ishihara 1996; Cubrinovski and Ishihara 2000). Nonetheless, this is not an easy task, since liquefaction-induced ground failure is a complex phenomenon governed by many interdependent factors such as sloping ground conditions, earthquake characteristics (shear stress ampli-

G. Chiaro (✉)
University of Canterbury, Christchurch 8041, New Zealand
e-mail: gabriele.chiaro@canterbury.ac.nz

T. G. Sitharam et al. (eds.), *Latest Developments in Geotechnical Earthquake Engineering and Soil Dynamics*, Springer Transactions in Civil and Environmental Engineering, https://doi.org/10.1007/978-981-16-1468-2_5

Fig. 5.1 **a** Lateral ground deformation up to 40 cm due to liquefaction of a micaceous silty sand deposit observed along the Trishuli River levee after the 2015 Gorkha Earthquake, Nepal (Chiaro et al. 2015); and **b** Liquefaction-induced extremely large deformation (600 m) of a volcanic soil deposit within a gentle slope observed after the 2016 Kumamoto earthquake, Japan (Chiaro et al. 2017, 2018)

tude and number of cycles), confining stress level and soil density. (Yoshimi and Oh-oka 1975; Castro and Poulus 1977; Vaid and Finn 1979; Tatsuoka et al. 1982; Vaid and Chern 1983; Hyodo et al. 1991; Hyodo et al. 1994; Vaid et al. 2001; Yang and Sze 2011a, b; Sivathayalan and Ha 2011; Chiaro et al. 2012, 2013a, b, 2015, 2017; Ziotopoulou and Boulanger 2016; Lee and Seed 1967; Seed 1968; Boulanger et al. 1991). To provide new understanding into such a challenging topic, in this paper, experimental results obtained using a state-of-the-art torsional shear testing device that is capable of reproducing the simple shear condition and achieving double amplitude shear strain (γ_{DA}) 100% are presented and discussed. Specifically, the post-liquefaction response of Toyoura sand specimens is examined in terms of observed failure modes, strain development characteristics, cyclic resistance up to 50% single amplitude shear strain and limiting shear strain to cause strain localization into the specimens.

5.1.1 Effects of Static Shear on Liquefaction Resistance of Sand

It is well-known that the undrained response during cyclic loading of saturated sand deposits within sloping ground is different from that of level ground condition because these sand deposits are subjected to a gravitational driving static shear stress on the horizontal plane or assumed failure surface. During earthquake shaking, such sand deposits are subjected to further shear stresses due to shear waves propagating vertically upward from the bedrock. The superimposition of the gravitational static shear and seismically induced cyclic shear stresses can have major effects on the undrained response of the sand, leading to liquefaction and extremely large ground deformation.

Compared with the large body of experimental data describing the undrained cyclic behavior of sands under level ground conditions, there have been limited studies focusing on the undrained cyclic response of sands under sloping ground conditions. Such studies, mainly based on triaxial tests and occasionally on simple shear, torsional shear or ring shear tests, have produced valuable data showing that the presence of initial static shear can have major effects on the cyclic response of sands. Nevertheless, contradictory views seem to exist with respect to these effects.

The magnitude of the initial static shear (τ_{static}) is often expressed in terms of static shear stress ratio ($\alpha = \tau_{static}/\sigma_c'$), which is the ratio of the τ_{static} to the effective consolidation stress (σ_c') on the plane of interest. The effects of α on the cyclic resistance ratio (*CRR*) of the sand are then usually expressed in terms of a K_α factor ($K_\alpha = CRR_\alpha/CRR_{\alpha=0}$) where *CRR* refers to the cyclic stress ratio ($CSR = \tau_{cyclic}/\sigma_c'$) required to trigger liquefaction (according to a specific failure criterion) in a specified number of cycles; CRR_α is the value of *CRR* for a given value of α; and $CCR_{\alpha=0}$ corresponds to the level ground condition ($\alpha = 0$). Experimental results on a range of sands at confining pressure less than 300 kPa showed that cyclic resistance would tend to increase with increasing α for dense sands and tend to decrease with increasing α for loose sands. These trends are reported in Fig. 5.2a showing values of K_α from simple shear tests results by Vaid and Finn (1979) and Boulanger et al. (1991) for Ottawa sand tested at $D_r = 35–68\%$ with $\sigma_{vc}' = 202$ kPa. On the other hand, typical results of triaxial tests are reported in Fig. 5.2b for Toyoura sand tested at $D_r = 10–70\%$ with $\sigma_{nc}' = 100$ kPa (Yang and Sze 2011). In such a case, it appears that the cyclic resistance would tend to increase with increasing α irrespective of the density of sand. Yet, for loose sand, the cyclic resistance appears to decrease with increasing α at higher α values.

Due to the scarcity of experimental data and the lack of convergence and consistency in the existing data, it has been strongly recommended that such experimental findings should not be used by non-specialists or in routine engineering practice (Youd et al. 2001). Evidently, there is a need for continued in-depth research to improve our understanding of the complicated effects of initial static shear on the cyclic resistance of sandy soils within sloping ground. As reported in

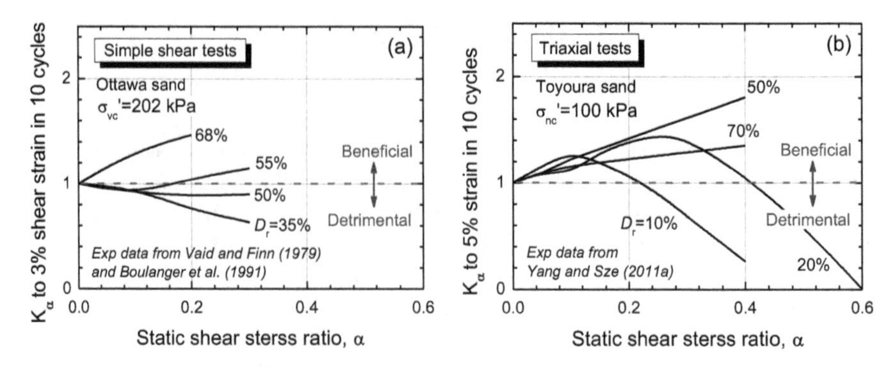

Fig. 5.2 Typical experimental relationships between K_α factor and static shear stress ratio: **a** simple shear tests and **b** triaxial tests

this paper, the development of a state-of-the-art large-strain torsional shear testing device may contribute to clarify a few ambiguities and obtain new insights on this challenging topic.

5.1.2 *Large Deformation Properties of Liquefied Sand Within Sloping Ground*

In most of the above-mentioned studies (Yoshimi and Oh-oka 1975; Castro and Poulus 1977; Vaid and Finn 1979; Tatsuoka et al. 1982; Vaid and Chern 1983; Hyodo et al. 1991, 1994; Vaid et al. 2001; Yang and Sze 2011a, b; Ziotopoulou and Boulanger 2016; Lee and Seed 1967; Seed 1968; Boulanger et al. 1991), due mainly to mechanical limitation of the employed testing devices and/or due to large extents of non-uniform deformation of the specimens at higher strain levels, as well as technical difficulties in correcting for the effects of membrane resistance during the tests, the shear strain levels were limited to the range of 10–20%. Such shear strain levels are, therefore, not able to fully describe the occurrence of extremely large liquefaction-induced ground deformations of 100% or more as often observed in the field in the case of sloping ground. Nevertheless, in torsional shear tests on hollow cylindrical specimens, it is nowadays possible to achieve higher strain levels by increasing the amount of torsional shear displacement that is applied to the soil specimen through the rotation of the top cap (Yasuda et al. 1992; Kiyota et al. 2008, 2010). For instance, Yasuda et al. (1992) investigated the properties of liquefied sand under undrained monotonic torsional shear conditions up to large strains levels of about 50%. Later, Kiyota et al. (2008) conducted undrained cyclic torsional simple shear tests up to double amplitude shear strain exceeding 50%. In the latter tests, a correction for the effects of membrane resistance on measured shear stress was carefully applied (Kiyota et al. 2008). Kiyota et al. (2010) reported that the maximum amounts of liquefaction-induced ground displacement observed in relevant model tests and field observations are consistent with the limiting value to initiate strain localization observed in torsional shear tests (Kiyota et al. 2008). Therefore, as long as the shear deformation remains uniform, the results of torsional shear tests can be effectively used to estimate the extent of large deformation that will occur in the field during earthquakes (Kiyota et al. 2010).

To clarify the uncertainties before mentioned on the effects of initial static shear on the undrained cyclic behavior of sand and overcome major technical limitation in reaching large cyclic post-liquefaction shear strain levels, Chiaro et al. (2012, 2013a, 2021) performed a number of large-strain cyclic undrained torsional simple shear tests with initial static shear using the same advanced testing apparatus employed by Kiyota et al. (2008, 2010). The tests were carried out on loose and medium dense Toyoura sand specimens (D_r = 25–49%) under different combinations of static and cyclic shear stress reproducing reversal stress and no-reversal stress conditions (Hyodo et al. 1991). This paper first provides an overview of the test apparatus and the procedures developed to carry out high-quality experimental tests. From the test results, the post-liquefaction behavior of Toyoura sand is assessed in terms of observed failure modes, strain development characteristics and cyclic strength up to 50% single amplitude shear strain. Lastly, the occurrence of strain localization in liquefied sand specimens and its effects on the cyclic strength evaluation is discussed. A comparison between cyclic strength typically obtained by triaxial tests (by other authors) and torsional shear tests (this study) is also reported with the aim of discussion on the appropriateness or not of using simple shear condition in the evaluation of liquefaction resistance of sand with special reference to tests with static shear stress.

5.2 Large-Strain Hollow Cylindrical Torsional Shear Apparatus

Torsional shear apparatus on hollow cylindrical specimens is recognized to be a good tool to properly evaluate liquefaction soil response (Kiyota et al. 2008, 2010; Chiaro et al. 2019, 2021; Arangelovski and Towhata 2004; Georgiannou et al. 2008). In particular, it offers the possibility to reproduce simple shear conditions that are a close representation of field stress conditions during earthquakes (Vaid and Finn 1979; Tatsuoka et al. 1982; Ziotopoulou and Boulanger 2016; Boulanger et al. 1991; Cappellaro et al. 2019).

To achieve double amplitude torsional shear strain levels exceeding 50%, the torsional shear apparatus on hollow cylindrical specimens shown in Fig. 5.3a was employed. The torque and axial loads were detected by using a two-component load cell, which is installed inside the pressure chamber. Difference in pressure levels between the cell pressure and the pore water pressure was measured by a high-capacity differential pressure transducer (HCDPT). Volume change during the consolidation process was measured by a low-capacity differential pressure transducer (LCDPT). A potentiometer with a wire and a pulley was employed to

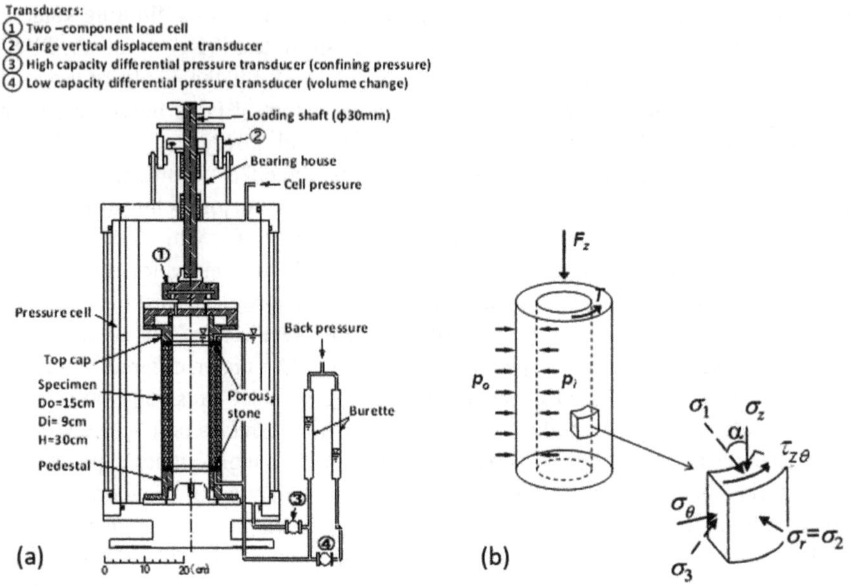

Fig. 5.3 **a** Torsional shear apparatus used in this study (Kiyota et al. 2008); **b** External forces and stress components acting on a hollow cylindrical specimen (Chiaro et al. 2017)

measure the rotation angle of the top cap and, thus, the torsional shear strains. The loading system was controlled by a computer (via a feedback-loop set onto the force readings from the load cell and the rotation angle from the potentiometer) that computes the stresses and strains applied on the soil specimen and in real time controls the device. Full details of the loading system are provided in Kiyota et al. (2008).

5.2.1 Stress and Strains Definition

The hollow cylinder torsional shear apparatus allows independent control of four loading components (Fig. 5.3b), namely vertical axial load (F_z), torque load (T), inner cell pressure (p_i) and outer cell pressure (p_o). The correspondent stress components [i.e., axial stress (σ_z), radial stress (σ_r), circumferential stress (σ_θ) and torsional shear stress ($\tau_{z\theta}$)] were calculated as follows (Hight et al. 1983):

$$\sigma_z = \frac{F_z}{\pi(r_o^2 - r_i^2)} + \frac{(p_o r_o^2 - p_i r_i^2)}{(r_o^2 - r_i^2)} \tag{5.1}$$

$$\sigma_r = \frac{(p_o r_o + p_i r_i)}{(r_o + r_i)} \tag{5.2}$$

$$\sigma_\theta = \frac{(p_o r_o - p_i r_i)}{(r_o - r_i)} \tag{5.3}$$

$$\tau = \tau_{z\theta} = \frac{3T}{2\pi(r_o^3 - r_i^3)} \tag{5.4}$$

where r_o and r_i are the outer and inner radii of the specimen, respectively.

The average principal stresses σ_1 (major), σ_2 (intermediate), σ_3 (minor) and the mean principal stress p are given by

$$\sigma_2 = \sigma_r \tag{5.6}$$

$$p = \frac{\sigma_1 + \sigma_2 + \sigma_3}{3} \tag{5.7}$$

$$p' = p - u \tag{5.8}$$

In Eq. (5.8), p' is the effective mean principal stress, and u is the pore water pressure. It should be noted that, in this study, p_i and p_o were kept equal to each other. Moreover, the measured shear stress and mean effective principal stress were corrected for the effects of the membrane force, as described in the next section.

Finally, the average torsional shear strain can be computed as

$$\gamma = \gamma_{z\theta} = \frac{2\theta\,(r_o^3 - r_i^3)}{3H\,(r_o^2 - r_i^2)} \tag{5.9}$$

where θ is the circumferential angular displacement, and H is the specimen height.

5.2.2 *Experimental Evaluation of Membrane Resistance and Its Correction*

Koseki et al. (2005) pointed out that in performing torsional shear tests on hollow cylindrical soil specimens, due to the presence of inner and outer membranes, the effect of membrane resistance on the measured torsional shear stress cannot be neglected (i.e., to calculate the shear stress effectively applied on the soil specimen, the total stress measured by the load cell needs to be corrected for the apparent shear stress induced by the presence of the membrane). In a similar manner, the mean effective principal stress requires amendments due to membrane resistance effects as well (Chiaro et al. 2013a). Usually, the membrane resistance is computed based on the linear elasticity theory, which assumes cylindrical deformation of

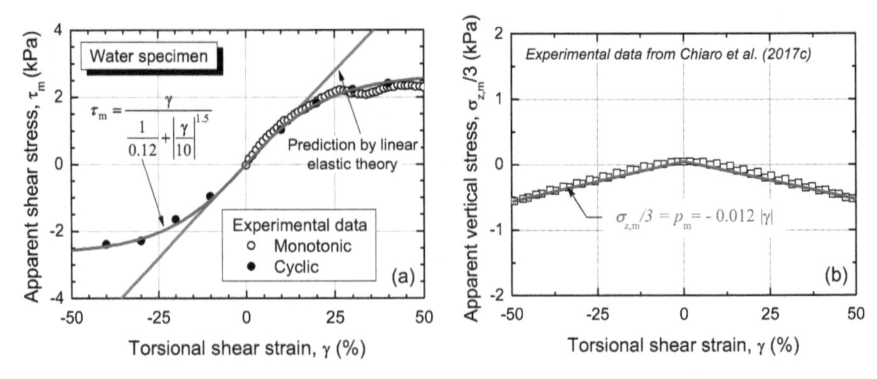

Fig. 5.4 Membrane force evaluation and correction: **a** apparent shear stress and **b** apparent vertical stress

specimen (Koseki et al. 2005). Nevertheless, experimental evidences clearly demonstrate that at large shear strains, deformation of a hollow cylindrical sand specimen is not uniform along the specimen height, and specimen shape is far from being perfectly cylindrical (Chiaro et al. 2013a; Kiyota et al. 2008). To experimentally evaluate the membrane force, a specific testing procedure was developed by Kiyota et al. (2008), which consists of shearing a hollow cylindrical water specimen from small to large shear strain levels in the torsional shear apparatus.

As shown in Fig. 5.4, the measured shear stress can be corrected for the effect of membrane force by employing the empirical hyperbolic correlation between the apparent shear stress due to the membranes (τ_m) and the shear strain (γ) (Chiaro et al. 2017). In contrast, the effective mean principal stress can be corrected by using the linear expression between the apparent vertical stress due to the membranes ($\sigma_m/3$) and the shear strain (γ) that was experimentally derived by Chiaro et al. (2013a). The corrected shear stress and effective mean principal stress take the form of

$$\tau_{cor} = \tau - \tau_m = \tau - \frac{\gamma}{\frac{1}{0.12} + \left|\frac{\gamma}{10}\right|^{0.15}} \tag{5.10}$$

$$p'_{cor} = \frac{\left[(\sigma_z - \sigma_{z,m}) + \sigma_r + \sigma_\theta\right]}{3} - u = \frac{\left[(\sigma_z - 0.036|\gamma|) + \sigma_r + \sigma_\theta\right]}{3} - u \tag{5.11}$$

5.3 Testing Material and Procedure

All the tests were performed on Toyoura sand, which is a uniform quartz sand with sub-angular particle shape and negligible fines content (G_s = 2.656; e_{max} = 0.992; e_{min} = 0.632; F_c = 0.1%). Its gradation curve and an optical microscope photo showing the typical particle shape are shown in Fig. 5.5.

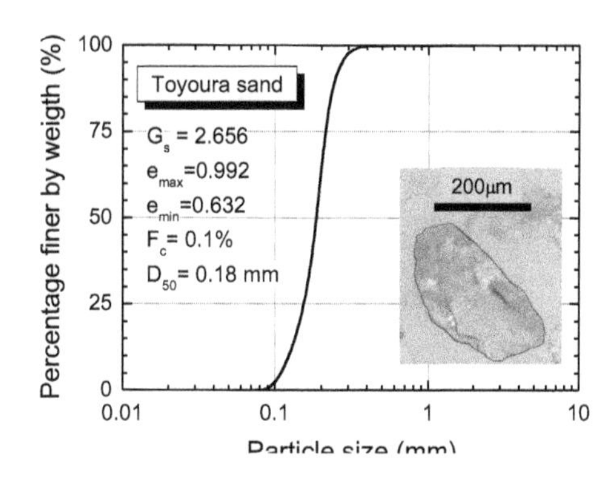

Fig. 5.5 Particle size distribution and optical microscope photo of Toyoura sand

Several medium-sized hollow cylindrical specimens with dimension of 150 mm in outer diameter, 90 mm in inner diameter and 300 mm in height were prepared by air pluviation method at a relative density (D_r) of 25–30% and 44–48%. To minimize the degree of inherent anisotropy in the radial direction of hollow cylindrical sand specimens, specimen preparation was carried out by carefully pouring air-dried sand particles into a mold while moving radially and at the same time circumferentially in alternative directions (i.e., first in clockwise and then anti-clockwise directions) the nozzle of the pluviator (De Silva et al. 2015). To achieve specimens with highly uniform density, the falling height of sand grains was kept constant throughout the pluviation process. To assure a high degree of saturation, the double vacuum method (Ampadu 1991) was employed, and de-aired water was circulated into the specimens. Finally, a back pressure of 200 kPa was applied. Skempton's B-value $\geq$ 0.96 was measured for all the tested specimens.

After saturation, specimens were isotropically consolidated by increasing the effective mean stress (p') to 100 kPa. Subsequently, a specific value of initial static shear was applied by drained monotonic torsional shearing, which replicates the gravitational shear stress component induced by slope inclination. Finally, to replicate seismic loading conditions, uniform cycles of undrained cyclic torsional shear stress amplitude were applied at a shear strain rate of 2.5%/min. The loading direction was reversed when the amplitude of shear stress reached the target value. During the process of undrained cyclic torsional loading, the vertical displacement of the top cap was prevented with the aim to simulate as much as possible the simple shear conditions that ground undergoes during horizontal excitation (Kiyota et al. 2010).

5.3.1 Stress Reversal and no-Stress Reversal Loading Conditions

As described schematically in Fig. 5.6, before an earthquake shaking, a soil element beneath sloping ground is subjected to an initial static shear stress (τ_{static}) induced

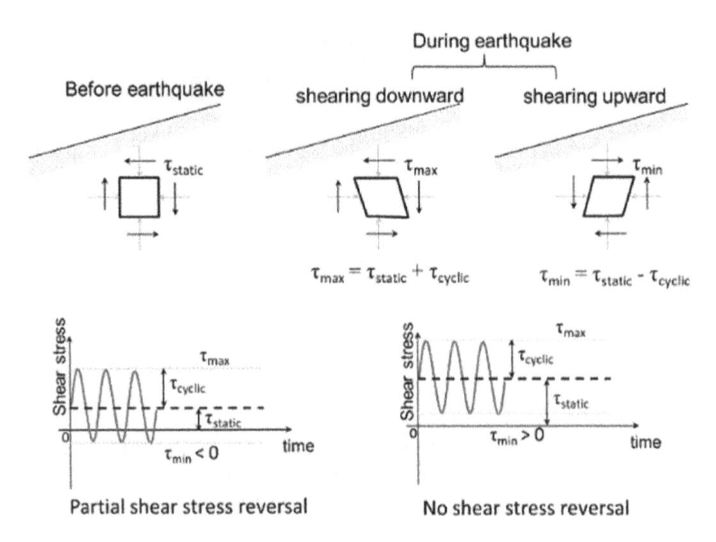

Fig. 5.6 Stress conditions in sloping ground during earthquakes (Chiaro et al. 2017, 2021)

by the slope inclination conditions. During the earthquake shaking, soil elements undergo partial or no shear stress reversal loading conditions due to the superimposition of seismically induced cyclic shear stress (τ_{cyclic}) to τ_{static}. When τ_{static} < $_{cyclic}$, the shear stress changes within the maximum positive value τ_{max} (=τ_{static} + $_{cyclic}$) > 0 and the minimum negative value τ_{min} (=$\tau_{static} - \tau_{cyclic}$) < 0 during each cycle of loading. This type of loading is known as stress reversal or two-way loading. On the other hand, when $\tau_{static} > \tau_{cyclic}$, the shear stress is always positive (i.e., $\tau_{max} > 0$ and $\tau_{min} > 0$). This condition is called no-stress reversal or one-way loading (Yoshimi and Oh-oka 1975; Hyodo et al. 1991). In view of the above, in this study, the influence of stress reversal and no-stress reversal was investigated by considering various combinations of τ_{static} and τ_{cyclic}. Tests were carried out over a range of τ_{static} varying from 0 to 30 kPa and three levels of τ_{cyclic} (i.e., 12, 16 and 20 kPa), as listed in Table 5.1.

5.4 Tests Results

5.4.1 Undrained Shear Strength

In Fig. 5.7, the effective stress paths and stress–strain relationships obtained by undrained monotonic torsional simple shear tests are shown for three tests, which were conducted under the same initial conditions of relative density (D_r = 27–30% or 44–48%) and effective mean principal stress (p_0' = 100 kPa) but by varying the initial static shear level (τ_{static} = 0–25 kPa).

Table 5.1 Undrained cyclic large-strain torsional shear tests

Test	D_r (%)	τ_{cyclic} (kPa)	τ_{static} (kPa)	Stress reversal	References
1	30	12	0	Yes	Chiaro et al. (2021)
2	28	12	5	Yes	Chiaro et al. (2021)
3	26	12	10	Yes	Chiaro et al. (2021)
4	26	12	15	No	Chiaro et al. (2021)
5	25	12	20	No	Chiaro et al. (2021)
6	28	12	25	No	Chiaro et al. (2021)
7	25	12	30	No	Chiaro et al. (2021)
8	25	16	0	Yes	Chiaro et al. (2021)
9	26	16	5	Yes	Chiaro et al. (2021)
10	26	16	10	Yes	Chiaro et al. (2021)
11	30	16	15	Yes	Chiaro et al. (2021)
12	28	16	20	No	Chiaro et al. (2021)
13	29	16	25	No	Chiaro et al. (2021)
14	26	16	30	No	This study
15	46	16	0	Yes	Chiaro et al. (2012, 2013a)
16	46	16	5	Yes	Chiaro et al. (2012, 2013a)
17	47	16	10	Yes	Chiaro et al. (2012, 2013a)
18	44	16	15	Yes	Chiaro et al. (2012, 2013a)
19	45	16	20	No	Chiaro et al. (2012, 2013a)
20	46	20	0	Yes	Chiaro et al. (2012, 2013a)
21	48	20	5	Yes	Chiaro et al. (2012, 2013a)
22	46	20	10	Yes	Chiaro et al. (2012, 2013a)
23	44	20	15	Yes	Chiaro et al. (2013a, 2013a)
24	47	20	20	Yes	Chiaro et al. (2012, 2013a)
25	46	20	25	No	Chiaro et al. (2012, 2013a)
26	49	10	15	No	This study

Effective mean principal stress, $p_0' \approx 100$ kPa

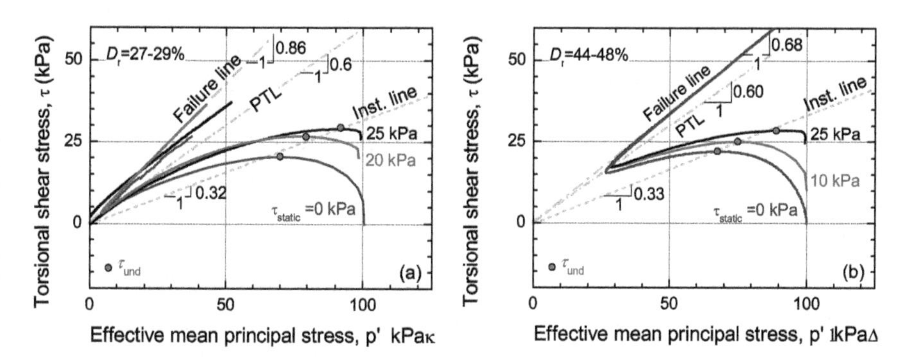

Fig. 5.7 Typical effective stress paths of medium dense Toyura sand specimens in monotonic torsional simple shear tests

All the specimens initially showed a contractive behavior (i.e., decrease in the p'), during which the shear stress (τ) steadily increased to a transient peak (here referred to as the undrained shear strength, τ_{und}). The τ_{und} marks the initiation of unstable behavior, as with further loading the shear stress drops to a transient minimum value (or quasi-steady state Verdugo and Ishihara 1996) during which the specimen deforms under a nearly constant shear stress. As soon as the shear stress reaches the phase transformation line (PTL; Ishihara et al. 1975), dilative behavior takes place, and the effective stress paths follow the failure envelope line.

Lade (1993) defined the instability line (IL) which connects the peak points of the effective stress paths to the origin of stress space. Further, Kramer (1996) termed this line as the flow liquefaction surface (FLS), since flow liquefaction behavior was observed in the tests in which the monotonic or cyclic loading stress path exceeds the undrained shear strength (τ_{und}). They showed that the slope of this line can be uniquely determined for specimens having similar relative density, irrespective of the initial effective stress level.

The experimental data reported in Fig. 5.7 show that under the same initial conditions of relative density and effective mean stress, the τ_{und} increases with increasing τ_{static}. However, it seems that IL line is unique despite the change in density.

5.4.2 *Failure Mechanisms and Development of Large Deformation*

In this section, three distinct deformation modes that loose and medium dense sand deposits in sloping ground may experience during earthquakes are identified: (i) potential development of large deformation due to cyclic liquefaction; (ii) abrupt development of large deformation due to rapid (flow) liquefaction and (iii) progressive cyclic accumulation of large residual deformation inducing shear failure. The case of no failure is reported for completeness.

Failure modes under stress reversal loading conditions. It has been observed in previous studies that under stress reversal conditions liquefaction is likely to occur, but its severity and consequent deformation development may vary significantly (Hyodo et al. 1991; Yang and Sze 2011). As described in detail hereafter, to gain a better understanding about the not unique sand response under reversal stress, the undrained response of several specimens undergoing reversal stress conditions is analyzed in terms of change in excess pore water pressure (PWP) and development of shear strain from cyclic and monotonic torsional shear tests.

Potential development of large shear deformation under cyclic loading
Figure 5.8 shows the results of a stress reversal test in which progressive excess pore water pressure generation (EPWP) besides a nearly zero shear strain development was observed until the full liquefaction state is reached at Δu = 100 kPa. Yet, significantly, only after this point (i.e., post-liquefaction state), sudden

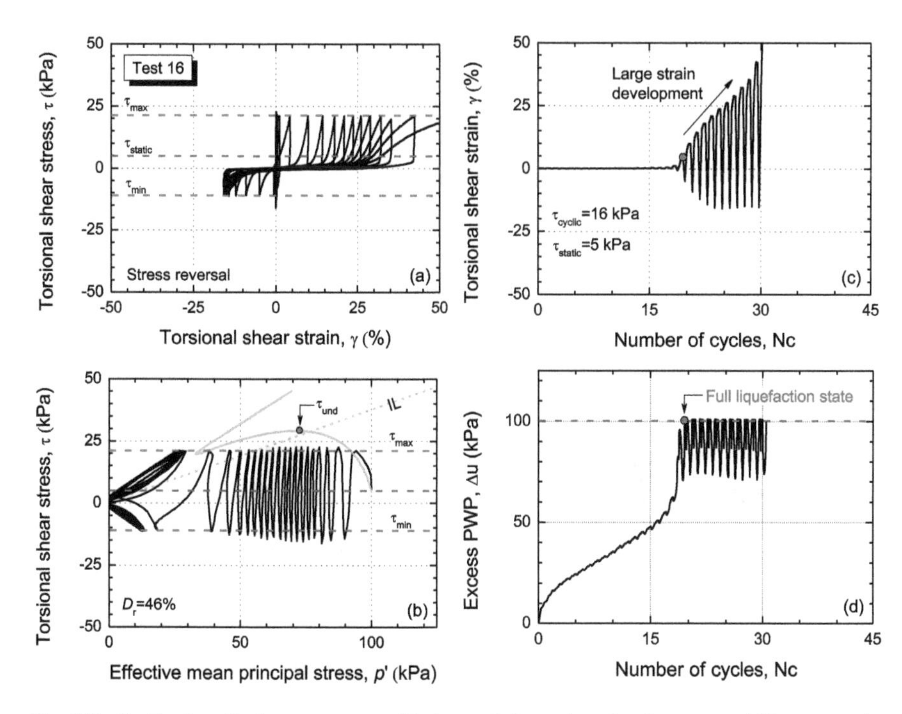

Fig. 5.8 Cyclic liquefaction response with large deformation development within ten cycles under stress reversal torsional shear loading

development of large shear strain clearly took place. This type of undrained cyclic response, which potentially may induce large post-liquefaction flow deformation, is referred to as cyclic mobility (Yang and Sze 2011) or cyclic liquefaction (Chiaro et al. 2012). It is commonly observed in loose/medium dense sand exhibiting a limited strain-softening behavior (i.e., no flow-type static liquefaction failure) under monotonic undrained shearing, and it may occur only under reversal stress conditions ($\tau_{min} < 0$) if the maximum shear stress does not exceed the undrained shear strength of sand ($\tau_{max} < \tau_{und}$).

From a practical viewpoint, under realistic earthquake conditions, cyclic liquefaction would take place and trigger flow deformation only if an earthquake produces a substantial number of cycles of large shear stress amplitude. For instance, if a moment magnitude (M_w) 7–7.5 earthquake is considered, then it is estimated that it will induce 10–15 shear stress cycles of uniform amplitude (Seed and Idriss 1982). Thus, as shown in Fig. 5.8a, under the stress conditions considered in Test 16, although the sand has the potential to liquefy, initial liquefaction will not take place, and large extent of strains will not develop within 10–15 cycles of loading. On the other hand, as shown in Fig. 5.9, under the stress conditions employed in Test 17, initial liquefaction will be achieved in less than ten loading cycles, and consequently, the rapid development of extremely large shear strain can be expected during the cyclic mobility phase.

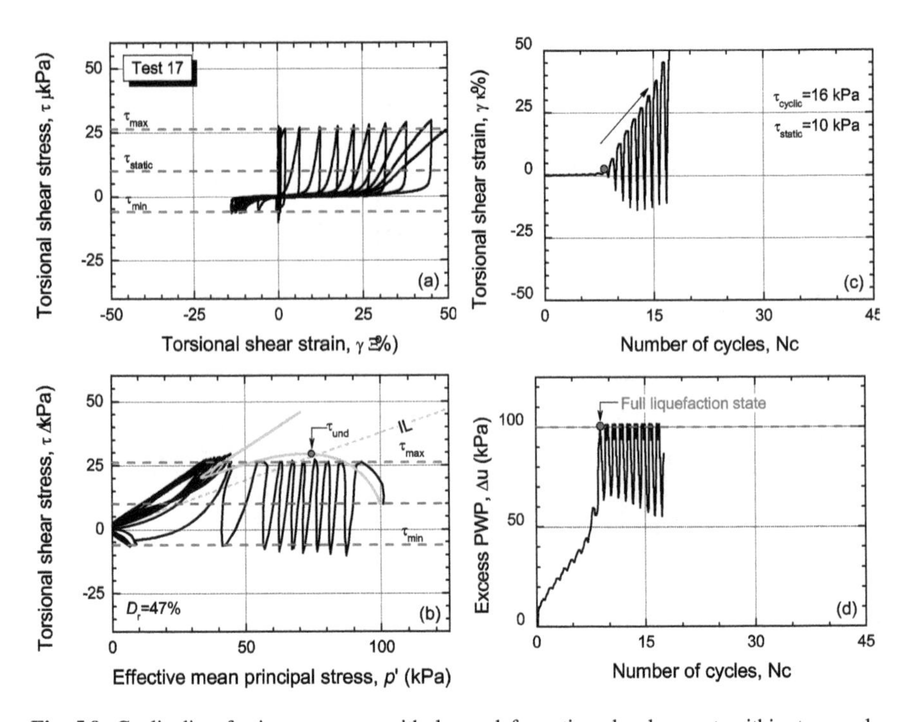

Fig. 5.9 Cyclic liquefaction response with large deformation development within ten cycles under stress reversal torsional shear loading

Liquefaction-induced abrupt development of large shear deformation
In contrast to the previous case, in the stress reversal test shown in Fig. 5.10, immediate EPWP generation was observed (i.e., full liquefaction state at Δu = 100 kPa was achieved in less than one cycle) and rapid development of extremely large shear strain in less than eight cycles. This type of undrained cyclic response is referred to as flow-type failure (Yang and Sze 2011) or rapid (flow) liquefaction (Chiaro et al. 2012) and similarly to the case previously examined, is also commonly observed in loose/medium dense sand exhibiting stable behavior under monotonic undrained loading. However, it only takes place under reversal stress conditions ($\tau_{min} < 0$) when the maximum shear stress (τ_{max}) exceeds the undrained shear strength of sand (τ_{und}).

The above described flow-type liquefaction behavior can be easily initiated by a strong earthquake with moment magnitude of M_w7 to M_w 8, which would correspond to 10 to 30 equivalent uniform cycles of large shear stress amplitude.

Failure modes under no-stress reversal loading conditions. It has been observed in previous studies that under non-stress reversal conditions, large deformation may bring loose sand to failure (Hyodo et al. 1991; Yang and Sze 2011). However, this is not always the case. In fact, under certain conditions, loose and medium dense sand may be very resistant against cyclic loading and will not experience liquefaction nor shear failure as described in detail hereafter.

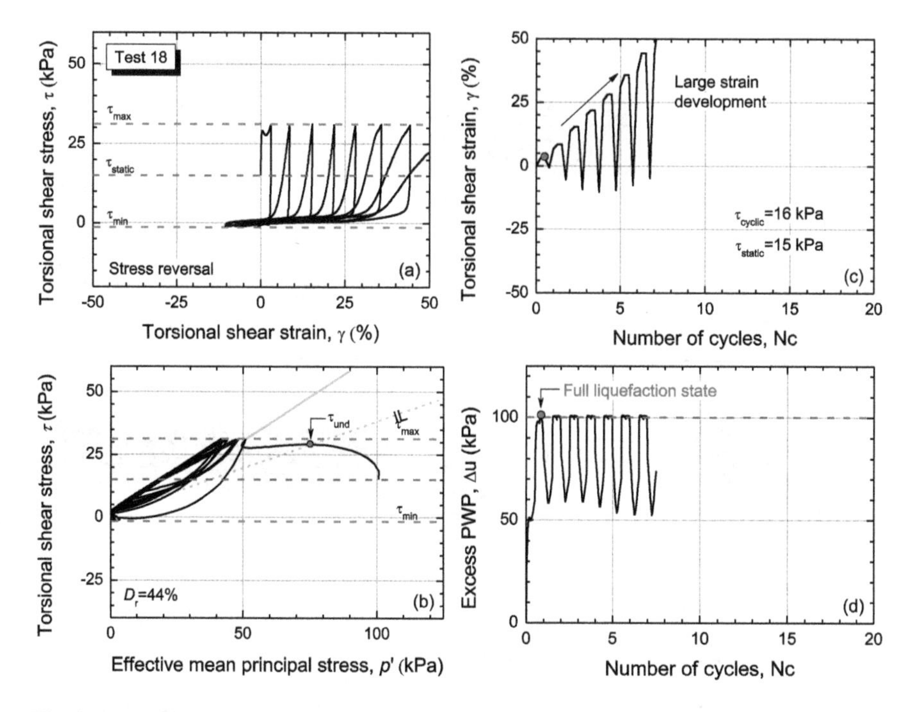

Fig. 5.10 Cyclic liquefaction with development of large deformation in less than ten cycles under stress reversal torsional shear loading

Shear failure due to accumulation of cyclic residual deformation

In the no-stress reversal test shown in Fig. 5.11, despite the gradual EPWP generation, full liquefaction state was not reached (i.e., $\Delta u \approx 90$ kPa after about 100 cycles of shear loading). However, progressive development of large single amplitude shear strain was observed. After ten cycles, the acculturated residual shear strain is almost 12%. At a large shear strain level of 21%, the specimen's failure was then observed as shown by formation of a shear band (Fig. 5.11c).

This type of soil response is referred to as plastic strain accumulation (Yang and Sze 2011) or residual deformation failure (Chiaro et al. 2012). It mostly occurs under no-stress reversal conditions when τ_{max} is higher than τ_{und}. However, as shown in Fig. 5.12, in some cases for loose sand, it may also occur where τ_{max} does not exceed τ_{und}, if the cyclic stress path intercepts the monotonic stress path.

Yet, under earthquake loading conditions, such an accumulation of large residual deformation would occur only if an earthquake produced a substantial number of uniform cycles with large stress amplitude and would result in severe serviceability problems.

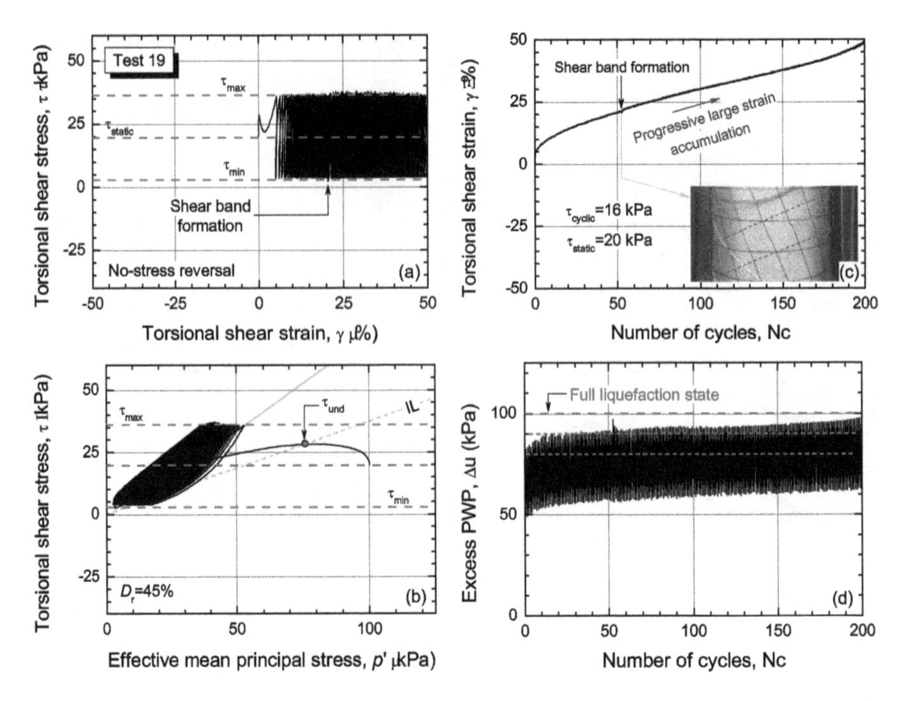

Fig. 5.11 Cyclic liquefaction with development of large deformation in less than ten cycles under no-stress reversal torsional shear loading

No-failure conditions

Figure 5.13 refers to the cyclic undrained response of a medium dense Toyoura sand specimen experiencing no-stress reversal loading conditions (Test 26) in which τ_{max} does not exceed τ_{und}. In this test, despite applying 100 cycles of loading, only very limited excess PWP is generated (i.e., ($\Delta u/p' \approx 20\%$). Significantly, it is also observed that the extent of shear strain is small ($\approx$0.1%). This type of no liquefaction and no failure behavior under cyclic loading has rarely been reported in the literature for sandy soils. As shown in the previous section, no failure will take place only if the cyclic stress path does not intercept the monotonic stress path.

It may be argued that, in these type of tests, failure could potentially be reached by applying a number of cycles higher than 100. However, it is expected that a M_w 7–8 earthquake will produce no more than 10–30 equivalent cycles of uniform-amplitude shear stress loadings. Consequently, it is rational to state that, a slope of loose or medium dense saturated sand will not experience failure in the field when shaken by an earthquake producing the same cyclic loading conditions here examined.

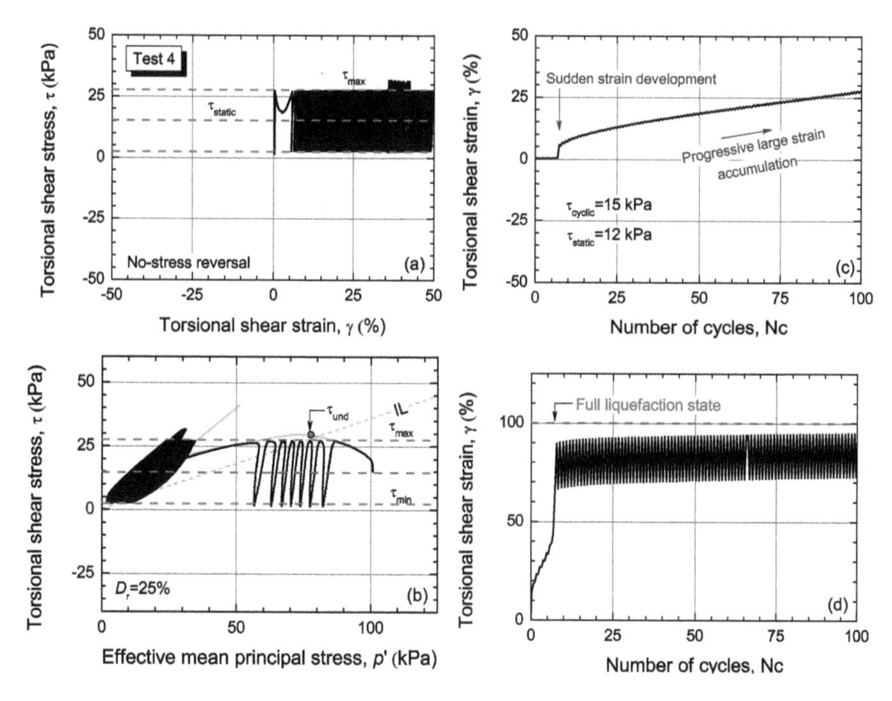

Fig. 5.12 Shear failure due to accumulation of cyclic residual deformation under no-stress reversal torsional shear loading

5.4.3 Cyclic Strength Against Large Deformation Accumulation

A key factor to evaluate the cyclic response of liquefiable soil in level ground ($\tau_{static} = 0$) is the resistance to cyclic strain accumulation (referred hereafter simply to as cyclic strength), which is usually expressed as the cyclic stress ratio ($CSR = \tau_{cyclic}/p_0'$) required to develop a specific amount of double amplitude shear strain (γ_{DA}) within a given number of uniform shear stress loading cycles. When considering sloping ground condition, the assessment of cyclic strength has to also account for the static shear ratio ($\alpha = \tau_{static}/p_0'$).

When τ_{static} is present, γ_{DA} tends to accumulate in the loading direction where the τ_{static} is applied (i.e., loading becomes unsymmetrical). Therefore, the use of single amplitude shear strain (γ_{SA}) rather than γ_{DA} has been recommended to estimate the cyclic strength of sand under sloping ground conditions (Chiaro et al. 2012, 2013a). The value of γ_{SA} can be expressed in terms of either the largest cyclic shear deformation of slopes during earthquakes γ_{max} (defined at $\tau = \tau_{max}$) or residual deformation of slopes just after earthquakes γ_{RS} (defined at zero cyclic shear stress, i.e., $\tau = \tau_{static}$ Tatsuoka et al. 1982). Nevertheless, Chiaro et al. (2012) found that γ_{max} and γ_{RS} almost coincide with each other in cyclic undrained torsional shear tests, so that both can be used interchangeably.

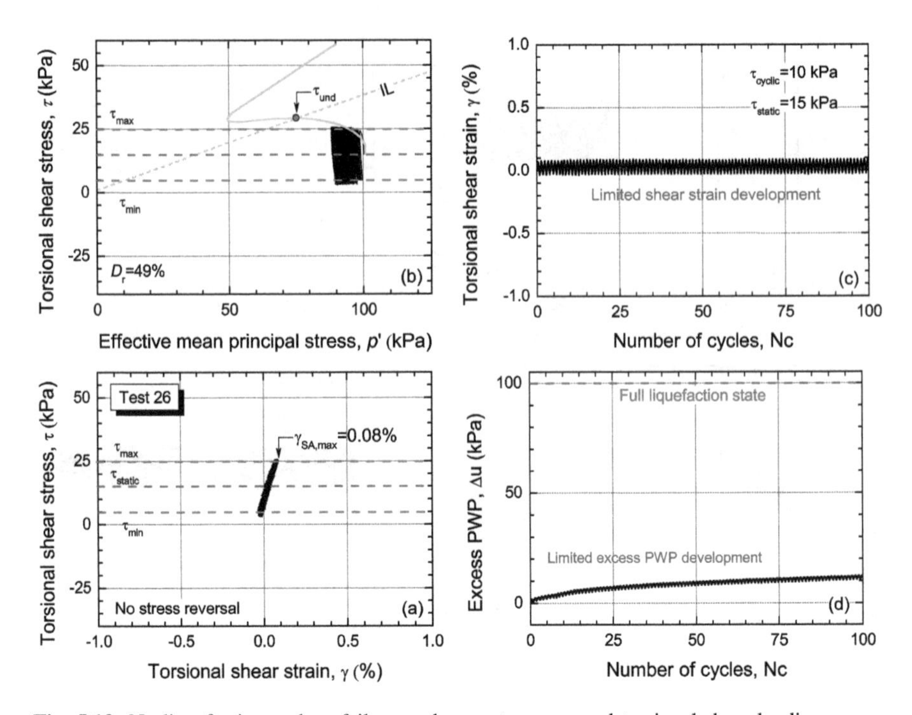

Fig. 5.13 No liquefaction and no failure under no-stress reversal torsional shear loading

In view of the above, henceforward, for any given value of α, the cyclic strength is defined as the value of *CSR* to achieve $\gamma_{\max} = \gamma_{SA} = 7.5$, 20 and 50%. The following sections discuss the effects of initial static shear, cyclic shear stress and relative density on the cyclic resistance of tested Toyoura sand specimens. Figure 5.14 shows the cyclic strength of loose (D_r = 25–30%) and medium dense (D_r = 44–48%) Toyoura sand specimens, respectively. In Fig. 5.15, to better capture the effects of τ_{static} on cyclic strength, the same data are plotted also in terms of variation of cyclic strength with α.

It is important to mention that although the cyclic strength is reported for γ_{SA} = 50%, such data should be taken only as a reference since these data can be affected by strain localization (i.e., the formation of shear band) that may take place during undrained torsional shear loading (Chiaro et al. 2012, 2013a; Kiyota et al. 2010), as described in details later in this paper.

Key factors influencing the cyclic strength of sand

Effect of initial static shear. Obviously in the case of 7.5% the sand resistance decreases with an increase in α. At larger strains of 20% and 50%, however, the cyclic strength first decreases and then increases. This change in strain accumulation response can be associated with change in failure mode from cyclic liquefaction to rapid liquefaction to shear failure. Further, for any given value of α, the

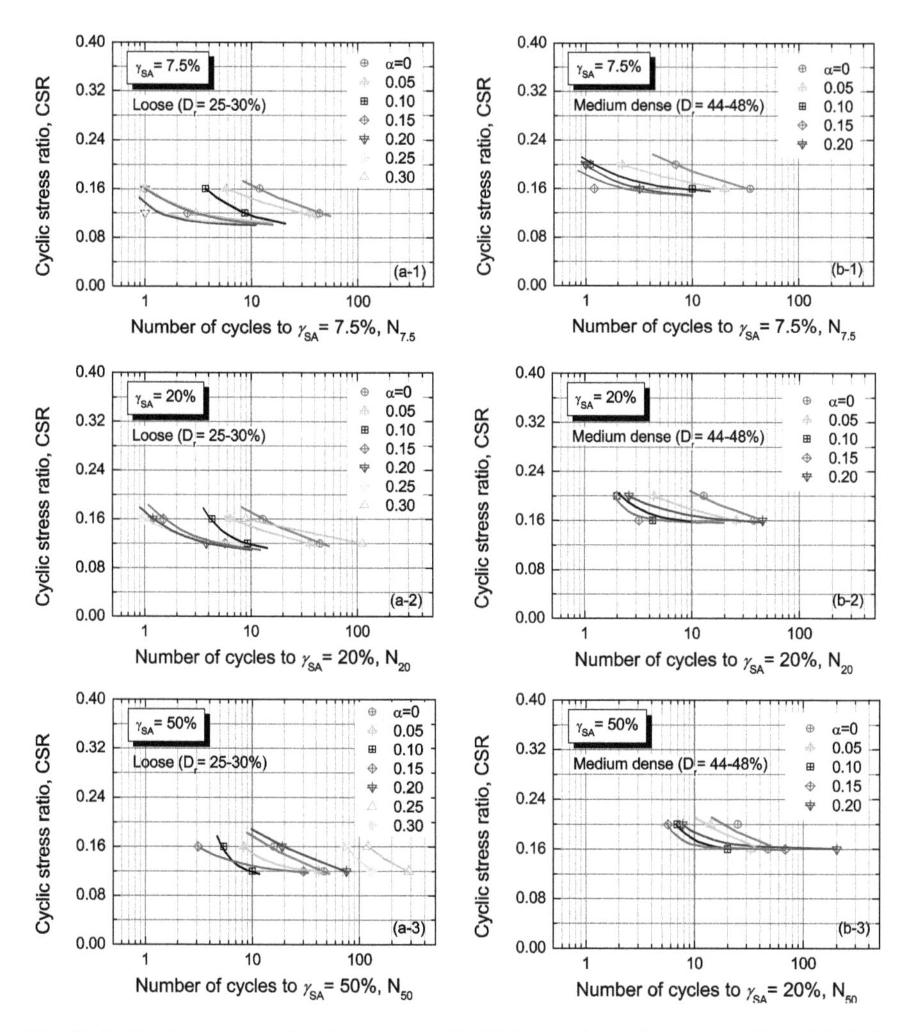

Fig. 5.14 Cyclic strength of **a** loose (D_r = 25–30%) and **b** medium dense (D_r = 44–48%) Toyoura sand under sloping ground conditions

cyclic strength significantly decreases with increasing *CSR*. This response is in accordance with those observed in previous studies (e.g., Hyodo et al. 1991).

Effect of relative density. In general, despite the two different levels of relative density tested in this study, Toyoura sand behavior is reasonably similar. In fact, under stress reversal loading, a drastic drop in the cyclic resistance is observed as α increases. On the other hand, under no-stress reversal loading, cyclic strength increases with the increase in α. However, as anticipated, it is obvious that the loose sand is much weaker against cyclic strain accumulation, so that large deformation is achieved in less number of loading cycles under the same *CSR* and α conditions.

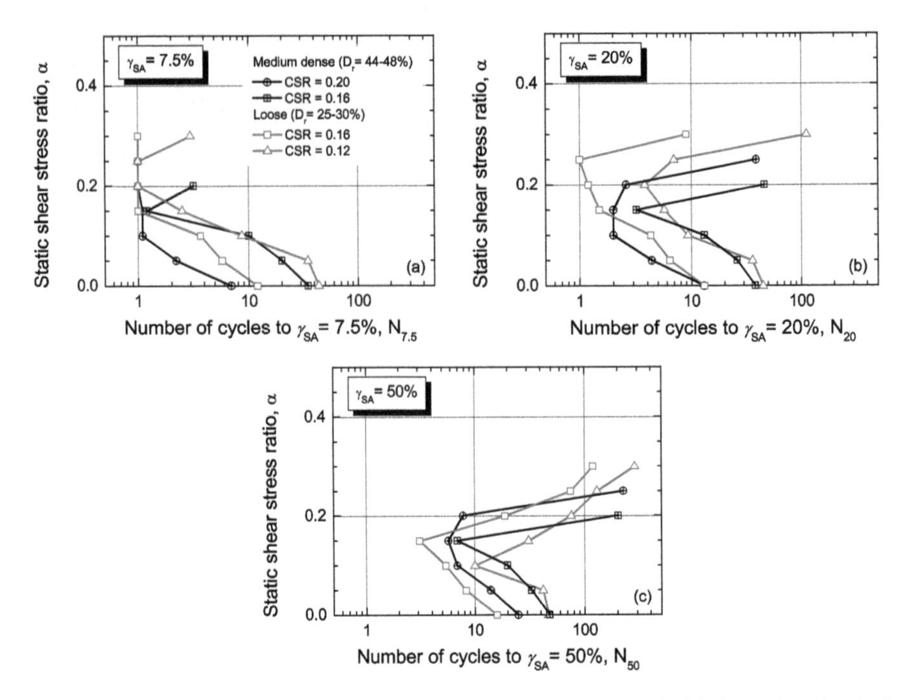

Fig. 5.15 Cyclic strength expressed in terms of static shear stress ratio (α) for various level of single amplitude shear strain levels

Effects of fabric. Laboratory observations have shown that soil fabric (i.e., spatial arrangement of sand particles and associated voids) plays an important role on sand response to cyclic loading (Ladd 1977; Mulilis et al. 1977) and its failure modes (Sze and Yang 2014). Therefore, the fabric should be regarded as a state parameter as important as density and stress state in describing soil behavior. However, due to the lack of a comprehensive experimental database for hollow cylindrical Toyoura sand specimens prepared with different methods (e.g., moist tamping or water sedimentation) that would provide different soil fabric, fabric effects are not discussed in this paper. This would need to be addressed by future and more inclusive undrained cyclic torsional shear tests with initial static shear.

Cyclic resistance ratio (CRR)

CRR. The cyclic resistance ratio (*CRR* = *CSR* at ten cycles of loading; the selection of ten cycles is consistent with the recommendation by Ishihara (1996) versus α relationships are illustrated in Fig. 5.16, for both loose and medium dense sand specimens. In both cases, the cyclic resistance first decreases (detrimental effect of τ_{static}) and then increases (beneficial effect of τ_{static}) upon the no-stress reversal loading line. These results and trends are reasonably consistent with experimental results from simple shear tests reported by Vaid and Finn (1979) and Boulanger et al. (1991).

K_α factor. The effect of α on the *CRR* is often expressed by the K_α factor (Seed 1983) that is defined as the ratio of CRR_α at any given α value to that of *CRR* for level ground conditions at $\alpha = 0$ (i.e., $K_\alpha = CRR_\alpha/CRR_0$). The value of $K\alpha$ indicates if the presence of τ_{static} is detrimental ($K_\alpha < 1$) or beneficial ($K_\alpha > 1$) to the sand cyclic resistance.

Figure 5.16 reports the variation of K_α with α obtained in this study for loose and medium dense Toyoura sand under sloping ground conditions and a mean effective stress of 100 kPa. It shows that for loose sand the presence of τ_{static} appears to be detrimental up to α = 0.15–0.20. This is also confirmed by the failure mechanisms analysis, since in the range of α = 0–0.15, a change from cyclic liquefaction to rapid flow liquefaction behavior was observed. However, beyond α = 0.15, due to an increase of CRR_α, K_α starts to increase and eventually becomes >1 (beneficial) for α = 0.2–0.25. This change from detrimental to beneficial can be associated with a change from rapid liquefaction with flow deformation development to shear failure with progressive accumulation of residual deformation.

In a similar manner, for medium dense sand, the presence of τ_{static} is detrimental up to α = 0.15. However, beyond this value, CRR_α increases, and K_α eventually becomes >1 (beneficial) for α = 0.22.

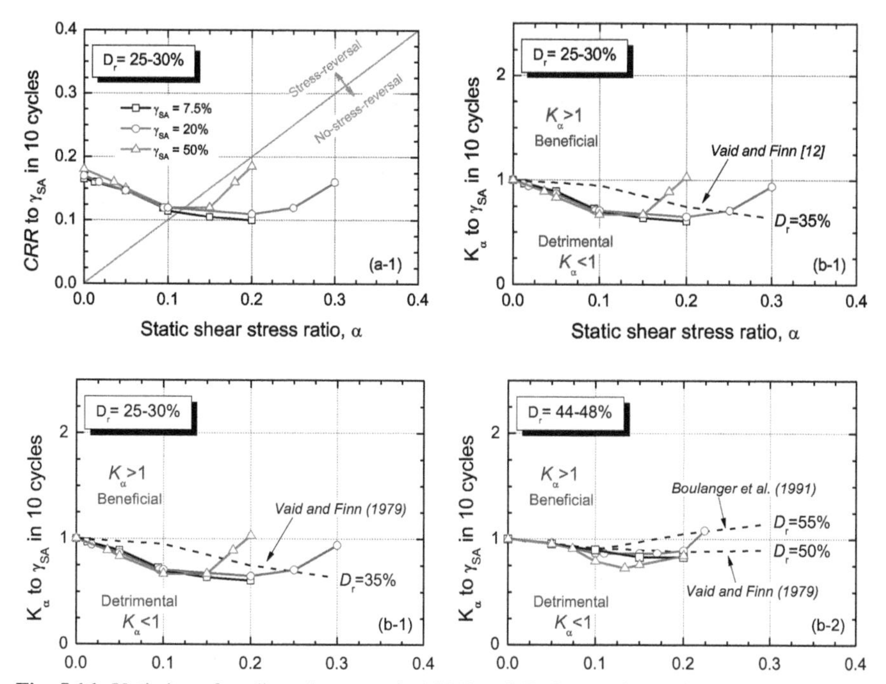

Fig. 5.16 Variations of cyclic resistance ratio (*CRR*) and K_α factor with static shear stress ratio (α) for loose and medium Toyoura sand

Cyclic strength of sand: simple shear versus triaxial tests with static shear
Figure 5.17 compares the cyclic strength of Toyoura sand specimens evaluated by undrained cyclic torsional simple shear tests (Chiaro et al. 2012), as reported in this study, with that obtained by Yang and Sze (2011) by undrained cyclic triaxial tests, under similar initial conditions of relative density and mean effective principal stress ($p_0' = 100$ kPa). Clearly, the cyclic strength of Toyoura sand measured in terms of single amplitude shear strain (i.e., $\gamma_{SA} = 7.5\%$ for torsional tests and $\varepsilon_{SA} = 5\%$ for triaxial tests) is in contrast to each other. Specifically, under torsional simple shear loading, the cyclic strength decreases with increasing α. As a result, the initial static shear has a detrimental effect on the liquefaction resistance and cyclic strain accumulation of sand. On the contrary, under triaxial shear loading, an opposite trend is observed, where the cyclic strength initially increases with increasing α. Hence, in this case, the initial static shear seems to be favorable to the liquefaction resistance of sand.

As well-known in cyclic triaxial tests, sand specimens are subjected to both extension and compression shearing loading phases within a single cycle of loading. For low values of initial static shear, the extension behavior is predominant, which may cause the sand to liquefy rapidly due to the effect of anisotropy. With the increase of the initial static shear on the triaxial compression side, the compression behavior predominates, and the sand turns out to be more resistant against liquefaction. Therefore, under such conditions, the initial static shear has a beneficial effect on the liquefaction resistance of sand. Moreover, as reported by Castro and Poulus (1977), the larger deformation observed in extension with respect to compression for a given deviator stress does not correspond to the field conditions. Therefore, generally cyclic triaxial tests overestimate the cyclic deformation that may be developed in the field due to liquefaction.

It is evident that the evaluation of the effect of the initial static shear on the liquefaction resistance of sand is significantly affected by the testing method employed and should be carefully addressed. In this regard, it is well recognized

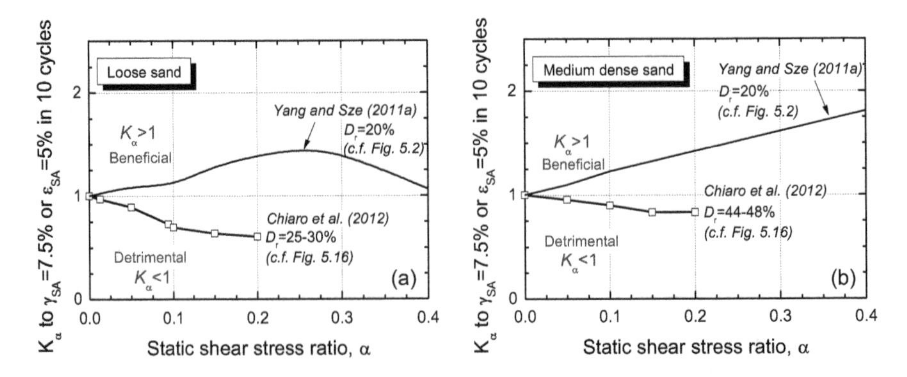

Fig. 5.17 Comparison of cyclic strength of loose and medium dense Toyoura sand evaluated by torsional simple shear and triaxial tests

that simple shear tests can simulate field stress conditions expected during earthquakes more accurately than triaxial tests. Hence, a torsional shear apparatus capable of reproducing simple shear conditions, as performed in this study, would be a useful tool for better understanding and evaluating the effect of initial static shear on the cyclic undrained behavior of sands, especially considering that extremely large shear strain can be achieved.

5.4.4 Strain Localization in Liquefied Sand Specimens

In all tests listed in Table 5.1, irrespective of specimen densities and applied stress conditions, γ_{SA} values progressively increased up to large shear strain levels during cyclic loading. However, non-uniform specimen deformation was observed at large level of γ_{SA} exceeding 25% or more. The initiation of strain localization could not be defined purely on the basis of visual observation. As described in detail hereafter, the shear strain limit to begin strain localization was, therefore, systematically evaluated based on the change in the vertical stress value during cyclic loading (Fig. 5.4).

Specimen deformation of sand in undrained cyclic torsional shear tests
Figure 5.18 shows typical specimen deformation observed at several loading stages during undrained cyclic torsional shear tests for the case of a medium dense Toyoura sand specimen (photos refer to Test 17 listed in Table 5.1). At $\gamma < 20\%$, although the outer membrane may appear slightly wrinkled, usually the deformation is almost uniform (as shown by the red dotted line) except for the areas close to the pedestal and the top cap that are affected by the end restraint. Then, at $\gamma = 20\text{–}30\%$, the outer membrane is noticeably wrinkled, and in the zone near the top cap, the deformation of the specimen started to localize (as shown by the blue dotted line) due probably to water film formation (Kokusho 1999). Next, at $\gamma = 30\text{–}40\%$, the localization of the specimen deformation is visibly developed in the upper part of the specimen. Alternatively, in the bottom part, the uniformity of the specimen deformation is generally maintained even if many wrinkles appear. Finally, at $\gamma > 40\%$, the specimen is almost twisted near the top cap.

Limiting value of shear strain to initiate strain localization
By keeping the vertical (σ_z') and radial (σ_r') effective stress values constant, Tatsuoka et al. (1986) carried out a series of drained monotonic torsional shear tests on hollow cylindrical Toyoura sand specimens. They reported that the vertical strain accumulated on the extension side due to the mobilization of positive dilatancy, and it reduced abruptly as the shear band formed in the specimen.

Later, Kiyota et al. (2008) found the change in stress (σ_z'–σ_r' here referred to as $\Delta\sigma$) response observed in undrained torsional shear tests, in which any vertical displacement of the top cap was prevented (i.e., to reproduce simple shear conditions), to be consistent with the behavior observed during drained monotonic torsional shear tests by Tatsuoka et al. (1986). Therefore, Kiyota et al. (2008) regarded

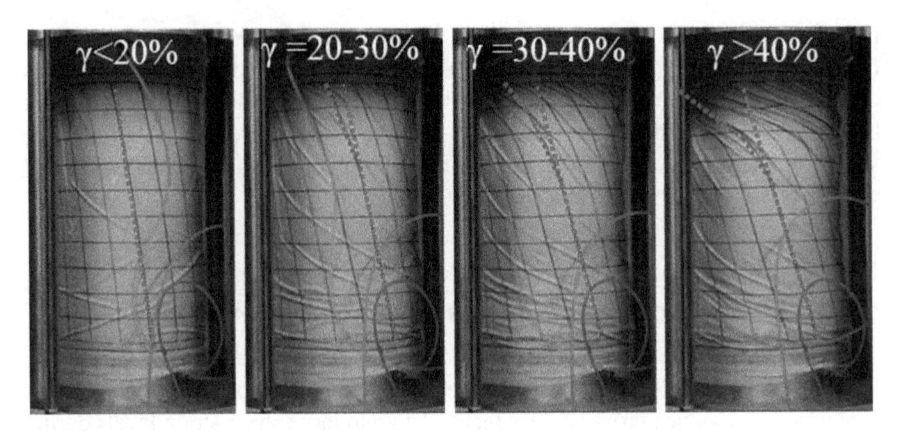

Fig. 5.18 Typical torsional deformation of a medium dense Toyoura sand specimen during undrained cyclic torsional shear tests (adopted from Chiaro et al. 2013)

the state at which the $\Delta\sigma$ value abruptly decreased (referred as state A henceforward) as the limiting state to initiate formation of shear band and, thus, strain localization. Furthermore, Kiyota et al. (2008) found that the drop in $\Delta\sigma$ at state A is usually accompanied by an increase in the increment of single amplitude shear strain ($\Delta\gamma_{SA}$) at state B. More recently, looking into the cyclic strain-softening behavior of Toyoura sand, Chiaro et al. (2013a) confirmed that the state A is effectively the beginning of shear band(s) formation into liquefied sand specimens and the state B is the beginning of the residual stress state after the full development of shear band(s).

Kiyota et al. (2008) defined the limit shear strain in terms of half of double amplitude shear strain $\gamma_{L(DA)}/2$, while Chiaro et al. (2013a) recommended that $\gamma_{L(SA)}$ is a more appropriate parameter when non-symmetric cyclic shear stress conditions are considered (i.e., an initial static shear in applied). Nevertheless, in the case of symmetric cyclic loading (i.e., zero static shear), $\gamma_{L(DA)}/2$ and $\gamma_{L(SA)}$ are well in accordance with each other (Chiaro et al. 2012).

In Fig. 5.19, typical experimental results are presented for Toyoura sand. It appears that during undrained cyclic simple torsional shear loading (where the vertical displacement is prevented) a sudden drop in $\Delta\sigma$ can be clearly observed at state A. According to Chiaro et al. (2012) and Kiyota et al. (2008), the state A can be regarded as the limit shear strain to initiate strain localization ($\gamma_{L(SA)}$). As anticipated, state A is closely followed by the increase of $\Delta\gamma_{SA}$ after state B (Fig. 5.19b, c). As a result, these features imply that the stress–strain characteristics of the sand specimen vary because of the formation of shear band(s) and the initiation of strain localization in the sand specimen.

Key factors affecting strain localization of liquefied sand
As described below, several factors affect the strain localization properties of loose and medium sand in undrained cyclic torsional shear tests with initial static.

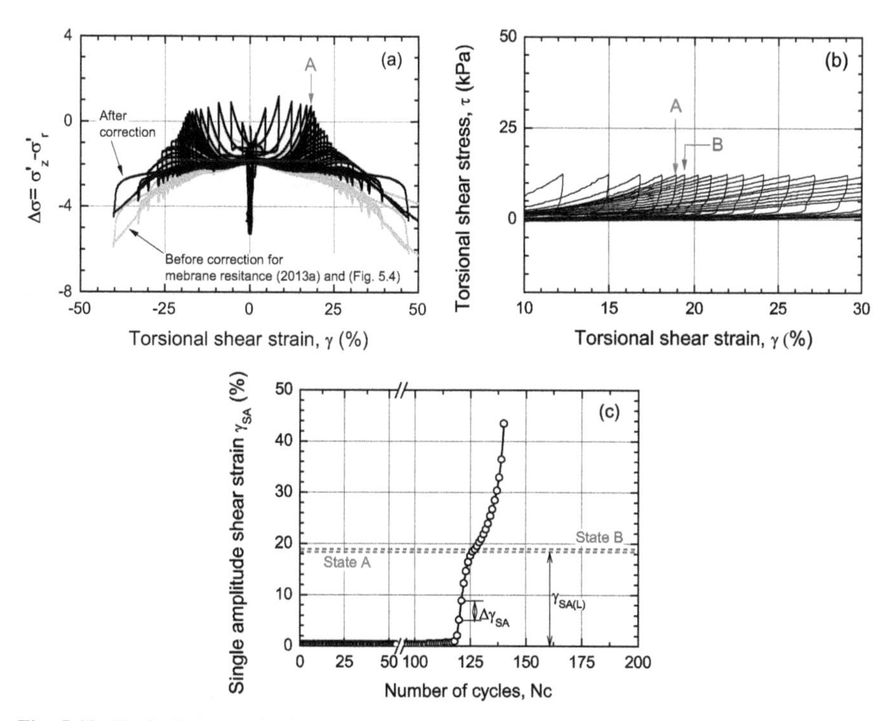

Fig. 5.19 Typical change in $\Delta\sigma$ and shear strain properties of Toyoura sand during undrained cyclic torsional shear loading

Effects of relative density. Density is indeed one of the key parameters that governs the cyclic undrained behavior of sand, including liquefaction resistance and large deformation development. Figure 5.20 shows that density also greatly affects the strain localization characteristics of liquefied sandy soils. In particular, the greater the relative density (D_r) is, the smaller the limit shear strain to cause strain localization is. In other words, dense sands tend to show strain localization at smaller γ_{SA} levels compared to loose sands. This is rational considering that dense sands usually show a dilative behavior during undrained shear loading, while loose sands show a contractive behavior.

Effect of *CSR*. Mean effective stress and cyclic shear stress are other two important factors that greatly affect the liquefaction behavior of sand. To reflect this aspect, the ratio between cyclic stress ratio and mean effective stress, namely cyclic stress ratio (*CSR*), is commonly employed to evaluate several features of the cyclic undrained soil behavior. However, it has been observed that $\gamma_{L(SA)}$ is not affected by *CSR* (Chiaro et al. 2013a, 2015).

Effect of combined cyclic and static shear stresses. In sloping ground, the presence of initial static shear and cyclic shear stress during an earthquake loading

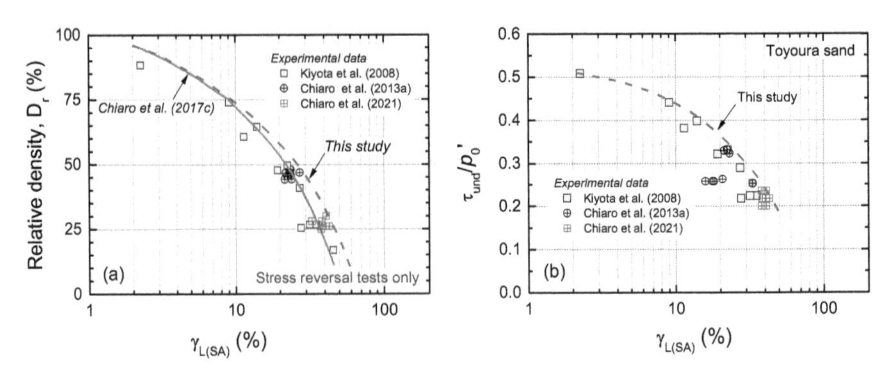

Fig. 5.20 Relationships between limiting shear strain and **a** relative density and **b** undrained shear strength for Toyoura sand

results into various degrees of reversal loading. In general, for the case of Toyoura sand, the smaller the reversal stress degree is, the smaller the limit shear strain to cause strain localization is (Chiaro et al. 2013a; 2015).

Use of undrained shear strength to define limiting shear strain
As described in the previous section, different factors influence the strain localization properties of liquefied sands. However, density, mean effective stress and static shear have a major effect on shear strain characteristics. In fact, contrary to cyclic shear stress, they directly affect the undrained shear behavior of sand (Chiaro 2010). With the aim of capturing the mutual effects of density, mean effective stress and static shear, in Fig. 5.20b, the experimental data shown in Fig. 5.20a are replotted in terms of undrained shear strength ratio (τ_{und}/p_0') versus $\gamma_{L(SA)}$ relationship. Note that, when feasible τ_{und} was experimentally evaluated, otherwise it was estimated by using the constitutive model for liquefiable sands developed by Chiaro et al. (2013b, 2017).

Thus, from Fig. 5.20b, it appears that $\gamma_{L(SA)}$ decreases with an increase in τ_{und}. This is rational since τ_{und} increases with the density, mean effective stress and static shear (Chiaro 2010). Moreover, an upper limit boundary seems to exist in spite of different stress conditions applied on loose to dense specimens. According to Kiyota et al. (2010), the maximum amounts of liquefaction-induced ground displacement observed in relevant model tests and field observations are consistent with the limiting value to initiate strain localization observed in torsional shear tests (Kiyota et al. 2008). Therefore, as long as the shear deformation remains uniform, the results of torsional shear tests can be effectively used to estimate the extent of large deformation that will occur in the field during earthquakes (Kiyota et al. 2010).

5.5 Summary and Conclusions

Due to the large number of factors that may affect the liquefaction resistance of sand, the mechanisms for the development of large liquefaction-induced deformation of sand within sloping are not fully understood yet. Compared with the large body of experimental data describing the undrained cyclic behavior of sands under level ground conditions, there have been very limited studies focusing on the undrained cyclic response of sands under sloping ground conditions. Previous studies, mainly based on triaxial tests and occasionally on simple shear, torsional shear or ring shear tests, have produced valuable data showing that the presence of initial static shear can have major effects on the cyclic response of sands. Nevertheless, contradictory views seem to exist with respect to these effects. Moreover, technical limitations of testing devices have prevented to entirely define the effects of sloping ground on the cyclic resistance of sands. As reported in this paper, however, the development and use of a state-of-the-art torsional shear testing device that is capable of reproducing the simple shear condition and achieving double amplitude shear strain (γ_{DA}) levels of 100%, have made it possible to obtain new insights on this challenging topic.

In this paper, the results of two series of large-strain undrained torsional simple shear tests carried out on loose (D_r = 25–30%) and medium dense (D_r = 44–48%) Toyoura sand specimens were reported, and the post-liquefaction response of the specimens was examined in terms of observed failure modes, strain development characteristics, cyclic resistance up to 50% single amplitude shear strain and strain localization characteristics. Key findings and recommendations derived from the study summarized in this paper are as follows:

Testing conditions. It is evident that the evaluation of the effect of the initial static shear on the liquefaction resistance of sand is significantly affected by the testing conditions employed (i.e., simple shear vs. triaxial shear) and should be carefully investigated. This study provides additional evidence that the use of a torsional shear apparatus capable of reproducing simple shear conditions for assessing the liquefaction properties of soils should be encouraged. It particular, it is an ideal tool for better understanding and evaluating the effect of initial static shear on the cyclic undrained behavior of sands, especially considering the extremely large shear strain level and large uniform specimen deformation extent that such devices can achieved.

Failure modes. Reversal stress plays a key role on the type of failure that loose and medium dense sand deposits within sloping ground may experience during undrained cyclic loading. It is confirmed in this study that liquefaction may take place only under reversal stress conditions. On the other hand, shear failure can occur only under non-stress reversal conditions.

Specifically, there exist three distinct failure modes: (i) potential development of large deformation due to cyclic liquefaction; (ii) abrupt development of large deformation due to rapid (flow) liquefaction and (iii) progressive cyclic accumulation of large residual deformation inducing shear failure. Among them, the rapid liquefaction followed by the sudden development of large deformation is the most detrimental since it occurs without any warning.

Development of large deformation. During undrained cyclic loading with static shear, sand can develop extremely large shear deformation under both stress reversal and non-stress reversal conditions. However, in the first case, the development of such deformation will be extremely rapid and have catastrophic effects only if liquefaction will take place in less than 10–15 cycles of loading. A more progressive accumulation of substantial deformation, leading to shear failure, is likely to occur in the latter case although liquefaction is not achieved. Nevertheless, in both the cases, the extent of large deformation and its unfavorable effects will be more severe if the maximum shear stress (i.e., combined static and cyclic shear tresses) exceeds the undrained shear strength of the sand.

Cyclic strength. Regardless of the relative density of the sand specimens (i.e., loose or medium dense) and the criterion used to define the cyclic strength (i.e., single amplitude shear strain (γ_{SA}) equal to 7.5, 20 or 50% in ten cycles), the cyclic resistance decreases with increasing static shear stress ratio (α) under reversal stress conditions and increases with increasing static shear stress ratio (α) under non-reversal stress conditions.

It is important to mention that even though the cyclic strength was reported for γ_{SA} = 50%, such data should be taken only as a reference since these data can be affected by strain localization (i.e., the formation of shear band) that may take place during undrained torsional shear loading.

Strain localization. Strain localization is likely to occur at large shear strain levels in liquefied sand specimens and may significantly affect the evaluation of the cyclic strength of sand. Therefore, for any given strain level used to define the strength criterion, it is important to confirm whether or not strain localization has occurred yet in the specimen.

The limiting shear strain to cause strain localization highly depends on the relative density of the sand and is affected also by the static shear and mean effective stresses. On the other hand, the cyclic shear stress appears to have no effects on the strain localization.

The undrained shear strength is found to be a useful factor to describe the strain localization properties of sand because it inherently captures the combined effects of relative density of the sand and is affected also by the static shear and mean effective stresses.

Acknowledgements I would like to acknowledge the contributions of Dr. Umar Muhammad (formerly Master and Ph.D. student) and Takeshi Sato (laboratory technician) for their assistance in the laboratory testing conducted at the University of Tokyo. Special thanks to my colleagues Professor Junichi Koseki and Professor Takashi Kiyota, for their contributions to these research studies, continuous support and invaluable recommendations provided over the years. This research conducted at the University of Tokyo was supported by the Ministry of Education,

Culture, Sport and Technology of Japan (MEXT Ph.D. Research Scholarship) and the Japanese Society for Promotion of Science (Research Fellowship Award No. P14056). Their support is gratefully acknowledged.

References

Ampadu SIK (1991) Undrained behavior of kaolin in torsional simple shear. PhD thesis, Department of Civil Engineering, University of Tokyo, Japan

Arangelovski G, Towhata I (2004) Accumulated deformation of sand with initial shear stress and effective stress state lying near failure conditions. Soils Found 44(6):1–16

Cappellaro C, Cubrinovski M, Chiaro G, Stringer ME, Bray JD, Riemer MF (2019) Effects of fines content, fabric, and structure on the cyclic direct simple shear behaviour of silty sands. In: 7th international conference on earth geotechnology engineering, pp 1588–1595

Castro G, Poulus SJ (1977) Factors affecting liquefaction and cyclic mobility. J Geotech Eng Div ASCE 103(GT6):501–551

Boulanger RW, Seed RB, Chan CK, BSH, Sousa JB (1991) Liquefaction behavior of saturated and under uni-directional and bi-directional monotonic and cyclic simple shear loading. B.S.H., University of California Berkley, USA

Chiaro G (2010) Deformation properties of sand with initial static shear in undrained cyclic torsional shear tests and their modeling. PhD thesis, Department of Civil Engineering, University of Tokyo, Japan, p 310

Chiaro G, Koseki J, Sato T (2012) Effects of initial static shear on liquefaction and large deformation properties of loose saturated Toyoura sand in undrained cyclic torsional shear. Soils Found 52(3):498–510

Chiaro G, Kiyota T, Koseki J (2013a) Strain localization characteristics of loose saturated Toyoura sand in undrained cyclic torsional shear tests with initial static shear. Soils Found 53(1):23–34

Chiaro G, Koseki J, De Silva LIN (2013b) A density- and stress-dependent elasto-plastic model for sands subjected to monotonic torsional shear loading. Geotech Eng J 44(2):18–26

Chiaro G, Kiyota T, Pokhrel RM, Goda K, Katagiri T, Sharma K (2015a) Reconnaissance report on geotechnical and structural damage caused by the 2015 Gorkha Earthquake, Nepal. Soils Found 55(5):1030–1043

Chiaro G, Kiyota T, Koseki J (2015b) Strain localization characteristics of liquefied sands in undrained cyclic torsional shear tests. Advances Soil Mech Geotech Eng, IOS Press 6:832–839

Chiaro G, Alexander G, Brabhaharan P, Massey C, Koseki J, Yamada S, Aoyagi Y (2017a) Reconnaissance report on geotechnical and geological aspects of the 2016 Kumamoto Earthquake, Japan. Bull New Zealand Soc Earth Eng 50(3):365–393

Chiaro G, De Silva LIN, Koseki J (2017b) Modeling the effects of static shear on the undrained cyclic torsional simple shear behavior of liquefiable sand. Geotech Eng J 48(4):1–9

Chiaro G, Kiyota T, Miyamoto H (2017) Liquefaction potential and large deformation properties of Christchurch liquefied sand subjected to undrained cyclic torsional simple shear loading. In: 19th international conference on soil mechanics geotechnology engineering, pp 1497–1500

Chiaro G, Umar M, Kiyota T, Massey C (2018) The Takanodai landslide, Kumamoto, Japan: insights from post-earthquake field observations, laboratory tests and numerical analyses. ASCE Geotech Special Publication 293:98–111

Chiaro G, Kiyota T, Umar M (2019) Undrained monotonic and cyclic torsional simple shear behavior of the Aso pumiceous soil deposits. In: 7th international conference on earth geotechnology engineering, pp 1754–1761

Chiaro G, Umar M, Kiyota T, Koseki J (2021) Deformation and cyclic strength characteristics of loose and medium-dense clean sand under sloping ground conditions: insights from cyclic undrained torsional shear tests with static shear. Geotech Eng J 52(2):in press

Cubrinovski M, Ishihara K (2000) Flow potential of sandy soils with different grain compositions. Soils Found 40(4):103–119
Cubrinovski M, Green RA, Allen J, Ashford SA, Bowman E, Bradley BA, Cox B, Hutchinson TC, Kavazanjian E, Orense RP, Pender M, Quigley M, Wotherspoon L (2010) Geotechnical reconnaissance of the 2010 Darfield (Canterbury) Earthquake. Bull New Zealand Soc Earth Eng 42(4):243–320
Cubrinovski M, Bray JD, Taylor M, Giorgini S, Bradley BA, Wotherspoon L, Zupan J (2011) Soil liquefaction effects in the Central Business Districts during the February 2011 Christchurch Earthquake. Seismol Res Lett 82(6):893–904
De Silva LIN, Koseki J, Chiaro G, Sato T (2015) A stress-strain description for saturated sand under undrained cyclic torsional shear loading. Soils Found 55(3):559–574
Georgiannou VN, Tsomokos A, Stavrou K (2008) Monotonic and cyclic behaviour of sand under torsional loading. Geotechnique 58(2):113–124
Hamada M, O'Rourke TD, Yoshida N (1994) Liquefaction-induced large ground displacement. In: Performance of ground and soil structures during earthquakes. 13th international conference on soil mechanics and foundation engineering, pp 93–108
Hight DW, Gens A, Symes MJ (1983) The development of a new hollow cylinder apparatus for investigating the effects of principal stress rotation in soils. Geotechnique 33:355–383
Hyodo M, Murata H, Yasufuku N, Fujii T (1991) Undrained cyclic shear strength and residual shear strain of saturated sand by cyclic triaxial tests. Soils Found 31(3):60–76
Hyodo M, Tanimizu H, Yasufuku N, Murata H (1994) Undrained cyclic and monotonic triaxial behavior of saturated loose sand. Soils Found 34(1):19–32
Ishihara K (1996) Soil behaviour in earthquake Geotechnics. Clarendon Press, Oxford
Ishihara K, Tatsuoka F, Yasuda S (1975) Undrained deformation and liquefaction of sand under cyclic stresses. Soils Found 15(1):29–44
Kiyota T, Sato T, Koseki J, Mohammad A (2008) Behavior of liquefied sands under extremely large strain levels in cyclic torsional shear tests. Soils Found 48(5):727–739
Kiyota T, Koseki J, Sato T (2010) Comparison of liquefaction-induced ground deformation between results from undrained cyclic torsional shear tests and observations from previous model tests and case studies. Soils Found 50(3):421–429
Kiyota T, Kyokawa H, Konagai K (2011) Geo-disaster report on the 2011 Tohoku-Pacific Coast Earthquake. Bull Earth Resistant Struct Res Center 44:17–27
Kokusho T (1999) Formation of water film in liquefied sand and its effects on lateral spread. J Geotech Geoenv Eng ASCE 125:817–826
Koseki J, Yoshida T, Sato T (2005) Liquefaction properties of Toyoura sand in cyclic torsional shear tests under low confining stress. Soils Found 45(5):103–113
Kramer SL (1996) Geotechnical earthquake engineering. Prentice Hall, New Jersey
Ladd RS (1977) Specimen preparation and cyclic stability of sands. J Geotech Eng Div ASCE 103 (6):535–547
Lade PV (1993) Initiation of static instability in the submarine Nerlerk berm. Can Geotech J 30 (6):895–904
Lee KL, Seed HB (1967) Dynamic strength of anysotropically consolidated sand. J Soil Mech Found Div ASCE 93(SM5):169–190
Mulilis JP, Arulanandan K, Mitchell JK, Chan CK, Seed HB (1977) Effects of sample preparation on sand liquefaction. J Geotech Eng Div ASCE 103(2):91–108
Seed HB (1968) Landslides during earthquakes due to soil liquefaction. J Soil Mech Found Div ASCE 94(SM5):1055–1122
Seed HB (1983) Earthquake-resistant design of earth dams. In: Symposium on seismic design Earth Dams and Caverns, New York, pp 41–46
Seed HB, Idriss IM (1982) Ground motions and soil liquefaction during earthquakes. Earthquake Engineering Research Institute Monograph, Oakland
Sivathayalan S, Ha D (2011) Effect of static shear stress on the cyclic resistance of sands in simple shear loading. Can Geotech J 48(10):1471–1484

Sze HY, Yang J (2014) Failure modes of sand in undrained cyclic loading: impact of sample preparation. J Geotech Geoenv Eng 140:152–169

Tatsuoka F, Muramatsu M, Sasaki T (1982) Cyclic undrained stress-strain behavior of dense sands by torsional simple shear stress. Soils Found 22(2):55–70

Tatsuoka F, Sonoda S, Hara K, Fukushima S, Pradhan TBS (1986) Failure and deformation of sand in torsional shear. Soils Found 26:79–97

Vaid YP, Chern JC (1983) Effects of static shear on resistance to liquefaction. Soils Found 23: 47–60

Vaid YP, Finn WDL (1979) Static shear and liquefaction potential. J Geotech Eng Div ASCE 105:1233–1246

Vaid YP, Stedman JD, Sivathayalan S (2001) Confining stress and static shear effects in cyclic liquefaction. Can Geotech J 38(3):580–591

Verdugo R, Ishihara K (1996) The steady-state of sandy soils. Soils Found 36(2):81–91

Yang J, Sze HY (2011a) Cyclic behavior and resistance of saturated sand under non-symmetrical loading conditions. Geotechnique 61(1):59–73

Yang J, Sze HY (2011b) Cyclic strength of sand under sustained shear stress. J Geotech Geoenv Eng 137(2):1275–1285

Yasuda S, Nagase N, Kiku H, Uchida Y (1992) The mechanisms and a simplified procedure for analyses of permanent ground displacement due to liquefaction. Soils Found 32(1):149–160

Yoshimi Y, Oh-oka H (1975) Influence of degree of shear stress reversal on the liquefaction potential of saturated sand. Soils Found 15(3):27–40

Youd TL, Idriss IM, Andrus RD et al (2001) Liquefaction resistance of soils: summary report from the 1996 NCREE and 1998 NCREE/NSF workshops on evaluation of liquefaction resistance of soils. J Geotech Geoenv Eng ASCE 127(10):817–833

Ziotopoulou K, Boulanger RW (2016) Plasticity modeling of liquefaction effects under sloping ground and irregular loading conditions. Soil Dyn Earth Eng 84:269–283

Chapter 6
High-Speed Trains with Different Tracks on Layered Ground and Measures to Increase Critical Speed

Amir M. Kaynia

6.1 Introduction

6.1.1 Early Studies on Moving Loads

The early studies in the field of moving loads on elastic media date back to the middle of the nineteenth century; however, interest in the theoretical solutions appears to have been triggered by Sneddon in 1951. Since then, numerous solutions with different constraints and varying degrees of sophistication have been reported in the literature. The complexity of the problem has largely confined the solutions to loads moving on homogeneous elastic half-space solids. The existing solutions can be broadly divided into 2D and 3D solutions. An important differentiation between the various solutions has been the range of load speeds. This is because the form of the solution depends primarily on the speed of the load with respect to the characteristic wave velocities of the propagating medium. It is common to refer to load speed as sub-seismic, super-seismic and trans-seismic, depending on whether the load speed is less than the Rayleigh wave velocity of the medium, greater than the longitudinal wave velocity or intermediate between those two velocities (Baron et al. 1967). The speed corresponding to the Rayleigh wave velocity is often called the *critical speed* (e.g., Madshus and Kaynia 2000), because below this speed, the ground response is primarily quasi-static, whereas for higher speeds, the ground response is largely characterized by dynamic features and larger magnitudes.

Two-dimensional solutions, i.e., solutions for a moving line load, have been presented by Sneddon (1951), Cole and Huth (1958), Ang (1960), Craggs (1960), Payton (1964), Eringen and Suhubi (1975). Sneddon developed a closed-form

A. M. Kaynia (✉)
Norwegian Geotechnical Institute (NGI), Oslo, Norway
e-mail: amir.m.kaynia@ngi.no; amir.kaynia@ntnu.no

A. M. Kaynia
Norwegian University of Science and Technology (NTNU), Trondheim, Norway

T. G. Sitharam et al. (eds.), *Latest Developments in Geotechnical Earthquake Engineering and Soil Dynamics*, Springer Transactions in Civil and Environmental Engineering, https://doi.org/10.1007/978-981-16-1468-2_6

solution using integral transforms; however, he focused his attention to the case of low speeds. Cole and Huth (1958) investigated the same problem and considered the trans-seismic and super-seismic speeds, as well. They used the Helmholtz decomposition technique to decouple the shear and compressional wave components, solved the resulting equations using the techniques of complex analytic functions and derived closed-form expressions for the displacements and stresses in a half-space. They defined Mach numbers $M_P = C_P/V$ and $M_S = C_S/V$ to represent the speed of the moving load, V, relative to the pressure wave velocity, C_P and shear wave velocity, C_S, of the medium. Eringen and Suhubi (1975) followed the same solution scheme and presented results for the trans-seismic and super-seismic speeds in addition to duplicating the Cole and Huth (1958) results for the sub-seismic load speed.

Other relevant 2D solutions are due to Ang (1960), Papadopoulos (1963) and Niwa and Kobayashi (1966) who used the Helmholtz decomposition technique to uncouple the dilatational (pressure) and distortional (shear) components. Ang [1] solved the equations for the sub-seismic case by using the Laplace transform and the Laplace inversions by a numerical technique, whereas Niwa and Kobayashi (1966) used the Fourier transform. A detailed review of the early studies on the subject can be found in Frýba (1973).

For layered half-space, the only analytical solutions appear to have been advanced by Sackman (1961) and Wright and Baron (1970). Sackman (1961) used for each layer the solutions for the dilatational and distortional potentials obtained by Cole and Huth (1958) and constructed the wave field from multiple reflections and refractions of plane waves from plane boundaries. Wright and Baron (1970) considered a soil layer over a half-space under a moving pressure pulse with a speed which is super-seismic with respect to the surficial layer and sub-seismic with respect to the half-space. They used the potential theory combined with the Fourier transform along the horizontal axis and derived formal integral solutions for stresses and velocities, which they in turn solved numerically. In this context, Achenbach et al. (1967) used the potential approach and by numerically evaluating the integrals studied the response of a plate, resting on a half-space, to a moving line load. Although they only considered the sub-seismic case, they were able to document large bending moments in the plate as the load speed approached the Rayleigh wave velocity in the half-space.

For general heterogeneous soil, researchers have had to resort to semi-analytical or purely numerical solutions. Several different semi-analytical models have been developed for this problem. These models consider the load either moving on a plate or track over a half-space (e.g., Pan and Atluri 1995; Krylov 1995), or directly moving over a layered half-space (e.g., Aubry et al. 1994; de Barros and Luco 1994). Krylov (1995) investigated the problem of a train moving on a track with sleeper periodicity d. He used a classical beam-on-elastic-foundation solution and calculated the contribution of each sleeper to the total pressure distribution. Then, by assuming each sleeper as a point source and convolving its pressure distribution by the half-space Green's functions, he derived formal solutions for the vertical component of the ground surface motions. He used, however, only the Rayleigh wave contribution to the Green's functions. Results of Krylov's model agree with theoretical solutions. More specifically, his results show the formation of Mach lines with accurate Mach angles for the trans-seismic speed and show that no radiation wave fields exist for speeds less than the Rayleigh wave velocity of the medium.

One of the most rigorous formulations and extensive numerical results has been presented by de Barros and Luco (1994). They developed a semi-analytical formulation for the ground vibrations due to a point load moving on or inside a layered viscoelastic half-space. The frequency-domain formulation of de Barros and Luco (1994) is based on double Fourier transform in the two horizontal spatial coordinates. For each layer, the transformed displacements and stresses at the layer boundaries were expressed in terms of the wave amplitudes in that layer through the so-called transmission and reflection matrices. The wave amplitudes were obtained by imposition of the appropriate condition at the layer interfaces. The inversion of the Fourier integrals from the wave number domain to the space domain was achieved by a quadrature algorithm (Filon 1928), and the response in the time domain was obtained through the inverse Fourier transform.

6.1.2 Studies on High-Speed Trains

The development of high-speed trains and their increased deployment in areas with low shear wave velocities, which bring the train speeds closer to the trans-seismic regime, has stimulated an extensive research on the subject since the 1990s. The solutions described above have inspired a host of new semi-empirical and numerical solutions and applications. They vary from 2D finite element solutions (e.g., Peplow and Kaynia 2007; Norén-Cosgriff et al. 2019) to 3D finite element (e.g., Hall 2003), combination of finite element and boundary element solutions (e.g., Andersen and Jones 2006 and Auersch 2005) and 2.5-D schemes (e.g., Coulier et al. 2013). The latter approach allows reducing the computational demand in models where the track is considered as invariant in the direction of the load passage. This allows the problem to be described with a 2D geometry while the loading is accounted for in 3D. This approach can also be combined with the boundary element method resulting in computationally efficient models (e.g., Karlström and Boström 2006).

An alternative solution is an efficient analytical–numerical 3D solution by use of the sub-structuring technique. In this solution, the dynamic response of the layered ground is represented by Green's functions, and the track is represented by finite elements (e.g., Kaynia et al. 2000). Two versions of this model are used in the present study. The next section presents the principles of these models, and the subsequent sections provide representative results of simulations for both models and the features of the responses. The main objective of this study is to investigate how to increase the critical speed in order to avoid excessive track responses associated with trans-seismic and super-seismic conditions.

6.2 Simulation Models

6.2.1 Track Cases

Two track concepts were considered in the present study. The first one is the conventional tracks consisting of the rail/ballast/sub-ballast placed either directly over the ground or over an embankment. This case is referred to as *surface track* in this study and is shown in Fig. 6.1a. For this case, two solutions for increasing the critical speed were considered: (1) stiffened track and (2) ground treatment, for example by replacing part of the natural soil by stiffer material or by grouting (Fig. 6.1c).

The second track concept consisted of the rail/ballast/sub-ballast placed on a deck (beam) resting on large-diameter piles. This case, which resembles a bridge, is portrayed in Fig. 6.2 and is referred to as *piled track*. While the deck could be initially in contact with the ground, the contact is weakened or possibly lost with time. This case can in practice be considered as a measure to increase the critical speed the same way that the soil improvement/replacement in Fig. 6.1c is considered.

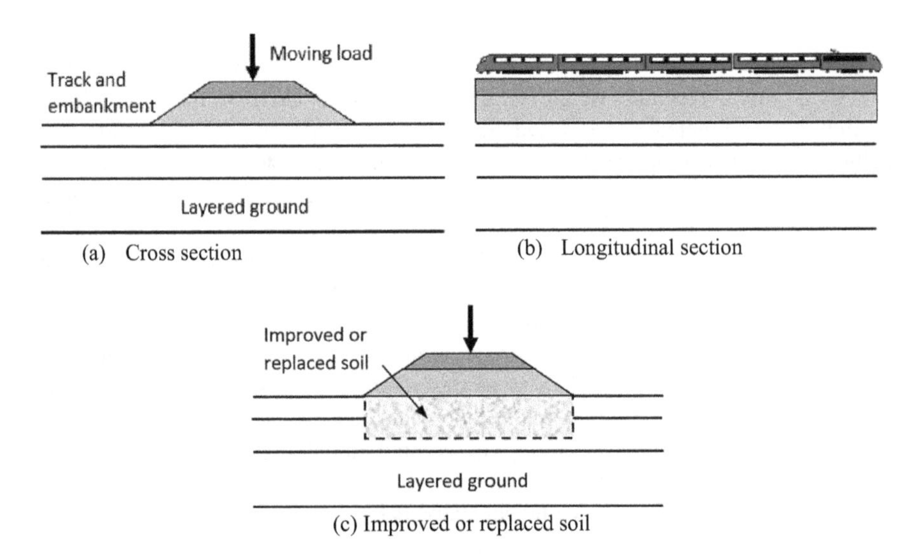

Fig. 6.1 Key features of surface track on layered ground: **a** cross section with idealized train load, **b** longitudinal section with schematics of train cars and **c** track with soil improvement or replacement

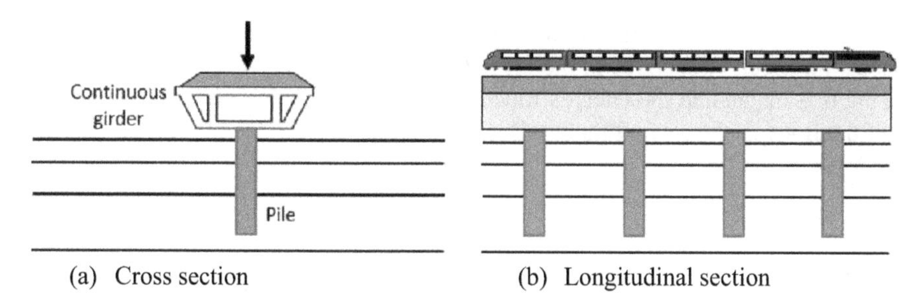

Fig. 6.2 Key features of piled track on layered ground: **a** cross section with idealized train load and **b** longitudinal section with piles and schematics of train cars

6.2.2 *Computational Tools*

For the surface track cases in Fig. 6.1, the numerical simulation tool *VibTrain* (Kaynia et al. 2000) was used. For the piled track concept in Fig. 6.2, the numerical model *VibTrain_pile*, which was recently developed by the author, was employed. A brief account of these models is presented in the following.

The solutions of both cases are based on sub-structuring in which the ground or the piles in the second track concept are represented by the impedance matrices at the track–support interface. For the surface track model, the interface nodes are defined by regularly placed points along the track. For the piled track concept, the interface nodes are the pile heads. If the vertical displacements and forces at these nodes are denoted by $\boldsymbol{W}$ and $\boldsymbol{P}$, one can write

$$\boldsymbol{P} = \boldsymbol{G}^{-1}\boldsymbol{W} = \boldsymbol{K}_S\boldsymbol{W} \tag{6.1}$$

where $\boldsymbol{K}_S$ is the impedance matrix of the ground nodes (or pile heads in the piled track model). A similar relationship can be established by considering the equilibrium of the track sub-structure as

$$\boldsymbol{F} - \boldsymbol{P} = \boldsymbol{K}_B\boldsymbol{W} \tag{6.2}$$

where $\boldsymbol{F}$ is the vector of applied nodal loads (i.e., moving loads) and $\boldsymbol{K}_B$ is the dynamic stiffness matrix of the track. Elimination of the interaction force vector, $\boldsymbol{P}$, from Eqs. (6.1) and (6.2) leads to following equations of the coupled system.

$$\boldsymbol{F} = (\boldsymbol{K}_B + \boldsymbol{K}_S)\boldsymbol{W} \tag{6.3}$$

By taking the Fourier transform of the applied loads, solving Eq. (6.3) for the Fourier frequencies, and combining the frequency contributions by the inverse Fourier transform, one can compute the time histories of the track response. See Kaynia et al. (2000) for details. This formulation was extended to cases where the excitation is due to rail roughness by Kaynia (2001) and to cases involving track defects, such as hanging sleepers, by Kaynia et al. (2017).

For computation of the ground impedance and pile impedance matrices, one needs the Green's functions of the layered soil for unit loads on the ground surface in the first model and the Green's functions for cylindrical loads representing pile–soil tractions in the second model. These functions are described in the following sections.

6.2.3 Green's Functions for Layered Viscoelastic Soil

Park and Kaynia (2018) recently developed a semi-analytical solution for the Green's functions for unit disk loads and pressure sources in layered anisotropic layered ground under water. The solution is an extension of the model proposed by Kausel and Roesset (1981) for isotropic soil media. The solution technique is based on the application of Fourier and Hankel transforms to the wave equations in each layer to reduce them to a series of ordinary differential equations. These equations are then solved by the imposition of the appropriate stress and kinematic boundary conditions at layer interfaces and the free surface. This is achieved through a stiffness matrix approach in which each layer is represented by a stiffness matrix that relates the Fourier transform of the stresses and displacements at the upper and lower surfaces of the layer. Stiffness expressions derived by Park and Kaynia (2018) are incorporated here for completeness. The reader is referred to the original reference for the details.

If λ_t and G_t are the complex *Lamé* constants related to wave motion propagating in the vertical direction (z), and λ and G are related to wave propagation in the two horizontal directions (r,θ), then the associated wave velocities can be expressed by

$$C_p = \sqrt{\frac{\lambda + 2G}{\rho}}; \quad C_s = \sqrt{\frac{G}{\rho}}; \quad C_{pt} = \sqrt{\frac{\lambda_t + 2G_t}{\rho}}; \quad C_{st} = \sqrt{\frac{G_t}{\rho}} \tag{6.4}$$

where ρ is the mass density, C_p and C_s are the P- and S-wave velocities in x-y plane (i.e., along the horizontal direction); C_{pt} and C_{st} are the corresponding velocities in the vertical direction. The anisotropy can be defined by introducing the following three parameters defining the ratios between these wave velocities.

$$\alpha = \frac{C_{pt}}{C_{st}}; \quad a = \frac{C_p}{C_{pt}}; \quad b = \frac{C_s}{C_{st}} \tag{6.5}$$

The 4 × 4 stiffness matrix for an anisotropic soil layer subjected to P-SV wave motion is given by the following formula.

$$\mathbf{K} = 2kG_t \begin{Bmatrix} \mathbf{K}_{11} & \mathbf{K}_{12} \\ \mathbf{K}_{21} & \mathbf{K}_{22} \end{Bmatrix} \tag{6.6}$$

where the sub-matrices in the above formula are given by

$$
\begin{aligned}
\mathbf{K}_{11} &= \frac{(\bar{k}_{z1}/\gamma_1 - \bar{k}_{z2}/\gamma_2)}{2D_t} \begin{Bmatrix} \gamma_2\left(C_1S_2 - \frac{\gamma_1}{\gamma_2}C_2S_1\right) & -\left[(1-C_1C_2) + \frac{\gamma_1}{\gamma_2}S_1S_2\right] \\ -\left[(1-C_1C_2) + \frac{\gamma_1}{\gamma_2}S_1S_2\right] & \frac{1}{\gamma_1}\left(C_2S_1 - \frac{\gamma_1}{\gamma_2}C_1S_2\right) \end{Bmatrix} \\
&\quad - \frac{1}{2}(1 + \bar{k}_{z2}/\gamma_2)\begin{Bmatrix} 0 & 1 \\ 1 & 0 \end{Bmatrix} \\
\mathbf{K}_{12} &= \frac{(\bar{k}_{z1}/\gamma_1 - \bar{k}_{z2}/\gamma_2)}{2D_t} \begin{Bmatrix} \gamma_2\left(\frac{\gamma_1}{\gamma_2}S_1 - S_2\right) & -(C_1 - C_2) \\ (C_1 - C_2) & \frac{1}{\gamma_1}\left(\frac{\gamma_1}{\gamma_2}S_2 - S_1\right) \end{Bmatrix} = \mathbf{K}_{21}^T \\
\mathbf{K}_{22} &= \frac{(\bar{k}_{z1}/\gamma_1 - \bar{k}_{z2}/\gamma_2)}{2D_t} \begin{Bmatrix} \gamma_2\left(C_1S_2 - \frac{\gamma_1}{\gamma_2}C_2S_1\right) & \left[(1-C_1C_2) + \frac{\gamma_1}{\gamma_2}S_1S_2\right] \\ \left[(1-C_1C_2) + \frac{\gamma_1}{\gamma_2}S_1S_2\right] & \frac{1}{\gamma_1}\left(C_2S_1 - \frac{\gamma_1}{\gamma_2}C_1S_2\right) \end{Bmatrix} \\
&\quad + \frac{1}{2}(1 + \bar{k}_{z2}/\gamma_2)\begin{Bmatrix} 0 & 1 \\ 1 & 0 \end{Bmatrix}
\end{aligned} \tag{6.7}
$$

The 2 × 2 stiffness matrix for an anisotropic half-space subjected to P-SV wave is expressed as

$$
\mathbf{K}_{\text{half}} = 2kG_t\left[\frac{(\bar{k}_{z1}/\gamma_1 - \bar{k}_{z2}/\gamma_2)}{2(1-\gamma_1/\gamma_2)}\begin{Bmatrix} \gamma_1 & 1 \\ 1 & 1/\gamma_2 \end{Bmatrix} - \frac{1}{2}(1 + \bar{k}_{z1}/\gamma_1)\begin{Bmatrix} 0 & 1 \\ 1 & 0 \end{Bmatrix}\right] \tag{6.8}
$$

Similarly, the stiffness matrices for an anisotropic layer and a half-space subjected to SH were derived and are defined as

$$
\mathbf{K} = kG_t\begin{Bmatrix} \mathbf{K}_{11} & \mathbf{K}_{12} \\ \mathbf{K}_{21} & \mathbf{K}_{22} \end{Bmatrix} = kG_t\frac{\bar{k}_{z3}}{S_3}\begin{Bmatrix} C_3 & -1 \\ -1 & C_3 \end{Bmatrix} \tag{6.9}
$$

$$
\mathbf{K}_{\text{half}} = kG_t\bar{k}_{z3} \tag{6.10}
$$

The parameters in the above formula are defined in the following.

$$
C_1 = \cos hk\bar{k}_{z1}h, \ S_1 = \sin hk\bar{k}_{z1}h \tag{6.11a}
$$

$$
C_2 = \cos hk\bar{k}_{z2}h, \ S_2 = \sin hk\bar{k}_{z2}h \tag{6.11b}
$$

$$
C_3 = \cos hk\bar{k}_{z3}h, \ S_3 = \sin hk\bar{k}_{z3}h \tag{6.11c}
$$

$$
D_t = 2(1 - C_1C_2) + \left(\frac{\gamma_1}{\gamma_2} + \frac{\gamma_2}{\gamma_1}\right)S_1S_2 \tag{6.11d}
$$

$$\bar{k}_{z1} \text{ or } \bar{k}_{z2} = \sqrt{\frac{\alpha^2(a^2-1)+2-(1+\alpha^{-2})\Omega_{st}^2 \pm \sqrt{\left[\alpha^2(a^2-1)-(1-\alpha^{-2})\Omega_{st}^2\right]^2+4(a^2-1)(\alpha^2-1)}}{2}} \tag{6.11e}$$

$$\bar{k}_{z3} = \sqrt{b^2-\Omega_{st}^2};\ \gamma_1 = \frac{\alpha^2 a^2 - \bar{k}_{z1}^2 - \Omega_{st}^2}{(\alpha^2-1)\bar{k}_{z1}};\ \gamma_2 = \frac{\alpha^2 a^2 - \bar{k}_{z2}^2 - \Omega_{st}^2}{(\alpha^2-1)\bar{k}_{z2}} \tag{6.11f}$$

$$\Omega_{pt} = \frac{\omega}{kC_{pt}} = \frac{\Omega_{st}}{\alpha};\ \Omega_{st} = \frac{\omega}{kC_{st}} \tag{6.11g}$$

As expected, by setting $a = 1$ and $b = 1$ in the above matrices one can recover the stiffness matrices for the isotropic medium developed by Kausel and Roesset (1981).

6.2.4 Green's Functions for Piles in Layered Soil

The Green's functions for cylindrical loads, representing pile–soil tractions, are those developed by Kaynia (1988). The solution is based on the stiffness matrix approach and in principle is similar to the one described above. For details, the reader is referred to Kaynia (1988) and Kaynia and Kausel (1991).

6.3 Soil and Load Data and Simulation for Base Case

The numerical results presented in the following sections correspond to the ground conditions established at a test site in Ledsgård, Sweden and for the Swedish passenger train X-2000. The soil at this site was characterized through a series of laboratory and field measurements. The parameters determined for the site for two speed regimes, namely sub-seismic and trans-seismic, were reported in Kaynia et al. (2000) and are summarized in Table 6.1. Figure 6.3 shows the bogie loads for the X-2000 train.

The track (including ballast and embankment) was represented by an equivalent beam with bending rigidity, $EI = 200$ MNm^2 for sub-seismic speed regime and 80 MNm^2 for trans-seismic condition (Kaynia et al. 2000). The different values of soil parameters (also in Table 6.1) reflect the soil nonlinearity. Mass of the track was taken 10,600 kg/m.

Figure 6.4 displays the computed time history of track displacements for the surface track base case (i.e., with no countermeasures) for train passages in different speed regimes, sub-seismic (100 and 140 km/h), critical speed (200 km/h) and

Table 6.1 Soil parameters at Ledsgård test site used in numerical simulations

Soil layer	Thickness [m]	Density [kg/m^3]	C_S [m/s]		C_P [m/s]	
			$V = 70$ [km/h]	$V = 200$ [km/h]	$V = 70$ [km/h]	$V = 200$ [km/h]
Crust	1.1	1500	72	65	500	500
Organic clay	3.0	1260	41	33	500	500
Clay	4.5	1475	65	60	1500	1500
Clay	6.0	1475	87	85	1500	1500
Half-space	–	1475	100	100	1500	1500

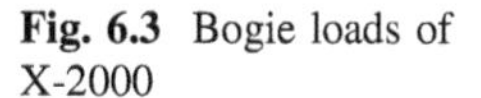

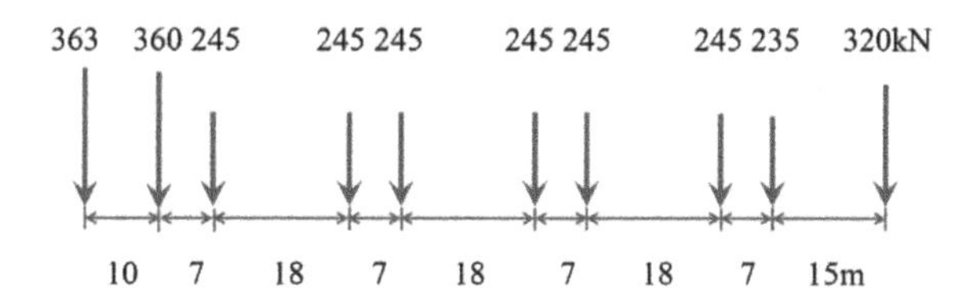

Fig. 6.3 Bogie loads of X-2000

trans-seismic (250 km/h). As the plots show, the response in the sub-seismic regime is quasi-static, while those at the critical speed and trans-seismic conditions are dynamic. The latter are characterized by upward and downward oscillations and considerably larger amplitudes. The relatively low critical speed (around 200 km/h) is primarily due to the soft peat layer at about 1.5 m depth. If this layer is replaced by, for example, sand/gravel or if it is stiffened by lime–cement columns, the Rayleigh wave velocity and critical speed will increase accordingly.

6.4 Measures to Increase Critical Speed

This section presents the results for effect of different countermeasures. Three cases were considered for this purpose: (1) stiffened track where the stiffness of the track is increased by for example improving the stiffness of the embankment or by placing a stiff girder under the track, (2) soil improvement under the track by for example replacing the natural soil with stiffer materials (Fig. 6.1c) and (3) use of piles under the track, namely piled track (Fig. 6.2). In each case, the time histories of the track displacements for the same train load (Fig. 6.3) were computed and compared with the displacements of the base case shown in Fig. 6.4.

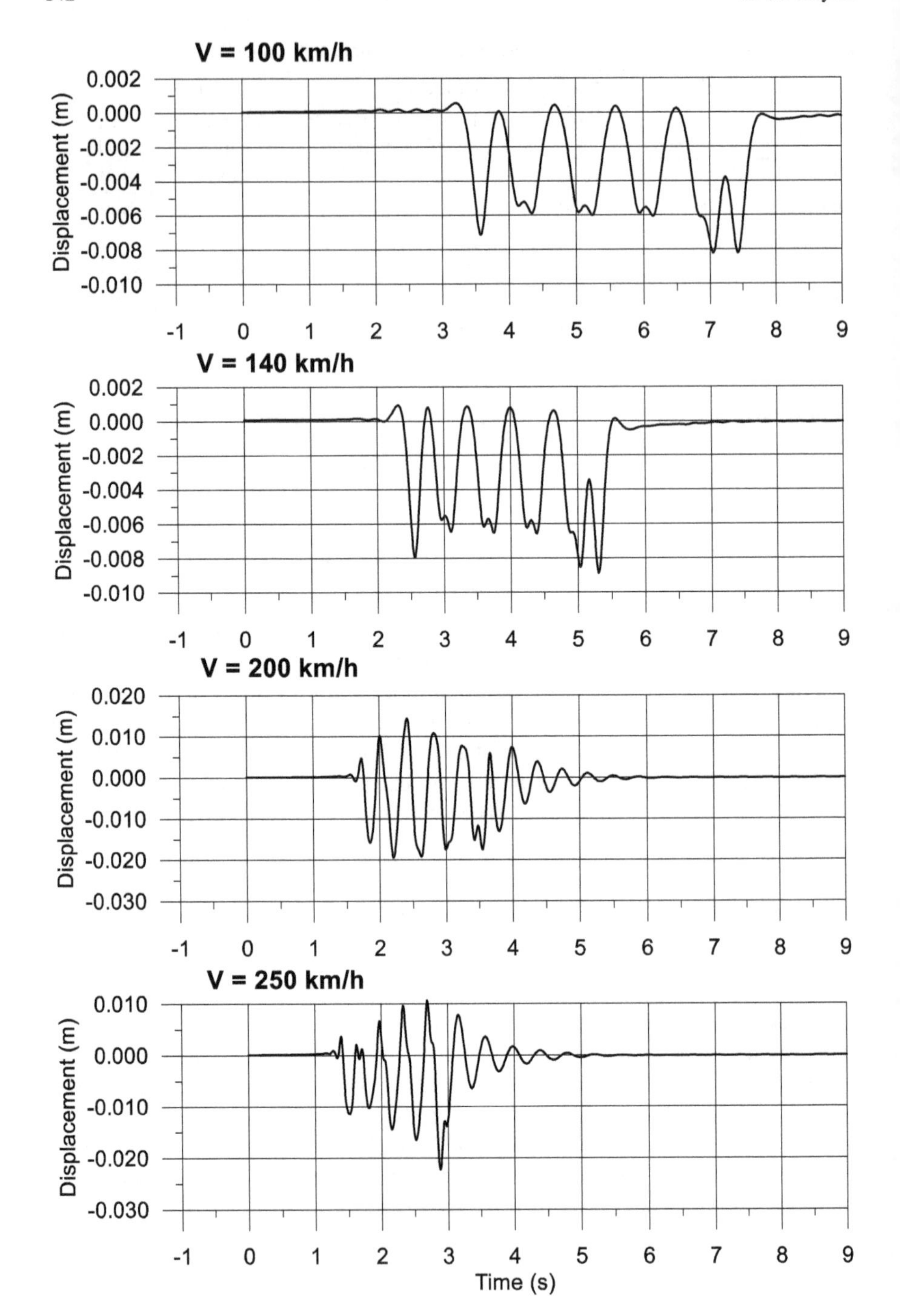

Fig. 6.4 Track displacements for low and high train speeds with track/ground model at Ledsgård

6.4.1 *Track Stiffening*

It is intuitive to think that stiffening the track, by for example use of a girder beam under the track, would reduce the vibrations and improve the performance of surface tracks. Figure 6.5 displays the track displacements for a case in which the total bending rigidity of the track is increased by a factor of 4 (to include a girder beam). Comparison of these results with those for the track in Fig. 6.4 indicates a considerable reduction of the response; however, the displacement oscillations for the speed 250 km/h indicates that the system is still in a trans-seismic regime. This is an undesirable condition for the track and should be avoided. While this measure is cable of reducing the track displacements, it will hardly affect the critical speed. To elucidate this point, Fig. 6.6 compares the trackside ground response for the base case and for the stiffened track for load speed 250 km/h. The figure demonstrates that the trackside vibration is only marginally reduced by stiffening the track, confirming the conclusion that track stiffening does not noticeably affect the critical speed although it could reduce the track response.

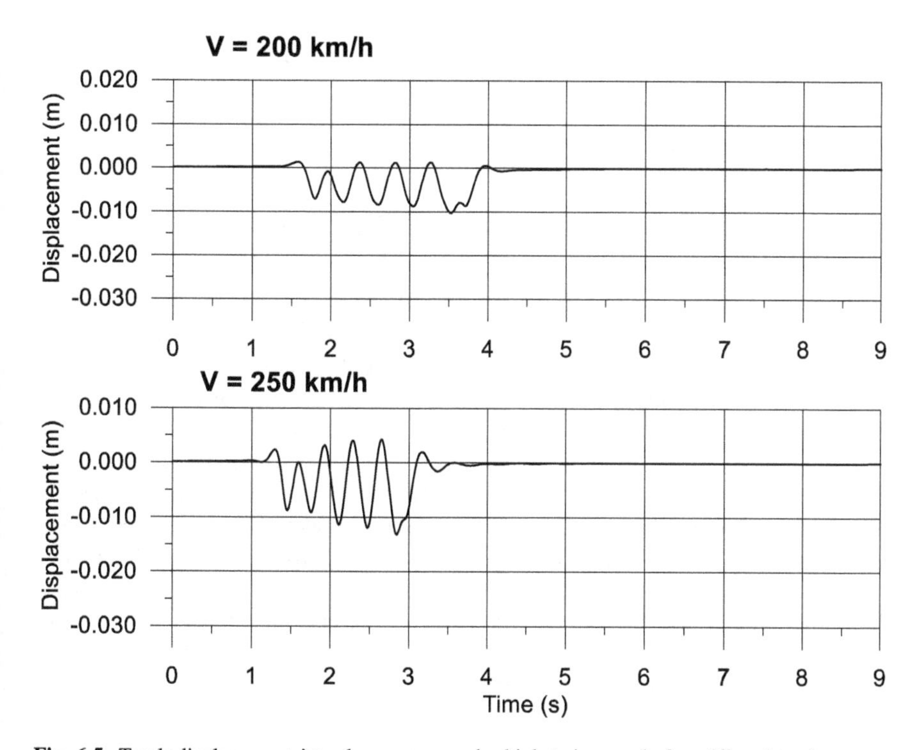

Fig. 6.5 Track displacements' track response under high train speeds for stiffened track

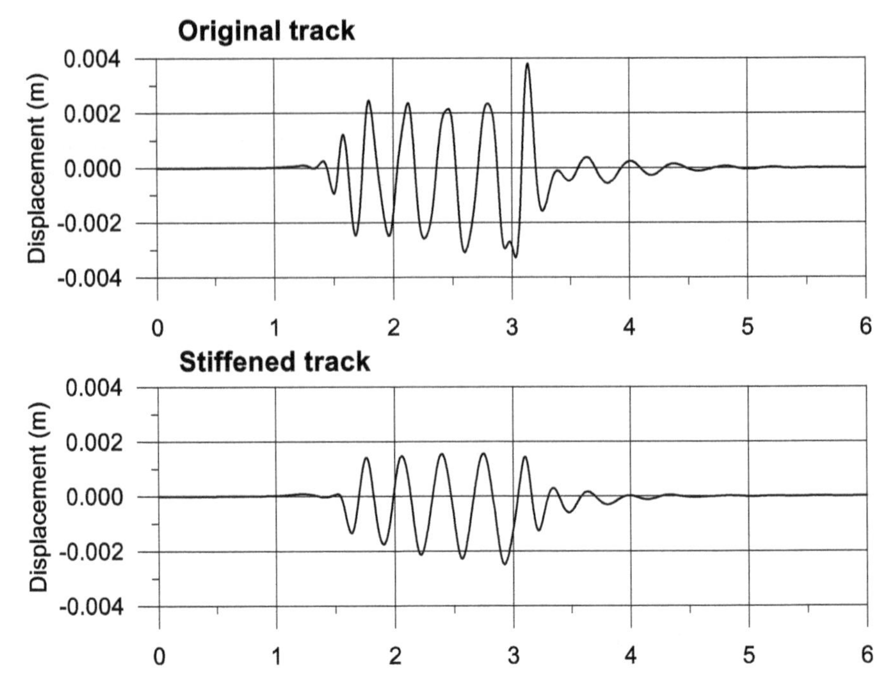

Fig. 6.6 Trackside ground displacements at 10 m distance for base case track (top) and stiffened track (bottom)

6.4.2 Ground Improvement and Soil Replacement

An effective solution for increasing the critical speed is stiffening the ground by either soil replacement or improvement techniques such as grouting or use of lime–cement columns (Norén-Cosgriff et al. 2019; Peplow and Kaynia 2007). Soil replacement is generally considered an expensive solution; however, it has the advantage that the properties of the new material are often more accurately known.

To assess the performance of ground improvement, the soil in the top two layers at Ledsgård site (Table 6.1) directly under the track was numerically replaced by gravel or crushed rock. It was assumed that the new material has an average shear wave velocity of 200 m/s. Moreover, for comparative purposes, the soil parameters of the track and the other soil layers were kept the same as in the base case. It should be noticed that this is strictly not correct because the soil parameters were selected by proper accounting for soil nonlinearity due to the induced stresses in the track and soil layers. The track response was then computed for the same loads (Fig. 6.3) using *VibTrain*.

The results of the numerical simulations are presented in Fig. 6.7. The plots in this figure clearly demonstrate that, although the track displacement increases with the train speed as expected, the responses for all the considered train speeds (in the

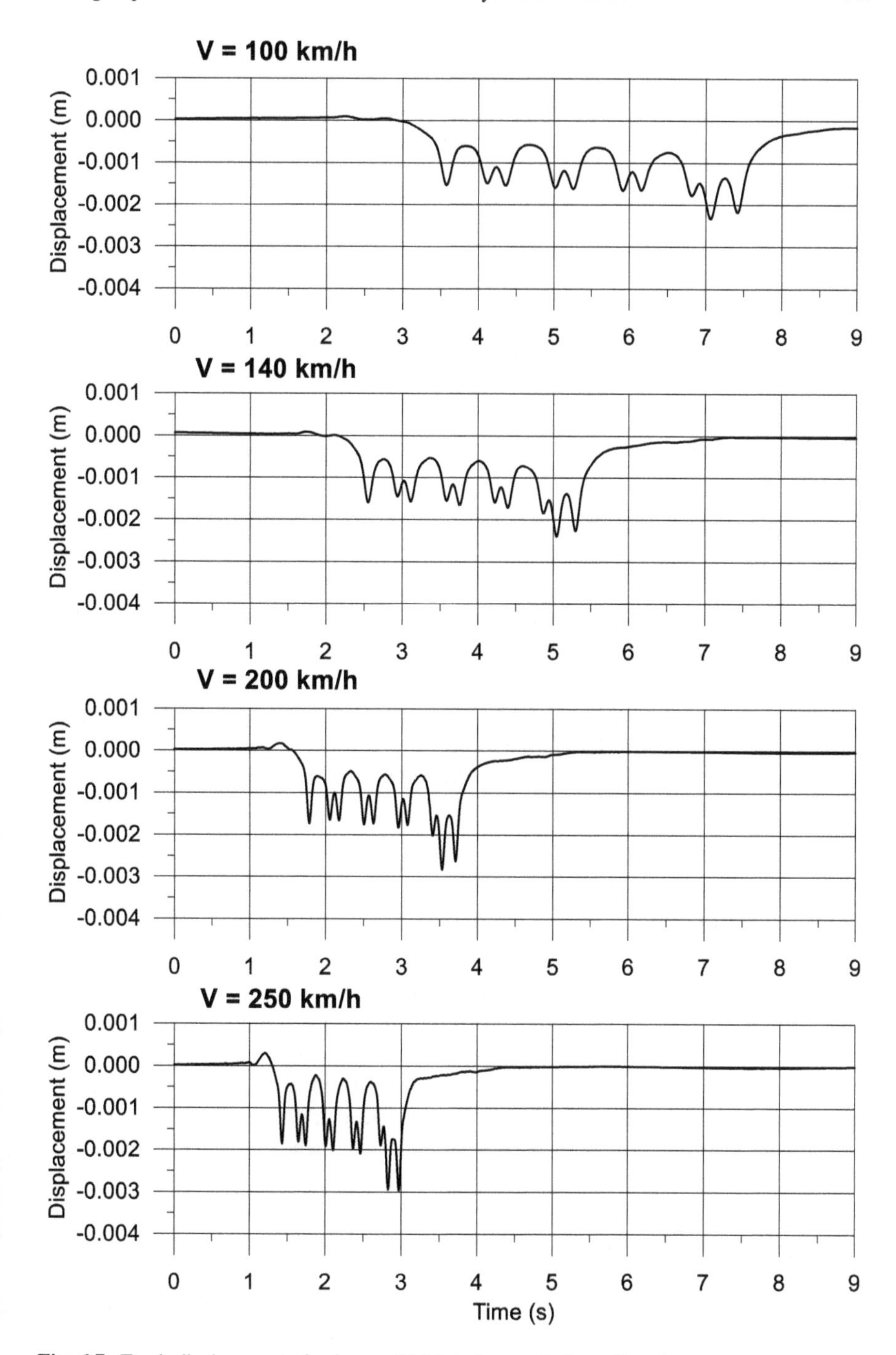

Fig. 6.7 Track displacements for low and high train speeds for soil replacement

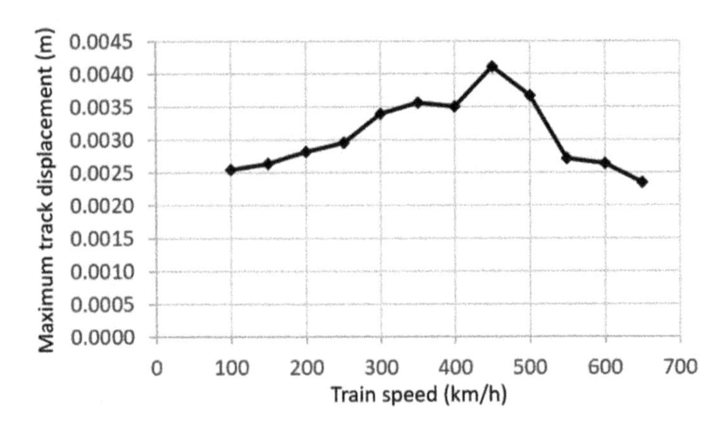

Fig. 6.8 Track displacements for low and high train speeds for soil replacement

range 100–250 km/h) are in the sub-seismic regime; that is, the responses are primarily quasi-static. These results confirm that soil replacement or ground improvement is a viable solution for increasing the critical speed.

For the above modified ground conditions, the critical speed was estimated by computing the maximum displacements of the track for increased train speed. Figure 6.8 shows a plot for the maximum track displacement as a function of the train speed in the range 100–700 km/h. The figure displays a peak displacement at a train speed of about 450 km/h. Therefore, the applied ground improvement has more than doubled the critical speed and has considerably reduced the displacements.

6.4.3 Piled Track

Another effective technique to reduce the critical speed is by use of piles under the track (Fig. 6.2). Piles have two functions in this regard. Firstly, they transfer the loads to deeper strata that are normally stiffer and hence contribute to generally stiffer track behavior. Secondly, the waves in the ground set up by one pile, as the load passes over the pile, do not excite the adjacent piles by the same extent as surface waves on the ground do. This is an important mechanism that leads to large track response when the load speed reaches the wave propagation velocity in the ground.

To investigate the performance of piled tracks, the same bending rigidity of the track in the stiffened track cased (Sect. 4.1) was used for the girder on the piles. The piles were assumed to have a diameter of 2 m and installed to the bottom of the third layer (i.e., about 9 m long, see Table 6.1). The spacing of the piles was taken as 12.0 m.

Figure 6.9 presents the results of simulations for the track response for the train speeds in the range 140–250 km/h. The plots show the time histories of the displacements for a section directly over the piles. The corresponding results for the mid-sections of the girder are displayed in Fig. 6.10. Comparing these results with those for the stiffened track on ground surface for the same speeds (Fig. 6.5), one can observe that while the track responses over the piles are lower, the response of the mid-sections is significantly higher. This large response is due to the vibration of the girder between the piles, which as the plots indicate, is amplified as the load speed increases.

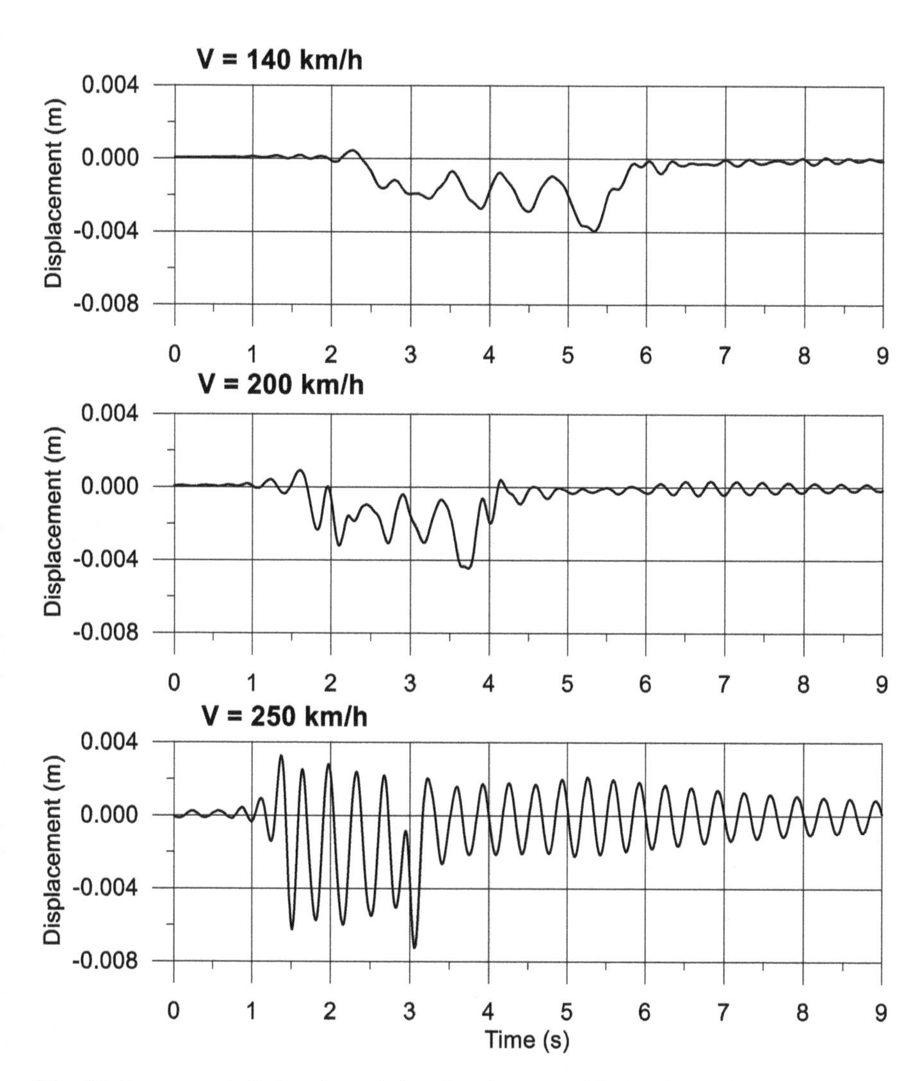

Fig. 6.9 Response of piled track at pile locations for low and high train speeds

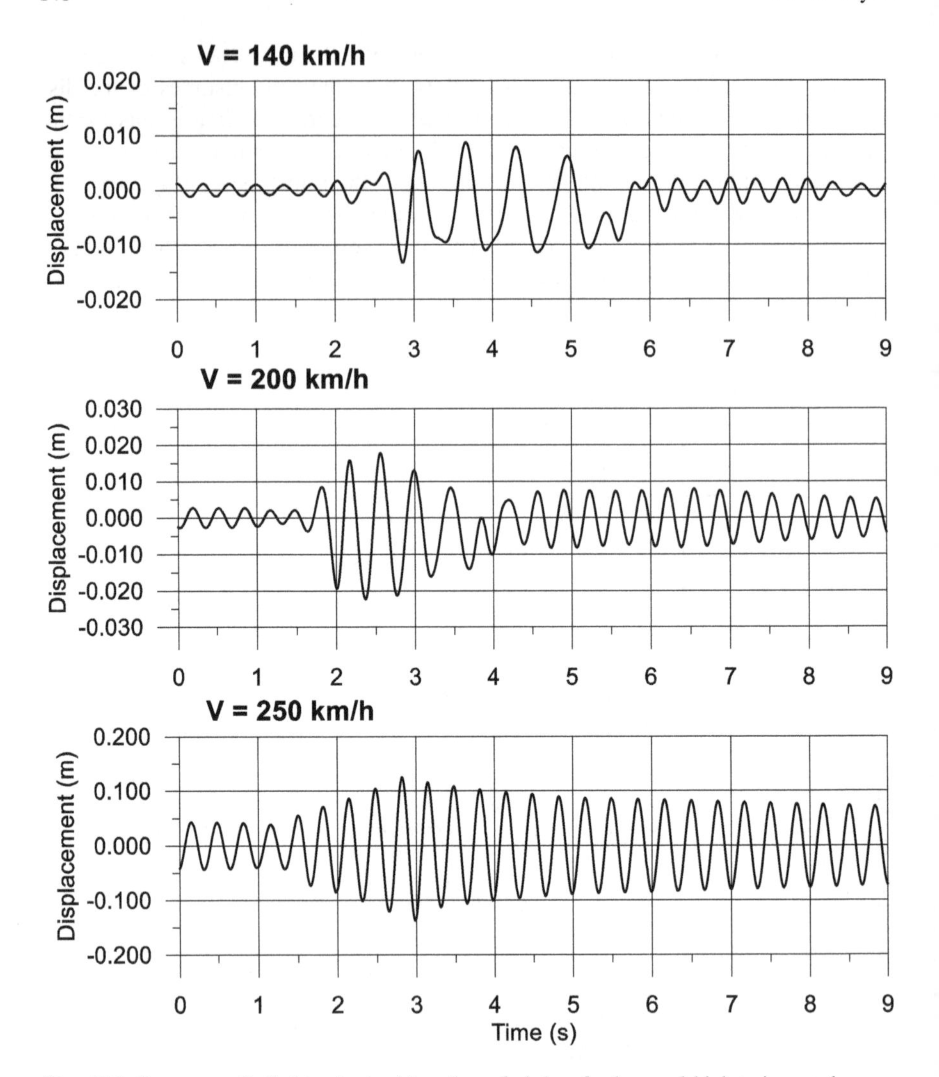

Fig. 6.10 Response of piled track at mid-section of girders for low and high train speeds

The results presented in this section indicate that while piled tracks could increase the critical speed and reduce the track response, one needs to perform sensitivity analyses on the bending rigidity of the girder to ensure acceptable vibration at mid-spans. The design challenge has therefore been moved from the critical speed issue to the proper structural design of the girder.

6.5 Conclusion

This paper presented an overview of the developments in the subject of moving loads over layered ground. Two analytical–numerical models based on the use of Green's functions in viscoelastic layered ground were presented. The first model is for conventional tracks on ground surface, and the second model is for elevated tracks on girders supported by piles. Typical results of simulations for both models were presented, and the responses in the speed regimes sub-seismic and trans-seismic were discussed. It was shown that one can reduce the track vibration near the critical speed by stiffening the track although not being able to noticeably alter the critical speed. Moreover, the trackside ground surface vibration is not significantly reduced.

Use of piles under tracks, namely piled tracks, appears to increase the critical speed; however, special attention should be given to design of the girders to avoid excessive track vibration between the piles.

Ground improvement or soil replacement under the track was shown to be a reliable solution for increasing the critical speed; however, it should be noticed that soil replacement is often a costly operation. Ground improvement techniques such as use of lime–cement columns could offer more economical alternatives.

References

Achenbach JD, Keshava SP, Herrmann G (1967) Moving load on a plate resting on an elastic half space. J Appl Mech 34:910–914

Andersen L, Jones CJC (2006) Coupled boundary and finite element analysis of vibration from railway tunnels—a comparison of two- and three-dimensional models. J Sound Vib 293:611–625

Ang DD (1960) Transient motion of a line load on the surface of an elastic half-space. Quart Appl Math 18:251–256

Aubry I, Clouteau D, Bonnet G (1994) Modeling of wave propagation due to fixed or mobile dynamic sources. In: Proceedings of workshop Wave'94, Ruhr University, Berg-Verlag, Bochum, pp 79–93

Auersch L (2005) The excitation of ground vibration by rail traffic: theory of vehicle-track-soil interaction and measurements on high-speed lines. J Sound Vib 284:103–132

Baron ML, Bleich HH, Wright JP (1967) Ground shock due to Rayleigh waves from sonic booms. J Engng Mech Div ASCE 93(5):135–162

Cole J, Huth J (1958) Stress produced in a half-space by moving loads. J Appl Mech 25:433–436

Coulier P, Francois S, Degrande G, Lombaert G (2013) Subgrade stiffening next to the track as a wave impeding barrier for railway induced vibrations. Soil Dyn Earthq Eng 48:119–131

Craggs JW (1960) Two-dimensional waves in an elastic half-space. Proc Cambridge Phil Society 56:269–275

de Barros FCP, Luco JE (1994) Response of a layered viscoelastic half-space to a moving point load. Wave Motion 19:189–210

Eringen AC, Suhubi ES (1975) Elastodynamics, volume II: linear theory. Academic Press, Inc., New York

Filon LNG (1928) On a quadrature formula for trigonometric integrals. Proc Roy Soc Edinburgh 49:38–47

Frýba L (1973) Vibration of solids and structures under moving loads. Noordhoff International Publishing, The Netherlands

Hall L (2003) Simulations and analyses of train-induced ground vibrations in finite element models. Soil Dyn Earthquake Eng 23:403–413

Karlström A, Boström A (2006) An analytical model for train induced ground vibrations from railways. J Sound Vib 292:221–241

Kausel E, Roesset JM (1981) Stiffness matrices for layered soils. Bull Seism Soc Am 71(6):1743–1761

Kaynia AM (1988) Characteristics of the dynamic response of pile groups in homogeneous and nonhomogeneous media. In: Proceedings 9th world conference earthquake engineering, vol 3, Tokyo-Kyoto, Japan, pp 575–580

Kaynia AM (2001) Measurement and prediction of ground vibration from railway traffic. In: Proceedings of the 15th international conference on soil mechanics and geotechnical engineering, vol 3. Istanbul, Turkey, 27–31 Aug, pp 2105–2109

Kaynia AM, Kausel E (1991) Dynamics of piles and pile groups in layered soil media. Soil Dyn Earthquake Eng 10(8):386–401

Kaynia AM, Madshus C, Zackrisson P (2000) Ground vibration from high-speed trains: prediction and countermeasure. J Geotech Geoenviron Eng 126(6):531–537

Kaynia AM, Park J, Norén-Cosgriff K (2017) Effect of track defects on vibration from high speed train. Proc Eng 199:2681–2686

Krylov VV (1995) Generations of ground vibrations from superfast trains. Appl Acoustics 44:149–164

Madshus C, Kaynia AM (2000) High-speed railway lines on soft ground: dynamic behaviour at critical train speed. J Sound Vibration 231(3):689–701

Niwa Y, Kobayashi S (1966) Stresses produced in an elastic half-space by moving loads along its surface. Mem Fac Engng Kyoto Univ 28:254–276

Norén-Cosgriff KM, Bjørnarå TI, Dahl BM, Kaynia AM (2019) Advantages and limitation of using 2-D FE modelling for assessment of effect of mitigation measures for railway vibrations. Appl Acoust 155:463–476

Pan G, Atluri SN (1995) Dynamic response of finite sized elastic runways subjected to moving loads: a coupled BEM/FEM approach. Int J Num Meth Eng 38:3143–3166

Papadopoulos M (1963) The elastodynamics of moving loads. J Australian Math Soc 3:9–92

Park J, Kaynia AM (2018) Stiffness matrices for fluid and anisotropic soil layers with applications in soil dynamics. Soil Dyn Earthq Eng 115:169–182

Payton RG (1964) An application of the dynamic Betti-Rayleigh reciprocal theorem to moving point load in elastic media. Quart Appl Math 21:299–313

Peplow A, Kaynia AM (2007) Prediction and validation of traffic vibration reduction due to cement column stabilization. Soil Dyn Earthq Eng 27:793–802

Sackman JL (1961) Uniformly moving load on a layered half-plane. J Eng Mech Div ASCE 87 (4):75–89

Sneddon IN (1951) Fourier transforms. McGraw-Hill Book Company, New York

Wright JP, Baron ML (1970) Exponentially decaying pressure pulse moving with constant velocity on the surface of a layered elastic material (superseismic layer, subseismic half-space). J Appl Mech 96:141–152

Chapter 7
Numerical Simulation of Coir Geotextile Reinforced Soil Under Cyclic Loading

Jayan S. Vinod, Abdullah Al-Rawabdeh, Ana Heitor, and Beena K. Sarojiniamma

7.1 Introduction

Coir fibers are permeable natural fibers that are developed from the husk of the coconut. The coir fibers degrade very slowly compared to other natural fibers (e.g., jute), and the longevity of these fibers in the field is around 2–3 years (Dutta and Rao 2008). The limited use of non-renewable resources and the low cost of coir geotextiles have attracted attention toward using them as an alternative to synthetic geotextiles for infrastructure development (Sarsby 2007; Subaida et al. 2008, 2009; Chauhan et al. 2008; Vinod and Minu 2010; Hejazi et al. 2012; Balan 2017). In the recent past, many studies have been carried out on the bearing capacity of soil using synthetic polypropylene materials (Unnikrishnan et al. 2002; Bueno et al. 2005; Hufenus et al. 2006; Basudhar et al. 2008; Rawal and Sayeed 2013). Unnikrishnan et al. (2002) showed that the efficacy of placing reinforcement in soil is related to the stress transfer from the soil to the reinforcement. Basudhar et al. (2008) developed a linear elastic model using the finite element method on synthetic geotextile reinforcing sand under strip loading. A parametric study was conducted to investigate the geotextile's reinforcement depth. Many studies are reported on the bearing capacity of coir geotextile reinforced soil during monotonic loading (Noorzad and Mirmoradi 2010; Bhandari and Han 2010; Lal et al. 2017; Sridhar and Prathap Kumar 2018). Rashidian et al. (2018) studied the effect of the depth of

J. S. Vinod (✉) · A. Al-Rawabdeh
School of Civil Mining and Environmental Engineering, University of Wollongong, Wollongong, NSW 2522, Australia
e-mail: vinod@uow.edu.au

A. Heitor
School of Civil Engineering, University of Leeds, Leeds, UK

B. K. Sarojiniamma
Division of Civil Engineering, School of Engineering, Cochin University of Science and Technology, Kochi, Kerala, India

T. G. Sitharam et al. (eds.), *Latest Developments in Geotechnical Earthquake Engineering and Soil Dynamics*, Springer Transactions in Civil and Environmental Engineering, https://doi.org/10.1007/978-981-16-1468-2_7

placement of coir geotextiles and reported that the bearing capacity varies with the geotextile's position. In addition, the number of layers has an insignificant effect on the bearing capacity of soil. Kurian et al. (1997) presented a 3D nonlinear finite element model of a sand foundation reinforced with coir rope. A significant reduction in the settlement was observed for the coir reinforced foundation compared to the unreinforced model.

Performance under cyclic loading is considered to be crucial for the design of infrastructure for transport and seismic loading. Many studies reported the cyclic behavior of geosynthetic reinforced soil (Das and Shin 1994; Naeini and Gholampoor 2014; Sreedhar and Goud 2011). Raymond and Williams (1978) conducted a repeated triaxial test and indicated that the deformation under repeated loading is higher than the magnitude of deformation under static loading. Cunny and Sloan (1962) studied the dynamic loading effect on footing to establish a criterion for designing foundations under cyclic loading. Vesic et al. (1965) have concluded that the bearing capacity of footing under cyclic loading is less than the bearing capacity under static loading. Brumund and Leonards (1972) studied the behavior of circular footing on sand under dynamic loading and presented a linear relationship between footing settlement and peak acceleration. Al-Qadi et al. (2008) presented the efficiency of using geogrid in low-volume flexible pavements under dynamic loads. They showed that placing the geogrid between the subbase and subgrade layers gives the best performance for thin-layered base courses, while placing the geogrid at the depth of one-third of the base layer is the best for thick-based layers. Perkins et al. (2011) developed a two-dimensional model for geosynthetic reinforced unpaved roads to study the rutting deformation of flexible pavements. Sridher and Prathap Kumar (2018) investigated the behavior of coir geotextile reinforced sand under cyclic loading and concluded that placing the coir geotextile improves the sand's bearing capacity and reduces its settlement. However, only limited research studies focused on the cyclic behavior of coir geotextile reinforced soil (Sridhar and Prathap Kumar 2018).

The main objective of this research is to evaluate the mechanical behavior of coir geotextile reinforced soil during cyclic loading using a finite element method. The effect of different parameters influencing the performance of coir geotextile reinforced soil under cyclic loading is investigated and reported.

7.2 Numerical Model for Coir Geotextile Reinforced Soil Under Cyclic Loading

The finite element model for cyclic loading was created using Plaxis 2D. The soil bed has two soil layers: The first layer (layer I) is classified as (GW) crushed stone with high-quality material based on the unified soil classification system, and the second layer (layer II) is classified as (CH) clay with low-quality material. These two soils layers with different strengths were selected to understand the interaction between various soils and coir geotextiles. The test tank model has dimensions of 800 mm length × 500 mm width. The model dimensions are similar to the laboratory experimental program reported by Subaida et al. (2009). Figure 7.1 shows the test conditions used for this study. The vertical load was applied on the left corner of the model by a 100-mm-diameter plate having a thickness of 25 mm. Only a half portion of the test bed was modeled considering the symmetry of the test bed. The hardening soil model with small strains (HSsmall) which is an advanced model designed by Schanz et al. (1999) has been considered for all soil layers. The material parameters used for the model were evaluated using laboratory tests reported by Subaida et al. (2009) and Sridhar and Prathap Kumar (2018). Table 7.1 presents the soil properties for layer I, layer II and sand soil for the hardening soil with small strains model (HSsmall).

The physical properties such as dry unit weight, saturated unit weight and Poisson's ratio (Bowles 1996) have been defined inclusive of strength parameters like lateral earth pressure, friction angle and dilatancy angles (Das et al. 2016) and the stiffness properties E_{oed}, E_{50} and Eur (Brinkgreve et al. 2014). In this study, the soil layers were considered to be dry.

The coir geotextile was modeled as a linear elastic plate element and assigned a bending stiffness (EI) value of 0.15E−9 kNm^2/m, and the elastic stiffness (EA) is 500 kN/m. The lateral deformation is restricted on the left and right boundary walls, and both lateral and vertical deformations are restricted for the bottom boundary of the model. The coir geotextile in the model represents woven coir geotextile with 1286.56 g/m^2 mass/unit area, 20.7 and 36 kN/m weft and warp tensile strength, respectively.

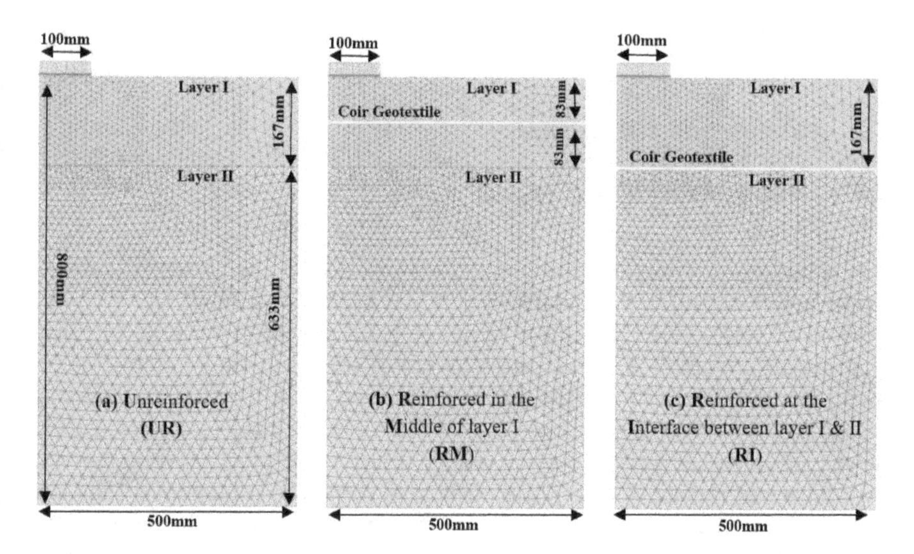

Fig. 7.1 Test models; **a** unreinforced soil (UR), **b** reinforcement in the middle of layer (I) and **c** reinforcement at the interface between layers (I) and (II)

Table 7.1 Layer I, layer II and sand soil properties (hardening soil model with small strains)

General, stiffness and strength parameters	Symbol	Unit	Value		
			Layer I	Layer II	Sand[a]
Dry unit weight	γ_d	kN/m^3	18	12	17.4
Saturated unit weight	γ_{sat}	kN/m^3	21	16.95	19.96
Void ratio	e	–	0.46	1.02	0.5
Tangent stiffness for primary oedometer loading	E_{oed}	kN/m^2	10,000	350	75,00
Secant stiffness in standard drained triaxial test	E_{50}	kN/m^2	12,500	700	10,000
Unloading/reloading stiffness	Eur	kN/m^2	45,000	2100	30,000
Power for stress-level dependency of stiffness	m	–	0.5	0.5	0.5125
Shear strain level, where the secant shear modulus reduced to 70% of G_0	$\gamma_{0.7}$	–	2×10^{-4}	0.1×10^{-3}	0.1×10^{-3}
Initial shear modulus	G_0	kN/m^2	140,000	8400	120,000
Coefficient of earth pressure	K_0	–	0.253	–	0.293
Friction angle	ϕ	0	48.3	–	45
Dilatancy angle	ψ	0	19	–	15
Cohesion	C	kN/m^2	–	19.5	–

[a]The sand soil layer used in the dynamic load validation

A typical sinusoidal cyclic loading with variant frequency and cyclic stresses (σ_c) is used to study the cyclic behavior of coir geotextile reinforced soil. In this investigation, number of cycles (N), cyclic stress amplitudes (σ_c = 50, 100, 150 kPa) and frequency (f = 0.5, 1, 1.5 Hz) were varied during cyclic loading.

7.3 Results and Discussion

7.3.1 Calibration of FE Model

Calibration of the finite element model for cyclic loading was carried out on a model having dimensions of 500 mm length and width (See inset of Fig. 7.2). The cyclic load was applied through a 50 mm circular footing (Sridhar and Prathap Kumar 2018). Figure 7.2 shows the settlement of sand with the number of cycles during cyclic loading. The model was subjected to a cyclic stress of 100 kPa and f = 0.5 Hz. It is evident from Fig. 7.2 that the numerical model captures the settlement of sand during cyclic loading similar to the laboratory experiment reported by Sridhar and Prathap Kumar (2018). The sand properties are found in Table 7.1 and the model geometry in Fig. 7.2. The footing settlement increases with the increase of the number of cycles, and the significant increase in the settlement of footing was observed for $N > 2000$.

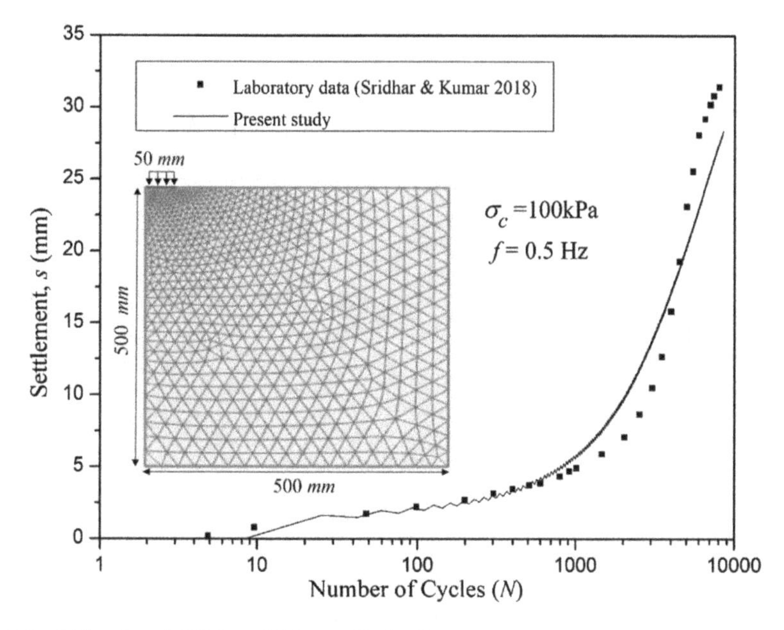

Fig. 7.2 Calibration of FEM model for cyclic loading

7.3.2 Behavior of Coir Geotextile Reinforced Soil During Cyclic Loading

Figure 7.3 shows the variation of *s/B* with *N* for UR, RI and RM for σ_c= 150 kPa and *f* = 0.5 Hz, where *s/B* is settlement over plate width ratio. It is evident from Fig. 7.3 that the inclusion of coir geotextiles reduces the *s/B* of soil during cyclic loading, and it is interesting to note that the initial stiffness increases with the inclusion of coir geotextiles. The coir geotextiles placed in the middle of layer I (RM) exhibits lower settlement compared to RI. For a particular value of *N* (say *N* = 10,000), the UR shows an *s/B* of 23% compared with 14% and 19% for RM and RI, respectively.

7.3.3 Effect of Cyclic Stress on the Settlement of Coir Geotextile Reinforced Soil

Figure 7.4 shows the effect of cyclic stress on the settlement behavior of UR, RI and RM. As expected, the settlement increases with the increase in cyclic stress. At 10,000 cycles, *s/B* increased from 5.5 to 24% when the cyclic stress increased from 50 to 150 kPa for UR. For RM, *s/B* decreased to 15.6% at 150 kPa cyclic stress and 3.4% at 50 kPa cyclic stress. For RI, s/*B* decreased to 20.1% at 150 kPa and 4.2%

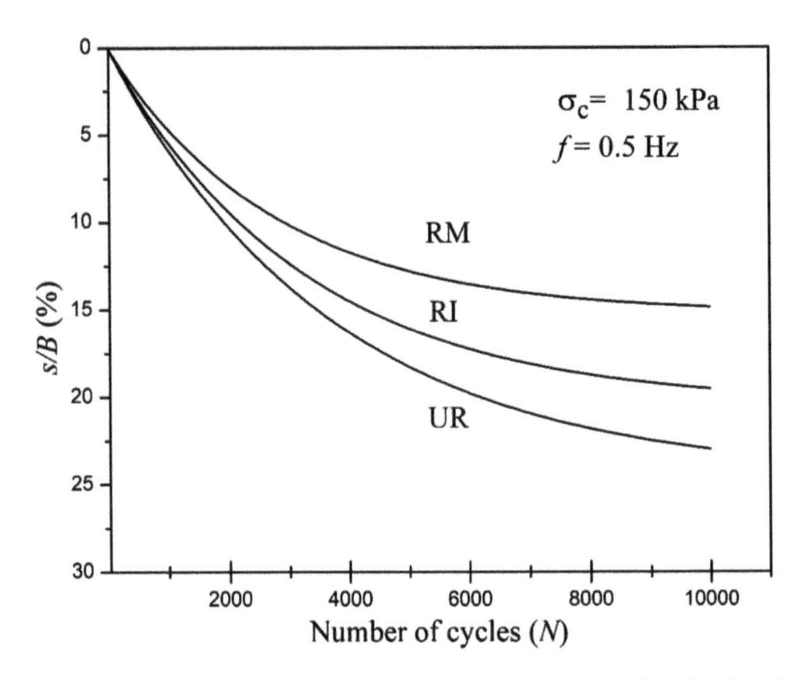

Fig. 7.3 Number of cycles—s/B relationship of unreinforced and reinforced soil with coir geotextiles

at 50 kPa. The maximum performance reduction in settlement was observed when the geotextile is placed in the middle of layer I (RM). The decrease in the footing settlement due to the placement of coir geotextiles at RI and RM for σ_c= 150 kPa was 16.3% and 35%, respectively. Nevertheless, for σ_c= 100 kPa, there is a 27.8 and 37.5% reduction in settlement for RI and RM, respectively. Table 7.2 summarizes the footing settlement under different σ_c for UR, RI and RM.

Figure 7.4 also shows the relationship between *s/B* and *N* for different frequencies (*f*) of cyclic loading. It is evident from Fig. 7.4 that the *s/B* increases with *f* irrespective of the location of coir geotextiles. In fact, for UR *s/B* increases from 14.3% to 48.3% when *f* increases from 0.5 to 1.5 Hz for *N* = 10,000. For RI, *s/B* decreased to 10.4% at 0.5 Hz and 46.2% at 1.5 Hz, and for RM, *s/B* decreased to 9.6% at 0.5 Hz and 33.3% at 1.5 Hz. The optimum performance of coir geotextiles in reducing settlement was observed for RM for the σ_c and *f* considered for this study.

The potential of the geotextile in controlling the settlements of the footing can be clearly seen in Table 7.2. Increasing the cyclic stress increases the footing settlement, while incorporating coir geotextiles at (RI) and in (RM) enhances soil performance and reduces settlement. The footing experienced a 23.6 and 38.2% reduction in settlement for RI and RM under 50 kPa and 27.8% and 37.5%

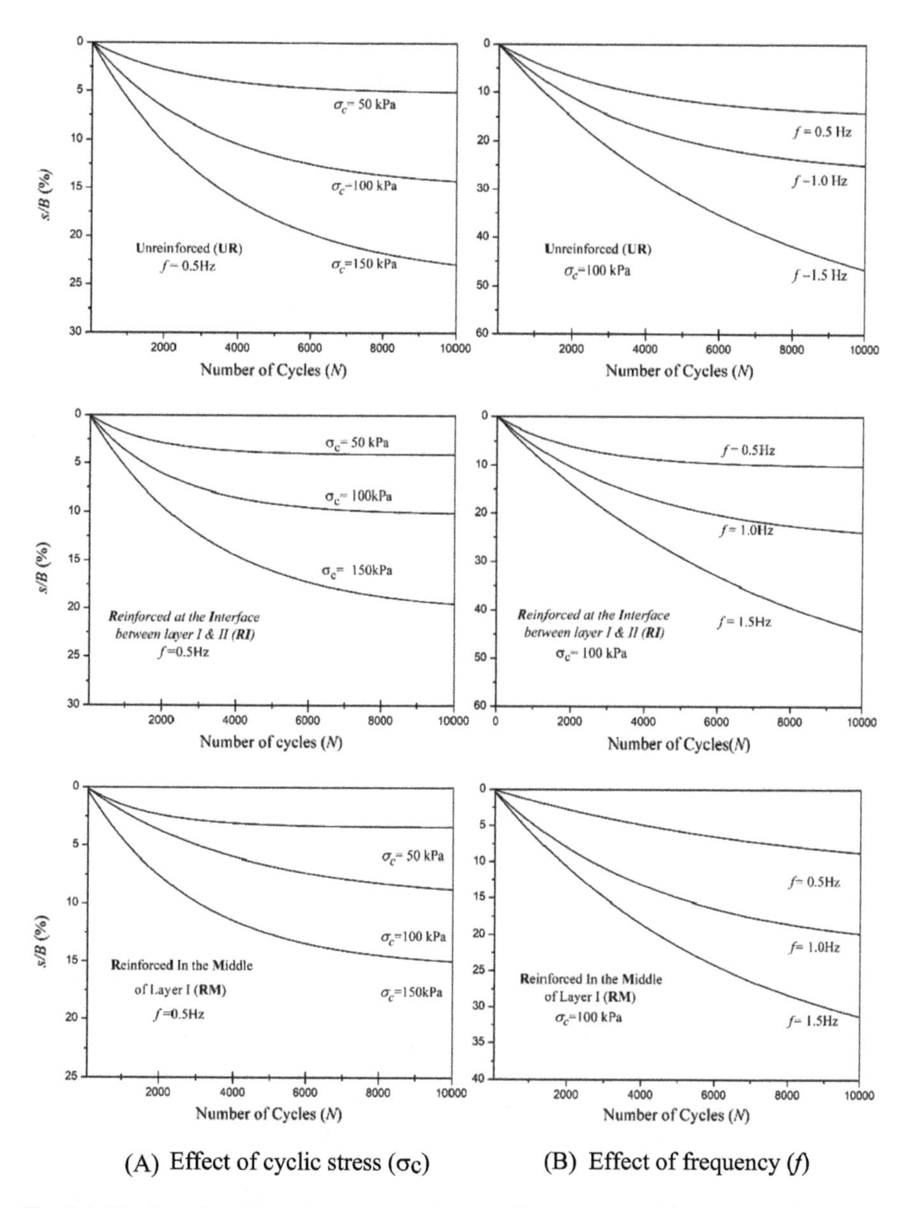

Fig. 7.4 Number of cycles—s/B curve; **a** various cyclic stresses and **b** various frequencies on unreinforced and reinforced soils

respectively, under 100 kPa cyclic stresses. While under 150 kPa cyclic stress, the coir geotextile at RI shows a 16.3% reduction in settlement and a 35% for RM. This shows that placing the coir geotextile in the middle of layer I results in an optimum performance of the geotextile in controlling settlements.

Table 7.2 Effect of cyclic stress on geotextile performance

Test	Amplitude (kPa)	Number of cycles (N)	s/B (%)	% Reduction in settlement (%)
	50		5.5	–
UR	100	10,000	14.4	–
	150		24	–
	50		4.2	23.6
RI	100	10,000	10.4	27.8
	150		20.1	16.3
	50		3.4	38.2
RM	100	10,000	9	37.5
	150		15.6	35

Table 7.3 Effect of frequency on geotextile performance

Test	Frequency (Hz)	Number of cycles (N)	s/B (%)	% Reduction in settlement (%)
UR	0.5		14.3	–
	1	10,000	26.4	–
	1.5		48.3	–
RI	0.5		10.4	27.3
	1	10,000	25.6	3
	1.5		46.2	4.3
RM	0.5		9.6	32.9
	1	10,000	21	20.5
	1.5		33.3	31.1

Table 7.3 summarizes the footing settlement under different f for UR, RI and RM. The ability of the geotextile to reduce the settlement of the footing can clearly be recognized, and the optimum performance occurred when the geotextile is placed in the middle of layer I (RM). The inclusion of a coir geotextile at RI and in RM enhanced the performance of soil and reduced settlement. For $f = 0.5$ Hz, a 27.3% and 32.9% reduction in s/B was observed for RI and RM, respectively. For $f = 1$ Hz, percentage reduction in s/B is 3% and 20.5% and for $f = 1.5$ Hz percentage reduction in s/B is 4.3% and 31.1% for RI and RM, respectively. These reductions in settlement indicate that placing the coir geotextile at RM has shown higher performance in terms of reducing settlement for the σ_c and f considered for this study.

7.3.4 Spatial Distribution of Stresses on Soil and Reinforcement During Cyclic Loading

Figure 7.5 shows the spatial stress distribution during cyclic loading for UR, RM and RI. The peak stress were captured when the assembly reaches a settlement of 40 mm. It is evident from Fig. 7.5 that peak stress for UR is 51.61 kN/m^2. However, for RM and RI the peak stress reaches 91.56 and 89.58 kN/m^2 respectively. The peak stress increases with the inclusion of coir geotextiles. The increase in the peak stress is mainly due to the additional axial tensile forces developed in the coir geotextile and has shown higher due to higher interface friction angle between soil and coir geotextiles.

Figure 7.6 shows the axial tensile force that developed in the geotextiles for RI and RM at N = 10,000 cycles. The maximum axial force for RM is 1.09 kN/m and for RI is 0.24 kN/m. The axial force observed in the coir geotextile for RM is about four times compared to RI. The frictional interaction between the coir geotextile and the soil generates interface shear stress in the coir geotextile. The axial tensile force developed in the coir geotextile is due to that interaction between the soil and the reinforcement during cyclic loading. The maximum axial force for RM generated at the middle of the footing decreases along the width of the footing. However, for RI, the axial force is found to distribute along the width of the footing. Moreover, a small amount of negative axial force is also seen to develop along the

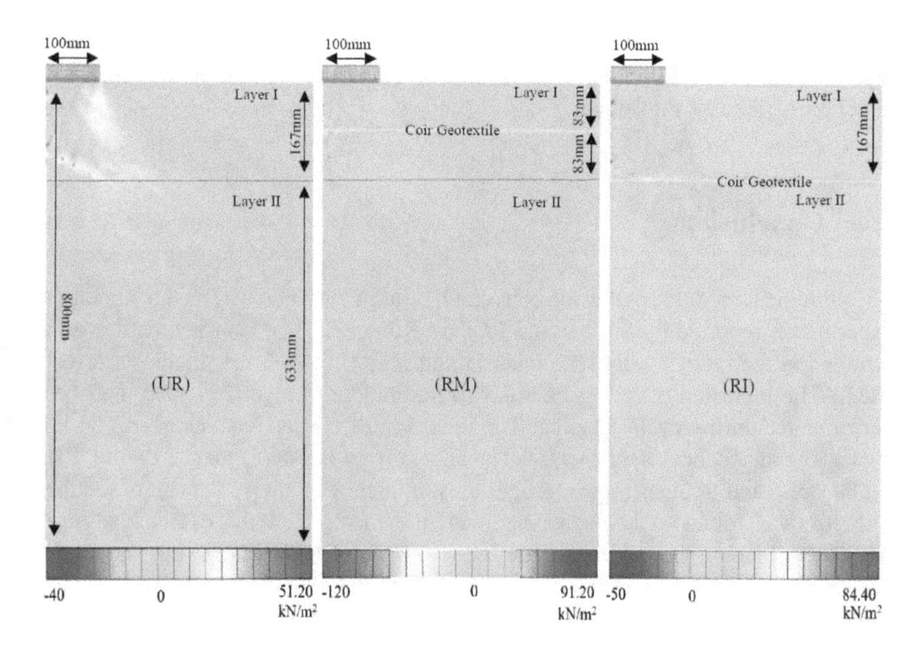

Fig. 7.5 Spatial stress distribution in unreinforced and reinforced soils with coir geotextiles under cyclic loading

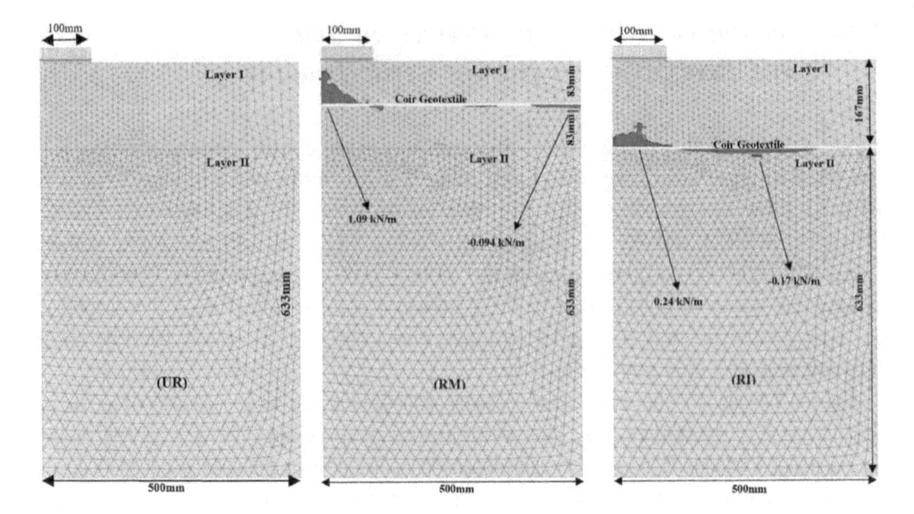

Fig. 7.6 Axial force developed in coir geotextile during cyclic loading

reinforcement for the case of RM. The negative value represents the generated shear force in coir geotextiles and soil in the direction opposite to the soil movement. The interface shear stress is mainly due to frictional interaction between the coir geotextile's surface and soil particles, thus aiding the geotextile to generate axial forces during cyclic loading as presented in Fig. 7.6. This observation is consistent with the experimental study, reported by Nguyen et al. (2013). Nguyen et al. (2013) report on the generation of shear stress in coir geotextile and soil in the transverse direction of the soil movement.

7.4 Conclusions

The numerical model captures the behavior of soil in the same way as the laboratory experiments reported by Sridhar and Prathap Kumar (2018). Coir geotextiles were installed at the interface between layers I and II (RI), and in the middle of layer I (RM). The inclusion of coir geotextiles in the middle of layer I yielded their best performance during cyclic loading. The inclusion of coir geotextiles in soil during cyclic loading reduces settlement and thus improves the performance of soil. The cyclic stress and frequency have a significant influence on the settlement of footing. The *s/B* was found to be increasing with the increase in the cyclic loads and frequency. Placing the coir geotextile in the middle of layer I (RM) has shown the best performance in term of the load carrying capacity of soil and may be due to the development of the interface shear stress in the coir geotextile. The interface friction between the soil and the coir geotextile generates an interface shear stress, and an axial force in the geotextile and consequently increases the load carrying capacity and reducing the settlement.

Acknowledgements The authors gratefully acknowledge the support of the Coir Board (Govt of India), University of Wollongong and Cochin University of Science and Technology (CUSAT), through EIS Partnership grant in carrying out this research. The first author also thanks the support from the Australian Academy of Science for funding Australia-India EMCR fellowship for this research project.

References

Al-Qadi IL, Dessouky SH, Kwon J, Tutumluer E (2008) Geogrid in flexible pavements: validated mechanism. Transp Res Rec 2045(1):102–109. https://doi.org/10.3141/2045-12

Balan K (2017) Coir geotextiles in infrastructure projects. Indian J Geosynthetics Ground Improv 6(2):8–16

Basudhar P, Dixit P, Gharpure A, Deb K (2008) Finite element analysis of geotextile- reinforced sand-bed subjected to strip loading. Geotext Geomembr 26(1):91–99. https://doi.org/10.1016/j.geotexmem.2007.04.002

Bhandari A, Han J (2010) Investigation of geotextile–soil interaction under a cyclic vertical load using the discrete element method. Geotext Geomembr 28(1):33–43. https://doi.org/10.1016/j.geotexmem.2009.09.005

Bowles L (1996) Foundation analysis and design. McGraw-hill, New York, pp 32–90

Brinkgreve R, Kumarswamy S, Swolfs W, Waterman D, Chesaru A, Bonnier P (2014) Plaxis 2014. Plaxis BV, The Netherlands

Brumund WF, Leonards GA (1972) Subsidence of sand due to surface vibration. J Soil Mech Found 98(1):27–42

Bueno BS, Benjamim CVS, Zornberg JG (2005) Field performance of a full-scale retaining wall reinforced with non-woven geotextiles. Slopes and retaining structures under seismic and static conditions. ASCE 1–9. https://doi.org/10.1061/40787(166)1

Chauhan MS, Mittal S, Mohanty B (2008) Performance evaluation of silty sand sub- grade reinforced with fly ash and fibre. Geotext Geomembr 26(5):429–435. https://doi.org/10.1016/j.geotexmem.2008.02.001

Cunny R, Sloan R (1962) Dynamic loading machine and results of preliminary small-scale footing tests. Symp Soil Dyn. https://doi.org/10.1520/STP44378S

Das B, Shin E (1994) Strip foundation on geogrid-reinforced clay: behavior under cyclic loading. Geotext Geomembr 13(10):657–667. https://doi.org/10.1016/0266-1144(94)90066-3

Das B, Sobhan K, Das B (2016) Principles of geotechnical engineering, 8th edn, SI edn. Cengage Learning, Boston

Dutta R, Rao GV (2008) Potential of coir based products as soil reinforcement. Int J Earth Sci Eng 1(2):71–79

Hejazi SM, Sheikhzadeh M, Abtahi SM, Zadhoush A (2012) A simple review of soil reinforcement by using natural and synthetic fibers. Constr Build Mater 30:100–116. https://doi.org/10.1016/j.conbuildmat.2011.11.045

Hufenus R, Rueegger R, Banjac R, Mayor P, Springman SM, Brönnimann R (2006) Full-scale field tests on geosynthetic reinforced unpaved roads on soft subgrade. Geotext Geomembr 24 (1):21–37. https://doi.org/10.1016/j.geotexmem.2005.06.002

Kurian NP, Beena K, Kumar RK (1997) Settlement of reinforced sand in foundations. J Geotech Geoenviron Eng 123(9):818–827. https://doi.org/10.1061/(ASCE)1090-0241(1997)123:9(818)

Lal D, Sankar N, Chandrakaran S (2017) Effect of reinforcement form on the behaviour of coir geotextile reinforced sand beds. Soils Found 57(2):227–236. https://doi.org/10.1016/j.sandf.2016.12.001

Naeini S, Gholampoor N (2014) Cyclic behaviour of dry silty sand reinforced with a geotextile. Geotext Geomembr 24(6):611–619. https://doi.org/10.1016/j.geotexmem.2014.10.003

Nguyen MD, Yang KH, Lee SH, Wu CS, Tsai MH (2013) Behavior of nonwoven-geotextile-reinforced sand and mobilization of reinforcement strain under triaxial compression. Geosynthetics Int 20(3):207–225. https://doi.org/10.1680/gein.13.00012

Noorzad R, Mirmoradi S (2010) Laboratory evaluation of the behavior of a geotextile reinforced clay. Geotext Geomembr 28(4):386–392. https://doi.org/10.1016/j.geotexmem.2009.12.002

Perkins S, Christopher B, Lacina B, Klompmaker J (2011) Mechanistic-empirical modeling of geosynthetic-reinforced unpaved roads. Int J Geomech 12(4):370–380. https://doi.org/10.1061/(ASCE)GM.1943-5622.0000184

Rashidian V, Naeini SA, Mirzakhanlari M (2018) Laboratory testing and numerical modelling on bearing capacity of geotextile-reinforced granular soils. Int J Geotech Eng 12(3):241–251. https://doi.org/10.1080/19386362.2016.1269042

Rawal A, Sayeed M (2013) Mechanical properties and damage analysis of jute/poly- propylene hybrid nonwoven geotextiles. Geotext Geomembr 37:54–60. https://doi.org/10.1016/j.geotexmem.2013.02.003

Raymond GP, Williams DR (1978) Repeated load triaxial tests on dolomite ballast. J Geotech Geoenviron Eng 104(7)

Sarsby RW (2007) Use of 'Limited Life Geotextiles' (LLGs) for basal reinforcement of embankments built on soft clay. Geotext Geomembr 25(4–5):302–310. https://doi.org/10.1016/j.geotexmem.2007.02.010

Schanz T, Vermeer PA, Bonnier PG (1999) The hardening soil model: formulation and verification. In: Brinkgreve RBJ (ed) Beyond 2000 in computation geotechniques. Rotterdam, the Netherlands, pp 281–290

Sreedhar M, Goud APK (2011) Behaviour of geosynthetic reinforced sand bed under cyclic load. In: Proceedings of Indian geotechnical conference, Kochi, India, pp 15–17

Sridhar R, Prathap Kumar MT (2018) Effect of number of layers on coir geotextile reinforced sand under cyclic loading. Geo-Eng 9(1):11. https://doi.org/10.1186/s40703-018-0078-y

Subaida E, Chandrakaran S, Sankar N (2008) Experimental investigations on tensile and pullout behaviour of woven coir geotextiles. Geotext Geomembr 26(5):384–392. https://doi.org/10.1016/j.geotexmem.2008.02.005

Subaida E, Chandrakaran S, Sankar N (2009) Laboratory performance of unpaved roads reinforced with woven coir geotextiles. Geotext Geomembr 27(3):204–210. https://doi.org/10.1016/j.geotexmem.2008.11.009

Unnikrishnan N, Rajagopal K, Krishnaswamy N (2002) Behaviour of reinforced clay under monotonic and cyclic loading. Geotext Geomembr 20(2):117–133. https://doi.org/10.1016/S0266-1144(02)00003-1

Vesic A, Banks D, Woodard J (1965) An experimental study of dynamic bearing capacity of footings on sand. In: Proceedings of VI international conference on soil mechanics and foundation engineerin, Canada, Montreal, (2), pp 209–213

Vinod P, Minu M (2010) Use of coir geotextiles in unpaved road construction. Geosynthetics Int 17(4):220–227. https://doi.org/10.1680/gein.2010.17.4.220

Chapter 8
Assessing the Effect of Aging on Soil Liquefaction Resistance

Ronald D. Andrus and Barnabas Bwambale

8.1 Introduction

The resistance of soil to liquefaction is often expressed by the cyclic resistance ratio (*CRR*) estimated from semi-empirical charts based on field tests, such as the cone penetration test (CPT), standard penetration test (SPT) or shear wave velocity (V_S) measurement (e.g., Youd et al. 2001; Idriss and Boulanger 2008; National Academies of Sciences 2016). Commonly used *CRR* charts are derived from primarily field case histories where liquefaction occurred in soil deposits that are less than a few thousand years old (Youd et al. 2001; Idriss and Boulanger 2008; National Academies of Sciences 2016; Hayati and Andrus 2009; Seed 1979; Bwambale and Andrus 2019). If these charts are applied without correction for the effect of aging, excessively conservative estimates of *CRR* might be obtained, leading to unnecessary and costly ground improvements. On the other hand, if older soil deposits are blindly assumed to be unsusceptible to liquefaction, less conservative assessments of the hazard might be obtained.

Aging (or diagenesis) is the post-depositional physical, chemical and biological processes that alter the structure of soil. As discussed by Boggs (2006), physical processes can include rearrangement and interlocking of soil particles, particles crushing and asperity shearing. Chemical processes can include precipitation of quartz, feldspar, carbonate cements, kaolinite or chlorite, and formation of pyrite or iron oxides. Biological processes include the reworking of sediments by living organisms, bacterial oxidation of organic matter and reduction of inorganic matter, and bacterial fermentation. The combination of these processes contributes to the

R. D. Andrus (✉)
Glenn Department of Civil Engineering, Clemson University, Clemson, SC 29634, USA
e-mail: randrus@clemson.edu

B. Bwambale
ECS Southeast, LLP, Fayetteville, NC 28304, USA
e-mail: bbwamba@clemson.edu

T. G. Sitharam et al. (eds.), *Latest Developments in Geotechnical Earthquake Engineering and Soil Dynamics*, Springer Transactions in Civil and Environmental Engineering, https://doi.org/10.1007/978-981-16-1468-2_8

net effect of diagenesis on liquefaction resistance, which can vary significantly even between locations in the same deposit.

The main objectives of this paper are: (1) to emphasize the importance of not blindly assuming older soil deposits to be unsusceptible to liquefaction by reviewing twelve cases of Holocene (<11.5 k years) liquefaction in Pleistocene (11.5 k to 2.6 M years) deposits; and (2) to review nine proposed relationships for estimating the effect of diagenesis on *CRR*. The 12 cases of Holocene liquefaction in Pleistocene deposits were compiled as part of the doctoral dissertation work of Bwambale (2018) and are published here for the first time. A comprehensive review of the proposed procedures for assessing the aging effect on soil liquefaction resistance is presented in the paper by Bwambale and Andrus (2019). Nine selected relationships for correcting commonly used *CRR* charts for the aging or diagenesis effect are summarized in this paper.

8.2 Holocene Liquefaction in Pleistocene Deposits

Although most cases of earthquake-induced liquefaction described in the literature involve soils deposited during the Holocene, several cases involving Pleistocene deposits have been reported. Summarized in Table 8.1 are 12 such cases. These cases are from Argentina, China, Israel, Lithuania, Republic of Karelia and the USA. The cases presented in Table 8.1 involve liquefaction of mainly Pleistocene alluvial/fluvial, beach and lacustrine sediments that are composed predominantly of sand, but also include some silt and silty sand. Sand boils (or sand blows) were observed at nearly all locations.

For the 12 cases summarized in Table 8.1, the liquefying events occurred from a few years to about 15,000 years ago. The time difference between the inferred geologic age and any documented liquefying event is, however, greater than 10,000 years, implying old deposits at the time of liquefaction. Some areas (e.g., South Carolina) have experienced liquefaction during multiple events, as indicated in Table 8.1.

Figure 8.1 presents a map of the world showing the geographical locations of the 12 cases of Holocene liquefaction in Pleistocene deposits summarized in Table 8.1. Half of the cases are from the USA. The other six cases are evenly distributed in the continents of Asia, Europe and South America.

Figure 8.2a, b show histograms of the case histories grouped according to deposit type and geologic age (i.e., time since deposition), respectively. As seen in the figures, 54% of the cases occurred in alluvial/fluvial deposits and 23% in each of the two other deposit types. About 70% of the cases involved deposits that are <100,000 years old; and 30% with age between 200,000 to 500,000 years old at the time of liquefaction. This observation suggests that liquefaction susceptibility varies significantly within Pleistocene deposits, often decreasing as the time since deposition increases. The 12 case histories summarized in Table 8.1 support the need for liquefaction assessments in Pleistocene deposits.

Table 8.1 Field case histories of Holocene liquefaction in Pleistocene deposits

Area	Deposit type	Geologic age (years)	Evidence of liquefaction	Year/time of liquefaction (years)	References
Charleston, South Carolina, USA	Beach, fluvial	33 k to >1,000 k	Sand blows, lateral spreading	1886 Charleston earthquake; 5 earlier events	Dutton (1889), Obermeier et al. (1985), Talwani and Cox (1985), Martin and Clough (1994), Lewis et al. (1999), Crone and Wheeler (2000), Talwani and Schaeffer (2001), Hu et al. (2002), Hayati and Andrus (2008), Heidari and Andrus (2012), Hasek and Gassman (2014)
Thousand Springs Valley, Idaho, USA	Alluvium fan low-energy stream channel fill	Probable 10–15 k	Lateral spread with fissures, buckled sod and sand boils	1983 Borah Peak earthquake	Andrus and Youd (1987), Andrus (1994), Andrus et al. (2004a)
Bluffton, South Carolina, USA	Beach?	Pleistocene	Sand blows	1.96 k	Crone and Wheeler (2000), Talwani and Schaeffer (2001), Obermeier et al. (1987)
Georgetown, South Carolina, USA	Beach	~450 k	Sand blows	1.64 and 5.04 k	Crone and Wheeler (2000), Talwani and Schaeffer (2001), Hu et al. (2002), Obermeier et al. (1987)
Marianna, Arkansas, USA	Fluvial	20 k	Sand blows and dike	5.5 k	Blum et al. (2000), Tuttle et al. (2006)
Mendoza and San Juan Provinces, Argentina	Fluvial, lacustrine and fluvio-lacustrine	Pleistocene to Holocene	Sand boils, sand dikes, cracks and fissures	1861–1997 earthquakes	Perucca and Moreiras (2006)

(continued)

Table 8.1 (continued)

Area	Deposit type	Geologic age (years)	Evidence of liquefaction	Year/time of liquefaction (years)	References
Dead Sea Basin, Israel	Lacustrine	15–70 k	Fluidization and injection of clastic dike	7–15 k	Porat et al. (2007), Jacoby et al. (2015)
Lake, Ladoga, Republic of Karelia	Alluvial and lacustrine	Pleistocene and Holocene	Diapir-like injections and dykes; flame and balls-and-pillows structures and breccia	Late Holocene	Biske et al. (2009)
Eastern Baltic Sea, Lithuania	Glacio-fluvial?	Middle Pleistocene	Sediment deformations	<13 k	Bitinas and Lazauskiene (2011)
Dandridge, Tennessee, USA	Fluvial terrace	203 k $\pm$ 13 k	Fluidization-filled fractures and dikes	<15 k	Hatcher et al. (2012), Hatcher (2015)
Xinglong Village, China	Alluvial	$\geq$ 12 k	Sand boiling, surface cracking	2008 Wenchuan earthquake	Li et al. (2013), Liu-Zeng et al. (2017)

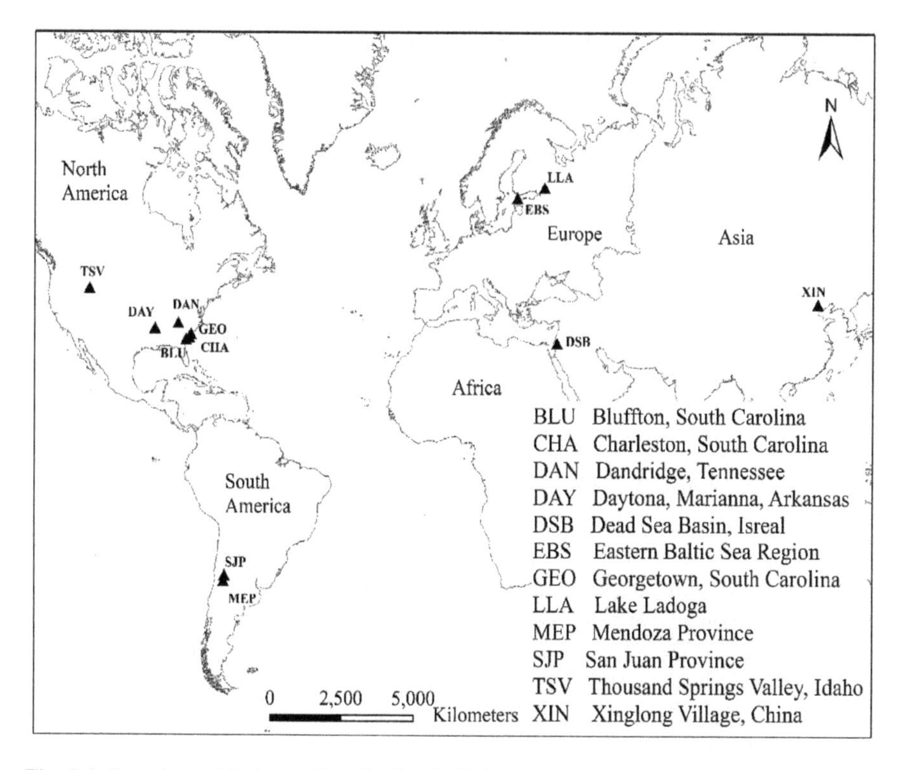

Fig. 8.1 Locations of Holocene liquefaction in Pleistocene deposits

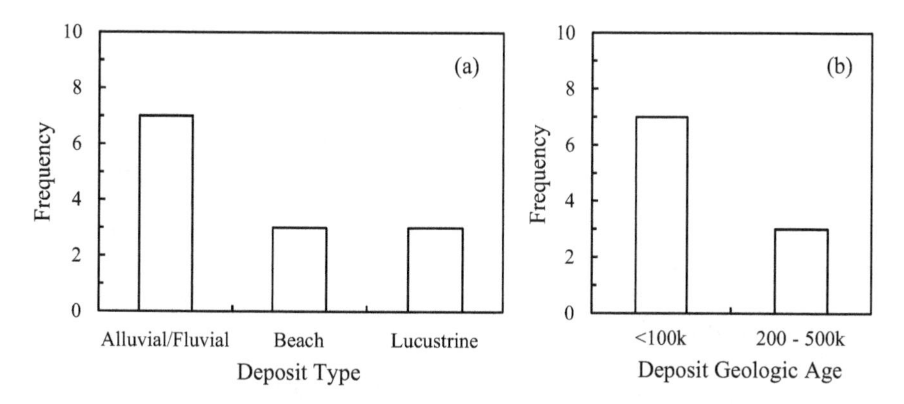

Fig. 8.2 Frequency of cases of Holocene liquefaction in Pleistocene deposits grouped by **a** deposit type and **b** deposit geologic age

8.3 Correcting *CRR* for Diagenesis

The correction of *CRR* for aging or diagenesis can be given by (Hayati and Andrus 2009; Seed 1979; Bwambale and Andrus 2019; Hayati et al. 2008; Arango et al. 2000):

$$CRR_{\text{corrected}} = K_{DR} CRR \tag{8.1}$$

where $CRR_{\text{corrected}}$ is the diagenesis-corrected *CRR*; and K_{DR} is the correction factor. Nine proposed relationships for estimating K_{DR} based on time and a ratio of measured to estimated shear wave velocity are reviewed below.

8.3.1 *Time-*K_{DR} *Relationships*

Because some processes (e.g., liquefaction during strong ground shaking; excavation and backfilling during construction of underground utilities) can cause the grain-to-grain contacts to be broken after deposition, 'time' in this section is defined as the period since the grain-to-grain contacts last formed. This definition of time is sometimes called the 'geotechnical age.' The geotechnical age can be less than the geologic age if an event causes the grain-to-grain contacts to be broken.

Figure 8.3 presents three proposed time-K_{DR} relationships that are primarily based on laboratory cyclic testing of intact and freshly deposited (or reconstituted) specimens composed of predominately silica-based sands. K_{DR} in Fig. 8.3 is defined as the *CRR* of the intact specimen divided by the *CRR* of the reconstituted specimen. The relationship by Seed (1979) is based on test results for five sands. The relationship by Arango et al. (2000) is based on the Seed (1979) relationship and test results for two sands in California and South Carolina. The relationship by Hayati and Andrus (2009) is based on test results for 13 sands in Japan, Taiwan, and the USA. All three relationships suggest a reference age (i.e., time when $K_{DR} = 1.0$) of <4 days, which is reasonable given that reconstituted laboratory test specimens are typically subjected to back-pressure saturation and consolidation over a period of a few days or less prior to cyclic testing.

Figure 8.4 presents four proposed time-K_{DR} relationships that use a penetration-based *CRR* chart as reference. K_{DR} in Fig. 8.4 is defined as the *CRR* of the intact material divided by the *CRR* from the chart for the given corrected field penetration resistance. The relationship by Hayati and Andrus (2009) is a refinement of the relationship by Hayati et al. (2008) and is based on 24 data points from sites in Canada, Japan, Taiwan and the USA. The relationship by Maurer et al. (2014) is based on the data compiled by Hayati and Andrus (2009) plus data from an area near Christchurch, New Zealand shaken by earthquakes in 2010 and 2011. The relationship by Towhata et al. (2017) is based on field data from Japan, as well as previous studies. The four relationships shown in Fig. 8.4 suggest a reference age

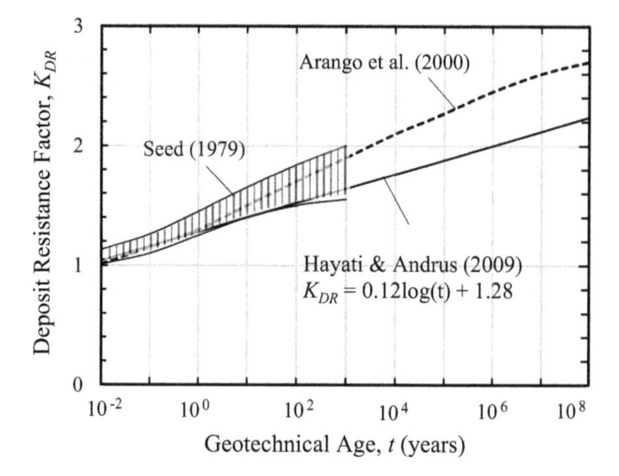

Fig. 8.3 Comparison of three time-K_{DR} relationships based on laboratory testing of silica-based sands (modified from Bwambale and Andrus 2019). K_{DR} is defined as *CRR* of the intact specimen divided by *CRR* of the reconstituted specimen

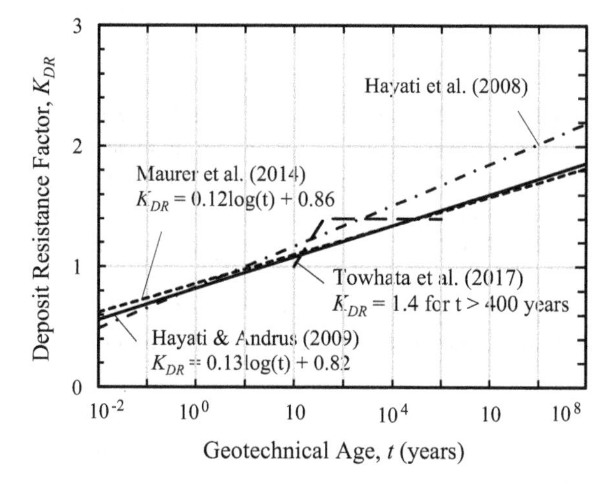

Fig. 8.4 Comparison of four time-K_{DR} relationships based on laboratory and field testing of silica-based sands (modified from Bwambale and Andrus 2019). K_{DR} is defined as *CRR* of the intact material divided by *CRR* from the chart based on penetration resistance

(i.e., time when $K_{DR} = 1.0$) of 10 to 100 years, which is reasonable given that many case histories used to develop the *CRR* charts are from post-earthquake investigations conducted 1 to 100 years after the liquefying events.

It is encouraging to observe the relationships shown in Figs. 8.3 and 8.4 which indicate similar rates of increase in K_{DR} (12–17%) per log cycle of time. The good agreement is likely because the regression data are from predominately silica-based sands. On the other hand, it should be noted that the results of tailings material reported by Troncoso et al. (1988) were omitted from the regression for the Hayati and Andrus (2009) relationship in Fig. 8.3 because the tailings data exhibit very high values of K_{DR}. Also, K_{DR} values obtained by Bwambale and Andrus (2017) for Pleistocene loess-colluvium near Christchurch, New Zealand plot well above the four relationships in Fig. 8.4. For these reasons, Bwambale and Andrus (2019) do not recommend using relationships based on time unless the rate of increase in resistance in a given deposit is well established.

8.3.2 MEVR-K_{DR} *Relationships*

The ratio of measured shear wave velocity to estimated shear wave velocity (*MEVR*) is a promising predictor variable for K_{DR}. Andrus et al. (2009) recommended using the following relationships for the estimated shear wave velocity (Andrus et al. 2004b, 2009):

$$V_{S1cs,E} = 62.6\left(q_{c1N,cs}\right)^{0.231} \tag{8.2}$$

$$V_{S1cs,E} = 87.6\left[(N_1)_{60,cs}\right]^{0.253} \tag{8.3}$$

where $V_{S1cs,E}$ is the estimated shear wave velocity corrected to a reference stress equal to 100 kPa and a clean-sand equivalent; $q_{c1N,cs}$ is the dimensionless overburden stress-corrected clean-sand equivalent cone tip resistance; and $(N_1)_{60,cs}$ is the dimensionless overburden stress-corrected clean-sand equivalent SPT blow count. Equations 8.2 and 8.3 correspond to deposits with average age of about 6 years (Andrus et al. 2009). In this section, *MEVR* is defined as the measured shear wave velocity ($V_{S1cs,M}$) divided by $V_{S1cs,E}$.

Figure 8.5 presents two proposed *MEVR*-K_{DR} relationships. The relationship by Bwambale and Andrus (2019) is an update to the relationship by Hayati and Andrus

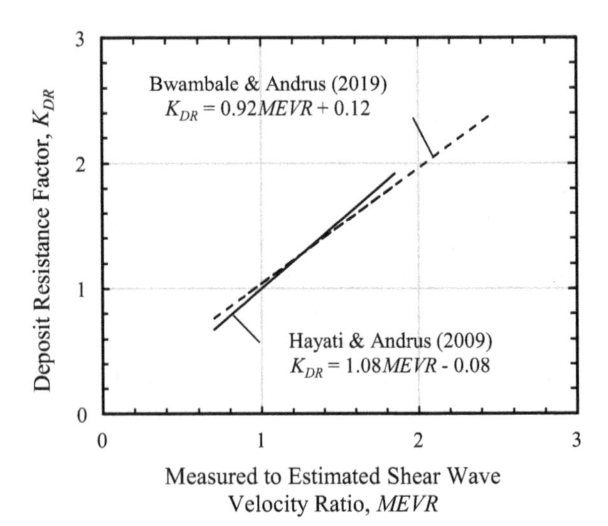

Fig. 8.5 Comparison of two *MEVR*-K_{DR} relationships based on laboratory and field testing, and ground behavior observations. *CRR* from the chart based on penetration resistance is used as reference in the calculation of K_{DR}

(2009). The update involved a critical re-evaluation of the 17 data points compiled by Hayati and Andrus (2009) and the addition of 11 new cases. Nine of the original 17 data points were excluded in the update because of insufficient information, a deficiency in testing procedures and/or a plasticity index too high to be liquefiable. The 20 data points used in deriving the Bwambale and Andrus (2019) relationship are from Canada, Japan, New Zealand, and the USA. *MEVR*-K_{DR} relationships are recommended, instead of time-K_{DR} relationships, because they provide higher coefficients of determination and lower root mean square errors.

8.4 Conclusions

Twelve cases of Holocene liquefaction in Pleistocene deposits were summarized in this paper. The 12 cases are from Argentina, China, Israel, Lithuania, Republic of Karelia and the USA. About 70% of the cases involved deposits that are less than 100,000 years old and 30% with geologic age between 200,000 to 500,000 years old at the time of liquefaction. These cases support the need for liquefaction assessments in Pleistocene deposits.

Nine proposed relationships for correcting the effect of age or diagenesis on *CRR* were reviewed. Although seven time-K_{DR} relationships exhibit similar rates of increase in K_{DR} (12–17%) per log cycle of time for predominately silica-based sands, test results for tailings material and loess-colluvium indicate that much greater rates of increase with time are possible. For this reason, relationships based on measured shear wave velocity to estimated shear wave velocity were recommended for correcting *CRR*.

References

Andrus RD (1994) In situ characterization of gravelly soils that liquefied in the 1983 Borah Peak Earthquake. PhD dissertation, University of Texas at Austin, Austin, TX, pp 246–290

Andrus RD, Youd TL (1987) Subsurface investigation of a liquefaction-induced lateral spread Thousand Springs Valley, Idaho. US Army Corps of Engineers Misc. Paper GL-87-8, Washington, DC

Andrus RD, Stokoe KH, Juang CH (2004a) Guide for shear-wave-based liquefaction potential evaluation. Earthq Spectra 20:285–308

Andrus RD, Piratheepan P, Ellis BS, Zhang J, Juang CH (2004b) Comparing liquefaction evaluation methods using penetration-V_S relationships. Soil Dyn Earthq Eng 24(9–10): 713–721

Andrus RD, Hayati H, Mohanan NP (2009) Correcting liquefaction resistance of aged sands using measured to estimated velocity ratio. J Geotech Geoenviron Eng 135(6):735–744

Arango I, Lewis MR, Kramer C (2000) Updated liquefaction potential analysis eliminates foundation retrofitting of two critical structures. Soil Dyn Earthq Eng 20(1–4):17–25

Biske YS, Sumareva IV, Sheetov MV (2009) Late Holocene paleoseismic event at southeastern coast of Lake Ladoga. I. Principles of research and deformation structures [Russian Source]. Vestnik Sankt-Peterburgskogo Universiteta, Seriya Geologiya i Geografiya 3-25 + 146 + 151

Bitinas A, Lazauskiene J (2011) Implications of the palaeoseismic events based on the analysis of the structures of the Quaternary deposits [Paleoseisminiu ivykiu prielaidos kvartero nuosedu teksturu tyrimo duomenimis]. Baltica 24:127–130

Blum MD, Guccione MJ, Wysocki DA, Robnett PC, Rutledge EM (2000) Late Pleistocene evolution of the lower Mississippi River Valley, southern Missouri to Arkansas. Geol Soc Am Bull 112:221–235

Boggs S Jr (2006) Principles of sedimentary stratigraphy, 4th edn. Pearson Prentice Hall, Upper Saddle River, NJ

Bwambale B (2018) Reducing uncertainty in the assessment of aging effects on soil liquefaction. PhD dissertation, Clemson University, Clemson

Bwambale B, Andrus RD (2017) Liquefaction resistance of pleistocene loess-colluvium deposits near Christchurch, New Zealand. In: 3rd international conference proceedings on performance-based design in earthquake geotechnical engineering (PBD-III), Vancouver, Canada, Paper No. 206

Bwambale B, Andrus RD (2019) State of the art in the assessment of aging effects on soil liquefaction. Soil Dyn Earthq Eng 125:

Crone AJ, Wheeler RL (2000) Data for quaternary faults, liquefaction features, and possible tectonic features in the Central and Eastern United States, east of the Rocky Mountain Front. United States Geological Survey, Denver, CO

Dutton CE (1889) The Charleston earthquake of August 31, 1886. Ninth annual report of the U.S. geological survey 1887–1888, Washington, DC, pp 203–528

Hasek MJ, Gassman SL (2014) Characterization of aged Coastal Plain soils at Hollywood, SC using petrography and microscopy. In: 2014 Geo-congress technical papers, Geo-characterization and Modeling for Sustainability 2088–2097

Hatcher RD Jr (2015) Written communication on the Dandridge liquefaction features to Andrus, R. D 2015/3/13

Hatcher RD Jr, Vaughn JD, Obermeier SF (2012) Large earthquake paleoseismology in the East Tennessee seismic zone: results of an 18-month pilot study. Spec Paper Geol Soc Am 493:111–142

Hayati H, Andrus RD (2008) Liquefaction potential map of Charleston, South Carolina based on the 1886 earthquake. J Geotech Geoenviron Eng 134(6):815–828

Hayati H, Andrus RD (2009) Updated liquefaction resistance correction factors for aged sands. J Geotech Geoenviron Eng 135(11):1683–1692

Hayati H, Andrus RD, Gassman SL, Hasek M, Camp WM, Talwani P (2008) Characterizing the liquefaction resistance of aged soils. Geotech Earthq Eng Soil Dyn IV, GSP 181:1–10

Heidari T, Andrus RD (2012) Liquefaction potential assessment of Pleistocene beach sands near Charleston, South Carolina. J Geotechn Geoenviron Eng 138(10):1196–1208

Hu K, Gassman SL, Talwani P (2002) In-situ properties of soils at paleoliquefaction sites in the South Carolina Coastal Plain. Seismol Res Lett 73:964–978

Idriss IM, Boulanger RW (2008) Soil liquefaction during earthquakes. Earthquake Engineering Research Institute Publication MNO-12, Oakland, CA

Jacoby Y, Weinberger R, Levi T, Marco S (2015) Clastic dikes in the Dead Sea basin as indicators of local site amplification. Nat Hazards 75:1649–1676

Lewis MR, Arango I, Kimball JK, Ross TE (1999) Liquefaction resistance of old sand deposits. In: 11th Pan American conference proceedings on soil mechanics and geotechnical engineering. Foz do Iguassu, Brazil, pp 821–829

Li LH, Fan LF, Deng XL, Hu RL, Zhang J, Wei X (2013) Engineering geological characteristics of sand liquefaction in Wenchuan earthquake. In: Wu, Qi (eds) Global view of engineering geology and the environment. Taylor & Francis Group, London, pp 733–738

Liu-Zeng J, Wang P, Zhang Z, Li Z, Cao Z, Zhang J, Xiaoming Y, Wang W, Xing X (2017) Liquefaction in western Sichuan Basin during the 2008 Mw 7.9 Wenchuan earthquake, China. Tectonophysics 694:214–238

Martin JR, Clough GW (1994) Seismic parameters from liquefaction evidence. J Geotech Eng 120:1345–1361

Maurer BW, Green R, Cubrinovski M, Bradley BA (2014) Assessment of aging correction factors for liquefaction resistance at sites of recurrent liquefaction. In: 10th national conference proceedings on earthquake engineering. Earthquake Engineering Research Institute, Anchorage, AK

National Academies of Sciences (2016) Engineering and medicine: state of the art and practice in the assessment of earthquake-induced soil liquefaction and its consequences. The National Academies Press, Washington, DC

Obermeier SF, Gohn GS, Weems RE, Gelinas RL, Rubin M (1985) Geologic evidence for recurrent moderate to large earthquakes near Charleston, South Carolina. Science, New Series 227:408–411

Obermeier SF, Weems RE, Jacobson RB (1987) Earthquake-induced liquefaction features in the Coastal South Carolina Region. In: Symposium on seismic hazards proceedings, ground motions, soil-liquefaction and engineering practice in Eastern North America, Sterling Forest Conference Center, Tuxedo, New York, pp 480–493

Perucca LP, Moreiras SM (2006) Liquefaction phenomena associated with historical earthquakes in San Juan and Mendoza Provinces, Argentina. Quat Int 158:96–109

Porat N, Levi T, Weinberger R (2007) Possible resetting of quartz OSL signals during earthquakes—evidence from late Pleistocene injection dikes, Dead Sea basin, Israel. Quat Geochronol 2:272–277

Seed HB (1979) Soil liquefaction and cyclic mobility evaluation for level ground during earthquakes. J Geotechn Eng Div 105(2):201–255

Talwani P, Cox J (1985) Evidence for recurrence of earthquakes near Charleston, South Carolina. Science, New Series 229:379–381

Talwani P, Schaeffer WT (2001) Recurrence rates of large earthquakes in the South Carolina Coastal Plain based on paleo-liquefaction data. J Geophys Res 106:6621–6642

Towhata I, Taguchi Y, Hayashida T, Goto S, Shintaku Y, Hamada Y, Aoyama S (2017) Liquefaction perspective of soil ageing. Géotechnique 67(6):467–478

Troncoso J, Ishihara K, Verdugo R (1988) Aging effects on cyclic shear strength of tailings materials. In: 9th world conference proceedings on earthquake engineering, vol III, Tokyo-Kyoto, Japan, pp 121–121

Tuttle MP, Al-Shukri H, Mahdi H (2006) Very large earthquakes centered southwest of the New Madrid Seismic Zone 5,000–7,000 years ago. Seismol Res Lett 77:755–770

Youd TL, Idriss IM, Andrus RD, Arango I, Castro G, Christian JT, Dobry R, Finn WDL, Harder LF Jr, Hynes ME, Ishihara K, Koester JP, Liao SSC, Marcusion WF III, Martin GR, Mitchell JK, Moriwaki Y, Power MS, Robertson PK, Seed RB, Stokoe KH (2001) II: Liquefaction resistance of soils: summary report from the 1996 NCEER and 1998 NCEER/NSF workshops on evaluation of liquefaction resistance of soils. J Geotechn Geoenviron Eng 127(10):817–833

Chapter 9
Uncertainties in Small-Strain Damping Ratio Evaluation and Their Influence on Seismic Ground Response Analyses

Sebastiano Foti, Mauro Aimar, and Andrea Ciancimino

9.1 Introduction

A proper evaluation of site effects is crucial to define the expected ground motion at the surface. Local site conditions modify the shaking characteristics due to variations of mechanical properties and basin/surface geometry. Site effects are therefore referred, respectively, as stratigraphic and geometrical amplification (e.g., Seed and Idriss 1982; Aki 1993; Kramer 1996).

Site response studies are usually performed to quantify the differences between the surface ground motion and the reference condition (i.e., flat rock-outcropping formation) where no amplification phenomena are expected. Different methods can be used to evaluate site effects: studies based on recorded ground motions (i.e., data-based approach) or numerical simulations (i.e., simulation-based approach). In the absence of a sufficient number of available records at the site, the latter represents the only feasible option (Olsen 2000; Rodriguez-Marek et al. 2014; Faccioli et al. 2015).

The complex phenomena affecting the seismic wave propagation can be represented through one-, two- and three-dimensional site response analyses, according to the specific features of the site. One-dimensional ground response analyses (hereafter, GRAs) are based on the assumption of vertical propagation of shear waves through a horizontally stratified medium. The applicability of this assumption is constrained by the geometry of the site: When no major basin or topographic effects are expected, GRAs are considered to be adequate to model the site response (Kramer 1996; Stewart and Kwok 2008). Several studies have addressed the actual capabilities of 1-D approaches in predicting the mean site response and, notwithstanding the well-known limitations, GRAs are still the primary choice for the

S. Foti (✉) · M. Aimar · A. Ciancimino
Politecnico di Torino, Turin, Italy
e-mail: sebastiano.foti@polito.it

T. G. Sitharam et al. (eds.), *Latest Developments in Geotechnical Earthquake Engineering and Soil Dynamics*, Springer Transactions in Civil and Environmental Engineering, https://doi.org/10.1007/978-981-16-1468-2_9

assessment of site effects (e.g., Stewart and Kwok 2008; Baturay and Stewart 2003; Assimaki et al. 2008a; Kwok et al. 2008; Li and Asimaki 2010; Asimaki and Li 2012; Kaklamanos et al. 2013, 2015; Stewart et al. 2014).

Leaving aside the dimensionality of the problem, an adequate simulation of the propagation of seismic waves should not disregard the actual stress–strain response of soils under cyclic loading. The reference parameters (termed as dynamic properties) adopted to describe the behavior of soils are usually the secant shear modulus (G_S) and the material damping ratio (D). The latter represents the energy internally dissipated by the soil as a consequence of friction between soil particles, nonlinear soil behavior and viscous effects. At very small strains, the soil response is practically linear and G_S assumes its maximum value G_0. The energy dissipation, given in this range mainly by friction and viscosity, is almost constant and equal to the small-strain material damping ratio (D_0). For larger shear strains, nonlinearity in the stress–strain soil behavior leads to a G_S decay and, consequently, to an increase of the energy dissipation. The relationships between G_S/G_0 and D along with the cyclic shear strain amplitude (γ_c) are usually termed as modulus reduction and damping (MRD) curves (Seed and Idriss 1970).

The evaluation of the dynamic soil properties is carried out through geotechnical laboratory tests together with geophysical in situ tests. The reliability of laboratory measurements is in fact constrained by sample disturbance effects, which alter the structure of the soil thus affecting the shear wave velocity V_S (and, then, the G_S) of the sample (Anderson and Woods 1975; Stokoe and Santamarina 2000). Therefore, the current state of practice is to evaluate the MRD curves in laboratory and adopt the V_S profile from specific in situ tests.

Different uncertainties and variabilities affect the results of GRAs due to both the approach adopted to model the complex nonlinear and inelastic response of soil and the selected model parameters (Foti et al. 2019). As a consequence, numerical simulations should be carried out within a probabilistic framework to identify, quantify and manage (i.e., IQM method Passeri 2019) all the uncertainties and variabilities involved in the analyses.

In the following, the main sources of uncertainties in GRAs are firstly analyzed to define the framework in which the uncertainties on D_0 are placed. The critical issues associated with the measurement of the small-strain damping are then identified to explain the differences observed between laboratory and in situ values. A review of approaches to evaluate D_0 from field data is then reported. Finally, two different applications are presented to highlight the influence of D_0 on the outcomes of GRAs. Firstly, a stochastic database of GRAs is used to analyze the average response over a wide range of soil profiles. Subsequently, a specific well-documented case study is considered to compare the impact of D_0 variability to the effects of the uncertainties on V_S profile and MRD curves.

9.2 Sources of Uncertainties in GRAs

Six factors can be identified as main sources of uncertainties in GRAs (after Idriss 2004; Rathje et al. 2010):

- Shear wave velocity profile;
- nonlinear approach;
- modulus reduction and damping curves;
- input motions;
- shear strength;
- small-strain damping ratio.

The V_S profile is the main parameter governing the wave propagation in the medium. It controls resonant frequencies and modifications of the motion at the interface. The V_S profile has to be based on adequate in situ geophysical measurements, and the specific uncertainties associated to the test typology have to be carefully evaluated (further details can be found in Foti et al. 2019 and Passeri 2019).

Different approaches can be used to model soil nonlinearity. A rigorous analysis should be based on a fully nonlinear (NL) approach, which allows the assessment of the actual stress–strain behavior in the time domain. However, frequency-domain equivalent linear (EQL) analyses are commonly used to approximate nonlinearity through an iterative approach based on the use of strain-compatible linear viscoelastic soil properties (Schnabel and Seed 1972). Finally, simplified linear viscoelastic (LE) analyses can be used to validate the model. The choice of an appropriate nonlinear approach depends on the expected shear strain level and/or on the possible development of excess pore water pressure due to the coupling of shear and volumetric strains. The EQL procedure has the advantage to be stable and straightforward, but its applicability is constrained when soil layers undergo excessive shear strains (e.g., Kaklamanos et al. 2013; Matasovic and Hashash 2012). Moreover, the time-independent assumption of the strain-compatible properties according to a predefined shear strain ratio is a further source of uncertainties (Kim et al. 2016). On the other hand, NL analyses are more rigorous and of general applicability. Their practical use is anyway limited by the complexity of the approach and the consequent need of expert users (e.g., Stewart and Kwok 2008; Kwok et al. 2008; Kaklamanos et al. 2015; Hashash et al. 2010). The choice of an appropriate constitutive model and the parameter calibration is, indeed, crucial to obtain reliable results (Régnier et al. 2018).

The MRD curves should be defined through specific laboratory tests. In absence of site-specific results, empirical models (e.g., Hardin and Drnevich 1972; Kokusho et al. 1982; Seed et al. 1986; Vucetic and Dobry 1991; Ishibashi and Zhang 1993; Darendeli 2001a; Menq 2003; Zhang et al. 2005; Senetakis et al. 2013; Vardanega and Bolton 2013; Ciancimino et al. 2019) can be used to predict the soil behavior as a function of different variables (e.g., soil type, plasticity index, mean confining pressure, overconsolidation ratio, loading frequency). The uncertainties on the empirical models related to the experimental variability of MRD curves and possible experimental errors can be quantified through the standard deviation provided along with the mean values (e.g., Darendeli 2001a; Zhang et al. 2005; Ciancimino et al. 2019; Akeju et al. 2017). Conversely, when laboratory tests are carried out, the main uncertainties are related to the experimental limitations and the natural randomness of the soil properties at the site scale, associated with the geological spatial variation (Park and Hashash 2005).

When very large strains occur, the MRD curves typically obtained with laboratory tests have to be corrected to cover the failure conditions (e.g., Yee et al. 2013; Zalachoris and Rathje 2015; Shi and Asimaki 2017). The main uncertainties in this regard are related to the randomness of the soil properties and to the specific tests performed to obtain the strength of the soil. Another source of uncertainties is related to the procedure adopted to merge the small- and large-strain behavior (Li and Asimaki 2010).

Real recorded ground motions are usually selected as input motions. The reference hazard condition is usually obtained with the probabilistic seismic hazard analysis (PSHA) (Cornell 1968), taking into account the source and path spatial variabilities through the spectral standard deviation. On the other side, the uncertainties related to the selection procedure are the choice of the hazard level, the type of reference spectrum, the spectral matching criterion and the type and number of inputs, along with the consistency with the reference condition (Stewart et al. 2014; Rathje et al. 2010; Passeri et al. 2018a).

Finally, the uncertainties on the soil small-strain damping ratio D_0 have to be considered. The D_0 obtained by laboratory tests is associated with material energy dissipation. Therefore, the main uncertainties related to D_0 are the same as previously specified for the MRD curves. However, the applicability of D_0 values obtained through laboratory tests for GRAs has been questioned by different authors (e.g., Stewart et al. 2014; Zalachoris and Rathje 2015; Thompson et al. 2012; Xu et al. 2019). Experimental evidence from back-analysis of downhole seismic arrays showed, in fact, small-strain damping ratios in the field larger than the values obtained through laboratory tests (note that the small-strain damping ratio in field is hereafter referred as $D_{0,\ site}$, while D_0 is adopted for the material small-strain damping ratio measured in the laboratory). These differences have to be interpreted taking into account the energy dissipation mechanisms acting at the site scale. Wave scattering effects can modify the propagating seismic waves due to heterogeneities in the soil profile (Thompson et al. 2009; Field and Jacob 1993). This phenomenon, which is relevant especially in the presence of large impedance contrasts (Zalachoris and Rathje 2015), causes additional energy dissipation to material dissipation and cannot be captured by laboratory tests. As a consequence, $D_{0,\ site}$ should be adopted as small-strain damping when GRAs are performed. When no measurements of $D_{0,\ site}$ are available, a procedure has to be adopted to correct D_0 according to the expected values on site.

Although the uncertainties related to D_0 are usually referred as secondary (Idriss 2004; Rathje et al. 2010; Cabas and Rodriguez-Marek 2018), the choice of adequate values can strongly influence the soil response, especially in the small-strain field (e.g., Thompson et al. 2012; Tao and Rathje 2019). For instance, Boaga et al. (Boaga et al. 2015) observed that D_0 affects the 1D amplification in presence of strong impedance contrasts and its effect is more relevant at high frequencies, whereas its impact is smaller in soil deposits with smooth variations of the mechanical properties. Indeed, for increasing impedance ratio, the 1D ground model exhibits a response closer to the theoretical case of homogeneous medium over a rigid bedrock, where the entity of the ground motion amplification is inversely proportional to D_0. Afshari and Stewart (Afshari and Stewart 2019) tested the effect of various approaches to estimate $D_{0,\ site}$, based on seismological relations and the site decay parameter (κ_0), respectively. The assessment compared the observed response for low-intensity ground motions and the predicted one. They stated that the κ_0-informed $D_{0,\ site}$ provides a better fit between predicted and observed amplification than alternative damping models. On the other side, there is no consensus about the best approach for its estimate and the proposed methods rely on data and resources that are often not available in common engineering applications. This difficulty has been highlighted by Stewart et al. (2014), who suggested to deal the discrepancy between the $D_{0,\ site}$ and the laboratory-based D_0 as an epistemic uncertainty and to run a sensitivity study by assuming different variations ΔD, ranging between zero and 5%.

9.3 Laboratory Tests

Laboratory tests are often carried out to obtain the dynamic properties of soils. The different tests can be grouped into two main categories: cyclic tests, performed at low frequencies, and dynamic tests, carried out at higher frequencies. The most common cyclic tests are the cyclic triaxial (CTx) test, the cyclic torsional shear (CTS) test and the cyclic direct simple shear (CDSS) test, along with its double-specimen (CDSDSS) variant. The stress–strain loops are directly used in cyclic tests to obtain the dynamic properties of the soil. On the other hand, the dynamic resonant column (RC) test analyzes the resonant conditions of the soil sample to obtain the MRD curves.

Results from laboratory tests were widely used in the past to identify the main parameters of the soil affecting, generally speaking, D and specifically D_0. For fine-grained soils, the D_0 is mainly influenced by the plasticity index (PI), the effective mean confining stress (σ'_m) and the overconsolidation ratio (OCR) (Darendeli 2001a). Conversely, the relevant parameters for granular materials are the uniformity coefficient (C_u) and σ'_m (Menq 2003). Additionally, an open issue is represented by the influence of the loading frequency (f) (e.g., Darendeli 2001a; Ciancimino et al. 2019; Shibuya et al. 1995; d'Onofrio et al. 1999; Stokoe et al. 1999; Matešić and Vucetic 2003; Rix and Meng 2005).

In the following, the main features of the tests are firstly described, along with critical issues associated with the experimental measurement of D_0. A final remark is given about the dependency of D_0 from the loading frequency.

9.3.1 RC Test

The RC test (ASTM D4015–15e1) is based on the theory of torsional waves propagation in the medium. The tests are usually carried out using modified versions of the free-fixed-type apparatus described by Isenhower (1979) and designed at the University of Texas at Austin.

Firstly, a cylindrical soil specimen is saturated, if required, through a back-pressure procedure. The consolidation phase takes then place, usually in isotropic conditions. Next, an electromagnetic driving system is used to excite the sample at the free top. The test is performed under loading control, applying torque loadings with increasing amplitudes. The bottom of the specimen is fixed to ensure adequate (i.e., well-defined) boundary conditions. For a given loading amplitude, several cycles are applied for variable frequencies over a wide range, in order to identify the resonance condition of the first torsional mode of the specimen (f_0) associated with the cyclic shear strain reached. The soil response is tracked through an accelerometer installed in the top cap. The conditions are generally undrained, and the pore water pressure build-up can be monitored. The test is able to investigate cyclic shear strain amplitudes ranging from 10^{-5} to 0.5%.

The response of the soil to the dynamic excitation can be represented in terms of rotation (θ) vs frequency curve, where the frequency associated with the maximum amplitude θ_{max} is the f_0 of the sample. The V_S of the soil is then obtained via the equation of motion for torsional vibrations (Richart et al. 1970):

$$\frac{I_\theta}{I_t} = \frac{2\pi f_0 h}{V_S} \cdot \tan\left(\frac{2\pi f_0 h}{V_S}\right) \tag{9.1}$$

where I_θ is the mass polar moment of inertia of the specimen, I_t is the driving system polar moment of inertia and h is the height of the specimen.

The G_S can then be obtained through the well-known relationship:

$$G_S = \rho V_S^2 \tag{9.2}$$

being ρ the mass density of the soil.

Two different methods can be applied to define the damping ratio, namely the half-power bandwidth and the free-vibration decay method. In the half-power bandwidth method, the connection between the shape of the frequency response curve and the dissipated energy is exploited (Fig. 9.1). It can be shown that, for small values of the damping ratio, the latter can be evaluated as:

$$D = \frac{f_2 - f_1}{f_0} \tag{9.3}$$

where f_1 and f_2 are the frequencies associated with a θ amplitude equal to $\sqrt{2}/2\theta_{\max}$. The soil is assumed to behave linearly: The method is therefore reliable only in the small-strain range.

Alternatively, the free-vibration decay method can be used to obtain the damping ratio from the amplitude decay of the torsional oscillations. At the end of the test, after the application of the forced vibrations, the input current is switched off and the damped free vibrations of the sample are recorded by the accelerometer. By knowing two successive peak amplitudes (z_n and z_{n+1} corresponding, respectively, to the n-th and n + 1-th cycle), the logarithmic decrement δ_{n+1} can be computed as:

$$\delta_{n+1} = \ln\left(\frac{z_n}{z_{n+1}}\right) \tag{9.4}$$

The logarithmic decrement is computed for different successive cycles, and then an average value (δ) is used to obtain the damping ratio as:

$$D = \frac{\delta}{2\pi} \tag{9.5}$$

The two methods are characterized by different advantages and disadvantages. When the free-vibration method is used in the small-strain range, the background noise recorded by the accelerometer is not negligible and a filtering procedure has to be applied to the output signals prior to amplitude interpolation (Fig. 9.1b). Moreover, given the small values of D_0, the difference between two consecutive peaks can be really small. As a consequence, the experimental standard deviation can be relatively high with respect to the average measured values.

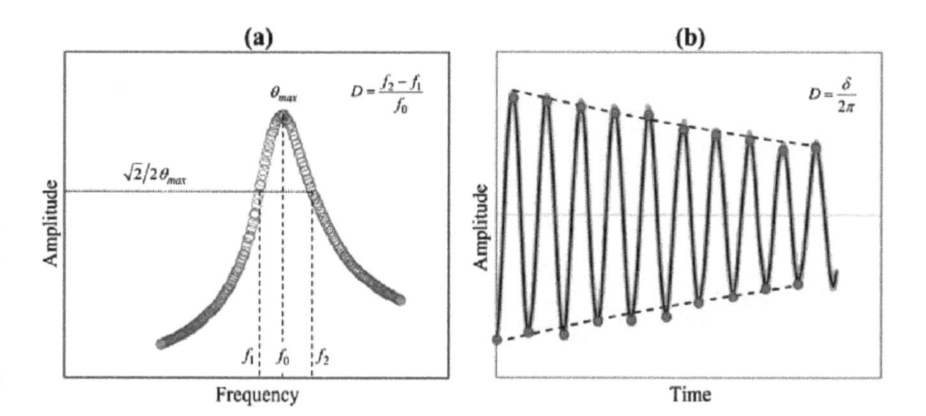

Fig. 9.1 Typical results of a RC test: **a** output amplitude vs frequency curve; **b** free-vibration decay plot

On the other side, a well-recognized source of error in RC measurements of D from forced vibrations arises from the use of an electromagnetic driving system to provide the torsional excitation (e.g., Kim 1991; Hwang 1997; Cascante et al. 2003; Wang et al. 2003; Meng and Rix 2003). The driving system is based on the interaction between the magnets and the magnetic field generated by the AC passing through the solenoids. The driving torque applied to the sample is given by the resulting motion of the magnets. Meanwhile, the motion of the magnets induces an electromagnetic force which is opposed to the motion.

The phenomenon results in equipment-generated damping which is added to the actual material damping. The bias can be substantial, especially in the small-strain range where small values of material damping are expected. Different studies suggested correcting the results of the RC test by subtracting the equipment-generated damping. The latter has to be obtained through a calibration procedure of the apparatus as a function of the loading frequency (e.g., Kim 1991; Hwang 1997; Wang et al. 2003). However, the extent of the bias is not yet totally understood.

9.3.2 CTS and C(DS)DSS Tests

Despite the different configurations, the cyclic tests are all based on the same concept, i.e., to measure the dynamic properties directly from the stress–strain response of the soil (Fig. 9.2a).

The G_S is obtained as the average slope of the loop, while D can be computed in analogy with the critical damping ratio of a single-degree-of-freedom system constituted by a mass connected to a linearly elastic spring and a viscous dashpot. The stress–strain path for soils subjected to cyclic shear strains is indeed similar to

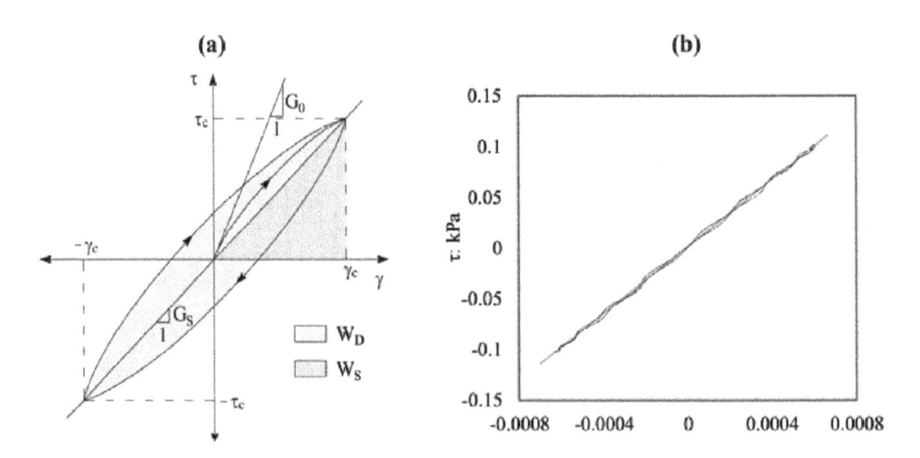

Fig. 9.2 Soil behavior under cyclic loadings: **a** idealized stress–strain loop; **b** real loop measured in a CTS test at small strains

the ellipse described by the SDOF system. Specifically, for a given loading–unloading cycle, D is evaluated as:

$$D = \frac{W_D}{4\pi W_S} \tag{9.6}$$

where W_D is the energy dissipated within one cycle and W_S is the maximum elastic strain energy.

In CTx tests (ASTM D3999/D3999M-11e1 and ASTM D5311/D5311M–13), a cylindrical specimen is firstly consolidated (either isotropically or anisotropically) in a standard triaxial cell. A cyclic deviator stress is then applied by keeping constant the cell pressure and changing the axial stress cyclically with a low loading frequency (about 1 Hz).

The test is commonly performed under load-controlled conditions, but some devices are also equipped to perform displacement-controlled tests. The stresses and the strains are used to compute G_S and D. The applicability of the CTx test is generally restricted to relatively high shear strains (greater than $10^{-2\%}$) because of bedding errors and system compliance effects (Kramer 1996). Local strain measurements can produce an increase of the accuracy of the device (e.g., Burland and Symes 1982; Ladd and Dutko 1985; Goto et al. 1991).

CTS tests can be performed in the same device used for RC tests. The driving system applies a fixed number of cycles for a given amplitude with a fixed loading frequency (usually between 0.1 and 0.5 Hz). The rotation of the specimen is measured through a couple of displacement transducers connected to the top cap. The shear strain is then obtained from the rotation and, by knowing the input applied, it is possible to draw the loading–unloading loops.

Finally, in a CDSS test, a cylindrical specimen is cyclically loaded under displacement control by a horizontal piston. The test is performed in undrained conditions and the specimen cannot, usually, be consolidated at horizontal to vertical stress ratios different than K_0. The applicability of the test in the small-strain range is limited mainly because of frictional problems. The range can anyway be increased adopting a double specimen configuration (Doroudian and Vucetic 1995). The CDSDSS device is a modified version of the standard device designed by the Norwegian Geotechnical Institute (Bjerrum and Landva 1966), and it can capture the soil behavior also at very small strains (Doroudian and Vucetic 1998).

The main issue regarding the cyclic tests is related to the measurement of the loops in the small-strain range. The stresses and the strains are indeed really small, and it becomes quite difficult to obtain a proper measurement even for the most apparatuses. For example, a loop measured during a CTS test, for a γ_c below the linearity threshold (i.e., in the almost linear branch of the stress–strain response) is showed in Fig. 9.2b. It is clear that although is quite straightforward to define the G_0 from the slope of the loop, the small area inside the loop can be affected by the accuracy of the measurement. Consequently, the experimental relative error on D_0 can be, again, substantial.

9.3.3 Frequency-Dependent Soil Behavior

The influence of the loading frequency on the material damping ratio is still an open issue. A number of experimental studies reported controversial results about the real extent of this dependency. Some studies also highlighted the possible impact of such dependency on GRAs (e.g., Park and Hashash 2008).

Kim (1991) presented the results of RC and CTS tests on undisturbed samples of cohesive soil with a PI of 20–30% showing that the small-strain damping ratio is almost linear for frequencies lower than 1 Hz but increases at higher frequencies. Subsequently, Shibuya et al. (1995) suggested the existence of three different branches. At low frequencies (<0.1 Hz) the damping ratio tends to decrease with increasing frequencies. In the medium range (between 0.1 and 10 Hz, the typical seismic bandwidth), the damping is almost constant, irrespectively of the loading frequency. Finally, for higher frequencies D increases with f because of viscous effects. A similar trend was reported also by other studies, highlighting anyway a rate dependency even in the seismic bandwidth (e.g., Stokoe and Santamarina 2000; Darendeli 2001a; Menq 2003; d'Onofrio et al. 1999; Matešić and Vucetic 2003; Rix and Meng 2005).

This trend is not clearly identifiable at high strain amplitudes, where nonlinearity partially covers these aspects. Darendeli (2001a) proposed then to model the damping ratio curves of fine-grained soils by adding a strain constant D_0 to the hysteretic damping ratio. The D_0 in the model depends on f. The latter has thus the effect of translating the damping curves. For granular dry materials, Kim and Stokoe (1994) suggested that f has a negligible impact. Conversely, the effects become relevant for saturated specimens (Menq 2003).

The motivation behind this behavior has to be found into the different mechanisms of energy dissipation taking place in soils during cyclic loadings (e.g., Shibuya et al. 1995; d'Onofrio et al. 1999). In the very low-frequency range, the application of the load is quasi-static and creep phenomena occur. As a consequence, the slower is the application of the load, the higher is the D_0. Conversely, in the medium range, the dissipation is given mainly by the hysteretic soil behavior, that is almost frequency-independent. In the small-strain range anyway, a substantial component of the energy dissipation is attributed to pore fluid viscosity. The relative movement between the water and the soil skeleton generates viscous damping that is, obviously, frequency-dependent. At high frequencies thus the D_0 dramatically increases with f. This effect exists independently from the strain level, but it becomes less relevant at high strains when the hysteretic damping increases.

Ciancimino et al. (2019) calibrated an empirical equation for predicting the D_0 of fine-grained soils from Central Italy. The equation, based on the model proposed by Darendeli (2001a), incorporates the dependency of D_0 from f in the range between 0.2 and 100 Hz:

$$D_0 = (\varphi_1 + \varphi_2 \cdot PI) \cdot \sigma_m^{'\varphi_3} \cdot [1 + \varphi_4 \cdot \ln(f)] \quad (9.7)$$

where φ_{1-4} are model parameters equal, respectively, to 1.281, 0.036, −0.274 and 0.134, PI is expressed in percentage, f in Hz, and $\sigma_m^{'}$ in atm. The equation is conceived to model the viscous component but neglects the creep effects at low frequencies.

The authors also suggested a possible application of this equation to correct the results of a laboratory test by taking into account the loading frequency. The procedure is relevant especially for RC tests, usually carried out at frequencies not representative of the typical seismic bandwidth.

However, it is worth noting that subtracting a constant value of D_0 from the damping curve obtained in a RC test is not completely correct. The RC test is in fact carried out at variable frequencies, according to the resonance conditions for different strain amplitudes. The steps below have to be followed to correct the experimental damping curve to a frequency of 1 Hz:

- the experimental $D_0(f_1)$ measured at the first shear strain amplitude is normalized to a frequency of 1 Hz, by dividing it by $[1 + \varphi_4 \cdot \ln(f_1)]$, where f_1 is the first loading frequency;
- from each i-th point of the damping curve is subtracted the corresponding $D_0(f_i)$, computed as $D_0(1Hz) \cdot [1 + \varphi_4 \cdot \ln(f_i)]$, where f_i is the i-th loading frequency;
- finally, $D_0(1Hz)$ is added to the $D - D_0$ curve previously computed in order to obtain the frequency-normalized damping curve $D(1Hz)$.

In Fig. 9.3 an example of the normalization procedure for a RC test is reported. Figure 9.3a shows the initial comparison between a RC and a CTS test carried out on the same sample (Massa Fermana BH1S1 sample, from Ciancimino et al. 2019). A marked difference is observed, especially in the small-strain range. The results of the two tests are then corrected to match a frequency of 1 Hz (Fig. 9.3b reports the procedure for the RC test). The correction is less significant for the CTS test performed at a constant frequency of 0.1 Hz. Finally, in Fig. 9.3c the normalized results are compared, showing good agreement.

The proposed approach can be applied also to other predictive models, provided that D_0 is given as a function of f (e.g., Darendeli 2001a). Anyway, it has to be pointed out that although the procedure is presented as a correction for the loading frequency, it is to some extent a correction for the type of test. As a matter of fact, tests performed at low loading frequencies are always cyclic tests, while at higher frequencies just RC tests can be carried out. As a consequence, the dependency of D_0 from f cannot be easily separated from other possible sources of discrepancies between cyclic and RC tests.

A possible alternative is given by the so-called non-resonance column (N-RC) method (Rix and Meng 2005; Lai and Rix 1998; Lai et al. 2001; Lai and Özcebe 2015). The method is based on the experimental measurement of the complex shear modulus $G_S^*(f)$ of a soil specimen, idealized as a linear viscoelastic medium. The

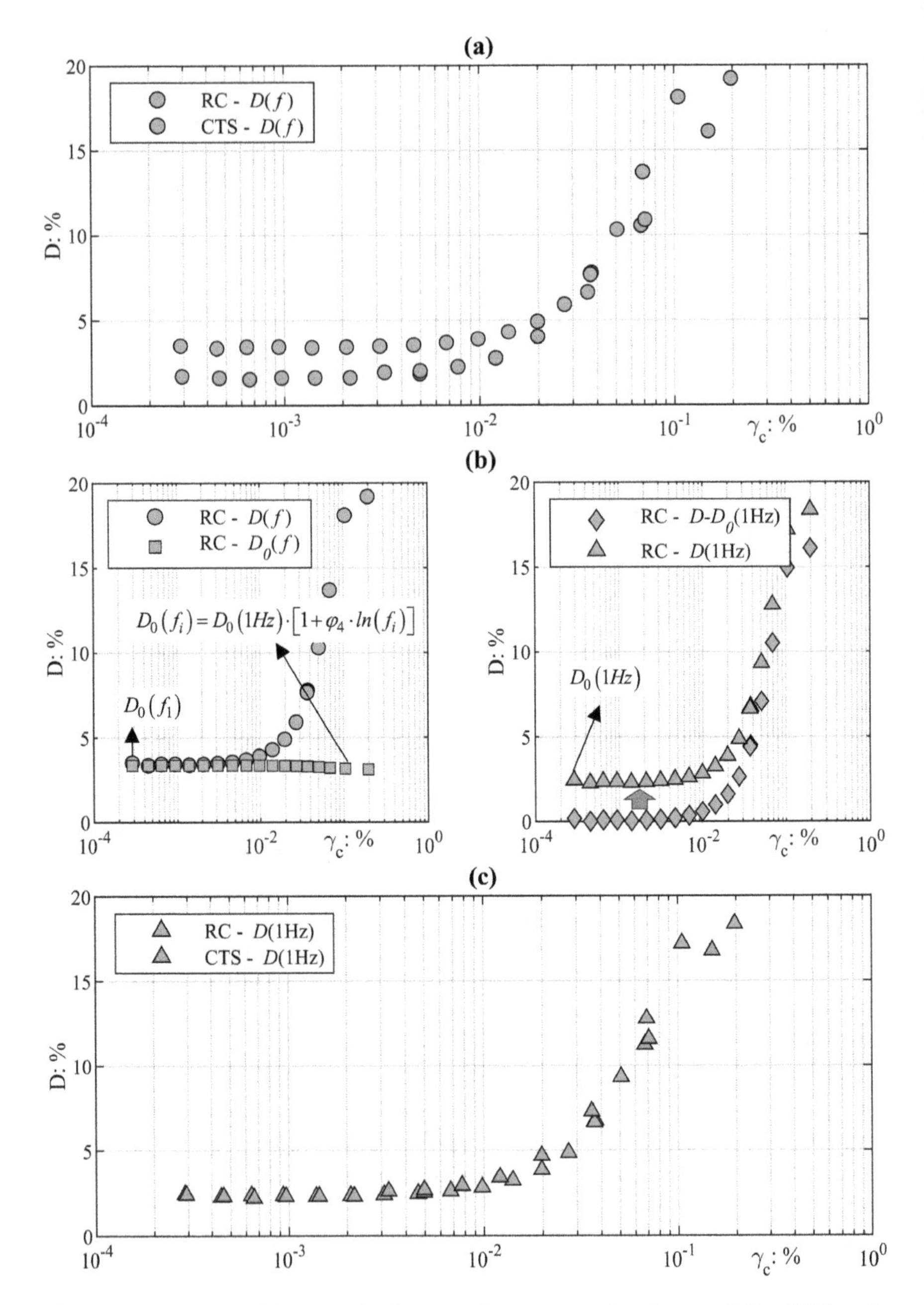

Fig. 9.3 Effectiveness of the normalization procedure: **a** comparison between RC and TC results; **b** normalization procedure for the RC test; **c** comparison of the normalized curves (modified from Ciancimino et al. 2019)

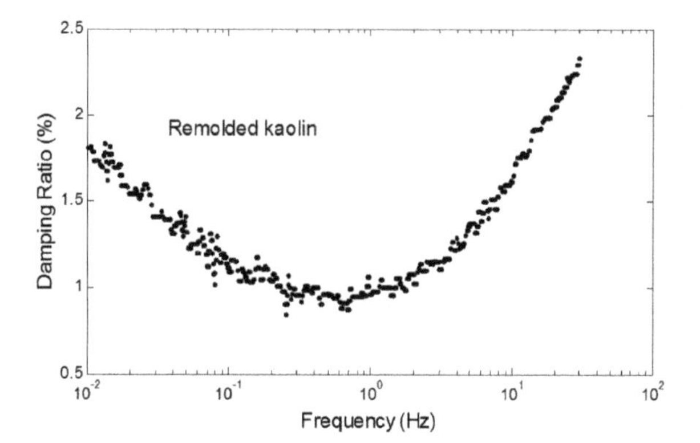

Fig. 9.4 Dependency of D_0 from f for a remolded kaolinite sample obtained applying the N-RC method (after Rix and Meng 2005)

latter is used to compute consistently V_S and D. The soil is thus assumed to be a dispersive medium and the frequency dependence is inherently taken into account by the method. Rix and Meng (Rix and Meng 2005), for instance, carried out a N-RC test on a remolded kaolinite sample, showing the 'U'-shaped dependence of D_0 from f in a wide range (i.e., 0.01–30 Hz) of frequencies (Fig. 9.4). The results confirmed the trends previously suggested by other authors (e.g., d'Onofrio et al. 1999). However, the application of the method in the current practice is still limited due to the complexity of the approach.

9.4 In Situ Tests

In situ tests are a common tool for site characterization, due to limited costs and the rapidity of execution. Moreover, they can provide a reliable estimate of the geotechnical parameters, since they assess the soil behavior in undisturbed conditions at a spatial scale compatible with the geotechnical application of interest.

Geophysical seismic tests are widely adopted for the determination of G_S on site. On the other hand, some methods have been proposed also for the estimation of D_0. The technical literature also includes some case studies of parameter estimation based on the interpretation of downhole arrays. This approach is less common, as it requires instrumented boreholes with seismic records, but it provides useful data for the assessment of the soil behavior in seismic conditions.

In addition, several seismological studies focus on the attenuation structure of the near surface. These studies provide an estimate of the dissipation properties, based on the high-frequency attenuation of seismic waves. Some of them propose empirical relationships with other geotechnical parameters—e.g., V_S—for specific geological formations. The next sections will focus on the geophysical approaches and the interpretation of borehole arrays.

9.4.1 Geophysical Tests

Geophysical seismic tests are generally classified as invasive and non-invasive. However, all the geophysical methods investigate the medium in its undisturbed natural state, but the sampled soil volume and the resolution are not the same. Therefore, they might provide different results, as a function of the degree of heterogeneity of the soil deposit (Foti et al. 2014). There are several interpretation techniques aimed at estimating the small-strain stiffness from the measured data, whereas the attempts of estimating the attenuation characteristics of the soil deposits are less numerous and often restricted to research. Moreover, they sometimes use strong assumptions which limit their applicability. The most important issue is the difficulty in separating geometric and intrinsic attenuation, i.e., the energy loss due to wavefront expansion and to wave scattering in heterogeneous media, on one side, and the one due to intrinsic material attenuation, on the other.

In the following, some applications of the invasive and non-invasive tests for the determination of D_0 are discussed, with focus on their assumptions and limitations.

Invasive tests. Invasive tests are a family of geophysical seismic tests for which a part of the instrumentation is installed in the ground. Typical methods are the cross-hole (CH) test (ASTM 2014), downhole (DH) test (ASTM 2017), the P–S suspension logging test, the seismic cone penetration Test (SCPT) (Campanella 1994), the seismic dilatometer test (SDMT) (Marchetti et al. 2008) and the direct-push cross-hole test (Cox et al. 2018). The technical literature involves many robust approaches for the determination of the shear wave velocity from the interpretation of the measured data. Conversely, the techniques aimed at estimating the dissipation characteristics of the soil deposit are limited to few attempts with limited applications outside the research field.

Techniques for the estimate of D_0 from CH data are the random decrement approach (Aggour et al. 1982), and the attenuation coefficient method (Hoar and Stokoe 1984; Mok et al. 1988; Michaels 1998; Hall and Bodare 2000). Lai and Özcebe (2015) observed that those methodologies rely on the hypothesis of frequency-independent (i.e., hysteretic) damping or on enforcing a specific constitutive model in the interpretation of the attenuation measurements. Moreover, they usually perform an uncoupled estimate of the low-strain parameters by using incompatible constitutive schemes: V_S is obtained according to a linear elastic model, whereas D_0 estimates are based on inelastic models. Therefore, these approaches may lead to inconsistent and biased estimates. To overcome those limitations, they applied the two-station interpretation scheme typical of the SASW method (Foti et al. 2014) to CH measurements, determining the S wave dispersion function from the unwrapped phase of the cross-power spectrum $G^S_{R_1R_2}$ of the S wave signal, detected at the two receivers.

$$V_S(\omega) = \frac{\omega \Delta L}{\arg G^S_{R_1R_2}} \tag{9.8}$$

In the equation, the terms R_1 and R_2 denote the distances between each receiver and the source, whereas ΔL is the interreceiver distance (Fig. 9.5a). The procedure then derives D_0 from the computed dispersion curves, by applying the solution of the Kramers–Kronig relation that relates stiffness and attenuation characteristics in a viscoelastic medium (Christensen 2012).

$$D_0(\omega) = \frac{\frac{2\omega V_S(\omega)}{\pi V_S(0)} \cdot \int_0^{\infty} \left(\frac{V_S(0)}{V_S(\tau)} \cdot \frac{d\tau}{\tau^2 - \omega^2} \right)}{\left[\frac{2\omega V_S(\omega)}{\pi V_S(0)} \cdot \int_0^{\infty} \left(\frac{V_S(0)}{V_S(\tau)} \cdot \frac{d\tau}{\tau^2 - \omega^2} \right) \right]^2 - 1} \tag{9.9}$$

where $V_S(0) = \lim_{\omega \to 0} V_S(\omega)$.

This interpretation method only requires measurements of velocity for determining either the stiffness or the damping parameters of the material; hence, an accurate tracking of particle motions is unnecessary. Moreover, the processing does not require a priori assumptions about the specific rheological behavior or the frequency-dependent nature of D_0. On the other side, broadband seismic sources are required to generate a wave signal with a wide frequency range. If not possible, some assumptions about the dispersive behavior of the soil parameters would be necessary to extrapolate the available data, introducing uncertainties in the estimate (Fig. 9.5b).

DH and SCPT-based techniques for the estimate of D_0 are theoretically more complex since they should account for the reflection and refraction phenomena at the layer interfaces in the computation of the attenuation. Some interpretation schemes are based on the attenuation coefficient method or on a simulation of the downward wave propagation in the DH testing. Actually, Campanella and Stewart (1991) stated several issues in the application of such approaches, due to the necessity of applying corrections to incorporate the effect of the wave passage

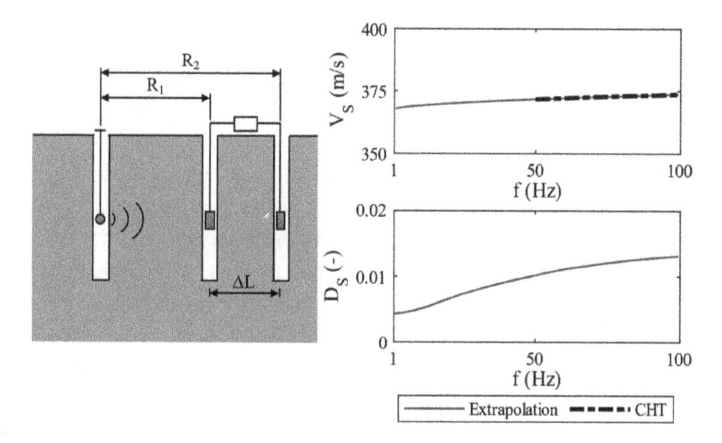

Fig. 9.5 **a** Scheme of the CH test layout; **b** resulting dispersion and damping curves from the interpretation of CH data (modified from Lai and Özcebe 2015)

through the layer interfaces. Furthermore, results were affected by large scatter, sometimes providing unphysical values. A popular method is the spectral ratio slope (SRS) method (Redpath et al. 1982; Crow et al. 2011). The approach provides a frequency-dependent estimate of D_0 at depth z_i by computing the second-order derivative of the wave amplitude spectral ratio (i.e., the amplitude ratio between the 1st and the *i*th receiver) with respect to the depth and the frequency.

$$D_0 = \frac{V_S}{2\pi} \cdot \frac{\partial^2}{\partial f \partial z} \ln\left(\frac{A_1}{A(z_i)}\right)\bigg|_{z=z_i} \tag{9.10}$$

This approach does not require interface corrections, and the scatter in the results is low. However, Badsar (2012) reported a low reliability of the SRS method in the determination of the damping profile, especially in the presence of complex stratigraphy, due to some simplifying assumptions for the geometrical damping.

A more robust approach is based on the spatial decay of the Arias intensity, developed by Badsar (2012): Once determined the V_S profile, the method calibrates the D_0 profile through an optimization algorithm minimizing the difference between the experimental evolution of the Arias intensity among the receivers and the theoretical one, computed for a vertical point force. This method properly considers all the phenomena of reflection and refraction and provides a good estimate of the D_0 profile. On the other side, its application requires an accurate modeling of the V_S profile and long computational time due to the multiple forward analyses.

Non-invasive tests. Non-invasive tests are geophysical seismic tests employing a source and a set of receivers on the ground surface. They include the seismic reflection survey (Schepers 1975), the seismic refraction survey (International ASTM 2011), surface wave testing (Foti et al. 2014) and the horizontal-to-vertical spectral ratio (Bard 2004). This section will focus on the techniques based on the measurement of surface waves generated from active sources.

Surface wave methods (SWM) rely on the dispersive behavior of Rayleigh waves in heterogeneous media, for which the phase velocity exhibits a dependence on frequency. Therefore, the procedure consists in acquiring the particle motion, processing the measured data to derive the experimental dispersion relationship and estimating the V_S profile with depth through an inversion scheme, where a theoretical soil model is calibrated to match the experimental data.

The SWM-based estimate of D_0 usually refers to the measurement of the spatial attenuation of surface waves along linear arrays with active sources. This quantity is linked to the geometrical spreading of the Rayleigh waves and the intrinsic dissipation properties of the material. The measurement requires precise tracking of the surface wave particle motion, since noise and amplitude perturbations might lead to wrong estimates. For this reason, the acquisition setup should guarantee an optimal coupling and verticality of each receiver, and a good sensor calibration for uniform response is required (Foti et al. 2014).

Rix et al. (2000) estimated the attenuation curves based on the regression of the displacement amplitude versus offset data, considering the equation for the

Rayleigh wave motion due to a harmonic point force (Lai and Rix 1998). The amplitude-offset regression provides an uncoupled estimate of the dispersion and attenuation curves, which is not mathematically robust and ignores the intrinsic relationship between velocity and attenuation in a linear viscoelastic material (Lai and Rix 1998).

An upgrade of the approach is the transfer function method (Rix et al. 2001; Lai et al. 2002). The technique is a multistation approach based on the estimate of the experimental displacement transfer function $T(r, \omega)$, i.e., the ratio between the measured vertical displacement at each sensor $u_z(r, \omega)$ with offset r and the input harmonic source $F \cdot e^{i\omega t}$ in the frequency domain:

$$T(r, \omega) = \frac{u_z(r, \omega)}{F \cdot e^{i\omega t}} \tag{9.11}$$

Then, for each frequency, the procedure jointly estimates the complex wavenumber $K(\omega)$ through the nonlinear fitting of the following expression (Lai and Rix 1998):

$$T(r, \omega) = Y(r, \omega) \cdot e^{-iK(\omega)r} \tag{9.12}$$

In the equation, $Y(r, \omega)$ is the geometrical spreading function, which is usually assumed as equal to $1/\sqrt{r}$ (e.g., Lai et al. 2002; Foti 2003). The complex wavenumber is defined as a combination of the real wavenumber $k(\omega)$ and the attenuation $\alpha(\omega)$. The latter are linked to the phase velocity and the phase damping of the Rayleigh waves:

$$K(\omega) = k(\omega) - i\alpha(\omega) = \frac{\omega}{V(\omega)} - i\frac{D(\omega) \cdot \omega}{V(\omega)} \tag{9.13}$$

The fitting of $T(r, \omega)$ can be performed in an uncoupled way, based on the separate fitting of its amplitude and phase (Lai et al. 2002). However, a coupled fitting of the transfer function in the complex domain is mathematically more robust and provides an estimate of the wave parameters compatible with amplitude phase data (Foti 2003) (Fig. 9.6).

Foti (2003, 2004) adopted a generalized version of the transfer function method by removing the effect of the input force, whose measurement is non-trivial and requires controlled sources. For this purpose, the author reformulated the displacement transfer function in terms of deconvolution of the seismic traces. The principle of this method consists in computing the experimental transfer function adopting the response of the closest receiver as the reference trace.

A limitation of the regression is the assumption that the wavefield is dominated by a single Rayleigh mode of propagation. Therefore, the result is an estimate of apparent Rayleigh phase dispersion and attenuation curves that can be affected by modal superposition when multiple propagation modes are relevant (Foti et al. 2014). For this reason, new advanced methods have been proposed, as the

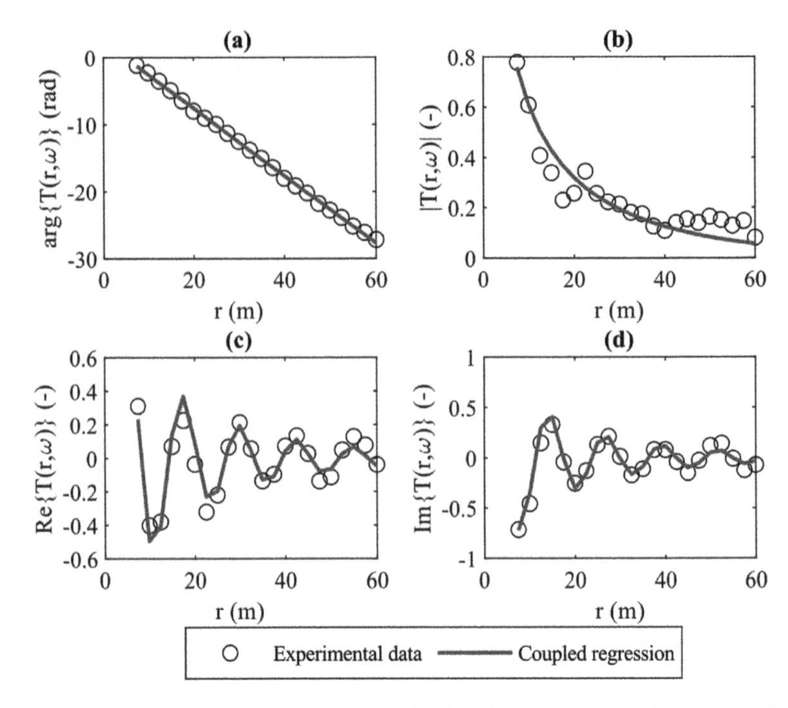

Fig. 9.6 Regression of the experimental transfer function for the coupled computation of dispersion and attenuation curves: **a** phase; **b** amplitude; **c** real part; **d** imaginary part. The data refer to the Pisa Leaning Tower site, at the frequency 11.5 Hz (modified from Foti 2003)

generalized multiple signal classification (Misbah and Strobbia 2014), wavelet decomposition methods (Bergamo et al. 2018) or sparse signal reconstruction (Mun and Zeng 2018). Moreover, Badsar et al. (2010) proposed a simplified method for the estimate of the attenuation curves, based on a generalization of the half-power bandwidth method, whereas Verachtert et al. (2017) introduced an alternative methodology for the determination of multimodal surface wave dispersion curves and attenuation curves, namely the circle fit method.

As for the inversion, a robust characterization method requires a joint inversion of the Rayleigh dispersion and attenuation curves into the V_S and D_0 profiles. The coupled inversion offers the advantages of accounting for the inherent relationships between the stiffness and the attenuation properties in the material, and it is a better-posed mathematical problem (Lai et al. 2002). Being the inversion procedure commonly based on the solution of multiple forward problems to fit the experimental dispersion and attenuation curves, it requires specific algorithms for the solution of the Rayleigh eigenvalue problem in linear viscoelastic media (Lai et al. 2002). Moreover, the coupled inversion requires the definition of a proper misfit function, which should be complex-valued to incorporate both stiffness and dissipation data. Most of the applications implemented this strategy into a constrained least squares algorithm, aiming at a smooth profile respecting the experimental data

(Lai et al. 2002; Foti 2003, 2004; Badsar et al. 2010; Verachtert et al. 2017). On the other side, there are also some attempts of application of the Monte Carlo technique in the joint inversion for the estimate of the uncertainties (Misbah and Strobbia 2014).

9.4.2 *Back-Analysis of Downhole Arrays*

Downhole instrumentations for the observation of ground motion represent a valuable tool for understanding the physics of the seismic amplification. The provided data are useful for the validation of theoretical models of amplification, highlighting issues due to the assumptions about the constitutive behavior of the soil deposit (e.g., Kwok et al. 2008; Kaklamanos et al. 2013) or the propagation model (e.g., Thompson et al. 2012). Moreover, borehole array data can be employed for the calibration of mechanical parameters by performing a back-calculation from observed ground motions. The literature reports several attempts at estimating the dissipation characteristics from the interpretation of earthquake records in instrumented boreholes (e.g., Shima 1962; Pecker 1995).

For instance, Assimaki et al. (2006, 2008b) implemented a seismic waveform inversion algorithm for the estimate of the small-strain parameters from weak motion records in downhole arrays. The procedure assumes a 1-D ground model, and it estimates the mechanical parameters, i.e., V_S, D_0 and density for each layer, through a two-step optimization algorithm, consisting in a genetic algorithm in the wavelet domain and a nonlinear least squares in the frequency domain. The stochastic optimization minimizes the misfit between the theoretical and the observed acceleration time histories, represented in the wavelet domain—rather than in the time domain—to ensure equal weighting of the information across all frequency bands. The local search process is a nonlinear least squares optimization algorithm in the frequency domain, minimizing the error between the theoretical transfer function and the empirical one. The combination of a stochastic search algorithm with a local search one provides the advantages of each technique, resulting in a robust search method.

A quite popular approach of D_0 estimation is based on site amplification synthetic parameters. The strategy consists in a search procedure where several GRAs are performed by keeping all the other model parameters (V_S, density, layer thickness) constant and iteratively adjusting D_0 to obtain a good level of consistency between the predicted and the observed response. The trial values of D_0 can be assumed a priori (Thompson et al. 2012), based on seismological relationships (Cabas et al. 2017) or from laboratory-based values (e.g., Zalachoris and Rathje 2015). The resulting value is compatible with the site behavior in seismic conditions, and it can be used for GRAs.

The downhole array data processing is not straightforward, and it incorporates some drawbacks. On one side, the computation of the empirical site response requires the selection of an adequate number of ground motion histories (Assimaki et al. 2008b). The only weak motions should be included, to avoid the rise of nonlinear phenomena and ensure the validity of viscoelastic behavior of the soil deposit (Zalachoris and Rathje 2015; Thompson et al. 2012; Xu et al. 2019; Tao and Rathje 2019; Cabas et al. 2017; Beresnev and Wen 1996). Moreover, a critical issue is the ambiguity about the wavefield conditions at the downhole sensors. Indeed, the downhole sensors record either upgoing or downgoing waves and, due to the impossibility of separating them, the modeling of their conditions is complex (Shearer and Orcutt 1987). Finally, the quality of the estimate strongly depends on the reliability of the available geotechnical information and the absence of lateral variabilities (Thompson et al. 2012).

A special remark about the role of the ground motion parameter adopted for measuring the site response should be pointed out. Common descriptors are frequency-domain parameters, as they best carry information about the frequency-dependent phenomenon of site amplification. We might refer to the acceleration transfer function (ATF), i.e., the ratio of the Fourier amplitude spectra between two locations (Borcherdt 1970), or the amplification function (AF), i.e., the ratio of the elastic response spectra (Thompson et al. 2012). The description might refer also to time-domain parameters, as the peak values of acceleration and velocity and the Arias intensity (Tao and Rathje 2019). Indeed, they synthesize the ground response of a broad range of frequencies, but the request of matching a time instant parameter might lead to physically unreliable data (Tao and Rathje 2019). An alternative approach adopts the high-frequency spectral attenuation κ_0, which describes the decay of the Fourier amplitude spectrum of the ground motion at high frequencies. The difference of attenuation $\Delta\kappa$ between surface and borehole records provides a measure of the attenuation along the borehole, and it is related to the small-strain parameters of the soil deposit (Hough and Anderson 1988).

$$\Delta\kappa = \int_0^z \frac{2D_0}{V_S} dz \tag{9.14}$$

In this way, by adjusting the damping parameters, we can identify a κ_0-informed D_0 that suits the observed high-frequency attenuation (e.g., Xu et al. 2019; Afshari and Stewart 2019; Cabas et al. 2017). Figure 9.7 shows the effect of the type of parameter on the damping correction for some sites with different geology.

There is no consensus about the best reference parameter. Several case studies performed an estimate in the frequency domain, based on the measured ATF (e.g., Kaklamanos et al. 2013; Zalachoris and Rathje 2015; Thompson et al. 2012). Tao and Rathje (2019) suggested keeping the time-domain parameters as a reference, since they capture the overall response of the site, whereas the calibration in the frequency domain may lead to an overestimation of the damping. On the other side, Afshari and Stewart (2019) stated that the κ_0-informed damping value best fits the observed site response.

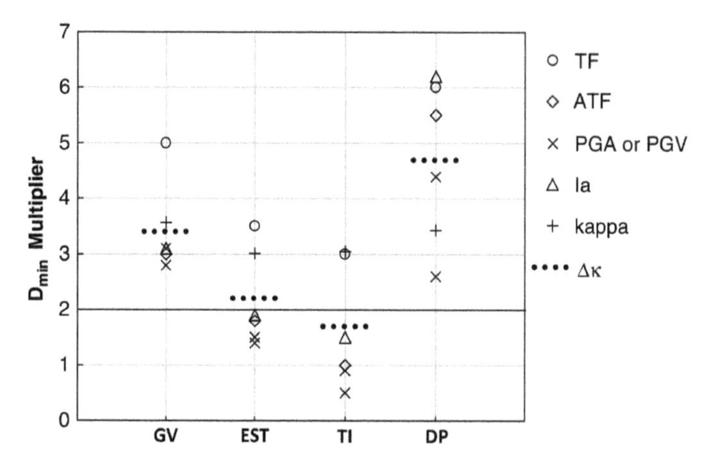

Fig. 9.7 Obtained damping multipliers for the Garner Valley (GV), EuroSeisTest (EST), Treasure Island (TI) and Delaney Park (DP) sites (after Tao and Rathje 2019)

9.5 Literature Approaches to Account for Wave Scattering Effects

The site characterization, i.e., the procedures and interpretations devoted to the formulation of a geotechnical model, is usually performed through laboratory tests and in situ surveys. Their combined use is strongly recommended by several guidelines, as they are complementary. Indeed, laboratory tests are performed on small-size soil specimens by applying an imposed stress/strain history with known hydro-mechanical boundary conditions, ensuring a controlled behavior and a rigorous estimate of the mechanical parameters. On the other side, in situ testing does not allow full control of the hydrogeologic and loading conditions, but the investigated soil volume is larger and closer to the representative volume for geotechnical applications. Moreover, in situ tests are free from effects due to sampling disturbance and also allow the characterization of hard-to-sample soils.

The complementary nature of such tests gives rise to discrepancies between laboratory-based D_0 values and the ones derived from in situ testing or inferred from DH arrays, as highlighted by several studies.

For instance, Foti (2003) compared the damping ratio obtained from the SWM and laboratory testing. He observed a slight overestimation of the dissipative properties in the former, due to the presence of additional attenuation mechanisms other than geometric and intrinsic attenuation, especially for shallow layers (Fig. 9.8).

The numerous studies based on DH arrays highlighted the differences between D_0 and $D_{0,\ \mathrm{site}}$. Some preliminary observations about this discrepancy are reported in Tsai and Housner (1970) and Dobry et al. (1971), where the calibrated $D_{0,\ \mathrm{site}}$ was much higher than the intrinsic attenuation properties. The reason for such

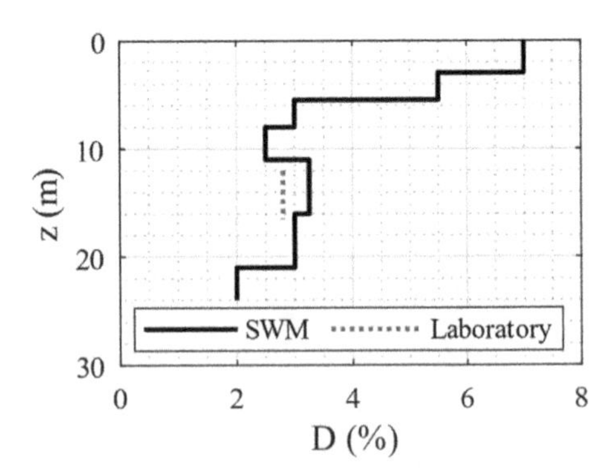

Fig. 9.8 Comparison between the SWM-based and the laboratory-based damping profile. The data refer to the Pisa Leaning Tower site (modified from Foti 2003)

difference was some bias due to plane wave assumption and the presence of other losses of energy (Joyner et al. 1976). Recent studies observed high $D_{0,\ \mathrm{site}}$ values (either in absolute terms or compared with laboratory data) in soft shallow layers that usually exhibit strong heterogeneities resulting in relevant scattering phenomena. For instance, Assimaki et al. (2006) stated a large increase of the attenuation in the shallow layers due to scattering, that is not accounted in the seismic waveform inversion algorithm, thus resulting in an additional energy loss. Zalachoris and Rathje (2015) corrected the D_0 profiles using the ATF-based approach that led to an increase of the damping ratio ranging between 2% and 5%. They also observed that the incompatibility is strong for deep arrays with no significant impedance contrast, mainly due to the modeled wavefield conditions at the downhole sensors (Fig. 9.9). Even the κ_0-informed $D_{0,\ \mathrm{site}}$ estimate is larger than the laboratory-based D_0, yet still lying close to the upper bounds of the statistical distribution of the D_0 data. However, some authors questioned the relation between κ_0 and the small-strain damping, due to the wave scattering phenomena that may be relevant in the presence of complex stratigraphy (Ktenidou et al. 2015).

Conversely, Kaklamanos et al. (2018) observed the necessity of decreasing the laboratory-based D_0, to reduce the high-frequency bias of the theoretical model with the observed amplification data. They recognized the limited physical background of such reduction, justifying it with some breakdowns of the 1D propagation assumption.

In summary, one of the most important factors affecting the in situ attenuation estimates is the presence of wave scattering phenomena, which is an additional dissipation mechanism not accounted in laboratory measurements. Seismic wave scattering is a phenomenon characteristic of the wave propagation in heterogeneous media, where the multiple reflections and refractions lead to a non-planar propagation and to the diffusion of the seismic energy.

Generally, the scattering depends on the relative size between the heterogeneity and the wavelength, and it holds only when they are compatible (Stein and Wysession 2003). For seismic applications, where the frequency content is close to 1–10 Hz, the material fluctuations usually present in the soil deposits have a significant effect on the wavefield; hence, the scattering is a relevant phenomenon

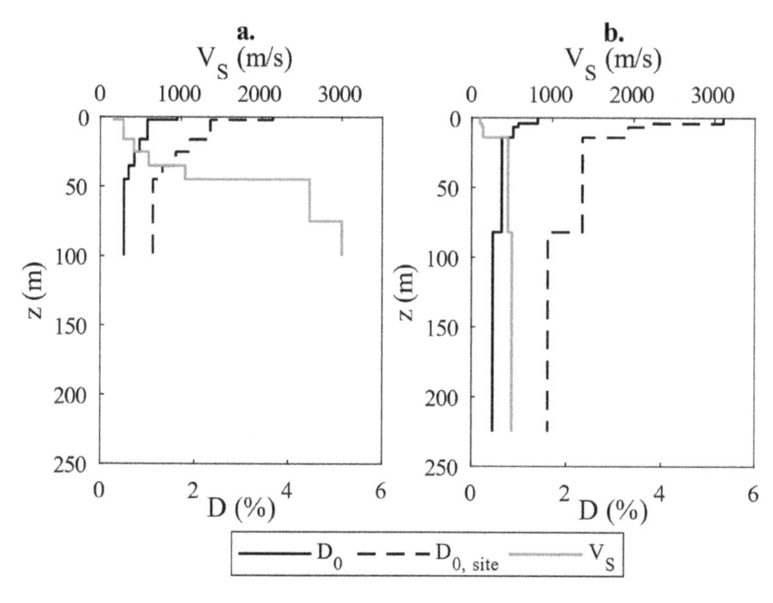

Fig. 9.9 D_0 profiles and obtained $D_{0,\ site}$ profiles (dotted line) **a** for a shallow site with a strong impedance contrast and **b** a deep site (modified from Zalachoris and Rathje 2015)

(Thompson et al. 2009). As pointed out by Zalachoris and Rathje (2015), the attenuation through wave scattering mechanisms is apparent. Indeed, the energy simply diffuses inside the medium instead of moving only toward the surface, which is perceived as an energy loss on the surface ground motion (Stein and Wysession 2003).

The role of wave scattering in the propagation phenomena—hence, in the site response analyses—limits the applicability of GRAs that rely on the assumption of a planar wavefield. For this reason, taxonomic schemes based on DH array data have been proposed to assess whether the 1D approach is reliable (Thompson et al. 2012; Tao and Rathje 2020). When the spatial variability of the mechanical parameters is relevant, specific 2D and 3D site response analyses (hereafter, SRAs) should be carried out. A common approach is based on generating soil models by assuming the material properties as spatially correlated random fields (e.g., Thompson et al. 2009). This approach considers a spatially correlated statistical distribution of the mechanical parameters, characterized by variance and range, i.e., the distance at which the correlation diminishes. Huang et al. (2019) observed that an increase of the variance in the distribution—namely the degree of heterogeneity—induces a reduction of the mean AF, due to the stronger wave scattering in the soil model. The variation in the AF is stronger at high frequencies, as the wave-length is small compared to the size of the heterogeneity. Yet, the performance of 2D or 3D analyses requires long computational time and involves a high degree of complexity in the model definition.

An alternative way to mimic lateral heterogeneities and wave scattering phenomena consists in performing a Monte Carlo simulation, by generating multiple 1D ground models with random V_S realizations and keeping the damping value constant and equal to D_0. Even though it does not explicitly consider lateral variabilities, the average response accounts for the wave scattering that happens in real soil deposits (Nour et al. 2003). A classical model for accounting for such variability is the Toro (Toro 1995) model that incorporates criteria for the V_S and the layering randomization. Several studies highlighted the limitations of the model (e.g., Rodriguez-Marek et al. 2014; Tao and Rathje 2019; Passeri et al. 2018b; Teague and Cox 2016), and some adjustments in the parameters or even in the framework have been proposed (Passeri 2019). The V_S randomization mimics the effects of damping at the site natural frequencies. Indeed, the realizations are soil profiles with different fundamental frequencies, and the average of the responses results in a smaller and smoother peak. On the other side, it does not have any significant effect on the Fourier amplitude spectrum at high frequencies; hence, it does not capture the behavior of κ (Tao and Rathje 2019). Moreover, it does not allow to incorporate complex phenomena as the presence of surface waves that introduce additional low-frequency oscillations (Baise et al. 2003).

9.6 Influence of D_0 Correction in GRAs

A critical step in conducting GRAs is the definition of the soil model and the choice of the parameters. Indeed, the key issue is the non-existence of a priori conservative values for the mechanical parameters. For this reason, GRAs should be carried out by considering the parameter uncertainties in an explicit way.

This section shows the influence of the uncertainties in D_0 in the seismic ground amplification, by reporting two case studies.

On one side, we assessed the effect of D_0 on a stochastic database of GRAs, consisting of the results of 3,202,500 EQL simulations performed over a collection of 91,500 1D soil models (Aimar et al. 2020) subjected to seismic inputs of different intensity. The aim of this specific study was to map the variations of D_0 on the seismic amplification of a subset of soil models of engineering interest. In this way, we figured out the role of D_0 on the response of generic soil models under seismic conditions.

On the other side, we assessed the influence of D_0 on a site-specific amplification study. For this purpose, we considered the site of Roccafluvione, in the Marche region, which was struck by the seismic sequence that started on the 24th of August 2016. The site was object of intense geological and geotechnical investigations, resulting in a detailed ground model. Foti et al. (2019) performed EQL GRAs over a statistical sample of ground models generated through a Monte Carlo simulation to capture the effects due to the variability in V_S and the MRD curves. In this paper, we analyzed the variations in the site response due to changes in the D_0 profile.

In both case studies, the definition of the $D_{0,\ \mathrm{site}}$ profile was a nontrivial operation. Indeed, as mentioned previously, there is not a common procedure for its computation. Moreover, no detailed information was available, as there were not DH arrays at the site and no specific geophysical studies for its determination were carried out. In these conditions, an estimate of $D_{0,\ \mathrm{site}}$ can be provided through the procedure prescribed by Stewart et al. (2014), who suggested dealing the difference between D_0 and $D_{0,\ \mathrm{site}}$ as an epistemic uncertainty. Therefore, for each ground model, we ran parallel analyses by assuming different $D_{0,\ \mathrm{site}}$ values, given as the sum of D_0—derived through empirical models—and a depth-independent additional damping ΔD, ranging between zero and 5%. This specific study considered three possible ΔD values: 0% (i.e., $D_{0,\ \mathrm{site}}$ coincides with D_0), 2.5% and 5%. The analyses were performed according to the EQL scheme, with the DEEPSOIL v7.0 software (Hashash and Park 2001; Hashash et al. 2017).

9.6.1 Stochastic Database of GRAs

Setting of the GRAs. The procedure for the generation of the 1D ground models consisted of a Monte Carlo simulation, which randomized a set of real soil profiles and assigned a V_S profile and the MRD curves to each ground model. The extraction of the V_S profiles with respect to depth was performed by means of a Monte Carlo procedure, according to the geostatistical model proposed by Passeri (2019), which represents an upgrade of the one introduced by Toro (1995). This new model provides a physically based population of soil models, compatible with the common geological features and the experimental site signatures. Then, the procedure derived the MRD curves from the literature models proposed by Darendeli (2001b), Rollins et al. (1998) and Sun and Idriss (1992). The reader can refer to Passeri (2019) and Passeri et al. (2018c) for further details about the model architecture and parameters for V_S profiles generation, whereas information about the MRD curves assignment is available in Aimar et al. (2020).

For the sake of simplicity, the study focused on some groups of profiles of engineering interest. Figure 9.10a shows the investigated regions, represented in the $V_{S,\ H}$-H domain, that are the time-weighted average of the V_S profile and the thickness of the soil column, respectively. On one side, we included a group of relatively stiff ground models, characterized by $V_{S,\ H}$ of 400 ÷ 450 m/s and bedrock depth close to 50 m, representative of gravelly soil deposits typical of the Alpine valleys (group A). On the other side, we considered three groups of soft soil deposits, with $V_{S,\ H}$ close to 250 m/s and sediment thickness ranging from 15 m (group B) to 50 m (group C) up to 120 m (group D). These groups represent different possible configurations of alluvial basins. Each set consisted of a population of 200 soil models, which can be considered reasonable for statistical purposes. Following the recommendations prescribed in Stewart et al. (2014), for each ground model we performed multiple GRAs, by computing $D_{0,\ \text{site}}$ as the sum of D_0, derived according to the above-mentioned literature models, and an additional damping ΔD equal to 0%, 2.5% and 5%.

The definition of the seismic input motions referred to two Italian sites, characterized by a small-to-moderate and high level of seismicity. The sites were Termeno sulla Strada del Vino and San Severo, characterized by expected values of maximum ground acceleration, a_g, of 0.54 m/s^2 and 2.07 m/s^2, respectively (return period of 475 years, Fig. 9.10b). For each site, seven natural accelerograms were selected from accredited ground motion databases, in compliance with the criteria of seismological compatibility and spectral compatibility with the reference uniform hazard elastic spectrum (Stewart et al. 2014; Ministero delle Infrastrutture e dei Trasporti 2018).

Results. For each ground model, results of the GRAs were averaged through logarithmic mean of the input motions, obtaining a representative response for every soil profile under the reference seismicity level. In order to describe the distribution of the results inside each group, the mean and the standard deviation of the spectral ordinates of the ground models were computed assuming a lognormal distribution of the data (Aimar et al. 2020). The procedure was applied for the four populations of ground models, for each level of additional damping.

Figure 9.11 shows the results in terms of AF for each group of soil models, for low and high seismicities. From the general viewpoint, all the soil models exhibited a smaller spectral amplification moving from high frequencies to periods ranging between 0.05 s and 0.2 s. Then, there was a peak at intermediate periods (i.e., 0.5 ÷ 1.1 s), corresponding to the average fundamental period of each soil group. These features were enhanced for increasing deformability and depth of the soil deposit. For an increasing level of seismicity, the amplification was smaller and the AF curves shifted toward higher vibration periods, due to the rise of nonlinear phenomena.

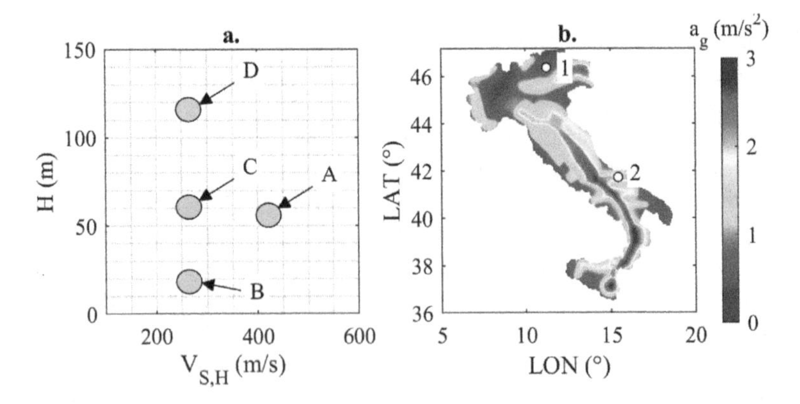

Fig. 9.10 **a** Distribution of the groups of soil deposits in the $V_{S,\ H}$-H domain; **b** position of the reference sites in the Italian seismic hazard map

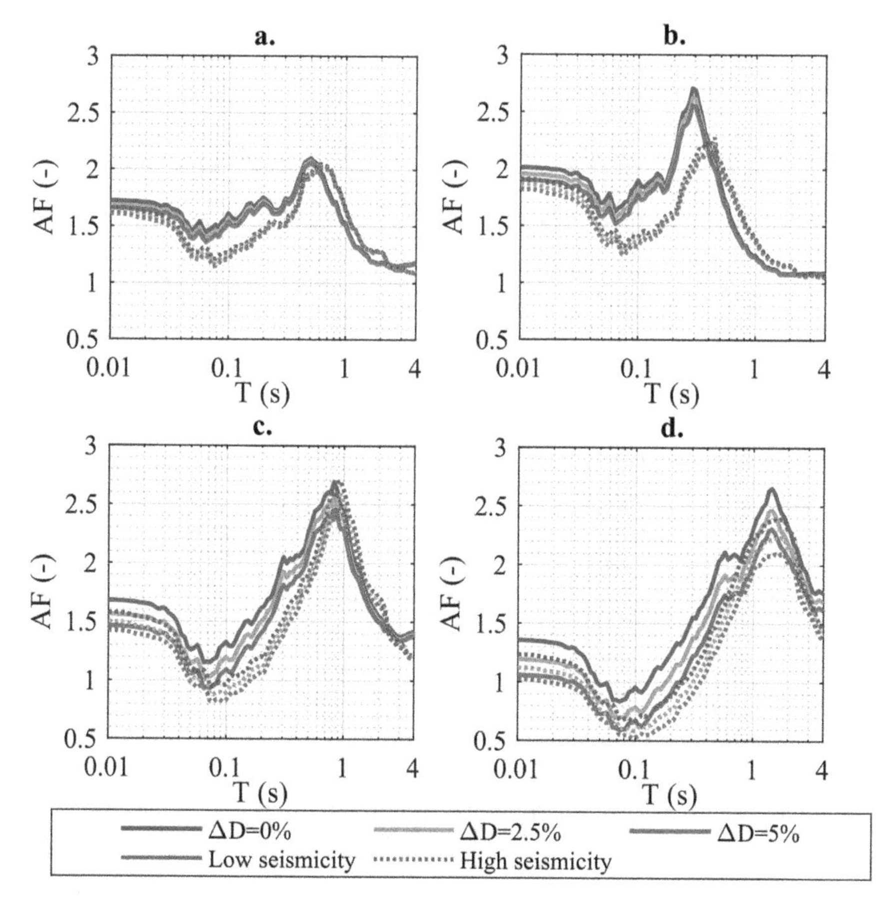

Fig. 9.11 Mean AFs for groups (**a**) A, (**b**) B, (**c**) C and (**d**) D as a function of ΔD and seismicity level

The impact of $D_{0,\ \mathrm{site}}$ depended on the deformability of the ground model and on the level of seismicity. Its increase induced a smaller amplification, but the difference was negligible in shallow soil models (Fig. 9.11a–b), with a maximum of 5% at high frequencies and at resonance just in case of soft deposits. This result was consistent with the findings of Stewart and Kwok (2008). Conversely, variations in $D_{0,\ \mathrm{site}}$ had a strong influence on the seismic amplification in deep and deformable soil deposits (Fig. 9.11c–d). For instance, very deep models underwent a reduction of the AF up to 15% at resonance and 35% at high frequencies for $\Delta D = 5\%$. Similar features were observed under strong seismic input motions, even though the effect was less relevant.

As for the variability in the stratigraphic amplification, Fig. 9.12 shows that it was higher in the range of intermediate periods, exhibiting a narrow peak close to the resonance period in case of shallow soil deposits. The increase of the $D_{0,\ \mathrm{site}}$

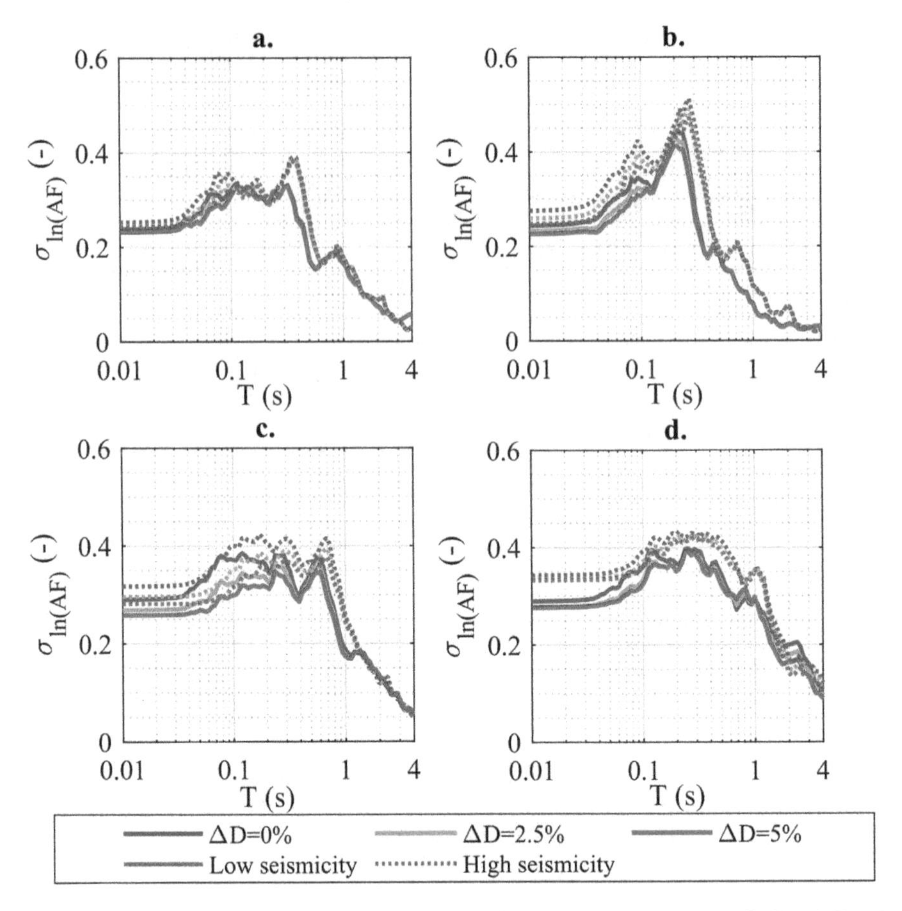

Fig. 9.12 Standard deviation (in logarithmic scale) of the AFs for groups (**a**) A, (**b**) B, (**c**) C and (**d**) D as a function of ΔD and seismicity level

induced a slight reduction in the variability of AF. This variation was negligible for low deformable soil models, whereas some differences were observed at short vibration periods in soft soil deposits. This kind of ground models, indeed, usually exhibits local variations—i.e., thin layers, in 1D conditions—that induce strong variability in the response.

On the other side, increasing $D_{0,\ \mathrm{site}}$ led to an overdamping of the high-frequency components of the wavefield that are more sensitive to such variations. Moreover, the effect of $D_{0,\ \mathrm{site}}$ on the response variability was observed on soft soil models under strong seismic inputs (e.g., Figure 9.12c). A possible reason might be the shifting of the D curve toward higher values at large strains due to the increase of $D_{0,\ \mathrm{site}}$, resulting in an additional attenuation of the high-frequency components of the wave.

9.6.2 The Roccafluvione Case Study

Setting of the GRAs. The stratigraphy of the Roccafluvione site is characterized by a 25-m-thick stratification of silty sands, lying over a formation of sands and gravels. A MASW survey provided an estimate of the V_S profile, shown in Fig. 9.13a. The study adopted the MRD curves proposed by Ciancimino et al. (2019), which is a specialized version of the Darendeli (2001a) model, adapted to capture the specific behavior of silty and clayey soils from the Central Italy area. More details about the stratigraphy and the parameter computation are available in Foti et al. (2019).

The uncertainties of $D_{0,\ site}$ were simulated through the approach suggested by Stewart et al. (2014). More specifically, the study focused on three soil models characterized by the V_S profile shown in Fig. 9.13a and $D_{0,\ site}$ computed as the sum of D_0 and an additional contribution ΔD, equal to 0%, 2.5% and 5%. The D_0 value was computed according to the model proposed by Ciancimino et al. (2019), for a reference frequency of 1 Hz to account for the rate dependence of this parameter.

The input motions consisted of unscaled seismologically and spectrum-compatible acceleration time histories, selected from accredited strong motion databases. The reference hazard levels corresponded to the target uniform hazard spectra for the return period of 50 and 475 years, provided by the National Institute of Geophysics and Volcanology (INGV) (Meletti and Martinelli 2008) (Fig. 9.13b). For each reference return period, ten time histories were selected.

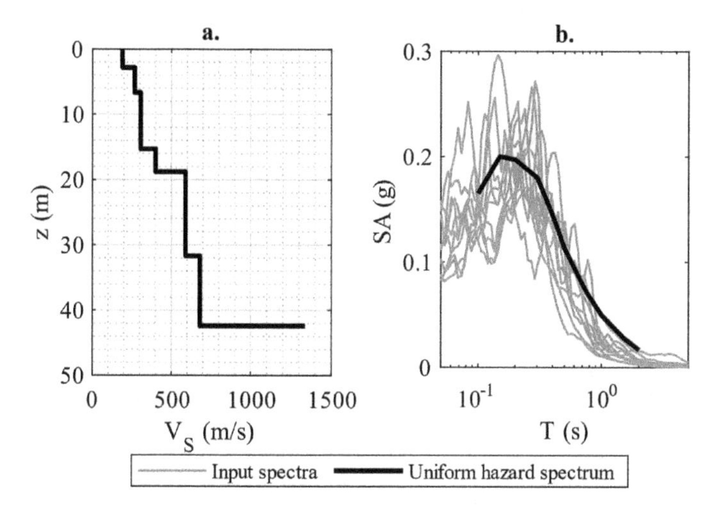

Fig. 9.13 **a** V_S profile obtained from the MASW survey; **b** comparison between the elastic response spectra of the input motions and the uniform hazard spectrum, for the return period of 50 years

Results. For each ground model, results were averaged through logarithmic mean with respect to the input motions, obtaining a representative response for every soil profile under the reference ground motion.

Figure 9.14 shows the AFs for each ground model for the weak motions (Fig. 9.14a) and the strong motions (Fig. 9.14b). The models exhibited a large amplification of the spectral ordinates for a wide range of vibration periods, especially at short periods and close to 0.25 s, which is the fundamental period of the soil deposit.

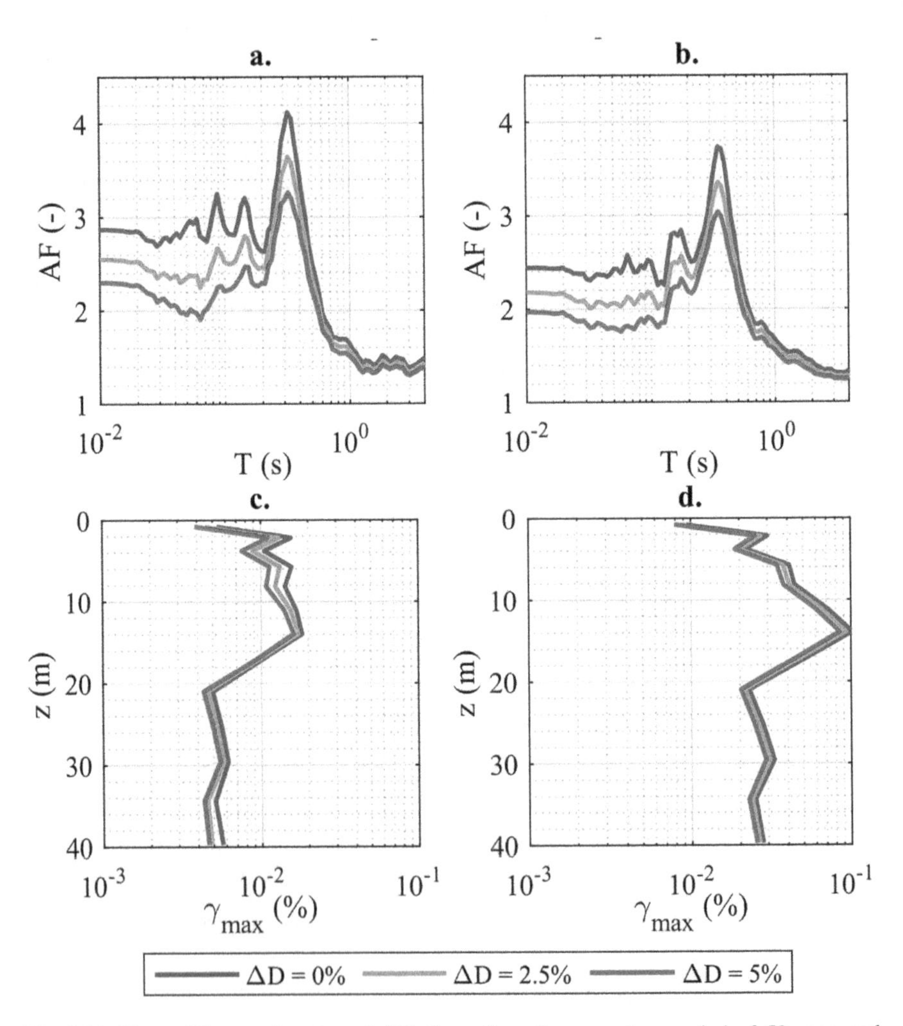

Fig. 9.14 Mean AFs as a function of ΔD, for **a** the reference return period of 50 years and **b** 475 years; maximum strain profiles for an input motion as a function of ΔD, for **c** the reference return period of 50 years and **d** 475 years

The effect of the variations in $D_{0,\,\text{site}}$ resulted in a deviation between the curves: For increasing dissipative properties, AF was smaller and the peaks were smoother. This effect was strong for short vibration periods, as the dissipative properties mainly affect the high-frequency components of the propagating wave, and at the resonance peak. For long vibration periods, the role of ΔD was negligible. When the seismicity level was higher, the role of $D_{0,\,\text{site}}$ was less significant, as the variation in terms of AF was smaller. Indeed, the maximum strain level increased of about one order of magnitude in all the ground model and the peak value shifted from $2 \times 10^{-2\%}$ to $10^{-1\%}$, as shown in Fig. 9.14c–d. Therefore, the nonlinear behavior strongly influenced the response of the soil deposit, and the small-strain parameters were less important.

It is interesting to compare the variations in the AF due to the epistemic uncertainty in the small-strain damping with the variability due to V_S and the MRD curves. Such variability was computed by keeping $D_{0,\,\text{site}}$ as equal to D_0 in Foti et al. (2019), resulting in a distribution of AF represented by the interval defined by the mean and one standard deviation (in logarithmic scale) in Fig. 9.15a–b. By overlapping the curves obtained as a function of ΔD with the AF distribution, we noticed that a change in $D_{0,\,\text{site}}$ led to a variation in the amplification which was not negligible if compared with the overall variability of the results. Indeed, for $\Delta D = 2.5\%$, the AF was close to the lower boundary of the distribution, whereas a value $\Delta D = 5\%$ led to a large reduction of the amplification, which lay completely below the bounds. This effect was relevant especially at high frequencies and close

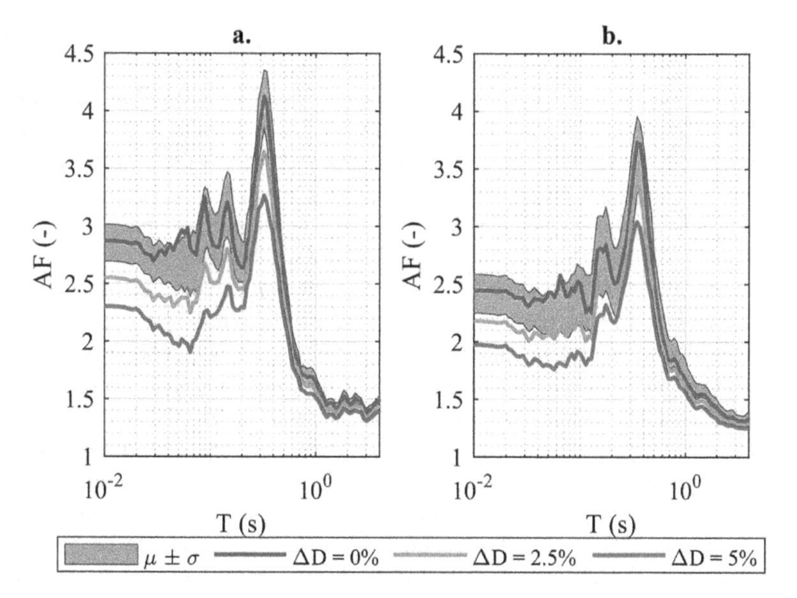

Fig. 9.15 Comparison between the AF distribution obtained by varying V_S and the MRD curves and the AF curves as a function of ΔD for **a** the reference return period of 50 years and **b** 475 years

to the resonance peak. This difference demonstrated that variations in $D_{0,\ site}$ may have an impact as strong as the ones in V_S and the MRD curves, and its proper quantification is necessary for a good prediction of the ground response in seismic conditions.

9.7 Final Remarks

The small-strain damping, along with the V_S profile, governs the response of a site in the almost linear range of the stress–strain behavior. Specifically, it affects the amplitude of the amplification functions.

In the present paper, the critical issues associated with the determination of the small-strain damping were firstly treated. Several tests can be carried out to measure the soil damping, either in laboratory or in situ through specific geophysical investigations. Each test is characterized by different advantages and disadvantages which may affect the measured values, but besides the differences, a main bias emerges: the small-strain damping observed in field is usually higher than the one measured in laboratory. These differences are generally attributed to wave scattering mechanisms that cannot be captured by laboratory tests. This bias affects the overall site response. A GRA performed neglecting wave scattering phenomena will in fact lead to an overestimation of the seismic motion at the surface, as shown by the comparisons of GRAs results with the actual motions measured by borehole arrays.

Although these differences are nowadays well-recognized, the difficulties associated with the determination of $D_{0,\ site}$ lead, usually, to the adoption of laboratory-based values, neglecting wave scattering phenomena. Moreover, the uncertainties associated with the small-strain damping are generally thought to be less relevant when compared, for instance, to the variability coming from the V_S profile and the MRD curves, especially when soil nonlinearity is involved.

A stochastic database of GRAs was then used in the present study to assess the actual influence of the small-strain damping on the seismic amplification. The study involved four different groups of 200 soil profiles subjected to seismic inputs of different intensity. Each group corresponds to a specific range of $V_{S,\ H}$ and bedrock depth. In order to study the influence of wave scattering phenomena, for each ground model, multiple GRAs were performed by computing $D_{0,\ site}$ as the sum of laboratory-based D_0 and a ΔD, assumed to be equal to 0%, 2.5% and 5%. The results show a moderate impact of the small-strain damping. The differences start anyway to be relevant when deep and deformable models are considered. For these models, a reduction of the average amplification function is observed up to 15% at resonance and to 35% at high frequencies, when ΔD is assumed to be 5%.

Subsequently, the influence of D_0 on a specific case study was assessed by considering the well-documented Roccafluvione site. Different input motions were selected to match the uniform hazard spectra of the site for the return periods of 50 and 475 years. A considerable impact of the small-strain damping was observed even when the soil profile is subjected to the higher seismicity level. For an increasing level of ΔD, the amplification functions get significantly reduced and the peaks become smoother. The variability coming from ΔD was then compared to the range of amplification functions defined considering the uncertainties related to MRD curves and V_S profile of the site. At least for this specific situation, ΔD seems to be even more relevant concerning the uncertainties in the other parameters. The amplification function at high frequencies falls in fact outside the range previously defined when at least a ΔD of 2.5% is taken into account. The role of ΔD becomes negligible just for periods higher than 0.5 s.

It is quite evident that the differences between site and laboratory values of the damping cannot be a priori neglected. The uncertainties related to D_0 have proven to be relevant in specific situations, leading to a modification of the expected ground motion at the surface. At the current state of practice anyway, there are not effective methods to take into account the differences between laboratory and site values of D_0. More efforts should then be devoted to improving our knowledge on the topic to develop specific and effective tools to manage the uncertainties related to the small-strain damping.

Acknowledgements Special thanks to Dr Federico Passeri, for letting us use his procedure of generation of soil profiles and for his cooperation on the subject of this paper. The study has been partially supported by the ReLUIS project, funded by the Italian Civil Protection Agency.

References

Afshari K, Stewart JP (2019) Insights from California vertical arrays on the effectiveness of ground response analysis with alternative damping models. Bull Seismol Soc Am 109(4):1250–1264

Aggour M, Yang J, Al-Sanad H (1982) Application of the random decrement technique in the determination of damping of soils. In: European conference on earthquake engineering, vol 7, pp 337–344

Aimar M, Ciancimino A, Foti S (2020) An assessment of the NTC18 simplified procedure for stratigraphic seismic site amplification prediction. Italian Geotechnical Journal—Rivista Italiana di Geotecnica 1

Akeju OV, Senetakis K, Wang Y (2017) Bayesian parameter identification and model selection for normalized modulus reduction curves of soils. J Earthq Eng 1–29

Aki K (1993) Local site effects on weak and strong ground motion. Tectonophysics 218(1–3): 93–111

Anderson DG, Woods RD (1975) Comparison of field and laboratory shear moduli. In: In Situ measurement of soil properties, Raleigh, N.C., ASCE, pp 66–92

Asimaki D, Li W (2012) Site-and ground motion-dependent nonlinear effects in seismological model predictions. Soil Dyn Earthq Eng 32:143–151

Assimaki D, Steidl J, Liu PC (2006) Attenuation and velocity structure for site response analyses via downhole seismogram inversion. Pure Appl Geophy 163(1):81–118

Assimaki D, Li W, Steidl J, Schmedes J (2008a) Quantifying nonlinearity susceptibility via site-response modeling uncertainty at three sites in the Los Angeles Basin. Bull Seismol Soc Am 98(5):2364–2390

Assimaki D, Li W, Steidl JH, Tsuda K (2008b) Site amplification and attenuation via downhole array seismogram inversion: a comparative study of the 2003 Miyagi-Oki aftershock sequence. Bull Seismol Soc Am 98(1):301–330

ASTM D3999/D3999M-11e1 (2011) Standard test methods for the determination of the modulus and damping properties of soils using the cyclic triaxial apparatus. American Society for Testing Material, West Conshohocken, Pennsylvania

ASTM D5311/D5311M–13 (2013) Standard test method for load controlled cyclic triaxial strength of soil. Amer, West Conshohocken, Pennsylvania

ASTM D4428/D4428M-14 (2014) Standard test methods for cross-hole seismic testing. american society for testing material, west Conshohocken, Pennsylvania

ASTM D4015–15e1 (2015) Standard test methods for modulus and damping of soils by fixed-base resonant column devices. American society for testing material, west Conshohocken, Pennsylvania

ASTM D7400-17 (2017) Standard test methods for downhole seismic testing. American Society for Testing Material, West Conshohocken, Pennsylvania

Badsar S (2012) In-Situ determination of material damping in the soil at small deformation ratios (in situ bepaling van de materiaaldemping in de grond bij kleine vervormingen). KU Leuven

Badsar SA, Schevenels M, Haegeman W, Degrande G (2010) Determination of the material damping ratio in the soil from SASW tests using the half-power bandwidth method. Geophys J Int 182(3):1493–1508

Baise LG, Dreger DS, Glaser SD (2003) The effect of shallow San Francisco Bay sediments on waveforms recorded during the Mw 4.6 Bolinas, California, earthquake. Bull Seismol Soc Am 93(1):465–479

Bard PY (2004) SESAME participants: the SESAME project: an overview and main results. In: 13th World conference on earthquake engineering, Vancouver, BC, Canada, pp 1–6

Baturay MB, Stewart JP (2003) Uncertainty and bias in ground-motion estimates from ground response analyses. Bull Seismol Soc Am 93:2025–2042

Beresnev IA, Wen K-L (1996) Nonlinear soil response—a reality? Bull Seismol Soc Am 86(6): 1964–1978

Bergamo P, Maranò S, Imperatori W, Fäh D (2018) Wavedec code: an application to the joint estimation of shear modulus and dissipative properties of the near-surface from multi-component, active surface-wave surveys. In: 36th General assembly of the european seismological commission, ESC2018, Valletta, Malta

Bjerrum L, Landva A (1966) Direct simple-shear tests on a Norwegian quick clay. Geotechnique 16(1):1–20

Boaga J, Renzi S, Deiana R, Cassiani G (2015) Soil damping influence on seismic ground response: a parametric analysis for weak to moderate ground motion. Soil Dyn Earthq Eng 79:71–79

Borcherdt RD (1970) Effects of local geology on ground motion near San Francisco Bay. Bull Seismol Soc Am 60(1):29–61

Burland J, Symes M (1982) A simple axial displacement gauge for use in the triaxial apparatus. Geotechnique 32(1):62–65

Cabas A, Rodriguez-Marek A (2018) Toward improving damping characterization for site response analysis. Geotechnical earthquake engineering and soil dynamics v: seismic hazard analysis, earthquake ground motions, and regional-scale assessment. American Society of Civil Engineers Reston, VA, pp 648–657

Cabas A, Rodriguez-Marek A, Bonilla LF (2017) Estimation of site-specific Kappa (κ 0)-consistent damping values at KiK-Net sites to assess the discrepancy between laboratory-based damping models and observed attenuation (of seismic waves) in the field. Bull Seismol Soc Am 107(5):2258–2271

Campanella RG (1994) Field methods for dynamic geotechnical testing: an overview of capabilities and needs. In: Dyn Geotech Testing II. ASTM Int

Campanella RG, Stewart WP (1991) Downhole seismic cone analysis using digital signal processing. In: Second international conference on recent advances in geotechnical earthquake engineering and soil dynamics, Saint Louis, Missouri, US, pp 77–82

Cascante G, Vanderkooy J, Chung W (2003) Difference between current and voltage measurements in resonant-column testing. Can Geotech J 40(4):806–820

Christensen R (2012) Theory of viscoelasticity: an introduction. Elsevier

Ciancimino A, Lanzo G, Alleanza GA, Amoroso S, Bardotti R, Biondi G, Cascone E, Castelli F, Di Giulio A, d' Onofrio A, Foti S, Lentini V, Madiai C, Vessia G (2019) Dynamic characterization of fine-grained soils in Central Italy by laboratory testing. Bull Earthq Eng (S.I.: SEISMIC MICROZONATION OF CENTRAL ITALY): 29. https://doi.org/10.1007/s10518-019-00611-6

Cornell CA (1968) Engineering seismic risk analysis. Bull Seismol Soc Am 58:1583–1606

Cox BR, Stolte AC, Stokoe II KH, Wotherspoon LM (2018) A direct-push crosshole test method for the in-situ evaluation of high-resolution P-and S-wave velocity. Geotech Test J 42

Crow H, Hunter JA, Motazedian D (2011) Monofrequency in situ damping measurements in Ottawa area soft soils. Soil Dyn Earthq Eng 31(12):1669–1677

Darendeli MB (2001a) Development of a new family of normalized modulus reduction and material damping curves. PhD Dissertation, University of Texas at Austin

Darendeli MB (2001b) Development of a new family of normalized modulus reduction and material damping curves. Doctoral Dissertation, University of Texas at Austin, Austin

Dobry R, Whitman RV, Roesset JM (1971) Soil properties and the one-dimensional theory of earthquake amplification. MIT Department of Civil Engineering, Inter-American Program

d'Onofrio A, Silvestri F, Vinale F (1999) Strain rate dependent behaviour of a natural stiff clay. Soils Found 39(2):69–82

Doroudian M, Vucetic M (1995) A direct simple shear device for measuring small-strain behavior. Geotech Test J 18(1):69–85

Doroudian M, Vucetic M (1998) Small-strain testing in an NGI-type direct simple shear device. In: Proceedings 11th Danube-European conferences on soil mechanics and Geotech. Engrg., Porec, Croatia, AA Balkema, pp 687–693

Faccioli E, Paolucci R, Vanini M (2015) Evaluation of probabilistic site-specific seismic-hazard methods and associated uncertainties, with applications in the Po Plain, northern Italy. Bull Seismol Soc Am 105:2787–2807

Field EH, Jacob KH (1993) Monte-Carlo simulation of the theoretical site response variability at Turkey Flat, California, given the uncertainty in the geotechnically derived input parameters. Earthq Spectra 9:669–701

Foti S (2003) Small-strain stiffness and damping ratio of Pisa clay from surface wave tests. Geotechnique 53(5):455–461

Foti S (2004) Using transfer function for estimating dissipative properties of soils from surface-wave data. Near Surf Geophy 2(4):231–240

Foti S, Lai C, Rix GJ, Strobbia C (2014) Surface wave methods for near-surface site characterization. CRC press

Foti S, Passeri F, Rodriguez-Marek A (2019) Uncertainties and variabilities in seismic ground response analyses. In: Earthquake geotechnical engineering for protection and development of environment and constructions: proceedings of the 7th international conference on earthquake geotechnical engineering, (ICEGE 2019), June 17–20, 2019, Rome, Italy, CRC Press, p 153

Foti S, Aimar M, Ciancimino A, Passeri F (2019) Recent developments in seismic site response evaluation and microzonation. In: Geotechnical engineering, foundation of the future, proceedings of the XVII ECSMGE, pp 223–248

Goto S, Tatsuoka F, Shibuya S, Kim Y, Sato T (1991) A simple gauge for local small strain measurements in the laboratory. Soils Found 31(1):169–180

Hall L, Bodare A (2000) Analyses of the cross-hole method for determining shear wave velocities and damping ratios. Soil Dyn Earthq Eng 20(1–4):167–175

Hardin BO, Drnevich VP (1972) Shear modulus and damping in soils: design equations and curves. J Soil Mechan Found Div 98(7):667–692

Hashash YMA, Park D (2001) Non-linear one-dimensional seismic ground motion propagation in the Mississippi embayment. J Eng Geol 62(1–3):185–206
Hashash Y, Phillips C, Groholski DR (2010) Recent advances in non-linear site response analysis. In: Paper presented at the fifth international conferences on recent advances in geotechnical earthquake engineering and soil dynamics, San Diego, California
Hashash YMA, Musgrove MI, Harmon JA, Okan I, Groholski DR, Phillips CA, Park D (2017) DEEPSOIL 7.0, user manual. University of Illinois at Urbana-Champaign
Hoar RJ, Stokoe KH (1984) Field and laboratory measurements of material damping of soil in shear. In: 8th World conference on earthquake engineering, San Francisco, pp 47–54
Hough SE, Anderson JG (1988) High-frequency spectra observed at Anza, California: implications for Q structure. Bull Seismol Soc Am 78(2):692–707
Huang D, Wang G, Du C, Jin F (2019) Seismic amplification of soil ground with spatially varying shear wave velocity using 2D spectral element method. J Earthq Eng 1–16
Hwang SK (1997) Dynamic properties of natural soils. Ph.D. Dissertation, University of Texas at Austin
Idriss IM (2004) Evolution of the state of practice. In: International workshop on the uncertainties in nonlinear soil properties and their impact on modeling dynamic soil response. Pacific Earthquake Engineering Research Center Richmond, Calif
International ASTM (2011) Standard guide for using the seismic refraction method for subsurface investigation—ASTM D5777. ASTM International
Isenhower WM (1979) Torsional simple shear/resonant column properties of San Francisco Bay mud. University of Texas at Austin
Ishibashi I, Zhang X (1993) Unified dynamic shear moduli and damping ratios of sand and clay. Soils Found 33(1):182–191
Joyner WB, Warrick RE, Oliver AA III (1976) Analysis of seismograms from a downhole array in sediments near San Francisco Bay. Bull Seismol Soc Am 66(3):937–958
Kaklamanos J, Bradley BA, Thompson EM, Baise LG (2013) Critical parameters affecting bias and variability in site-response analyses using KiK-net downhole array data. Bull Seismol Soc Am 103:1733–1749
Kaklamanos J, Baise LG, Thompson EM, Dorfmann L (2015) Comparison of 1D linear, equivalent-linear, and nonlinear site response models at six KiK-net validation sites. Soil Dyn Earthq Eng 69:207–219
Kaklamanos J, Bradley BA, Brandenberg SJ, Manzari MT (2018) Insights from KiK-net data: what input parameters should be addressed to improve site response predictions? In: Geotechnical earthquake engineering soil dynamics V, Austin, pp 454–464
Kim DS (1991) Deformational characteristics of soils at small to intermediate strains from cyclic test. PhD dissertation, University of Texas at Austin
Kim D-S, Stokoe K (1994) Torsional motion monitoring system for small-strain (10–5 to 10–3%) soil testing. Geotech Test J 17(1):17–26
Kim B, Hashash YMA, Stewart JP, Rathje EM, Harmon JA, Musgrove MI, Campbell KW, Silva WJ (2016) Relative differences between nonlinear and equivalent-linear 1-D site response analyses. Earthq Spectra 32:1845–1865
Kokusho T, Yoshida Y, Esashi Y (1982) Dynamic properties of soft clay for wide strain range. Soils Found 22(4):1–18
Kramer SL (1996) Geotechnical earthquake engineering. International series in civil engineering and engineering mechanics, New Jersey
Ktenidou O-J, Abrahamson NA, Drouet S, Cotton F (2015) Understanding the physics of kappa (κ): insights from a downhole array. Geophys J Int 203(1):678–691
Kwok AOL, Stewart JP, Hashash YMA (2008) Nonlinear ground-response analysis of Turkey Flat shallow stiff-soil site to strong ground motion. Bull Seismol Soc Am 98:331–343
Ladd RS, Dutko P (1985) Small strain measurements using triaxial apparatus. In: Advances in the art of testing soils under cyclic conditions. ASCE, pp 148–165

Lai C, Özcebe A (2015) Non-conventional methods for measuring dynamic properties of geomaterials. In: 6th International conference on earthquake geotechnical engineering. christchurch, New Zealand
Lai CG, Rix GJ (1998) Simultaneous inversion of rayleigh phase velocity and attenuation for near-surface site characterization
Lai C, Pallara O, Presti DL, Turco E (2001) Low-strain stiffness and material damping ratio coupling in soils. In: Advanced laboratory stress-strain testing of geomaterials. Routledge, pp 265–274
Lai CG, Rix GJ, Foti S, Roma V (2002) Simultaneous measurement and inversion of surface wave dispersion and attenuation curves. Soil Dyn Earthq Eng 22(9–12):923–930
Li W, Asimaki D (2010) Site-and motion-dependent parametric uncertainty of site-response analyses in earthquake simulations. Bull Seismol Soc Am 100:954–968
Marchetti S, Monaco P, Totani G, Marchetti D (2008) In situ tests by seismic dilatometer. In: Symposium honoring Dr. John H. Schmertmann for his contributions to civil engineering at research to practice in geotechnical engineering, New Orleans, Louisiana, US
Matasovic N, Hashash Y (2012) Practices and procedures for site-specific evaluations of earthquake ground motions, vol Project 20–05 (Topic 42-03)
Matešić L, Vucetic M (2003) Strain-rate effect on soil secant shear modulus at small cyclic strains. J Geotech Geoenviron Eng 129(6):536–549
Meletti C, Martinelli F (2008) I dati online della pericolosità sismica in Italia. esse1.mi.ingv.it
Meng J, Rix G (2003) Reduction of equipment-generated damping in resonant column measurements. Géotechnique 53(5):503–512
Menq FY (2003) Dynamic properties of sandy and gravelly soils. University of Texas, Austin, USA
Michaels P (1998) In situ determination of soil stiffness and damping. J Geotech Geoenviron Eng 124(8):709–719
Ministero delle Infrastrutture e dei Trasporti (2018) DM 17/01/2018—Aggiornamento delle "Norme Tecniche per le Costruzioni"
Misbah AS, Strobbia CL (2014) Joint estimation of modal attenuation and velocity from multichannel surface wave data. Geophysics 79(3):EN25-EN38
Mok YJ, Sanchez-Salinero I, Stokoe KH, Roesset JM (1988) In situ damping measurements by crosshole seismic method. In: Earthquake engineering and soil dynamics II—recent advances in ground-motion evaluation, Park City, Utah, US, pp 305–320
Mun S-C, Zeng S-S (2018) Estimation of Rayleigh wave modal attenuation from near-field seismic data using sparse signal reconstructions. Soil Dyn Earthq Eng 107:1–8
Nour A, Slimani A, Laouami N, Afra H (2003) Finite element model for the probabilistic seismic response of heterogeneous soil profile. Soil Dyn Earthq Eng 23(5):331–348
Olsen K (2000) Site amplification in the Los Angeles basin from three-dimensional modeling of ground motion. Bull Seismol Soc Am 90(6B):S77–S94
Park D, Hashash YM (2005) Evaluation of seismic site factors in the Mississippi Embayment. II. Probabilistic seismic hazard analysis with nonlinear site effects. Soil Dyn Earthq Eng 25(2): 145–156
Park D, Hashash YM (2008) Rate-dependent soil behavior in seismic site response analysis. Can Geotech J 45(4):454–469
Passeri F (2019) Development of an advanced geostatistical model for shear wave velocity profiles to manage uncertainties and variabilities in Ground Response Analyses. Ph. D. dissertation, Politecnico di Torino
Passeri F, Bahrampouri M, Rodriguez-Marek A, Foti S (2018a) Influence of the uncertainty in bedrock characteristics on seismic hazard: a case study in Italy. Paper presented at the geotechnical earthquake engineering and soil dynamics V
Passeri F, Aimar M, Foti S (2018b) Modelli geostatistici per la valutazione delle incertezze e delle variabilità nei profili di Vs. In: Incontro Annuale dei Ricercatori di Geotecnica, Genova, Italy. Associazione Geotecnica Italiana

Passeri F, Foti S, Rodriguez-Marek A (2018c) Geostatistical models for the assessment of the influence of shear wave velocity uncertainty and variability on ground response analyses. Paper presented at the 7th International conference of earthquake geotechnical engineering, Roma, Italy

Pecker A (1995) Validation of small strain properties from recorded weak seismic motions. Soil Dyn Earthq Eng 14(6):399–408

Rathje EM, Kottke AR, Trent WL (2010) Influence of input motion and site property variabilities on seismic site response analysis. J Geotech Geoenviron Eng 136:607–619

Redpath BB, Edwards RB, Hale RJ, Kintzer FC (1982) Development of field techniques to measure damping values for near-surface rocks and soils. NSF

Régnier J, Bonilla LF, Bard PY, Bertrand E, Hollender F, Kawase H, Sicilia D, Arduino P, Amorosi A, Asimaki D (2018) PRENOLIN: international benchmark on 1D nonlinear site-response analysis—validation phase exercise. Bull Seismol Soc Am 108(2):876–900

Richart FE, Hall JR, Woods RD (1970) Vibrations of soils and foundations

Rix GJ, Meng J (2005) A non-resonance method for measuring dynamic soil properties. Geotech Test J 28(1):1–8

Rix GJ, Lai CG, Jr Wesley Spang A (2000) In situ measurement of damping ratio using surface waves. J Geotechn Geoenviron Eng 126(5):472–480

Rix GJ, Lai CG, Foti S (2001) Simultaneous measurement of surface wave dispersion and attenuation curves. Geotech Test J 24(4):350–358

Rodriguez-Marek A, Rathje E, Bommer J, Scherbaum F, Stafford P (2014) Application of single-station sigma and site-response characterization in a probabilistic seismic-hazard analysis for a new nuclear site. Bull Seismol Soc Am 104(4):1601–1619

Rollins KM, Evans MD, Diehl NB, Daily WD III (1998) Shear modulus and damping relationships for gravels. J Geotech Geoenviron Eng 124(5):396–405

Schepers R (1975) A seismic reflection method for solving engineering problems. J Geophy

Schnabel PB, Seed HB (1972) SHAKE: a computer program for earthquake response analysis of horizontally layered sites, vol 72-12. University of California, Berkeley, CA

Seed H, Idriss I (1970) Soil moduli and damping factors for dynamic response analyses, Report no. EERC 70-10. Earthquake Engineering Research Center, University of California, Berkeley, California

Seed HB, Idriss IM (1982) Ground motions and soil liquefaction during earthquakes. Earthquake engineering research insititute

Seed HB, Wong RT, Idriss I, Tokimatsu K (1986) Moduli and damping factors for dynamic analyses of cohesionless soils. J Geotech Eng 112(11):1016–1032

Senetakis K, Anastasiadis A, Pitilakis K (2013) Normalized shear modulus reduction and damping ratio curves of quartz sand and rhyolitic crushed rock. Soils Found 53(6):879–893

Shearer PM, Orcutt JA (1987) Surface and near-surface effects on seismic waves—theory and borehole seismometer results. Bull Seismol Soc Am 77(4):1168–1196

Shi J, Asimaki D (2017) From stiffness to strength: formulation and validation of a hybrid hyperbolic nonlinear soil model for site-response analyses. Bull Seismol Soc Am 107(3): 1336–1355

Shibuya S, Mitachi T, Fukuda F, Degoshi T (1995) Strain rate effects on shear modulus and damping of normally consolidated clay. Geotech Test J 18(3):365–375

Shima E (1962) Modifications of seismic waves in superficial soil layers as verified by comparative observations on and beneath the surface. Bull Earthq Res Inst 40:187–259

Stein S, Wysession M (2003) An introduction to seismology, earthquakes and earth structure

Stewart JP, Kwok AOL (2008) Nonlinear seismic ground response analysis: code usage protocols and verification against vertical array data. Geotechnical earthquake engineering and soil dynamics IV

Stewart JP, Afshari K, Hashash YMA (2014) Guidelines for performing hazard-consistent one-dimensional ground response analysis for ground motion prediction. In: PEER Report 2014

Stokoe K, Santamarina JC (2000) Seismic-wave-based testing in geotechnical engineering. In: ISRM international symposium. International society for rock mechanics

Stokoe K, Darendeli M, Andrus R, Brown L (1999) Dynamic soil properties: laboratory, field and correlation studies. In: 2nd International conference of earthquake geotechnical engineering, pp 811–846

Sun J, Idriss IM (1992) User's manual for SHAKE91: a computer program for conducting equivalent linear seismic response analyses of horizontally layered soil deposits. Center for Geotechnical Modeling, Department of Civil Engineering, University of California, Davis, California

Tao Y, Rathje EM (2019) Insights into modeling small-strain site response derived from downhole array data. J Geotech Geoenviron Eng 145(7):04019023

Tao Y, Rathje EM (2020) Taxonomy for evaluating the site-specific applicability of one-dimensional ground response analysis. Soil Dyn Earthq Eng 128:

Teague DP, Cox BR (2016) Site response implications associated with using non-unique vs profiles from surface wave inversion in comparison with other commonly used methods of accounting for vs uncertainty. Soil Dyn Earthq Eng 91:87–103

Thompson EM, Baise LG, Kayen RE, Guzina BB (2009) Impediments to predicting site response: seismic property estimation and modeling simplifications. Bull Seismol Soc Am 99(5):2927–2949

Thompson EM, Baise LG, Tanaka Y, Kayen RE (2012) A taxonomy of site response complexity. Soil Dyn Earthq Eng 41:32–43

Toro GR (1995) Probabilistic models of site velocity profiles for generic and site-specific ground-motion amplification studies. Brookhaven National Laboratory, Upton, New York

Tsai NC, Housner GW (1970) Calculation of surface motions of a layered half-space. Bull Seismol Soc Am 60(5):1625–1651

Vardanega P, Bolton M (2013) Stiffness of clays and silts: normalizing shear modulus and shear strain. J Geotechn Geoenviron Eng 139(9):1575–1589

Verachtert R, Lombaert G, Degrande G (2017) Multimodal determination of Rayleigh dispersion and attenuation curves using the circle fit method. Geophys J Int 212(3):2143–2158

Vucetic M, Dobry R (1991) Effect of soil plasticity on cyclic response. J Geotech Eng 117(1): 89–107

Wang Y-H, Cascante G, Santamarina JC (2003) Resonant column testing: the inherent counter emf effect. Geotech Test J 26(3):342–352

Xu B, Rathje EM, Hashash Y, Stewart J, Campbell K, Silva WJ (2019) κ 0 for soil sites: observations from Kik-net sites and their use in constraining small-strain damping profiles for site response analysis. Earthq Spectra 2019:0000–0000

Yee E, Stewart JP, Tokimatsu K (2013) Elastic and large-strain nonlinear seismic site response from analysis of vertical array recordings. J Geotechn Geoenviron Eng 139(10):1789–1801

Zalachoris G, Rathje EM (2015) Evaluation of one-dimensional site response techniques using borehole arrays. J Geotech Geoenviron Eng 141(12):04015053

Zhang J, Andrus RD, Juang CH (2005) Normalized shear modulus and material damping ratio relationships. J Geotech Geoenviron Eng 131(4):453–464

Chapter 10
Large Deformation Analysis of Coseismic Landslide Using Material Point Method

Gang Wang, Kewei Feng, and Duruo Huang

10.1 Introduction

Earthquake-induced landslides are one of the most catastrophic effects of earthquakes, as evidenced by many recorded historic events in the past. For example, during the 2008 M_w 7.9 Wenchuan earthquake in China, more than 15,000 incidences of earthquake-induced landslides, rock falls and debris flows increased the death toll by 20,000 (Yin et al. 2009). At Tangjiashan district, the earthquake-induced landslide of over 20.37 million m^3 blocked the main river channel, posing a significant threat to the lives and properties downstream. Among other recent earthquakes, the 2016 M_w 7.0 Kumamoto earthquake in Japan triggered 3460 landslides within an area of about 6000 km^2 (Xu et al. 2018); the 2016 M_w 7.8 Kaikōura earthquake in New Zealand generated more than 10,000 landslides, where the largest one exceeded a volume of 20 million m^3 (Massey et al. 2018).

Over the past decade, significant progress has been achieved in the development of analytical methods to predict earthquake-induced landslides. Of the many landslide assessment methods, the Newmark sliding mass model and its variants have been extensively used to estimate earthquake-induced displacements in natural slopes, earth dams and landfills since the 1960s (Jibson 2007; Du and Wang 2016). Newmark analysis assumes that the slope behaves as a rigid block and the material does not lose strength throughout the earthquake shaking. The slope sliding occurs on a predefined slip surface when the ground acceleration exceeds a critical value and the permanent displacement accumulates. The Newmark type analysis has also been recently integrated with a physics-based numerical method to assess landslides

G. Wang (✉) · K. Feng
Hong Kong University of Science and Technology, Hong Kong SAR, China
e-mail: gwang@ust.hk

D. Huang
Tsinghua University, Beijing, China

T. G. Sitharam et al. (eds.), *Latest Developments in Geotechnical Earthquake Engineering and Soil Dynamics*, Springer Transactions in Civil and Environmental Engineering, https://doi.org/10.1007/978-981-16-1468-2_10

in a regional scale (Huang et al. 2020). However, Newmark analysis provides only a simple index for the triggering of shallow landslides. The complicated seismic landslide failure process and the post-failure performance of a sliding mass cannot be modeled using this method, leading to large uncertainty for landslide prediction (Du et al. 2018a, b). Advanced nonlinear dynamic analyses, such as the finite element method or finite-difference method, can account for significant soil nonlinearity, hydraulic conditions or strength softening behavior during earthquake loading. However, these mesh-based methods often suffer from mesh distortion issues and consequently ill conditioning of the solution scheme when large deformation of the soil is encountered.

Recent advances in material point method (MPM) have made it possible to predict the seismic landslides performance. The MPM was initially proposed by Sulsky et al. (1994) and was further extended (Bardenhagen and Kober 2004). In MPM, the material domain is divided into number of material points, and the computational domain is discretized using a background mesh connected by nodes. The numerical procedure takes advantage of both Lagrangian and Eulerian methods to make it possible to model large deformation failure mechanics of geomaterials. Over the last decade, MPM has a widely application in geomechanics simulation (Andersen and Andersen 2010; Więckowski 2004; Soga et al. 2016; Liang and Zhao 2019), including soil–water interaction (Abe et al. 2013; Bandara and Soga 2015; Wang et al. 2018a), such that retrogressive slope failure can be captured due to rainfall infiltration and change of pore water pressure. However, the MPM has rarely been used in investigating dynamic landslides phenomena. Among the limited literature, Abe et al. (2012) used a modified Cam-Clay model in MPM to simulate seismic slope deformation in shaking table tests. All these examples demonstrate the promise of using MPM in modeling the coseismic landslides. Yet, many gaps still remain in the scientific understanding of earthquake-induced landslides. Important factors, such as soil properties, ground motion, topographic amplification effects and so on, have not yet been investigated.

In the study, we present a numerical model based on material point method to simulate earthquake-induced landslides. Case studies will be performed to demonstrate capacity of the proposed numerical model. The numerical model has been applied to predict the complex failure process of seismic landslides, including the landslides triggering failure and runoff and deposition process. The influence of the soil residual shear strength on the slope kinematic performance is also evaluated in this study.

10.2 Material Point Method

The computational scheme of the MPM method can be summarized in Fig. 10.1 as follows (Soga et al. 2016). First, the entire mass is divided into a set of material points. At same time, the computational domain is discretized using a background mesh connected by nodes, similar to the finite element method (FEM). (1) When computation starts, the information carried by material points (position, velocity, mass, volume, stress) is projected onto the background mesh. (2) Equilibrium equations are solved at the background nodes. (3) The acceleration and velocity field obtained at the background nodes are mapped to material points and are used to calculate strain and strain rate at the material points to update stress and history variables using a continuum constitutive model. (4) Finally, the positions of the material points are updated, and a new iteration step starts. It can be seen that the numerical procedure takes advantage of both Lagrangian and Eulerian methods to overcome difficulties encountered with a conventional FEM. Pure Lagrangian methods (e.g., FEM) typically result in severe mesh distortion and consequently ill conditioning of the solution scheme, while MPM has a distinct advantage in modeling large deformation failure mechanics of geomaterials because Lagrangian material points move within a Eulerian background mesh. MPM can easily incorporate advanced history-dependent soil constitutive models to handle soil dynamic problems. More importantly, prescribing boundary conditions in MPM is more straightforward than other mesh-free methods due to the use of a background mesh. The velocity of the MPM boundary particles can be used to directly prescribe an earthquake wave (Bhandari et al. 2016).

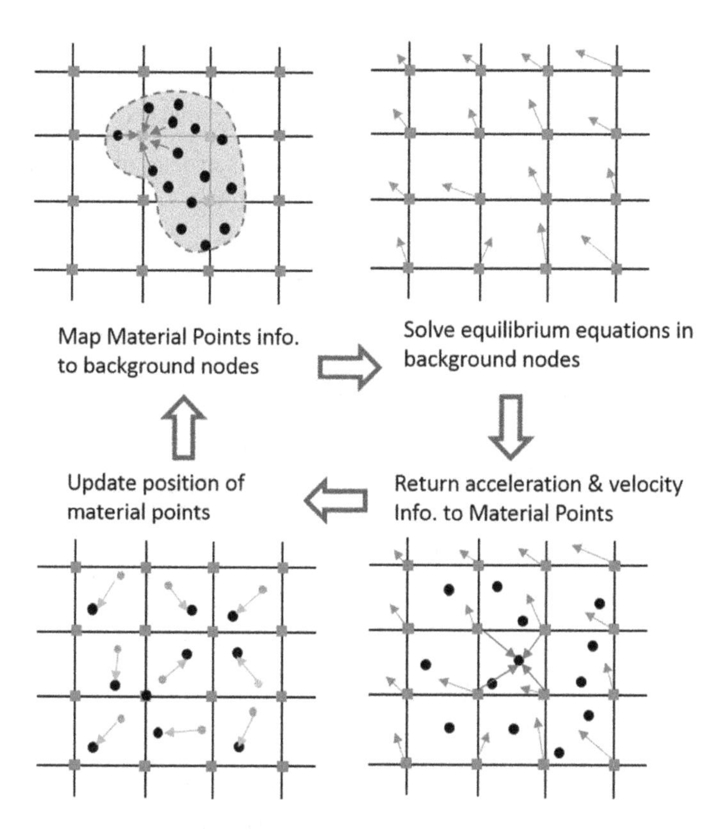

Fig. 10.1 MPM computational scheme (adapted from Soga et al. 2016)

10.3 Numerical Simulation of Dynamic Slope Failure

10.3.1 Model Setup

The failure mechanism of seismic landslides is analyzed in this section, including triggering, evolution of shear deformation and post-failure behavior. The schematic of the slope geometry is presented in Fig. 10.2a, where a slope is inclined at 45°, and the slope height is 20 m.

For the soil overlying bedrock, a strain-softening Mohr–Coulomb constitutive model is implemented to describe the soil behavior. Softening is accounted for by reduction of soil shear strength at large strains, including soil cohesion c, friction angle ϕ and dilation angle ψ, with respect to the accumulated deviatoric plastic strains ε_q^p according to the softening rules Eqs. (10.1–10.3).

$$c = c_r + (c_p - c_r)e^{-\eta\varepsilon_q^p} \tag{10.1}$$

$$\phi = \phi_r + (\phi_p - \phi_r)e^{-\eta\varepsilon_q^p} \tag{10.2}$$

$$\psi = \psi_r + \left(\psi_p - \psi_r\right)e^{-\eta\varepsilon_q^p} \tag{10.3}$$

where c_r, ϕ_r, ψ_r are the residual strength parameters, c_p, ϕ_p, ψ_p are the peak strength parameters, η is the strain-softening parameter, ε_q^p is accumulated deviatoric plastic strains defined as $\varepsilon_q^p = \sqrt{\frac{2}{3}e_{ij}^p e_{ij}^p}$, and e_{ij}^p is the deviatoric part of the plastic strain tensor.

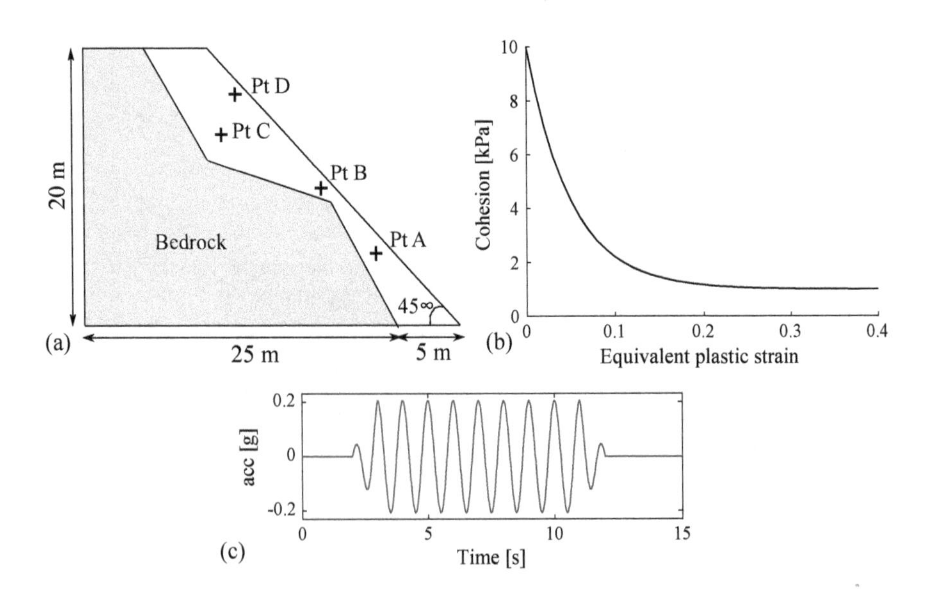

Fig. 10.2 **a** Schematics of slope geometry, **b** strain-softening behavior, **c** input motion

Table 10.1 Material properties for landslide simulation

Property	Value	Property	Value
Young's modulus E (MPa)	187	Peak dilation angle ψ (°)	10
Poisson's ratio v	0.3	Residual dilation angle ψ_r (°)	0
Density ρ_s (kg/m^3)	1800	Peak cohesion c (kPa)	10
Shear wave velocity Vs (m/s)	200	Residual cohesion c_r (kPa)	1
Peak friction angle ϕ (°)	40	Strain-softening parameter (η)	20
Residual friction angle ϕ_r (°)	33		

Parameters for modeling the overlying soil are summarized in Table 10.1. Figure 10.2b shows the softening behavior of soil cohesion as an example. A rough Coulomb friction boundary condition is applied at the interface between the overlying soil and the base of rigid bed rock. Initially, the slope remains in equilibrium, and the initial stress state is obtained by linearly increasing the gravity under the assumption that soil materials behave as a linearly elastic media. Critical damping 5% is applied in the landslides analysis in order to reduce numerical instabilities. A harmonic sinusoidal wave with the predominant frequency of 1 Hz is input at the base of the slope, as shown in Fig. 10.2c.

10.3.2 Dynamic Slope Failure Process

Figure 10.3 shows the progressive failure process of the landslide. At 4 s, a shear band is initiated in the upper part of the soil slope. Under continued shaking, the shear band intensifies, and the soil mass starts to move downward along the slip band, as shown in Fig. 10.3b. The shear zone spreads gradually under seismic shaking, in which the soil mass loses its shear strength following the softening rules and reaches to its residual state quickly. As shown in Fig. 10.3b, c, the upper sliding mass continues to slide on top of the lower soil layer and exerts a relatively high impact pressure on the lower soil layer, which induces generation of the lower shear band and subsequently the second landslide. Large mass flow and interaction between two sliding masses are well captured in the runoff process in Fig. 10.3b, c, which would be extremely difficult for an FEM modeling. Finally, the upper and the lower sliding masses mix and flow together over the base bed rock. Owing to the frictional resistance from the rigid base, the soil mass deposits with a final travel distance at 10 m at the end of the simulation, as shown in Fig. 10.3d.

Several material points in the slope are selected for monitoring during the landslides process (see Fig. 10.4a). Point D is located above the main slip surface in the upper layer of the slope. During the slope failure process, it experiences a longest travel distance (14 m in x direction), then interacts with lower soil layer and finally deposits in the lower part of the slope. Point C lies below the main slip surface, which almost has the same movement as the rigid bed rock. Point A starts

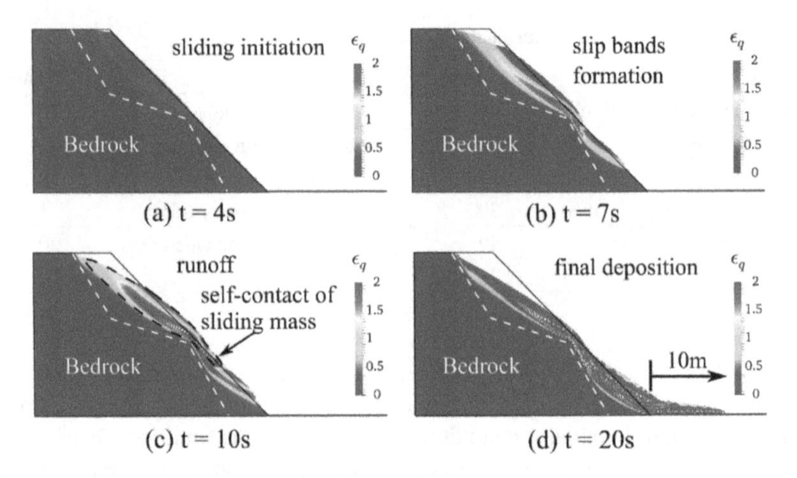

Fig. 10.3 Landslide process (showing deviatoric plastic strain)

to accumulate its displacement later than Point B, when the second sliding mass is induced by the impact pressure from the upper soil mass. The impact pressure exerted on the location Point E is recorded throughout the simulation (see Fig. 10.4b). Initially, the mean effective pressure 14 kPa is obtained due to gravity loading. This soil pressure increases rapidly and achieves a large value of about 44 kPa at 7 s, corresponding to the generation of the second slip surface shown in Fig. 10.3b. During the shaking, the impact pressure fluctuates but keeps a relative high value. Finally, the impact pressure gradually reduces when the sliding mass deposits on the base plate.

It is worth mentioning that simulating soil mass interaction requires a robust self-contact algorithm. When material points are projected onto the same background grid nodes, no-slip contact constraint is inherently reinforced between the material points. Figure 10.3 demonstrates that clusters of material points can generate self-contact during runoff, which is extremely helpful in simulating the complex landslide phenomena.

10.3.3 Effects of Residual Soil Strength

In order to evaluate the influence of residual soil cohesion on the failure mechanism of the landslides, two different residual soil cohesions of 2 and 4 kPa have also been analyzed in this study, while all of the other material properties are kept the same as in Table 10.1. Figure 10.5 demonstrates the equivalent plastic strain contours for these two cases. Similar to the residual soil cohesion $c_r = 1$ kPa case discussed above (Fig. 10.3), shear deformation is initially generated in the upper soil layer in all cases. A second shear band is also developed in the $c_r = 2$ kPa case, with a shorter run-out distance of 6 m in the end. However, owing to the increase in

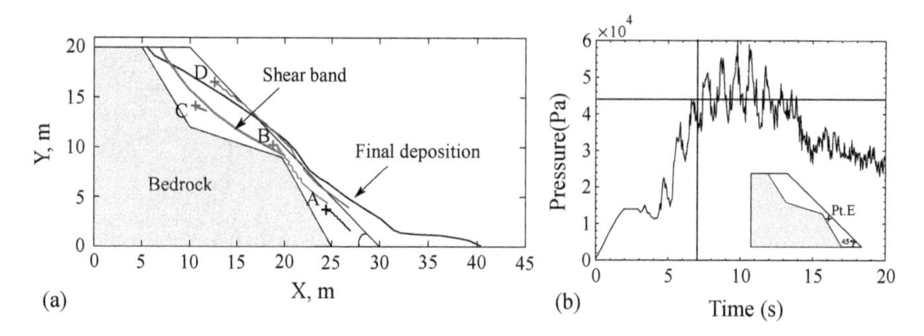

Fig. 10.4 **a** Displaement trajectory and **b** impact pressure on point E

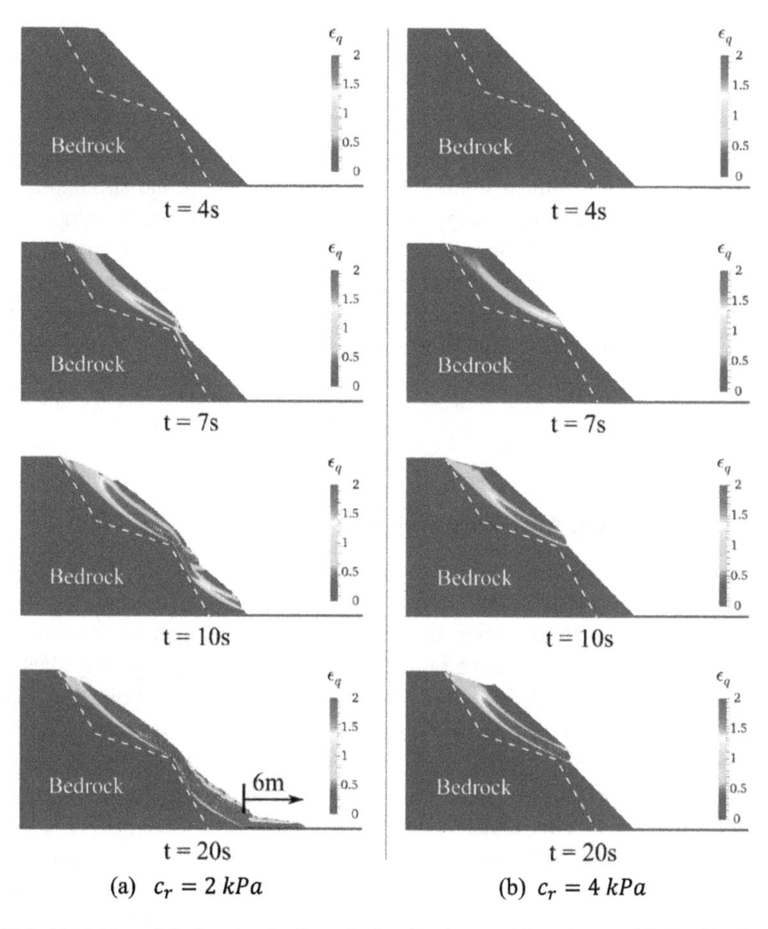

Fig. 10.5 Evolution of deviatoric plastic strain for the slopes with various residual soil cohesions, **a** $c_r = 2$ kPa, **b** $c_r = 4$ kPa

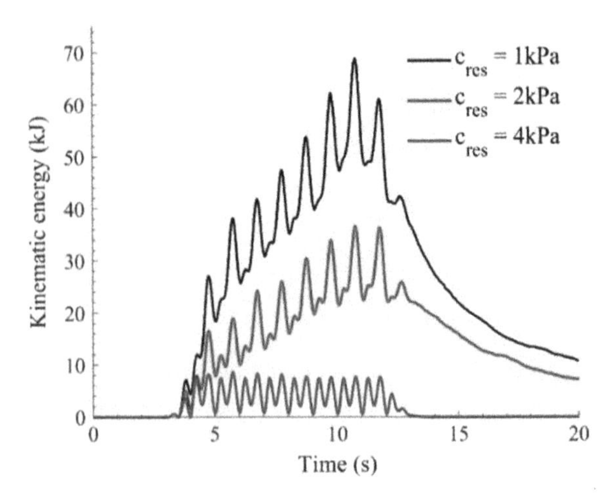

Fig. 10.6 Time histories of kinematic energy for the overlying sediments with various residual soil cohesions

residual soil strength, the landslide behaves differently in the c_r = 4 kPa case. Only one main shear zone is developed in the upper slope during the shaking, and the lower part of the slope still maintains its stability throughout the simulation.

The evolution of kinematic energy for the overlying soil during the slopes failure process is further explored, as shown in Fig. 10.6. At the onset of the ground shaking, the kinematic energy of the soil layer rapidly increases to reach a peak before dropping to a residual value. A maximum kinematic mobility can be found when the soil material has a minimum soil residual strength. For c_r = 4 kPa case, the kinematic energy remains at a lower level and dies down quickly after the ground excitation.

10.4 Conclusions and Discussions

This paper presents a numerical study of coseismic landslides based on material point method. The method demonstrated significant advantage in capturing large deformation and entire progressive failure process, including triggering, shear band formation, runoff and final deposition. It is also interesting to point out that the model has capability in capturing complicated interaction between soil masses during the failure process. The simulation also showed that residual soil cohesion is important in the post-failure stage, in which a higher residual soil strength can prohibit a large runoff after slope failure is triggered.

The present example is limited to dry granular soil simulation. Modeling water–soil interaction is also an important consideration in dynamic analysis of slope and embankment. The development of pore water pressure may influence the mechanical properties of unsaturated soils by changing the soil effective stress, thus resulting in the decrease of soil shear strength or even liquefaction. Recently, we developed a multiphase hydromechanically coupled MPM formulation based on a u-U formulation. Two set of material point layers are adopted to describe the true displacement field of solid skeleton and pore water, respectively. The proposed MPM formulation has been applied to model unsaturated or fully saturated soils involving large soil deformation and to better understand the complex soil–water interaction behavior in pre- and post-failure process of slope failure. In addition, advanced nonlinear soil model (Wang et al. 1990; Wang and Xie 2014) for liquefaction has been implemented in MPM framework, which can be used to realistically assess earthquake-induced liquefaction and large deformation in slopes and embankment dams.

Recent efforts have also been devoted to develop a multiscale, multiphase numerical method by coupling MPM and spectral element method (SEM) for large deformation analysis of coseismic landslides. On the regional scale, SEM is efficient in modeling a 3D wavefield in complex topography on a scale up to hundreds of kilometers (Huang et al. 2020; Wang et al. 2018b). On the local scale, the progressive slope failure process and post-failure large deformation behavior will be studied by MPM. The multiscale MPM-SEM method will be the first numerical tool of its kind for integrated study of the complete coseismic landslide process. These results will be reported in the future.

Acknowledgements This study is supported by General Research Fund 16214519 from Hong Kong Research Grants Council and research fund 2019-KY-02 from State Key Laboratory of Hydroscience and Engineering.

References

Abe K, Shinoda M, Watanabe K, Sanagawa T, Nakajima S, Nakamura S, Kawai T, Murata M, Nakamura H (2012) Numerical simulation of landslides after slope failure using MPM with SYS Cam-clay model in shaking table tests. In: Proceedings of 15th world conference on earthquake engineering

Abe K, Soga K, Bandara S (2013) Material point method for coupled hydromechanical problems. J Geotech Geoenviron Eng 140(3):04013033

Andersen S, Andersen L (2010) Modelling of landslides with the material-point method. Computat Geosci 14(1):137–147

Bandara S, Soga K (2015) Coupling of soil deformation and pore fluid flow using material point method. Comput Geotech 63:199–214

Bardenhagen SG, Kober EM (2004) The generalized interpolation material point method. Comput Model Eng Sci 5(6):477–496

Bhandari T, Hamad F, Moormann C, Sharma KG, Westrich B (2016) Numerical modelling of seismic slope failure using MPM. Comput Geotech 75:126–134

Du W, Wang G (2016) A one-step Newmark displacement model for probabilistic seismic slope displacement hazard analysis. Eng Geol 205:12–23

Du W, Huang D, Wang G (2018a) Quantification of model uncertainty and variability in Newmark displacement analysis. Soil Dyn Earthq Eng 109:286–298

Du W, Wang G, Huang D (2018b) Influence of slope property variabilities on seismic sliding displacement analysis. Eng Geol 242:121–129

Huang D, Wang G, Du C, Jin FS, Feng K, Chen Z (2020) An integrated SEM-Newmark model for physics-based regional coseismic landslide assessment. Earthq Eng Soil Dyn 132:106066

Jibson R (2007) Regression models for estimating coseismic landslide displacement. Eng Geol 91:209–218

Liang W, Zhao J (2019) Multiscale modeling of large deformation in geomechanics. Int J Numer Anal Meth Geomech 43(5):1080–1114

Massey C, Townsend D, Rathje E et al (2018) Landslides triggered by the 14 November 2016 Mw 7.8 Kaikōura Earthquake, New Zealand. Bull Seismol Soc Am 10(3B):1630–1648

Soga K, Alonso E, Yerro A, Kumar K, Bandara S (2016) Trends in large-deformation analysis of landslide mass movements with particular emphasis on the material point method. Géotechnique 66(3):248–273

Sulsky D, Chen Z, Schreyer HL (1994) A particle method for history-dependent materials. Comput Methods Appl Mech Eng 118(1–2):179–196

Wang G, Xie Y (2014) Modified bounding surface hypoplasticity model for sands under cyclic loading. J Eng Mech ASCE 140(1):91–101

Wang ZL, Dafalias YF, Shen CK (1990) Bounding surface hypoplasticity model for sand. J Eng Mech 116(5):983–1001

Wang B, Vardon PJ, Hicks MA (2018a) Rainfall-induced slope collapse with coupled material point method. Eng Geol 239:1–12

Wang G, Du C, Huang D, Jin F, Koo RCH, Kwan JSH (2018b) Parametric models for 3D topographic amplification of ground motions considering subsurface soils. Soil Dyn Earthq Eng 115:41–54

Więckowski Z (2004) The material point method in large strain engineering problems. Comput Methods Appl Mech Eng 193(39–41):4417–4438

Xu C, Ma S, Tan Z, Xie C, Toda S, Huang X (2018) Landslides triggered by the 2016 Mj 7.3 Kumamoto, Japan, earthquake. Landslides 15:551–564

Yin YP, Wang FW, Sun P (2009) Landslide hazards triggered by the 2008 Wenchuan earthquake, Sichuan, China. Landslides 6:139–152

Chapter 11
The State of Art on Equivalent State Theory for Silty Sands

Md Mizanur Rahman

11.1 Introduction

Traditionally, the void ratio (e) has been used as a state variable for the critical state soil mechanics (CSSM) framework. This is because an e represents a state of the force skeleton structure that corresponds to the observed behavior of sand. However, when sufficiently small fine particles are added to the sand particles matrix, the force skeleton structure of the sand-fines mixtures and their behaviors do not change significantly, but their e changes significantly. Therefore, an idea of assuming inactive fine particles as a void to an equivalent void ratio that corresponds to the force skeleton structure was developed. In 1976, Mitchell (1976), perhaps, was the first to propose an equation to calculate the maximum fraction of inactive fine particles in a coarse particles matrix, which later developed the foundation of skeleton void ratio (e_{SK}) where all fine particles are assumed as void. In about a quarter-century, it was realized that the assumption of all fine particles as inactive is not correct as a fraction of fine particles participates in the force skeleton structure of coarse particles. Therefore, the e_{SK} was further modified to the equivalent granular void ratio (e^*) where the active fraction of fine particles is considered via a b parameter (Thevanayagam et al. 2002a). The e^* concept is valid up to a threshold fines content (f_{thr}) where the force skeleton structure is dominated by coarse particles. This simple idea of the equivalent state produced many theories that were evaluated and evolved in the last 50 years via a wide range of experimental settings and discrete element method (DEM) simulations. These include the theories and equations for e_{SK}, e^*, b, equivalent granular critical state line (EG-CSL), equivalent granular state parameter (ψ^*) and a few fully developed constitutive models for monotonic and complicated cyclic loading conditions in a

M. M. Rahman (✉)
University of South Australia, Mawson Lakes, Adelaide, Australia
e-mail: Mizanur.Rahman@unisa.edu.au

T. G. Sitharam et al. (eds.), *Latest Developments in Geotechnical Earthquake Engineering and Soil Dynamics*, Springer Transactions in Civil and Environmental Engineering, https://doi.org/10.1007/978-981-16-1468-2_11

single framework for a range of $f_c < f_{thr}$. All of these theories, collectively, termed as the equivalent state theory hereafter. This article presents a state-of-art development of the equivalent state theory.

11.2 Equivalent State Theory (EST)

11.2.1 Equivalent Granular Void Ratio, e*

The early conceptual development of e^* came from the effort of determining the non-active clay content in the granular phase structure by Mitchell (1976). Shen et al. (1977), Kenney (1977) and Troncoso and Verdugo (1985) who studied natural sand-fines mixtures and tailings, respectively, reported that fine particles did not participate in the force skeleton structure, i.e., granular material's strength remained the same with the addition of fine particles, but their e reduced. Kuerbis et al. (1988) treated fine particles as voids to obtain the skeleton void ratio (e_{sk}), i.e., e_{sk} represents the force skeleton structure of the sand particles. Kuerbis et al. (1988) and others found consistent behavior with e_{sk} for sand-fines mixtures up to a low f_c (Georgiannou et al. 1990; Chu and Leong 2002). The e_{sk} can be written as in Eq. 11.1 when fine and coarse particles share an identical specific gravity.

$$e_{sk} = \frac{e + f_c}{1 - f_c} \tag{11.1}$$

However, it was reported that the quality of correlations between characteristic behaviors with e_{sk} was deteriorated for higher f_c (Zlatovic and Ishihara 1995). Pitman et al. (1994) presented a scanning electron microscope (SEM) imagery showing that a fraction of fine particles comes in between sand particles at $f_c =$ 0.20. Therefore, the assumption that all fine particles are inactive in the force skeleton structure was not correct at least for higher f_c. Therefore, the fraction of f_c contributes to the force skeleton structure was considered via a b parameter, and e_{sk} was modified to e^* as shown in Eq. 11.2 by Thevanayagam (2000):

$$e^* = \frac{e + (1 - b)f_c}{1 - (1 - b)f_c} \tag{11.2}$$

When $b = 0$, the Eq. 11.2 for e^* reduces to Eq. 11.1 for e_{sk}. The major micromechanical assumption in Eq. 11.2 is $(1 - b)$ represents the fraction of inactive fine particles that do not contribute to the force skeleton structure (i.e., to deviatoric stress, q), where b satisfies a condition of $1 \geq b \geq 0$. These assumptions are often challenged, criticized and require further evaluation. In addition to this criticism, the development of a consistent and widely acceptable method for estimating b parameter is a challenge.

11.2.2 Discrete Element Method (DEM) Evidence for Active/Inactive Fine Particles and Their Contribution

Acknowledging the technical limitation on visualizing the contribution of fine particles in the force skeleton structure, a systematic increase of f_c in the matrix of sand particles was simulated using DEM at the University of South Australia (Barnett et al. 2020). The results show that for nearly the same e of 0.675–0.685, the increase of f_c increases the number of particles (N) in the soil matrix but reduces the force skeleton structure at CS as shown in Fig. 11.1. This is due to the inactive contribution of fine particles in the force skeleton structure. However, the active contribution of fines was evaluated via the proportion of fine particles in strong contacts (PR_{SC}). Figure 11.1 shows that the active fraction of fine particles, PR_{SC}, increases with increasing f_c, i.e., b is an increasing function with f_c for the same sand-fines mixture.

11.2.3 Estimation of b

Experimental studies. After the introduction of Eq. 11.2 in 2000, many researchers claimed in their initial studies that b is a constant value for a given sand-fines mixture and loading condition, irrespective of $f_c < f_{thr}$. For example, Thevanayagam and Martin (2002) reported a b of 0.35 for Ottawa sand with silt mixture, and Ni et al. (2004) reported a b of 0.25 for Toyoura sand with silt mixture.

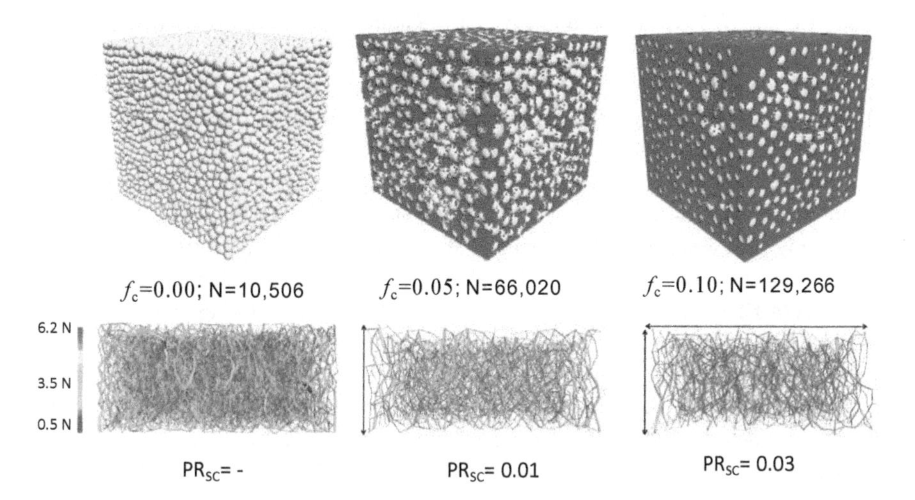

Fig. 11.1 DEM simulation for a systematic increase of f_c in the sand to evaluate active/inactive contribution of f_c in the force skeleton structure (modified after Barnett et al. 2020)

A couple of years later, Thevanayagam et al. (2002b) also suggested that the b values for different sand-fines mixtures may be correlated to $C_{u,c}C_{u,f}^2/R_d$, where $C_{u,c}$ = uniformity coefficient of the coarse particle (sand), $C_{u,f}$ = uniformity coefficient of fine particle, and R_d is the diameter ratio defined as D_{50}/d_{50}, where D is the diameter of the coarser particle, and d is the diameter of the fine particle, and subscript '50' denotes the particle size for 50% finer on particle size distribution (PSD) curve. Ni et al. (2004) also suggested that b values for different sand-fines mixtures decrease with the particle diameter ratio defined by $\chi = D_{10}/d_{50}$ of different soils, where the subscript '10' denotes the particle size for 10% finer on PSD curve.

Almost a decade later in 2009, Rahman et al. (2008) first suggested that b may be a function of χ and f_c, i.e., $f(\chi, f_c)$ and provided a semi-empirical equation for estimating b as shown in Eq. 11.3a, along with a prediction equation for f_{thr} as shown in Eq. 11.3b. The inclusion of f_c in their equation means that b is a variable, not a constant, for the same sand-fines mixtures. In fact, b is an increasing function of f_c. The predicted b value for decreasing f_c ($b \rightarrow 0$) is consistent with the well-established findings for e_{sk} where $b = 0$ for small f_c.

$$b = \left[1 - \exp\left\{-0.30\frac{1}{(1-\chi^{-0.25})}\right\}\right]\left(\frac{f_c/f_{thr}}{\chi}\right)^{1/\chi} \tag{11.3a}$$

$$f_{thr} = 0.40\left\{\frac{1}{1+\exp(0.50-0.13\chi)} + \frac{1}{\chi}\right\} \tag{11.3b}$$

Equations 11.3a and 11.3b have been evaluated with new emerging datasets by independent researchers, and general acceptability was observed (Mohammadi and Qadimi 2015; Lashkari 2014; Chen et al. 2020) and used in many other studies (Barnett et al. 2020; Rahman and Lo 2014; Rahman and Lo 2012). A generalized version of Eq. 11.3a is presented in Table 11.1 along with chronological development of different approaches of estimating b. The generalized equation has been adapted, simplified and used in many studies (Mohammadi and Qadimi 2015; Chen et al. 2020; Rahmani and Abolhasan Naeini 2020).

However, further analysis for the equivalent state theory within this state-of-art lecture is conducted with Eqs. 11.3a and 11.3b, along with Eq. 11.2. It is worth mentioning that there are other forms of equations found in the literature for estimating b as shown in Table 11.1 (Lashkari 2014; Chang and Deng 2019).

Discrete Element Method (DEM) studies. A very few studies have evaluated the critical state behavior of sand-fines mixtures using DEM (Barnett et al. 2020; Ng et al. 2016; Zhou et al. 2018) due to the large computational demand for modelling sand-fines mixtures in DEM. Zhou et al. (2018) reported that the CSL moves downward in the e-log (p') space with increasing f_c, for $f_c < f_{thr}$. This is consistent with many earlier experimental studies (Zlatovic and Ishihara 1995; Murthy et al. 2007). Studying binary mixtures of ellipsoids, Ng et al. (2016) observed e at CS is a decreasing function of f_c. Shire et al. (2016) evaluated the

Table 11.1 Approaches for estimating b from experimental studies

Year of development	Mathematical forms	Explanation/Mathematical expression	References
2000–2002	Constant or $f\left(C_{u,c}C_{u,f}^2/R_d\right)$	b remains constant for same soil and loading condition nonlinear correlation of $C_{u,c}C_{u,f}^2/R_d$	Thevanayagam et al. (2002)
2004	$f(\chi)$; $\chi = D_{10}/d_{50}$	b decreases with χ	Ni et al. (2004)
2008	$f(\chi, f_c)$	$b = \left[1 - \exp\left\{-m\frac{(f_c/f_{thr})^n}{(1-\chi^{-0.25})}\right\}\right]\left(\frac{f_c/f_{thr}}{\chi}\right)^{1/\chi}$	Rahman et al. (2008, 2009)
2008	$f(\chi, f_c)$; simplified	$b = \left[1 - \exp\left\{-0.30\frac{f_c/f_{thr}}{(1-\chi^{-0.25})}\right\}\right]\left(\frac{f_c/f_{thr}}{\chi}\right)^{1/\chi}$ $f_{thr} = 0.40\left\{\frac{1}{1+\exp(0.50-0.13\chi)} + \frac{1}{\chi}\right\}$	Rahman and Lo (2008)
2014	$f(\chi, f_c, r)$	$b = \left[\left(1.93 + 0.04\langle r-1\rangle^2\right)\left\{1 + 3.2\langle r-1\rangle^2\right\}\exp(-22f_c)\right]f_c\chi^{-0.2}$; $\langle X\rangle = X$ when $X > 0$ or 0 and r is roundness ratio of coarse and fine particles	Lashkari (2014)
2015	$f(\chi, f_c)$; simplified	$b = \left[1 - \exp\left\{-0.30\frac{1}{(1-\chi^{-0.25})}\right\}\right]\left(\frac{f_c/f_{thr}}{\chi}\right)^{1/\chi}$	Mohammadi and Qadimi (2015)
2019	$f(\chi, f_c, v_r)$	$b = v_r\left\{1 - (1 - 1/R_d)^S\right\}$; v_r is specific volume ratio of coarse and fine particles, $S = 2, 4, 8$	Chang and Deng (2019)

influence of a stress reduction factor, α on e_{sk}, and their observation suggested that the stress participation of fine particles could be related to the active/inactive participation of fine particles in the force skeleton structure.

Recently, the DEM study of Nguyen et al. (2017) suggested that the inactive fraction of fines $(1 - b)$ can be expressed as the integral of the probability density function of the normal distribution of coordination number (CN), from 0 to 3.5. A mathematical manipulation gives an expression of b as shown in Table 11.2.

Recently, Barnett et al. (2020) presented a comprehensive DEM study for a sand-fines mixture where deviatoric stress components for sand to sand contacts $(s - s)$, sand to fines contacts $(s - f)$ and fines to fines contacts $(f - f)$ were separated, and a b parameter was defined based on the relative contribution of these contacts to the total deviatoric stress (q) as shown in Table 11.2. Barnett et al. (2020) also found that Eqs. 11.3a and 11.3b also achieve the EG-CSL for their DEM data. These two studies are encouraging; however, the first expression of b by Nguyen et al. (2017) depends on CN which is known to vary with particle shapes and PSDs, and the second expression b by Barnett et al. (2020) needs the measurement of q for different contact types which are not possible in practice.

11.2.4 Philosophy of the Equivalent State Theory and a Few Experimental Databases for Evaluation

The anchor concept of the equivalent state theory is that the e^* eliminates the inconsistency in density state, i.e., e, for sand with a range of f_c when they are compared within a single framework. Therefore, many CSSM frameworks for a range of f_c may coalesce to a single CSSM framework within the equivalent state theory. This is simply achieved by merely substituting e^* for e into the established equations for a CSSM framework.

The database in Table 11.3 has been used to evaluate the philosophy of the equivalent state theory using e^* obtained by Eqs. 11.2, 11.3a and 11.3b. The footnote below the table shows a wide range of application of the equivalent state theory.

Table 11.2 Approaches for estimating b from DEM studies

Year of development	Mathematical forms	Explanation/Mathematical expression	References
2017	$f(R_d, f_c, e)$	$b = 1 - \frac{1}{\sigma\sqrt{2\pi}} \int_{x=0}^{x=3.5} \exp\left\{-\frac{(x-C_s)^2}{2\sigma^2}\right\} dx$ $C_s = 4.7(1 - f_c + 2f_c R_d)/\{R_d(e + f_c)\}$	Nguyen et al. (2017)
2020	$f(q)$; q = deviatoric stress	$b = \frac{q_{s-f} + q_{f-f}}{q_{s-s}}$	Barnett et al. (2020)

Table 11.3 Data from literature used to evaluate the equivalent state theory using Eqs. 11.3a, 11.3b and 11.2

Year	Sand	Fines	f_c	Range of e_0	Range of e^*_0	Range of p'_0 (kPa)	References
1977[1]	Iruma Z1	Iruma X1	0–0.14	0.38–0.76	0.55–0.78	50–200	Iwasaki and Tatsuoka (1977)
1977[1]	Iruma W	Iruma X1	0–0.11	0.46–0.56	0.56–0.61	50–200	Iwasaki and Tatsuoka (1977)
1994[3]	Brenda 20/200	Non-plastic	0–0.21	0.44–0.80	0.66–0.80	350	Vaid (1994)
1995[245]	Toyoura sand	Toyoura sand	0–0.30	0.45–0.92	0.83–1.04	50–500	Zlatovic and Ishihara (1995)
1998[1]	Yun-Ling	Yun-Ling	0–0.30	0.79–0.91	0.90–1.12	100–350	Chien and Oh (1998)
1999[3]	Yalesville	Yalesville silt	0–0.37	0.57–1.09	0.75–1.14	–	Polito (1999)
2000[1]	Ottawa	106 Sil-co-Sil	0–0.20	0.36–0.69	0.57–0.75	20–500	Salgado et al. (2000)
2001[3]	Monterey 0/30	Yalesville silt	0–0.25	0.40–0.83	0.67–1.02	–	Polito and Martin (2001)
2001[1245]	Foundary	Sil-co-Sil	0–0.30	0.36–0.86	0.61–0.86	100–350	Thevanayagam and Liang (2001)
2002[345]	OS00 #55	Sil-co-Sil	0–0.25	0.36–0.80	0.60–0.87	100	Thevanayagam and Martin (2002)
2004[2]	Mai Liao	Mai Liao	0–0.30	0.53–0.89	0.76–1.02	50–500	Huang et al. (2004)
2004[2]	Old Allivium	Old Allivium	0–0.09	0.44–0.72	0.63–0.81	215	Ni et al. (2004)
2005[1]	Volcanic soil	Volcanic soil	0.03–0.17	2.47–4.36	3.05–4.58	49–196	Sahaphol and Miura (2005)
2006[2]	Hokkasund	Chengbei	0–0.30	0.52–0.83	0.80–0.91	–	Yang et al. (2006)
2006[1]	Ottawa	Sil-co-Sil	0–0.15	0.38–0.66	0.51–0.69	100	Liu and Mitchell (2006)
2008[2345]	Sydney sand	Majura fines	0–0.30	0.46–0.82	0.74–0.93	100–1300	Rahman et al. (2008, 2014a)
2017[12]	Hostun sand	Quartz fines	0–0.30	0.65–0.89	0.70–1.17	55–200	Goudarzy et al. (2017, 2016)
2020[234]	Ahmedabad	Quarry dust	0–0.30	0.40–0.54	0.44–0.84	100	Rahman and Sitharam (2020)

[1]small strain stiffness; [2]equivalent granular critical state line (EG-CSL); [3]cyclic loading behavior; [4]equivalent granular state parameter; [5]constitutive modelling

11.2.5 Small Strain Stiffness Within the Equivalent State Theory

The most important elastic parameter that can be reliably measured for soil is the small strain shear modulus (G_{max}), which is widely used for mechanical response analysis of geo-materials. Hardin and Black (1966), arguably, were the first to propose one of the most widely used empirical relations in Eq. 11.4 to estimate G_{max}.

$$G_{max} = G_0 f(e)\left(\frac{p'}{p_a}\right)^n p_a \tag{11.4}$$

where G_0 is a material constant which depends on soil type, p_a is atmospheric pressure (≈100 kPa), p' is mean effective stress, and n is an exponent. The $f(e)$ is a function of e, commonly found as $(c - e)^2/(1 + e)$ by Hardin and Black (1966) and as e^d by Jamiolkowski et al. (1995), where c depends on the angularity of soil particles (e.g., $c = 2.97$ for angular sands and 2.17 for rounded sands), and d is a fitting constant. The equivalent state theory suggests that if the fitting constant in Eq. 11.4, i.e., G_0, n, c or G_0, n, d is found for clean sand or sand with an f_c, then substituting e^* for e into Eq. 11.4 should predict G_{max} for sand with other $f_c < f_{thr}$ (Rahman et al. 2012). The modified equation is presented in Eq. 11.5, where the predicted predict G_{max} from equivalent state theory is presented by G^*_{max}.

$$G^*_{max} = G_0 f(e^*)\left(\frac{p'}{p_a}\right)^n p_a \tag{11.5}$$

The G^*_{max} for Hostun sand with up to f_c of 0.30 is predicted by Eq. 11.5 and compared with measured G_{max} by a free–free resonant column device at Ruhr-Universität Bochum as shown in Fig. 11.2a (Goudarzy et al. 2016, 2017). Other six available datasets for sand with a range of f_c as presented in Table 11.3 are presented in Fig. 11.2b. Both figures show excellent agreement.

11.2.6 Equivalent Granular Critical State Line for the Equivalent State Theory

The equivalent state theory suggests that substituting e^* for e into CS data may coalesce CSLs for sand with a range of f_c ($<f_{thr}$) to a single equivalent granular critical state line (EG-CSL) (Rahman and Lo 2008). The EG-CSL has been verified with many datasets; however, only a few of them are reported in Table 11.3 that were verified with Eq. 11.3a. The EG-CSL for Ahmedabad sand with up to 0.30 f_c and Sydney sand with up to 0.30 f_c are presented in Fig. 11.3a, b, respectively.

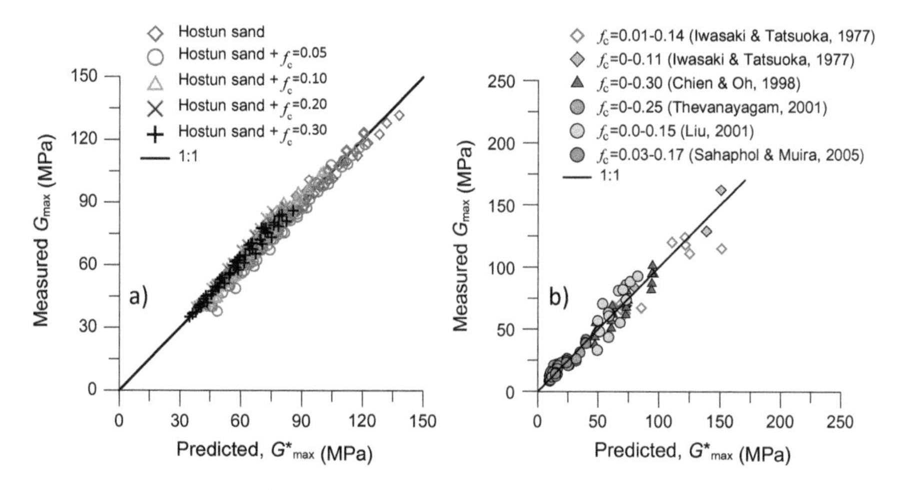

Fig. 11.2 Comparison of measured G_{max} and predicted G^*_{max}: **a** Hostun sand with 0–0.30 f_c, **b** other six available datasets for sand with fines

11.2.7 The Equivalent Granular State Parameter for the Equivalent State Theory

The CSL in e-log (p') space is used as a reference line for a soil's relative state. Soil with a state above the CSL exhibits contractive behavior and below the CSL exhibits dilative behavior. Been and Jefferies (1985) presented a mathematical expression of such relative state by the state parameter, $\psi = e - e_{cs}$, where e_{cs} is the e on the CSL at the same p'. Therefore, a state of $\psi > 0$ corresponds to contractive behavior, and a state of $\psi < 0$ corresponds to dilative behavior (Rabbi et al. 2019a, b; Nguyen et al. 2020a, b; Bobei et al. 2009, 2013). The substitution of e^* for e gives the equation of the equivalent granular state parameter as shown in Eq. 11.6.

$$\psi^* = e^* - e_{cs} \tag{11.6}$$

For the unique EG-CSL, $e_{cs} = e^*_{cs}$. The ψ^* at the beginning of a test is called initial equivalent granular state parameter (ψ^*_0), which correlates well with characteristic features of sand-fines mixtures such as the undrained instability stress ratio, η_{IS} (Rahman and Lo 2012), phase transformation and other characteristic states of cyclic/static shear behavior (Baki et al. 2014; Rahman et al. 2014b).

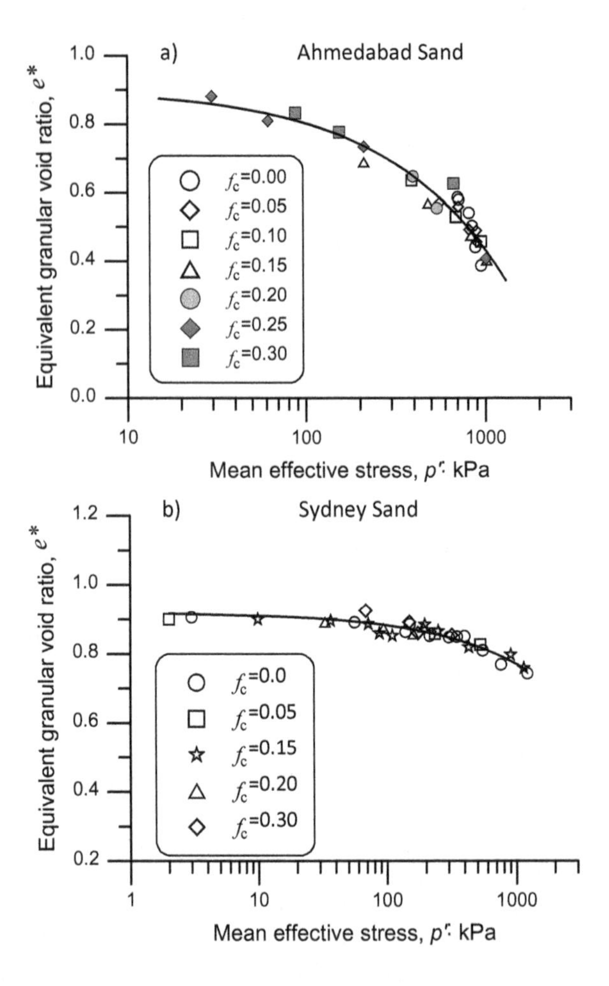

Fig. 11.3 EG-CSL obtained by substituting e^* for e: **a** Ahmedabad sand with 0–0.30 f_c, **b** Sydney sand with 0–0.30 f_c

11.2.8 Static Liquefaction/Instability Within the Equivalent State Theory

The onset of static liquefaction is defined as the onset of deviatoric strain-softening or instability which is often characterized by the effective stress ratio, $\eta_{IS} = q_{IS}/p'$, where $\eta_{IS} = q_{IS}/p'$, and the subscript 'IS' denotes the onset of instability (Chu and Leong 2002; Nguyen et al. 2018; Rabbi et al. 2018; Zhang et al. 2018; Yang 2002). Figure 11.4a shows a trend of decreasing η_{IS} with increasing ψ^*_0, irrespective of f_c. Figure 11.4b also shows a similar trend for other datasets (Rahman and Lo 2012). The scatter of η_{IS} is noticeable, and such a scatter is often presented by instability zone (Baki et al. 2012, 2014 Rahman et al. 2014b).

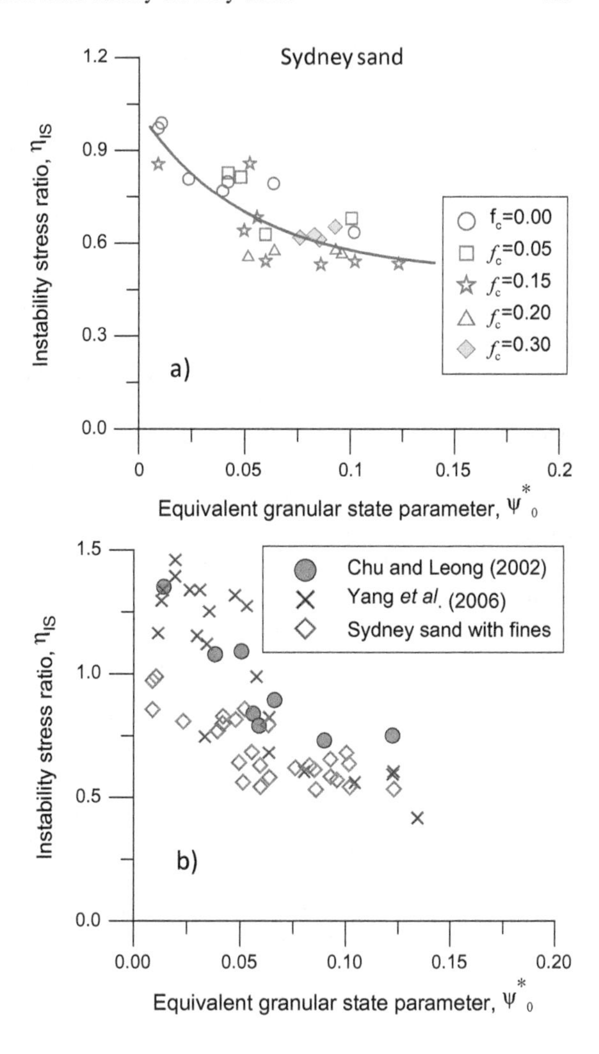

Fig. 11.4 Relation between η_{IS} and ψ^*_0: **a** Sydney sand with 0–0.30 f_c, **b** data from the literature

11.2.9 Cyclic Liquefaction Within the Equivalent State Theory

The cyclic resistance of a soil element depends on its e, p', the average magnitude of cyclic stress ratio (CSR) and the number of cycles it may be exposed due to an expected earthquake. This requires a large number of element tests on undisturbed specimens which is not practical in practice. Alternatively, reconstitute specimens were used (Ishihara 1993), and correction applied for the difference between undisturbed and reconstitution. The current practice of testing procedure for cyclic resistance was demonstrated in Seed and Lee (1966) where they applied equal amplitude of axial stress to a saturated and consolidated soil element until they deform a certain level of peak-to-peak axial strain. It was found in many laboratory studies that the pore water pressure (PWP), Δu, developed to 90–95% of p' at a 5% double amplitude (DA) axial strain, and it was become a common practice to consider 5% DA axial strain as the initiation of liquefaction (Ishihara 1993). However, the PWP development of 90–95% of p' was considered as initial liquefaction for Ahmedabad sand with up to f_c of 0.30 as sued in Fig. 11.5a. The number of cycles required to cause initial liquefaction (N_L) also depends on the applied CSR. For the same e and p', at least a few tests are required to develop a CSR vs N_L relation, which is referred to as cyclic resistance (CR) curve (Rahman and Sitharam 2020). The cyclic resistance ratio (CRR) of a soil can be found from the CR curve for an expected number of cycles, N_L, from an earthquake, e.g., CRR_{xx} is CRR on CR curve for N_L = xx cycles. The relation between CRR_{20} for N_L of 20 and ψ^*_0 for Ahmedabad sand with up to f_c of 0.30 is shown in Fig. 11.5a, and a similar trend is found for other datasets as shown in Fig. 11.5b. These relations are independent of f_c and suggest the applicability of the equivalent state theory. A procedure for liquefaction assessment, along with the performance of a case site during the Bhuj earthquake in 2001 in India, is presented below, and the details can be found in Rahman and Sitharam (2020).

The relation between CRR_{20} and ψ^*_0 for Ahmedabad sand in Fig. 11.5 can be presented by Eq. 11.7, which allows the prediction of liquefaction potential from the field test.

$$CRR_{20} = 0.05\ \exp(-5\psi^*) + 0.04 \tag{11.7}$$

Raju et al. (2004) reported a soil exploration and standard penetration test (SPT) data from a non-liquefied site near Sabarmati Rivers. The soil layers contained a variable fines content from f_c = 0.13 to 0.17 up to the top 14.5 m and the corresponding corrected SPT $(N_1)_{60}$ are shown in Fig. 11.6a. The ψ values were estimated from $(N_1)_{60}$ as discussed in Jefferies and Been (2006) and Rahman and

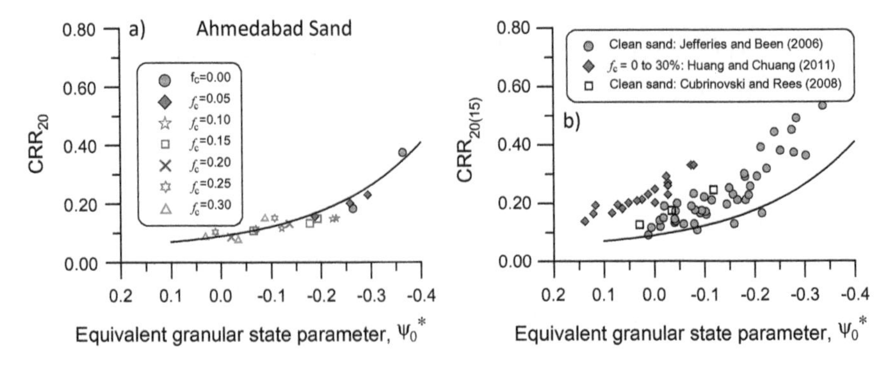

Fig. 11.5 Relation between CRR_{XX} and ψ^*_0: **a** Ahmedabad sand with 0–0.30 f_c, **b** data from the literature

Sitharam (2020). The ψ-values were then converted to a range of ψ*-values for different f_c using the Eq. (11.8).

$$\psi^*/\psi = 1 - (1 - b)f_c \tag{11.8}$$

The range of ψ* was presented by the bar in Fig. 11.6b against CSR values for a 7.5 Mw earthquake (equivalent to Bhuj earthquake in 2001). The (ψ*, CSR) data points fall on the right of CRR curve, i.e., Eq. (11.7). Therefore, the prediction was 'no liquefaction' which matched with the site observation. These data points were also compared with the boundary line by Youd et al. (2001). Again these data points fall on the right of the boundary curve, i.e., 'no liquefaction'. This provides strong evidence that EST and CS approach is applicable for liquefaction assessment.

Several limitations such as specimen preparation methods, the effect of reconstitution and the differences in consolidation state in laboratory experiment and field condition were not considered in this exercise; however, this shows that the CS framework can be used for liquefaction screening of sand with fines. Further research is needed to overcome these limitations in future studies. It is worth noting that the above case study is in line with the recent development of constitutive models for static loading behavior of sand with fines (Lashkari 2014; Rahman et al.

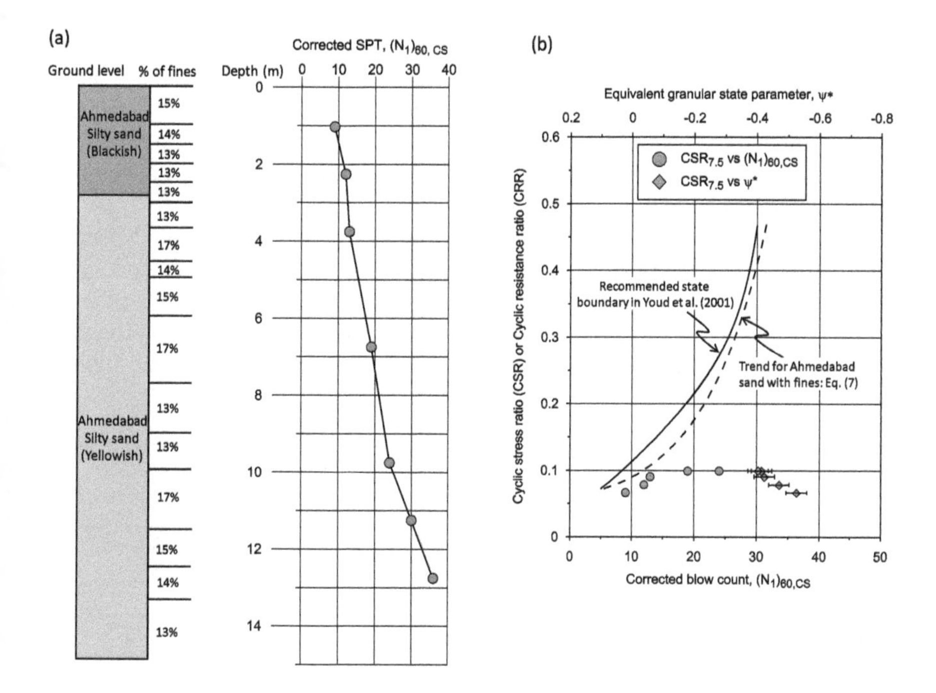

Fig. 11.6 Case study of a non-liquefied site near Sabarmati Rivers: **a** soil exploration and SPT data (modified after Raju et al. 2004), **b** liquefaction screening using EST and CS approach

2014a) to cyclic loading capability. Both models used e^* and ψ^* as state indexes in their constitutive formulations. These are discussed in the next section.

11.3 Constitutive Models Within the Equivalent State Theory

In 2009, Rahman (2009) showed the proof-of-concept that a simple substitution of e^* and ψ^* for e and ψ into state-dependent constitutive models (e.g., SANISAND family of models by Li and Dafalias 2000) works very well for the prediction of drained and undrained behavior under triaxial loading condition. The advantage of this equivalent state constitutive model is that it only requires one set of model parameters either for clean sand or sand with a f_c. In 2014, both Rahman et al. (2014a) and Lashkari (2014) showed that a single set of parameters works very well for sand with a range of f_c for monotonic and cyclic loading behavior, respectively. In 2019, Xu et al. (2019) used the equivalent state theory for modelling monotonic and cyclic loading behavior for Sydney sand with up to f_c of 0.30 and F55 foundry sand with up to f_c of 0.25.

The key constitutive relations after the substitution of e^* and ψ^* for e and ψ into a state-dependent constitutive model by Li and Dafalias (2000) is shown in Table 11.4. The triaxial test data for loose Sydney sand with f_c of 0.15, 0.20 and 0.30 are shown in Fig. 11.7a, and the corresponding model predictions are shown by dotted lines in Fig. 11.7b. An excellent agreement is observed. The triaxial tests data and model prediction for dense Sydney sand with f_c of 0 and 0.20 are shown in Fig. 11.7c and d, respectively. Slight deterioration of model prediction is observed for the dense Sydney sand with fines.

The simple substitution of e^* and ψ^* for e and ψ into state-fabric-dependent constitutive models by Li and Dafalias (2012) was also evaluated. This model is an extension of the model presented in Table 11.4. The model consists of a series of constitutive equations for elastic and plastic strain increment components. The yield function for this model is $f(q,p') = q - \eta p' = 0$. The additive decomposition of

Table 11.4 Key constitutive relations within the equivalent state theory and a single set of model parameters for Sydney sand with up to 0.30 f_c

Description	Mathematical expression	Values
Elasticity (G, v)	$G^* = G_0 \frac{(2.17-e^*)^2}{1+e^*}\left(\frac{p'}{p_a}\right)^{0.7} p_a$	$G_0 = 186$, $v = 0.30$
Yield function	$f(q,p') = q - p' = 0$	–
CSL	$e^* = e_{\lim} - \lambda\left(\frac{p'}{p_a}\right)^{\zeta}$	$e_{\lim} = 0.92$, $\lambda = 0.0375$ $\xi = 0.60$, $M = 1.305$
Dilatancy	$d^* = d_0/M[M\exp(m\psi^* - \eta)]$	$d_0 = 1.06$, $m = 0.05$
Plastic modulus	$K_p^* = C[M\exp(m\psi^* - \eta)]$ $C = (h_1 - h_2 e)G^* \exp(n\psi^*)$	$h_1 = 1.30$, $h_2 = 0.30$ $n = 0.95$

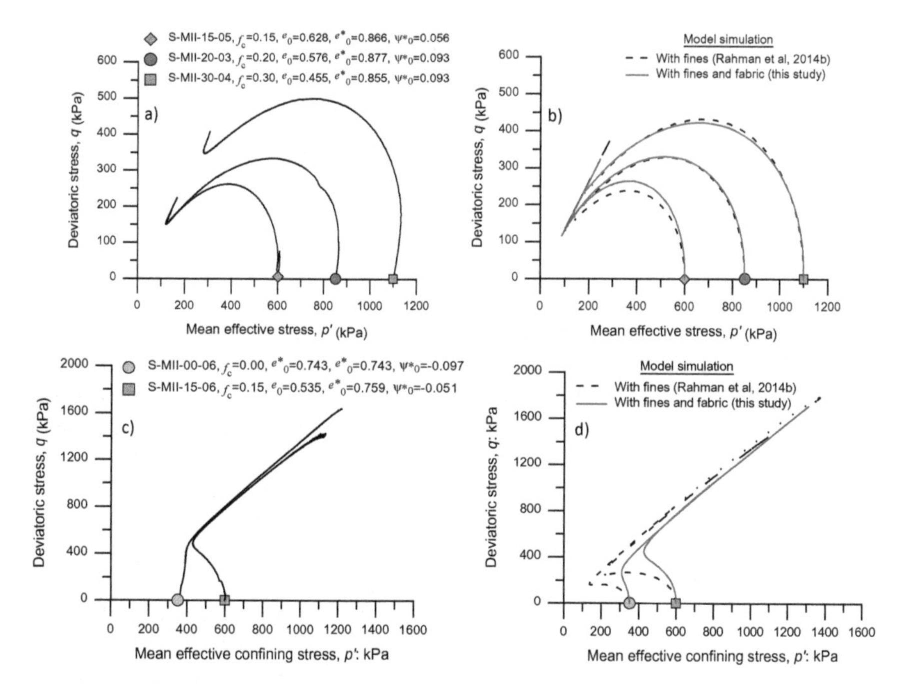

Fig. 11.7 Model prediction of experimental behavior: **a** loose Sydney sand with 0.15–0.30 f_c, **b** model prediction, **c** dense Sydney sand with 0–0.15 f_c, **d** model prediction

strain increment, as in one of SANISAND model (Li and Dafalias 2000), is often considered as

$$d\varepsilon_q = d\varepsilon_q^e + d\varepsilon_q^p = \frac{dq}{3G} + \frac{d}{H} \tag{11.9}$$

$$d\varepsilon_v = d\varepsilon_v^e + d\varepsilon_v^p = \frac{dp'}{K} + d\left|\frac{d\eta}{H}\right| \tag{11.10}$$

where the superscripts 'e' and 'p' denote elastic and plastic components, respectively; G and K are elastic shear and bulk moduli; H is the plastic hardening modulus; d is dilatancy. Note, in triaxial condition, ε_1 and ε_3 are axial and lateral strains, respectively, whereas σ'_1 and σ'_3 are major and minor principal effective stresses, respectively. Other triaxial variables are deviatoric strain ($\varepsilon_q = 2/3[\varepsilon_1 - \varepsilon_3]$), volumetric strain ($\varepsilon_v = \varepsilon_1 + 2\varepsilon_3$), deviatoric stress ($q = \sigma'_1 - \sigma'_3$) and mean effective stress ($p' = [\sigma'_1 + 2\sigma'_3]/3$).

It should be noted that H is a variant of K_p, which is also the plastic hardening modulus from the simple triaxial model in Li and Dafalias (2000). In that model, K_p is also used to calculate the incremental plastic deviatoric strain as the following equation:

$$d\varepsilon_q^p = \frac{p' d\eta}{K_p} \tag{11.11}$$

For dilatancy, Li and Dafalias (2000) proposed an exponential relation between d and the state parameter (ψ) as follows:

$$d = d_0/M[M\exp(m\psi) - \eta] \tag{1.12}$$

The model predictions are shown by blue lines in Fig. 11.7b, d, which are excellent irrespective of density and f_c. Further, the equivalent state theory was combined with a fabric dependent model by Dafalias and Manzari (2004) to evaluate cyclic loading behavior, particularly the link between monotonic instability (static liquefaction) and cyclic instability. This model is also an extension of the model presented in Table 11.4, and thus, the details of this model are not presented here. The triaxial data and model prediction for undrained monotonic triaxial loading are presented in Fig. 11.8a. The instability zone for an initial condition of (e^* of 0.880, p' of 850 kPa) predicted from Fig. 11.4a and shown by the shaded zone in Fig. 11.8a. Both the experimental and model instability fall within the range of instability zone. The model was also used to predict two one-way cyclic loading conditions (different q), and cyclic instability was observed with rapid pore water pressure generation as shown in Fig. 11.8b. Interestingly, cyclic instability triggered around the predicted instability zone from monotonic loading. This suggests the applicability of the equivalent state theory for evaluating linkage between static and cyclic instability (Baki et al. 2012, 2014).

For a more complete triaxial model addressing cyclic loading, the yield function is represented by a cone in multiaxial or a wedge in triaxial stress space, as shown in Manzari and Dafalias (1997) and Dafalias and Manzari (2004). The background theory of cyclic model was developed based on the SANISAND monotonic model. To capture the change in loading direction, Dafalias and Manzari (2004) added the fabric parameter to capture the change of fabric during cyclic loading. The yield function for this model is $f = |\eta - \alpha| - m = 0$, where α and m are both stress ratio

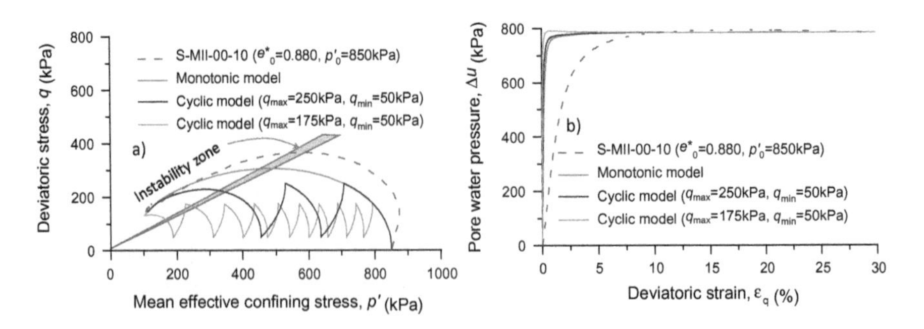

Fig. 11.8 Model prediction of for cyclic instability: **a** q–p' behavior of loose Sydney sand and model prediction, **b** pore water pressure, Δu for loose Sydney sand and model prediction

Table 11.5 Key constitutive relations within the equivalent state theory for cyclic loading and a single set of model parameters for Sydney sand with up to 0.30 f_c

Description	Mathematical expression	Values
Elasticity (G, v)	$G^* = G_0 \frac{(2.17-e^*)^2}{1+e^*}\left(\frac{p'}{p_a}\right)^{0.7} p_a$	$G_0 = 186$, $v = 0.30$
Yield function	$f = \lvert\eta - \alpha\rvert - m = 0$	$m = 0.01$
CSL	$e^* = e_{\lim} - \lambda\left(\frac{p'}{p_a}\right)^{\zeta}$	$e_{\lim} = 0.92$, $\lambda = 0.0375$ $\xi = 0.60$, $M = 1.305$
Dilatancy	$d = A_d(M^d - s\eta)$ $A_d = A_0(1 + \langle sz \rangle)$ $M^d = Me^{n^d\psi}$	$A_0 = 1.06$, $n^d = 0.5$
Plastic modulus	$H = h(M^b - s\eta)$ $h = b_0/\lvert\eta - \eta_{in}\rvert$ $M^b = Me^{-n^b\psi}$	$h_0 = 20$, $c_h = 0.90$ $n^b = 1.06$
Fabric	$dz = -c_z\langle -d\varepsilon_v^p \rangle(sz_{max} + z)$	$z_{max} = 6$, $c_z = 600$

quantities. Therefore, the dilatancy and hardening rules in the cyclic model have changed accordingly. Only CS and elastic parameters remained the same in the cyclic model (Table 11.5).

The plastic modulus controlling the hardening behavior was simulated based on the peak stress ratio (M^b), also known as bounding stress ratio, in the drained test. The plastic hardening modulus (H) is controlled by M^b and can be expressed as the following equation:

$$H = h(M^b - s\eta) \tag{11.13}$$

where s is loading direction (± 1), and h is the function of the state variables:

$$h = \frac{b_0}{\lvert\eta - \eta_{in}\rvert} \tag{11.14}$$

where b_0 can be defined as

$$b_0 = G_0 h_0 (1 - c_h e)(p'/p_a)^{-0.7} \tag{11.15}$$

Dilatancy relates to the difference of current η from the dilatancy stress ratio (M^d). The definition of M^d is similar to the phase transformation line in Ishihara et al. (1975). The dilatancy equation in this model, hence, can be written as

$$d = A_d(M^d - s\eta) \tag{11.16}$$

where A_d is the function of the state and can be expressed as

$$A_d = A_0(1 + \langle sz \rangle) \tag{11.17}$$

where A_0 is the model parameter; z is fabric-dilatancy internal variable; the Macauley brackets $\langle \ldots \rangle$ mean that $\langle sz \rangle = sz$ if $sz > 0$ and $\langle sz \rangle = 0$ if $sz \leq 0$. According to Dafalias and Manzari (2004), the change of fabric in this model influences the dilatancy behavior. The incremental fabric change can be defined as

$$dz = -c_z \langle -d\varepsilon_v^p \rangle (sz_{\max} + z) \tag{11.18}$$

where c_z and $z_{\max}$ are both model parameters.

11.4 Conclusions

The inactive contribution of finer particles in the matrix of coarser particles was long recognized. Perhaps the equation of inactive fraction of fines in the coarse matrix as presented in Mitchell (1976) was the initial form of today's equivalent granular void ratio (e^*). The assumption of the early e^* was that all fine particles are inactive and thus considered as a void in force skeleton structure, i.e., $e^* = e_{SK}$. The next 25 years, e_{SK}, have been used to explain the behavior of sand with fines in a single framework, although the active contribution of fine particles was noticed. In 2000, Thevanayagam (2000) proposed a mathematical equation of e^* where the active fraction of fine particles is considered via a b parameter. In the next 10 years, a constant b value was either back-analyzed or assumed for sand with a range of f_c to calculate e^*. In 2009, Rahman and Lo (2008) first suggested that b is a function of particle size ratio (χ) and f_c, i.e., b is not a constant but a variable for a sand-fines mixture, and presented a generalized equation to calculate b without back-analysis/assumption. This equation is adapted and simplified by many researchers. There are other b equations proposed using experimental and DEM data. However, which approach is used to obtain b and e^* that does not affect the equivalent state theory when substitution of e^* for e gives an equivalent granular critical state line (EG-CSL). This study uses the approach proposed by Rahman and Lo (2008). The major conclusions are as follows:

1. The substitution of e^* for e in the small strain stiffness ($G_{\max}$) model for clean sand or sand with a f_c and the equation can predict $G_{\max}$ for sand with other f_c.
2. The substitution of e^* for e into CS data coalesces CSLs for sand with a range of f_c to a single EG-CSL. This also modifies the state parameter (ψ) to equivalent granular state parameter (ψ^*).
3. The substitution of ψ^* for ψ gives a single relation between characteristic behavior and ψ^* irrespective f_c.

4. The triggering of onset instability during monotonic loading gives a single relation with ψ^* irrespective f_c and the cyclic resistance ratio for 20 cycles to initial liquefaction (CRR_{20}) during conventional cyclic liquefaction tests also gives a single relation with ψ^* irrespective f_c.
5. The substitution of e^* and ψ^* for e and ψ into state-dependent/state-fabric-dependent constitutive models was verified for monotonic (drained/undrained) and cyclic loading conditions by many researchers.

Acknowledgements The author has been working on this topic for nearly 15 years. The development and understanding of the equivalent state theory were achieved in collaboration with many renowned colleagues, to name a few—my Ph.D. supervisor A/Professor Robert Lo (UNSW, Canberra), Professor Yannis Dafalias (University California, Davis), Professor Misko Cubrinovski (University Canterbury, Christchurch), Professor TG Sitharam (Indian Institute of Technology, Guwahati), Late Professor Tom Schanz & Dr Meisam Goudarzy (Ruhr-Universität Bochum) and Dr. Antonio Carraro (Imperial College London). Many former and current graduate students should be acknowledged, particularly Dr Abdul Baki, Dr Khoi Nguyen and Mr. Nick Barnett.

References

Baki MAL, Rahman MM, Lo SR, Gnanendran CT (2012) Linkage between static and cyclic liquefaction of loose sand with a range of fines contents. Can Geotech J 49(8):891–906

Baki MAL, Rahman MM, Lo SR (2014) Predicting onset of cyclic instability of loose sand with fines using instability curves. Soil Dyn Earthq Eng 61–62:140–151

Barnett N, Rahman MM, Karim MR, Nguyen HBK, Carraro JAH (2020) Equivalent state theory for sand with non-plastic fine mixtures: a DEM investigation. Géotechnique p. Accepted

Been K, Jefferies MG (1985) A state parameter for sands. Géotechnique 35(2):99–112

Bobei D, Lo S, Wanatowski D, Gnanendran C, Rahman M (2009) Modified state parameter for characterizing static liquefaction of sand with fines. Can Geotech J 46(3):281–295

Bobei DC, Wanatowski D, Rahman MM, Lo SR, Gnanendran CT (2013) The effect of drained pre-shearing on the undrained behaviour of loose sand with a small amount of fines. Acta Geotech 8(3):311–322

Chang CS, Deng Y (2019) Revisiting the concept of inter-granular void ratio in view of particle packing theory. Géotechnique Letters 9(2):121–129

Chen G, Wu Q, Zhao K, Shen Z, Yang J (2020) A binary packing material-based procedure for evaluating soil liquefaction triggering during earthquakes. J Geotechn Geoenviron Eng 146 (6):04020040

Chien LK, Oh YN (1998) Influence on the shear modulus and damping ratio of hydraulic reclaimed soil in West Taiwan. Int J Offshore Polar Eng 8(3):228–235

Chu J, Leong WK (2002) Effect of fines on instability behaviour of loose sand. Géotechnique 52 (10):751–755

Dafalias Y, Manzari M (2004) Simple plasticity sand model accounting for fabric change effects. J Eng Mech 130(6):622–634

Georgiannou VN, Burland JB, Hight DW (1990) The undrained behaviour of clayey sands in triaxial compression and extension. Géotechnique 40(3):431–449

Goudarzy M, Rahman MM, König D, Schanz T (2016) Influence of non-plastic fines content on maximum shear modulus of granular materials. Soils Found 56(6):973–983

Goudarzy M, Rahemi N, Rahman MM, Schanz T (2017) Predicting the maximum shear modulus of sands containing non-plastic fines. J Geotech Geoenviron Eng 143(9):1–5

Hardin BO, Black WL (1966) Sand stiffness under various triaxial stresses. J Soil Mech Found Div ASCE 92(2):27–42
Huang Y-T, Huang A-B, Kuo Y-C, Tsai M-D (2004) A laboratory study on the undrained strength of silty sand from Central Western Taiwan. Soil Dyn Earthq Eng 24:733–743
Ishihara K (1993) Liquefaction and flow failure during earthquakes. Géotechnique 43(3):351–415
Ishihara K, Tatsuoka F, Yasuda S (1975) Undrained deformation and liquefaction of sand under cyclic stresses. Soils Found 15(1):29–44
Iwasaki T, Tatsuoka F (1977) Effects of grain size and grading on dynamic shear moduli of sands. Soils Found 17(3):19–35
Jamiolkowski M, Lancellotta R, Lo Presti DCF (1995) Remarks on the stiffness at small strains of six italian clays. In: Proceedings of Pre-failure deformation of geomaterials: the international symposium, pp 817–836. AA Balkema, Aapporo, Japan
Jefferies M, Been K (2006) Soil liquefaction: a critical state approach. Taylor & Francis, London
Kenney TC (1977) Residual strength of mineral mixture. In: Proceedings of 9th international conference of soil mechanics and foundation engineering, Tokyo
Kuerbis R, Negussey D, Vaid YP (1988) Effect of gradation and fine content on the undrained response of sand. In: Hydraulic fill structure, Geotechnical Special Publication 21, ASCE, New York
Lashkari A (2014) Recommendations for extension and re-calibration of an existing sand constitutive model taking into account varying non-plastic fines content. Soil Dyn Earthq Eng 61–62:212–238
Li XS, Dafalias YF (2000) Dilatancy for cohesionless soils. Géotechnique 50(4):449–460
Li X, Dafalias Y (2012) Anisotropic critical state theory: role of fabric. J Eng Mech 138(3):263–275
Liu N, Mitchell JK (2006) Influence of nonplastic fines on shear wave velocity-based assessment of liquefaction. J Geotech Geoenviron Eng 132(8):1091–1097
Manzari MT, Dafalias YF (1997) A critical state two-surface plasticity model for sands. Géotechnique 47(2):255–272
Mitchell JK (1976) Fundamental of soil behaviour, 1st edn, pp 170–172. John Wiley & Sons, Inc
Mohammadi A, Qadimi A (2015) A simple critical state approach to predicting the cyclic and monotonic response of sands with different fines contents using the equivalent intergranular void ratio. Acta Geotech 10(5):587–606
Murthy TG, Loukidis D, Carraro JAH, Prezzi M, Salgado R (2007) Undrained monotonic response of clean and silty sands. Géotechnique 57(3):273–288
Ng T-T, Zhou W, Chang X-L (2016) Effect of particle shape and fine content on the behavior of binary mixture. J Eng Mech 143(1):C4016008
Nguyen T-K, Benahmed N, Hicher P-Y (2017) Determination of the equivalent intergranular void ratio—application to the instability and the critical state of silty sand. Powders Grains, EPJ Web of Conferences 140(02019):1–4
Nguyen HBK, Rahman MM, Fourie AB (2018) Characteristic behavior of drained and undrained triaxial compression tests: DEM study. J Geotech Geoenviron Eng 144(9):04018060
Nguyen HBK, Rahman MM, Fourie AB (2020a) How particle shape affects the critical state and instability triggering of a granular material: results from a DEM study. Géotechnique 1–18. https://doi.org/10.1680/jgeot.18.p.211
Nguyen HBK, Rahman MM, Fourie A (2020b) Effect of particle shape on constitutive relation: A DEM study. J Geotech Geoenviron Eng 146(7):04020058
Ni Q, Tan TS, Dasari GR, Hight DW (2004) Contribution of fines to the compressive strength of mixed soils. Géotechnique 54(9):561–569
Pitman TD, Robertson PK, Sego DC (1994) Influence of fines on the collapse of loose sands. Can Geotech J 31(5):728–739
Polito CP (1999) The effects of non-plastic and plastic fines on the liquefaction of sandy soils, p 274. The Virginia Polytechnic Institute and State University, Blacksburg, USA
Polito CP, Martin JR (2001) Effects of nonplastic fines on the liquefaction resistance of solids. J Geotech Geoenviron Eng 127(5):408–415

Rabbi ATMZ, Rahman MM, Cameron DA (2018) Undrained behavior of silty sand and the role of isotropic and K0 consolidation. J Geotech Geoenviron Eng 144(4):04018014

Rabbi ATMZ, Rahman MM, Cameron DA (2019a) Instability of natural silty sand under undrained and constant shear drained path: a critical state study. Int J Geomech 19 (8):04019083

Rabbi ATMZ, Rahman MM, Cameron DA (2019b) The relation between the state indices and the characteristic features of undrained behaviour of silty sand. Soils Found 59(4):801–813

Rahman MM (2009) Modelling the influence of fines on liquefaction behaviour. In: Department of Civil Engineering. University of New South Wales at Australian Defence Force Academy, Canberra, Australia

Rahman MM, Lo SR (2008) The prediction of equivalent granular steady state line of loose sand with fines. Geomech Geoeng 3(3):179–190

Rahman MM, Lo SR (2012) Predicting the onset of static liquefaction of loose sand with fines. J Geotechn Geoenviron Eng 138(8):1037–1041

Rahman MM, Lo SR (2014) Undrained behaviour of sand-fines mixtures and their state parameters. J Geotechn Geoenviron Eng 140(7):04014036

Rahman MM, Sitharam TG (2020) Cyclic liquefaction screening of sand with non-plastic fines: critical state approach. Geosci Front 11(2):429–438

Rahman MM, Lo SR, Gnanendran CT (2008) On equivalent granular void ratio and steady state behaviour of loose sand with fines. Can Geotech J 45(10):1439–1456

Rahman MM, Lo SR, Gnanendran CT (2009) Reply to the discussion by Wanatowski and Chu on "On equivalent granular void ratio and steady state behaviour of loose sand with fines". Can Geotech J 46(4):483–486

Rahman MM, Cubrinovski M, Lo SR (2012) Initial shear modulus of sandy soils and equivalent granular void ratio. Geomech Geoeng 7(3):219–226

Rahman MM, Lo S-CR, Dafalias YF (2014a) Modelling the static liquefaction of sand with low-plasticity fines. Géotechnique 64(11):881–894

Rahman M, Baki M, Lo S (2014b) Prediction of undrained monotonic and cyclic liquefaction behavior of sand with fines based on the equivalent granular state parameter. Int J Geomech 14 (2):254–266

Rahmani H, Abolhasan Naeini S (2020) Influence of non-plastic fine on static iquefaction and undrained monotonic behavior of sandy gravel. Eng Geol 275:105729

Raju LG, Ramana GV, Rao CH, Sitharam TG (2004) Site specific ground response analysis. Geotech Earthq Hazaeds 87(10):1354–1362

Sahaphol T, Miura S (2005) Shear moduli of volcanic soils. Soil Dyn Earthq Eng 25(2):157–165

Salgado R, Bandini P, Karim A (2000) Shear strength and stiffness of silty sand. J Geotech Geoenviron Eng 126(5):451–462

Seed HB, Lee KL (1966) Liquefaction of saturated sands during cyclic loading. J Soil Mech Found Div ASCE 92(SM6):105–134

Shen CK, JL, Vrymoed, Uyeno CK (1977) The effect of fines on liquefaction of sands. In: The 9th International conference on soil mechanics and foundation engineering, pp 381–385. Tokyo

Shire T, O'Sullivan C, Hanley KJ (2016) The influence of fines content and size-ratio on the micro-scale properties of dense bimodal materials. Granular Matter 18(3):1–10

Thevanayagam S (2000) Liquefaction potential and undrained fragility of silty soils. In: Proceedings of 12th world conference earthquake engineering. New Zealand Society of Earthquake Engineering, Auckland, New Zealand

Thevanayagam S, Liang J (2001) Shear wave velocity relations for silty and gravely soils. In: Prakash S (ed) 4th international conference on soil dynamics and earthquake engineering. University of Missouri Rolla, San Diego, CA, pp 1–15

Thevanayagam S, Martin GR (2002) Liquefaction in silty soils-screening and remediation issues. Soil Dyn Earthq Eng 22(9–12):1035–1042

Thevanayagam S, Shenthan T, Mohan S, Liang J (2002a) Undrained fragility of clean sands, silty sands, and sandy silts. J Geotech Geoenviron Eng 128(10):849–859

Thevanayagam S, Kanagalingam T, Shenthan T (2002b) Contact density-confining stress-energy to liquefaction. In: 15th ASCE engineering mechanics conference. Columbia University, New York

Troncoso JH, Verdugo R (1985) Silt content and dynamic behaviour of tailing sands. In: Proceedings of 11th international conference on soil mechanics and foundation engineering

Vaid YP (1994) Liquefaction of silty soils in ground failure under seismic conditions. In: Prakash S, Dakoulas P (eds) Geotech Spl publ. No. 44, pp 1–16

Xu L-Y, Zhang J-Z, Cai F, Chen W-Y, Xue Y-Y (2019) Constitutive modeling the undrained behaviors of sands with non-plastic fines under monotonic and cyclic loading. Soil Dyn Earthq Eng 123:413–424

Yang J (2002) Non-uniqueness of flow liquefaction line for loose sand. Géotechnique 52(10):757–760

Yang SL, Lacasse S, Sandven RF (2006) Determination of the transitional fines content of mixtures of sand and non-plastic fines. Geotech Test J 29(2):102–107

Youd TL et al (2001) Liquefaction resistance of soils: summary report from the 1996 NCEER and 1998 NCEER/NSF workshops on evaluation of liquefaction resistance of soils. J Geotech Geoenviron Eng 127(10):817–833

Zhang J, Lo S-CR, Rahman MM, Yan J (2018) Characterizing monotonic behavior of pond ash within critical state approach. J Geotech Geoenviron Eng 144(1):04017100

Zhou W, Wu W, Ma G, Ng T-T, Chang X (2018) Undrained behavior of binary granular mixtures with different fines contents. Powder Technol 340:139–153

Zlatovic S, Ishihara K (1995) On the influence of nonplastic fines on residul strength. In: Ishihara K, Balkema AA(eds) Proceedings of IS-TOKY0'95/the first international conference on earthquake geotechnical engineering/Tokyo/14–16 November 1995, pp 239–244. Rotterdam, Tokyo, Japan

Chapter 12
Forensic Evaluation of Long-Distance Flow in Gently Sloped Ground During the 2018 Sulawesi Earthquake, Indonesia

Hemanta Hazarika, Divyesh Rohit, Takashi Kiyota, Mitsu Okamura, Siavash Manafi Khajeh Pasha, and Sukiman Nurdin

12.1 Introduction

The Sulawesi Island of the Indonesian archipelago was struck by a powerful earthquake of moment magnitude (Mw) 7.5 on September 28, 2018, at 18:02:44 local time. The earthquake had a hypocentral depth of 20 km with epicenter at 0.256°S (latitude) and 119.846°E (longitude) in the Donggala Regency of Minahasa Peninsula, about 70 km north of the provincial capital of Palu (USGS 2018). The event had a peak intensity of 8.5 on MMI scale observed in Palu region, which experienced the devastating damage. The mainshock was preceded by a major foreshock of magnitude Mw 6.1 and followed by multiple aftershocks of magnitude Mw > 5.5 for many days. The tremors of the earthquake were felt till the Eastern Kalimantan region of Borneo Islands as well as Tawau district in Malaysia. The event triggered extensive flow-slides in ground with very gentle gradient (about 1–5%) in some areas, such as Balaroa, Petobo, Jono Oge and Sibalaya. Numerous landslides, in the coastal areas of Palu, Donggala and Mamuju, also caused tsunami. According to the National Agency of Disaster Management, Indonesia, it caused 2,101 causalities, 4438 injuries and with 1373 still missing (BNPB 2018). The

H. Hazarika (✉) · D. Rohit
Kyushu University, Fukuoka 819-0395, Japan
e-mail: hazarika@civil.kyushu-u.ac.jp

T. Kiyota
Tokyo University, Tokyo 113-8654, Japan

M. Okamura
Ehime University, Ehime 790-8577, Japan

S. M. K. Pasha
IMAGEi Consultant, Tokyo 102-0083, Japan

S. Nurdin
Tadulako University, Palu 94148, Indonesia

T. G. Sitharam et al. (eds.), *Latest Developments in Geotechnical Earthquake Engineering and Soil Dynamics*, Springer Transactions in Civil and Environmental Engineering, https://doi.org/10.1007/978-981-16-1468-2_12

locations of mainshock, foreshock and aftershocks are shown in Fig. 12.1 (Hazarika et al. 2020). The data presented in Fig. 12.1 are for the duration between September 28 and October 8, 2018. The occurrence and magnitude of foreshock, mainshock and aftershocks signify the amount of energy accumulated across the Palu–Koro fault (PKF). The earthquake was triggered by the strike-slip movement of PKF which straddles the Palu city. The event created a rupture of length about 200 km in 30 s at a speed of 4–4.1 km/s, which lies in between the P and S wave propagation velocities, also termed as super-shear rupture (Bao et al. 2019).

The horizontal and vertical motions in EW, NS and UD directions are depicted in Fig. 12.2a, b along with their power spectrum recorded at the Japan International Cooperation Agency (JICA) seismological station in Palu (BMKG and JICA 2018). The measured peak ground accelerations (PGAs) in each (EW, NS and UD) direction were 199 Gals, 200 Gals and 335 Gals, respectively. The dominant frequency of strong motion is from 0.2 to 0.6 Hz, which can be attributed to many geotechnical damages.

The extraordinary and extensive large-scale flow failures in some areas of Palu city, Donggala and Sigi Regency of Central Sulawesi inflicted widespread damage to public and private infrastructural facilities including irrigation structures, houses, government buildings and bridges. Also, the event caused loss of thousands of lives, injuries to many more and thousands of missing. Balaroa, Jono Oge, Petobo

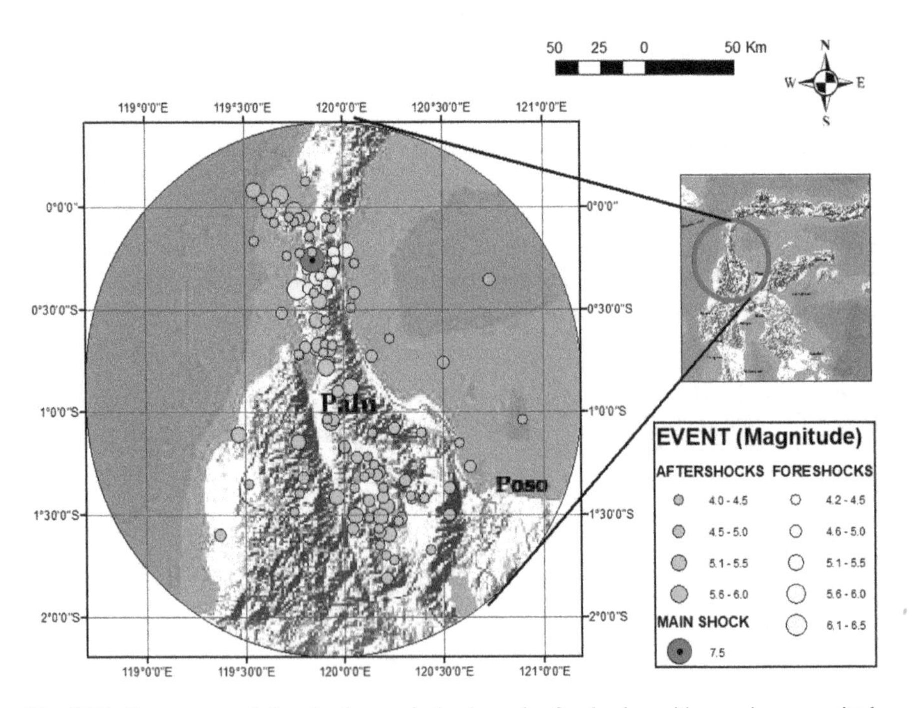

Fig. 12.1 Occurrence of foreshocks, mainshock and aftershocks with varying magnitude (Hazarika et al. 2020)

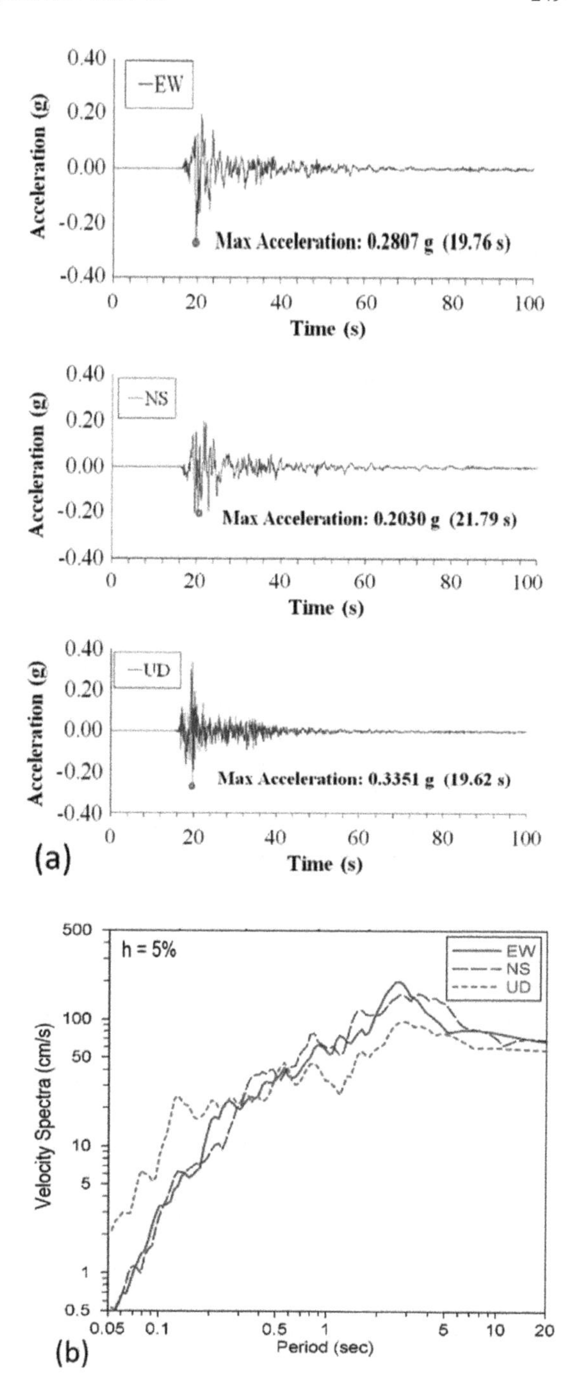

Fig. 12.2 Strong ground motion: **a** E–W, N–S and U–D components and **b** velocity response spectrum (modified from JICA report 2019)

and Sibalaya of Palu city were severely affected due to the large-scale flow-slides and mud flows. It was first time in the history of recorded seismic events that such large-scale flow failures were initiated by an earthquake, sweeping away thousands of houses, and that too, in grounds with gentle gradients. According to Kiyota et al. (2020), these long-distance flow-slides were caused by multiple factors such as presence of less-permeable cap layer leading to the formation of interlayer water called water film, liquefaction of underlying sandy layer, presence of confined aquifer and overall geology of the area.

The geological as well as seismological characteristics of Central Sulawesi region, the findings from the aerial, geotechnical and geological investigations and subsequent data analyses, performed by the authors aftermath the Palu earthquake for each flow-slide site, are described in the following sections.

12.2 Geological and Seismological Characteristics of Central Sulawesi Region

Sulawesi Island has an intricate tectonic cluster that segregates the converging Eurasian, Indo-Australian and Philippines Sea Plate (Kadarusman et al. 2011; Bellier et al. 2001; Watkinson et al. 2011). Due to the intense collision and interaction of the three plates, Sulawesi Island has multiple active faults, among which Palu–Koro fault (PKF) is the most seismically active. The sinistral PKF straddles the Palu region of the Central Sulawesi Province. On September 28, 2018, several meters of horizontal crustal displacement were observed across the PKF, which was established through satellite image analysis (Heidarzadeh et al. 2019). All the flow-slide sites were located across or near the fault zone.

The PKF with an average slip rate of 40 mm/year and an extension rate of 11–14 mm/year is the most active fault on earth, which is quite contrasting for a simple strike-slip fault (Socquet et al. 2019). The greater Palu area is surrounded by 10-km-wide mountains and lowland on eastern and western banks. The earthquake ground motions triggered four large-scale flow-slides in Balaroa, Petobo, Jono Oge and Sibalaya, with multiple resembling features in the flow-slide and the surrounding terrain.

The geomorphology of Palu city and vicinity is contrasting in eastern and western side of the lowland region along Palu River (van Leeuwen 2005). The underlying rocks on the eastern side are mainly Triassic to Jurassic metamorphic rocks (Palu Complex—Gumbasa Complex), while those on the western side are sedimentary rocks of the Upper Cretaceous, and there are penetrating rocks (Palu Granite/Kambuno Granite) in the underlying rocks. The plains along Palu River are overlaid by Holocene deposits of mainly alluvial fans, while the stratigraphy of the catchment area of Balaroa, Petobo, Jono Oge and Sibalaya, where the extensive flow-slides occurred, consists of granites and granitic metamorphic rocks (gneissose granite) which tend to be weathered into sandy soil with small uniformity

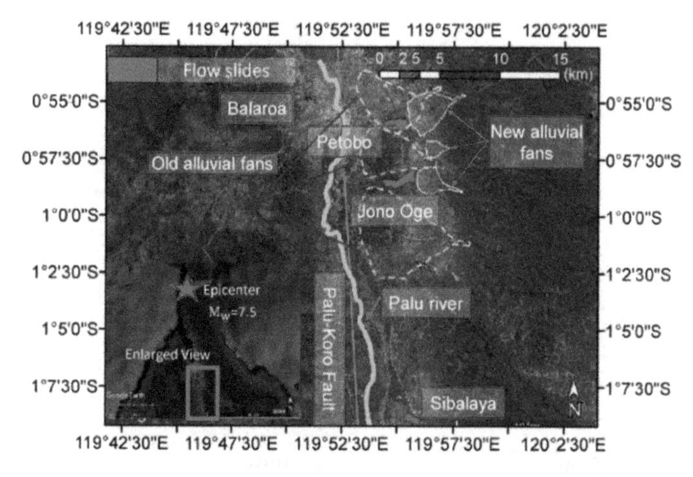

Fig. 12.3 Distribution of alluvial fans and the locations of long-distance flow-slide (marked in red) in different areas of Palu (Hazarika et al. 2020)

coefficient of particle size. Uniquely, in the western side of Palu Valley, there was no flow-slide observed with area overlain by sedimentary rocks, which are vastly distributed around the flow-slide area of Balaroa (Kiyota et al. 2020).

As shown in Fig. 12.3, the flow-slide sites are all situated along the edges of Palu Valley in the low-lying regions, where the new alluvial fans converge with the old alluvial fan deposits of Palu River (Hazarika et al. 2020). Further, according to Kiyota et al. (2020) the recent alluvial fans situated at higher elevation are comparably small and with a gradient of 5% had groundwater at lower level due to the permeable coarse gravel layers. On the other hand, old alluvial fans located at lower elevation are comparably large with a gradient of 2–5%. Due to shallow water table and easy access to groundwater, these areas have been utilized for growing paddy, with formation of some human settlements. Extensive flow-slides occurred in the floodplain with loose overlying sediment layers, located in the lowland area between the large fans.

12.3 Earthquake-Induced Flow-Slides and the Resulting Damage

Figure 12.4 shows an outline of the flow-slide consisting the collapsed part (main scarp), flow part and sedimentation part. The devastated areas have very gentle gradient with an average of 1–5%, and they are located directly below a knick line on a new alluvial fan. Furthermore, surface water was present in all the damaged areas despite scanty rainfall prior to the survey. The subsections below discuss the findings of the field investigation at all the major flow-slide sites by the three

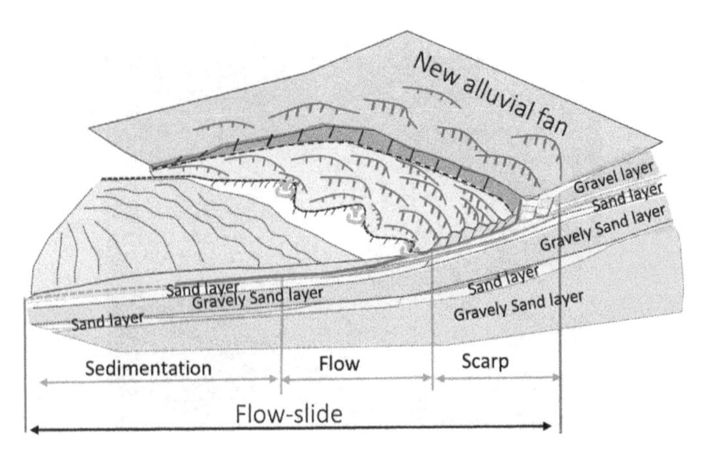

Fig. 12.4 Outline of long-distance flow-slide (JICA 2019)

research groups led by the authors from Kyushu University, University of Tokyo and Ehime University, Japan.

12.3.1 Flow-Slides at Jono Oge

Flow-slide extent and ground profile

The flow-slide at Jono Oge was the longest with respect to the debris flow and ground movement among all the flow-slides, which were triggered due to the 2018 Sulawesi earthquake. The origin of the flow-slide was observed at the crown, just besides the irrigation channel in eastern edge of Palu Valley (Fig. 12.5). The flow-slide was spread over an area of 1.5 km^2 of area with maximum displacement of 2 km. (Hazarika et al. 2020). The location of flow-slide was at the bottom of alluvial fan created at the valley mouth in the space between old alluvial fans with gentle gradient, low elevation and shallow groundwater table (Kiyota et al. 2020). The flow-slide damaged approximately 500 units of houses in the area with many being swept away or buried under the soil. It should be noted that 90% of the flow-slide area was utilized for agricultural activities. Shallow groundwater table and subsurface water in the overlying sandy layers, due to the presence of irrigation channel, probably would have aggravated the weak ground conditions and contributed significantly to the extensive flow-slides stretching over a few kilometers.

Characteristics of ground movement and soil profile.

Based on the observations made by Hazarika et al. (2020), the flow-slide profile in Jono Oge is categorized into four zones (I, II, III and IV) according to ground movement mechanism, surface features and maximum displacement (Fig. 12.6).

(a)

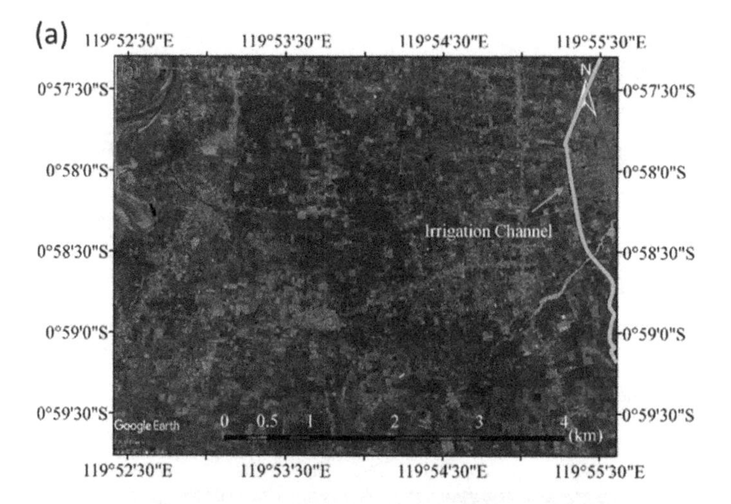

(b)

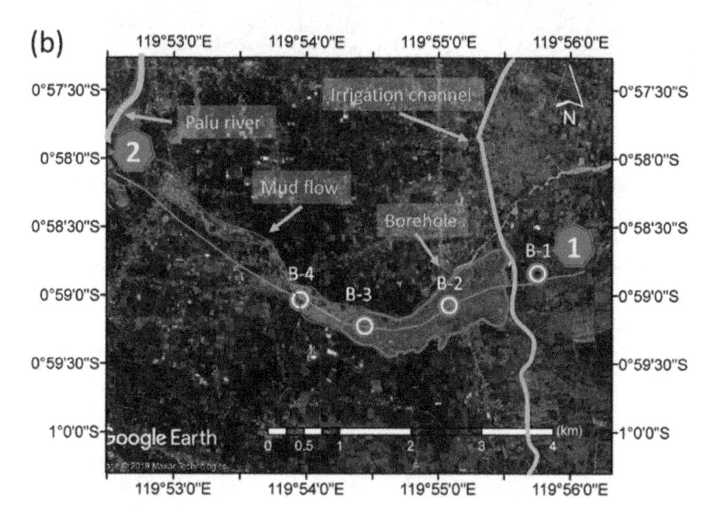

Fig. 12.5 Satellite imagery and ground profile of Jono Oge area **a** before earthquake, **b** after earthquake and **c** cross section at profile 1–2 (Hazarika et al. 2020)

Zone I includes the crown or head scarp, which extends from northeast side of failure zone (parallel to the western boundary of irrigation channel) to the south-eastern side of failure zone where it meets the channel. The overlying soil strata of the main scarp were highly stratified with an alternating layer of sand, silt and organic soil as seen in Fig. 12.7a which implies the area was a wetland at the lower end of the old fan (Kiyota et al. 2020). This zone also is characterized by scouring and erosion of surface sediments due to the breach of irrigation channel which discharged large volume of the water (Fig. 12.7b). The maximum displacement of soils in this zone ranges from 2 to 50 m.

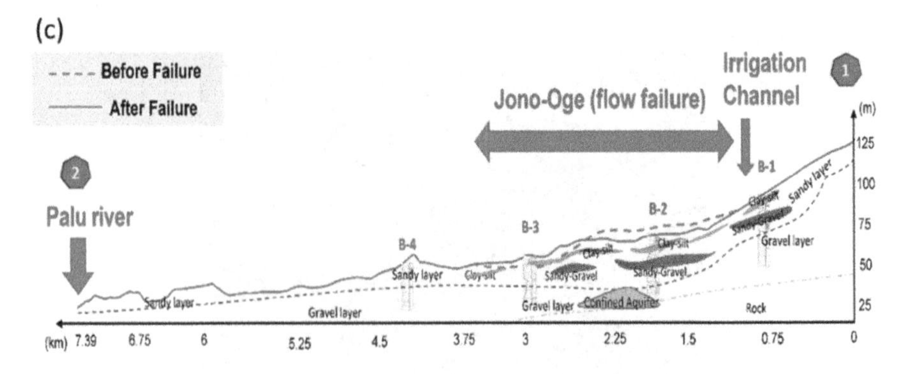

Fig. 12.5 (continued)

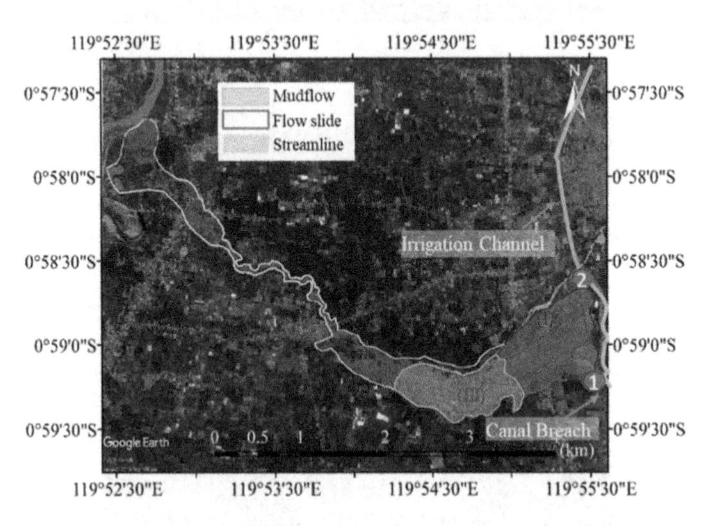

Fig. 12.6 Flow-slide area categorized into zones based on ground displacement in flow-slide area (Hazarika et al. 2020)

Zone II, where the flow-slide initiated after the onset of possible liquefaction in the sandy and sandy-silty layers below the ground surface, was the surface soil layers started moving toward the southwest direction of failure zone. The surface features show large extensional cracks with drastic change in soil morphology (Fig. 12.8a). Figure 12.8b shows the displacement of a house which moved approximately 1200 m due to the flow-slide. The maximum displacement in this region varies from 300 to 1300 m. The residuals of the displaced house after the cease of flow-slide can be seen in Fig. 12.8c.

Zone III is the region, where most of the disaster debris, including house rubbles and uprooted trees, got deposited. As shown in Fig. 12.8b, the debris of the structures and mudflow from Zone II piled up in the Zone III.

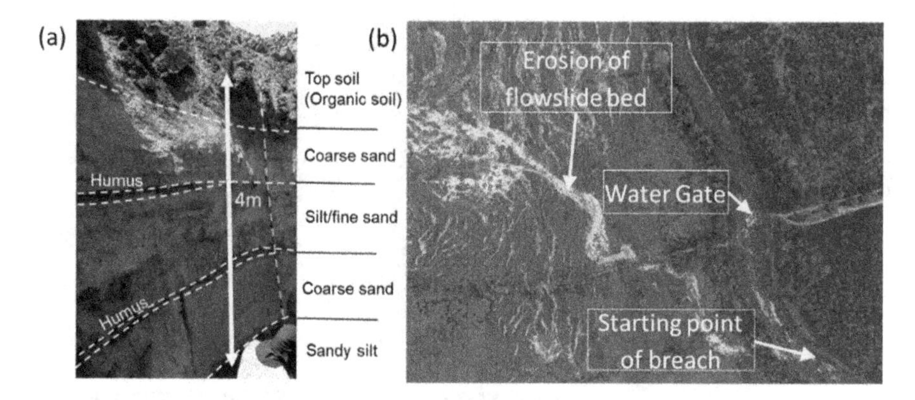

Fig. 12.7 **a** Soil stratification observed at the crown (Kiyota et al. 2020) and **b** UAV imagery of channel breaching and water gate destruction in Zone I (Hazarika et al. 2020)

Zone IV is located at the toe of the flow-slide and close to Palu River. The breached irrigation channel in Zone I discharged huge amount of water, which then flowed into Zone IV along with the debris. This debris flow which then confluence the stream flowing on the west bank of the flow-slide and finally ended up flowing into the Palu river (Fig. 12.9).

Damage to infrastructures

The ground motion caused damage to housing, transportation and irrigation structures. The damage due to flow-slide in Jono Oge was further aggravated due to the breach of irrigation channel, as shown in Fig. 12.10, which shows the completely dry state of the channel. According to eyewitness, the embankment of the irrigation channel breached 2 min after the earthquake, which implies that the damage to the embankment of irrigation channel was not caused by the earthquake motion, but it might have been caused by the dragging force due to flow-slide.

According to another eyewitness, living near the downstream of the flow-slide, after the ground motion ceased, the thick mud water flowed down for approximately an hour, and the houses along the river bank were inundated with the mud and debris, as shown in Fig. 12.11. From the mud inundation mark on the houses, it could be deduced that the height of mudflow at the downstream area was about 7 m above the current river water level. It is to be noted that such mud slide was not observed in other areas, such as Petobo and Sibalaya, where similar damage occurred in the irrigation channels. The origin of the water that caused the flood in Jono Oge could be attributed to a confined aquifer beneath the affected area of flow-slide (Kiyota et al. 2020).

In situ tests

In order to evaluate the probable mechanism and progression of long-distance flow-slide at Jono Oge, the research team from Kyushu University performed in situ tests using the portable dynamic cone penetrometer test (called hereafter PDCPT) at

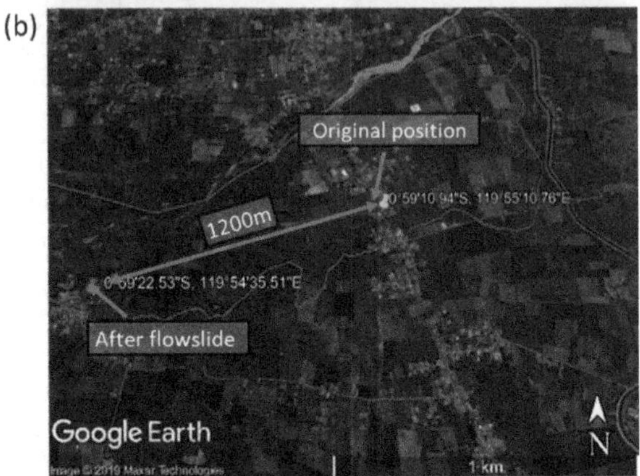

Fig. 12.8 Zone II **a** surface deformation with extensional cracks, **b** displacement of house structure from Zone II to Zone III due to the flow-slide and **c** residues of the displaced house after the flow-slide (Hazarika et al. 2020)

multiple locations in and outside the failure zone. One test (PDCPT1) was conducted outside the flow zone, and the other tests (PDCPT2 $\sim$ 8) were conducted inside the flow-slide zone in two arrays, to assess the subsoil profile in two different directions as shown in Fig. 12.12 by two dotted lines.

Fig. 12.9 Intersection of mudflow and stream (Hazarika et al. 2020)

Fig. 12.10 Damage to irrigation channel and control gate in Jono Oge

Fig. 12.11 Mud inundation mark on the house depicting the height of mud flow passing the area (Kiyota et al. 2020)

Fig. 12.12 Locations of the PDCPT and collected soil samples at Jono Oge (Hazarika et al. 2020)

Figure 12.13 represents the plot for converted N_{SPT} value versus depth for PDCPT1 at the location, which is outside the failure zone, far from extensional cracks. As observed in Fig. 12.13, there is a decrease in the N value for the soil layers below 2.5 m, indicating the presence of loose liquefiable layers. The risk of liquefaction is also high due to the presence of groundwater table at a shallow depth (water level at the time of testing G.W.L = 2.3 m). Furthermore, to correlate the soil lithology at location PDCPT1, a pictorial image of the cliff nearest to the extensional crack is shown in Fig. 12.14. As shown in Fig. 12.14, the overlying layer consists of 10–20 cm of organic soil, followed by 30–40 cm of clay and silty clay layer, while the underlying 10–20-cm-thick silty sand layer is sandwiched between the upper clayey layer and lower 10–20-cm-thick clayey silt layer.

Fig. 12.13 PDCPT conducted near the crown of flow-slide and the test result (modified from Hazarika et al. 2020)

Fig. 12.14 Profile of exposed soil layer due to the extensional cracks near PDCPT1 (Hazarika et al. 2020)

Furthermore, beneath the silty layer, there were several layers of sandy and gravelly sand, which appeared to be influenced by probable liquefaction in the underlying layers (Hazarika et al. 2020).

The PDCPTs 4, 5, 6 and 8 were conducted alongside the damaged road as discussed before in Fig. 12.13. PDCPT4 was conducted beside the collapsed bridge near the stream (Fig. 12.15a). PDCPT5 was conducted within the flow-slide Zone (I) near the reconstructed road as shown in Fig. 12.15b. PDCPT6 was conducted in the Zone (II) alongside the reconstructed road as depicted in Fig. 12.15c. PDCPT8 was conducted few meters away from the red house which experienced no displacement during the flow-slide (Fig. 12.15d). The results of the PDCPTs conducted are converted in terms of SPT N values and are plotted alongside the locations of the tests (Fig. 12.15a–d). Here, in Fig. 12.13a, the overall soil profile up to the depth of 5 m shows weak soil, and this was due to its presence near the river, where the bridge collapsed during the earthquake. As seen from the test results, the N values for all the locations are less than 10 up to a depth of 2.5 m which signifies very loose to loose soil below the ground surface. It can be observed in PDCPT 8 that the soil below 4 m is of medium strength as against those observed in PDCPTs 4, 5 and 6. The deposits at the PDCPT locations are believed to be loosely deposited sand and silty sand layers which were carried over from the upstream of flow-slide region (Hazarika et al. 2020).

The red house shown in Fig. 12.15d was the only structure that did not move from its original location, as compared to all other structures in the flow-slide region. This could be attributed to the massive mat foundation over which the house was built. The self-weight of the mat foundation along with the house restrained its

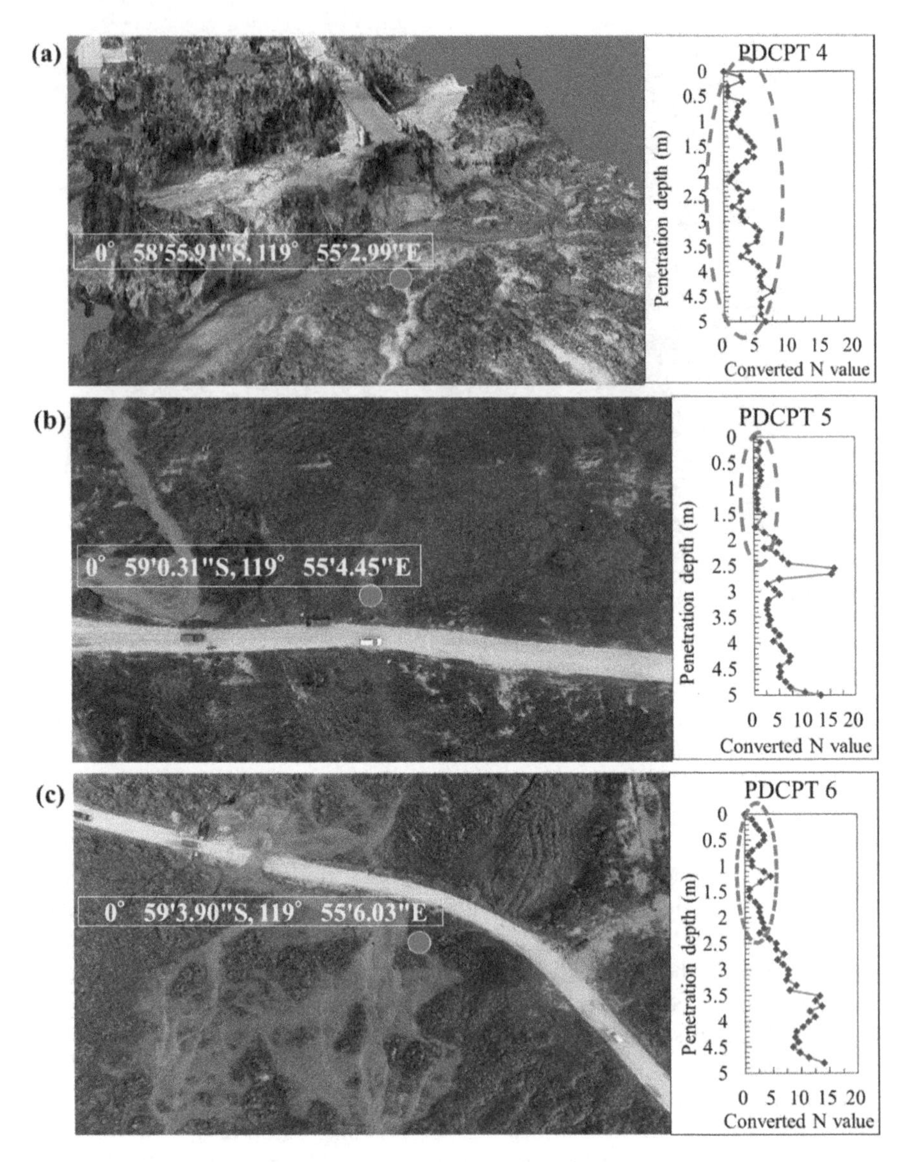

Fig. 12.15 Aerial imagery of PDCPT locations and converted SPT N value versus depth (modified from Hazarika et al. 2020)

lateral displacement even though some structural elements of the house were partially damaged due to the impact of debris. Figure 12.16 displays the aerial view of the unmoved red house along with other houses, which moved from few meters to few hundred meters.

Fig. 12.15 (continued)

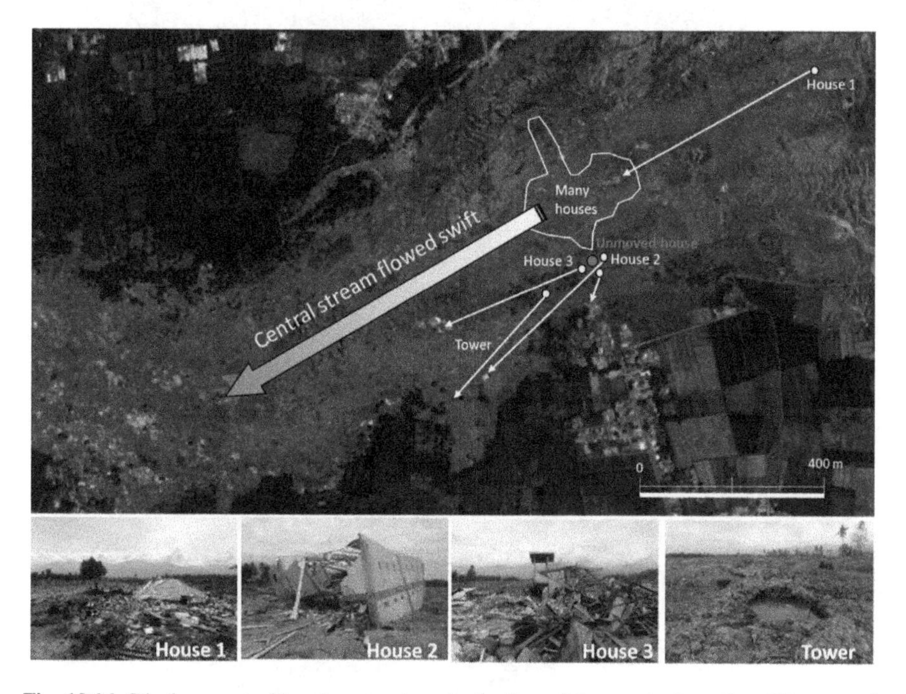

Fig. 12.16 Displacement of housing structures in the flow-slide zone in Jono Oge (Kiyota et al. 2020)

12.3.2 Flow-Slide at Sibalaya

Flow-slide extent and ground movement

The failed area of flow-slide in Sibalaya is situated on the east side of Palu Valley, 18 km south of Jono Oge area. The watershed of the failed area is approximately

5.3 km^2. Here, the flow-slide took place at the bottom of alluvial fan at the valley mouth. The alluvial fan at the higher elevation has a steep slope of about 8%, while, in the region of flow-slide, the slope gradient is approximately 4% in the upper part and 2% in the middle and bottom parts. Figure 12.17a shows the Google Earth image of the flow-slide in Sibalaya after the earthquake. The flow-slide area could be divided into three zones, and each zone flowed 200–400 m in the northwest direction from their respective positions (Fig. 12.18b). Sibalaya has an irrigation channel above the main scarp, which suffered significant damage similar to the ones in Jono Oge and Petobo.

The normal strain contour map of the ground movement shown in Fig. 12.18a provides a quantitative description of overall ground displacement characteristic. As

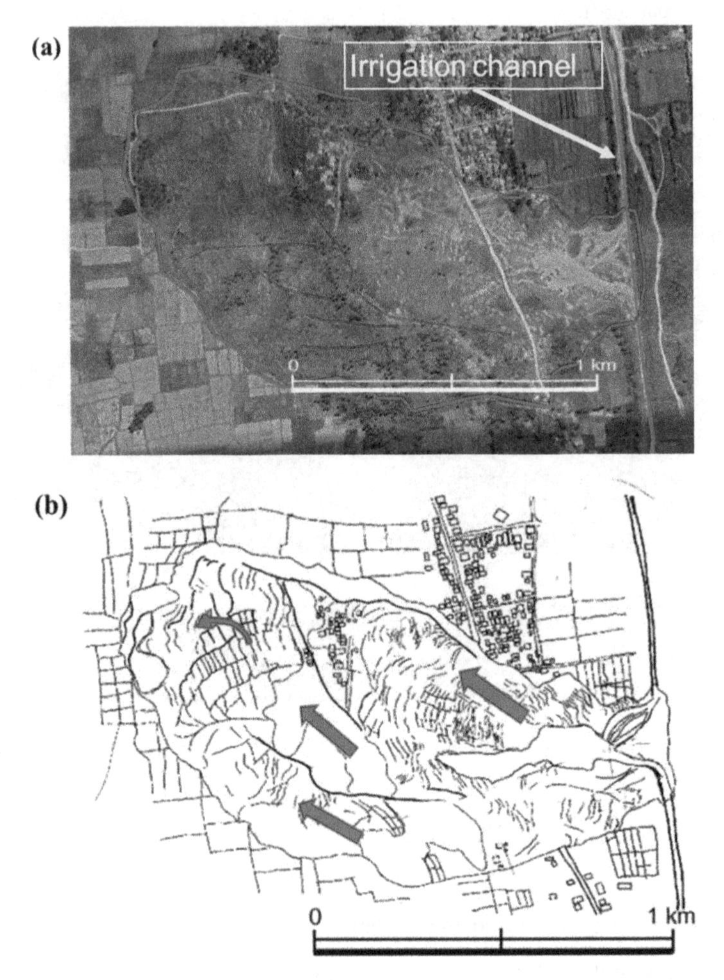

Fig. 12.17 Flow-slide at Sibalaya, **a** area after the earthquake as captured by the UAV and **b** ground movement during the flow-slide in different zones (Kiyota et al. 2020)

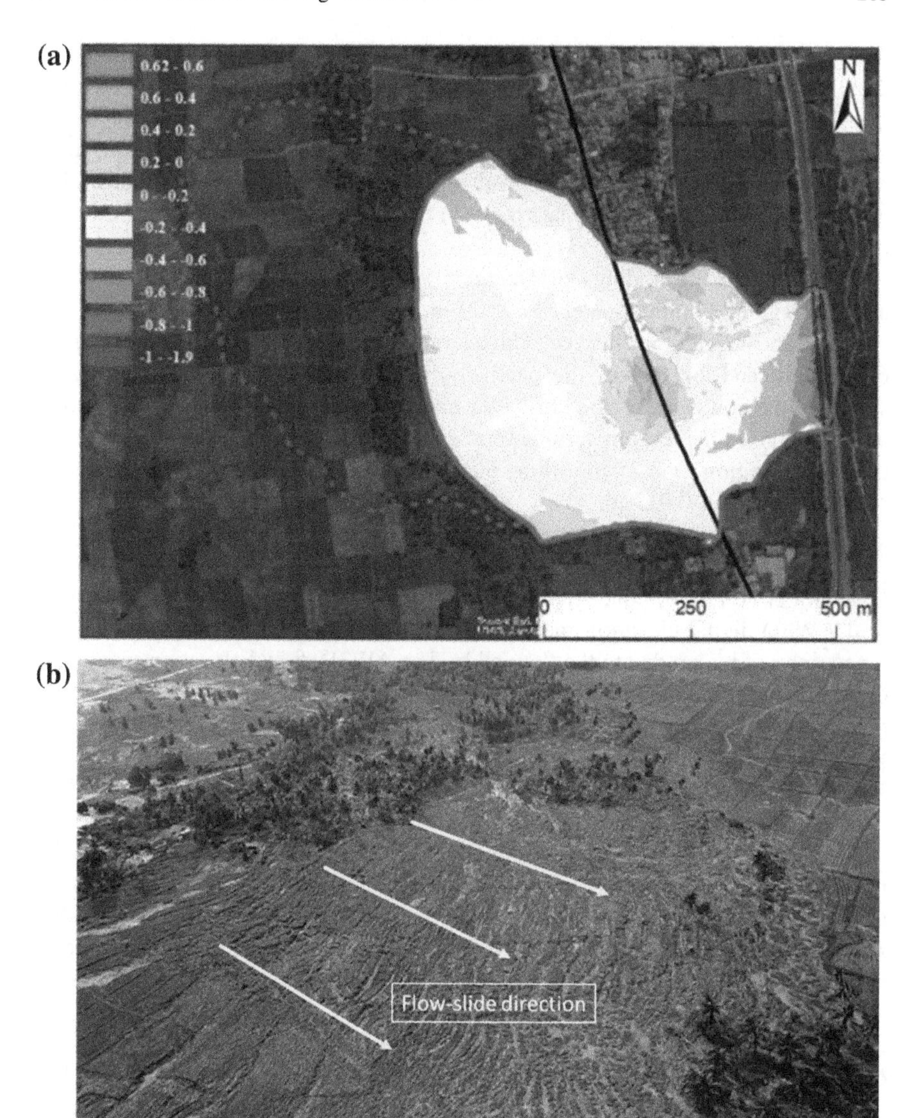

Fig. 12.18 Sibalaya flow-slide region: **a** normal strain in east–west direction, ε_x, and **b** flow-slide direction and extensional cracks in downstream zone (Okamura et al. 2020)

observed from the figure, extremely large tensile strain is generated in the narrow zone of the upstream region. Similarly, in the upstream region (paddy fields), which is just below the crown or main scarp, the tensile strain ranges from 20% to 60%. In this zone, all the houses and roads, along with paddy fields and trees, were displaced by about 350 m. In contrast, the tensile strain is relatively small in the area between the upstream and downstream paddy fields, including the original road that was passing through this area, and compressive strain was observed (Okamura et al. 2020). Here, the negative tensile strain implies that the ground moved downstream from its original position causing extensional cracks, while the positive tensile strain relates to the compression of existing soil surface due to thrust from accumulating or depositing soil mass from upstream. The aerial view of flow-slide direction showing the extensional cracks in the downslope region of the flow-slide is shown in Fig. 12.18b.

According to the findings of Okamura et al. (2020), the flow-slide occurred across a region of approximately 534,700 m^2, in which around 339,200 m^2 of area subsided and 195,500 m^2 heaved. The overall volume of subsidence and heaving is 609,900 m^3 and 745,900 m^3, respectively, which was evaluated from the terrain survey of the flow-slide region during the site investigation and the elevation profile of the region available from various sources. The average ground subsidence across the failure region is 0.25 m, while the area outside the edge of the flow-slide subsided by about 0.013 m.

Terrain profile of the flow-slide region

Okamura et al. (2020) conducted terrain survey of the flow-slide zone and generated AW3D digital elevation maps with a 0.5 m resolution to analyze change in topography of region due to the flow-slide. The AW3D digital elevation maps were developed with the help of Remote Sensing Technology Center of Japan, and the maps were constructed using satellite imageries acquired by Geo-Eye-1 and WorldView-1, 2, 3 and 4.

The terrain elevation from the AW3D maps is shown in Fig. 12.19a–f. In all the figures, the chainage of the irrigation channel is shown by a vertical dashed line set at 1340 m. Figure 12.19a–c shows the ground elevation profile along the flow-slide mainstream (section A–A') before and after the earthquake. Figure 12.19d–f shows the terrain elevation profile, ground slope and change in height of the flow-slide zone across sections A–A', B–B', C–C' and D–D'. All the sections were parallel to each other. It can be implied from Fig. 12.19a–c that the flow-slide caused an abrupt change in terrain elevation and slope profile in the damaged area. Also, the change in height (subsidence and upheaval) was caused due to the tension and compression strain in ground, as discussed before. Although the topographies of all these sections prior to the earthquake were similar, the post-earthquake deformations were significantly different. This could be attributed to the spatial variability of soil lithology, the mechanical properties of soils and the location of groundwater table.

Soil lithology of the flow-slide area

To understand the geological profile of the failure zone, multiple trenches were excavated in Sibalaya to examine the soil lithology and arrive at probable cause of the failure. Figure 12.20 shows the soil profile from one such trench located within the failure zone (refer Fig. 12.17a). As shown in Fig. 12.20, the subsurface soil lithology is highly stratified with layers of sandy gravel, silty sand and clay, sandy gravel and sandy soil. The silty sand and clay layer, which is sandwiched between

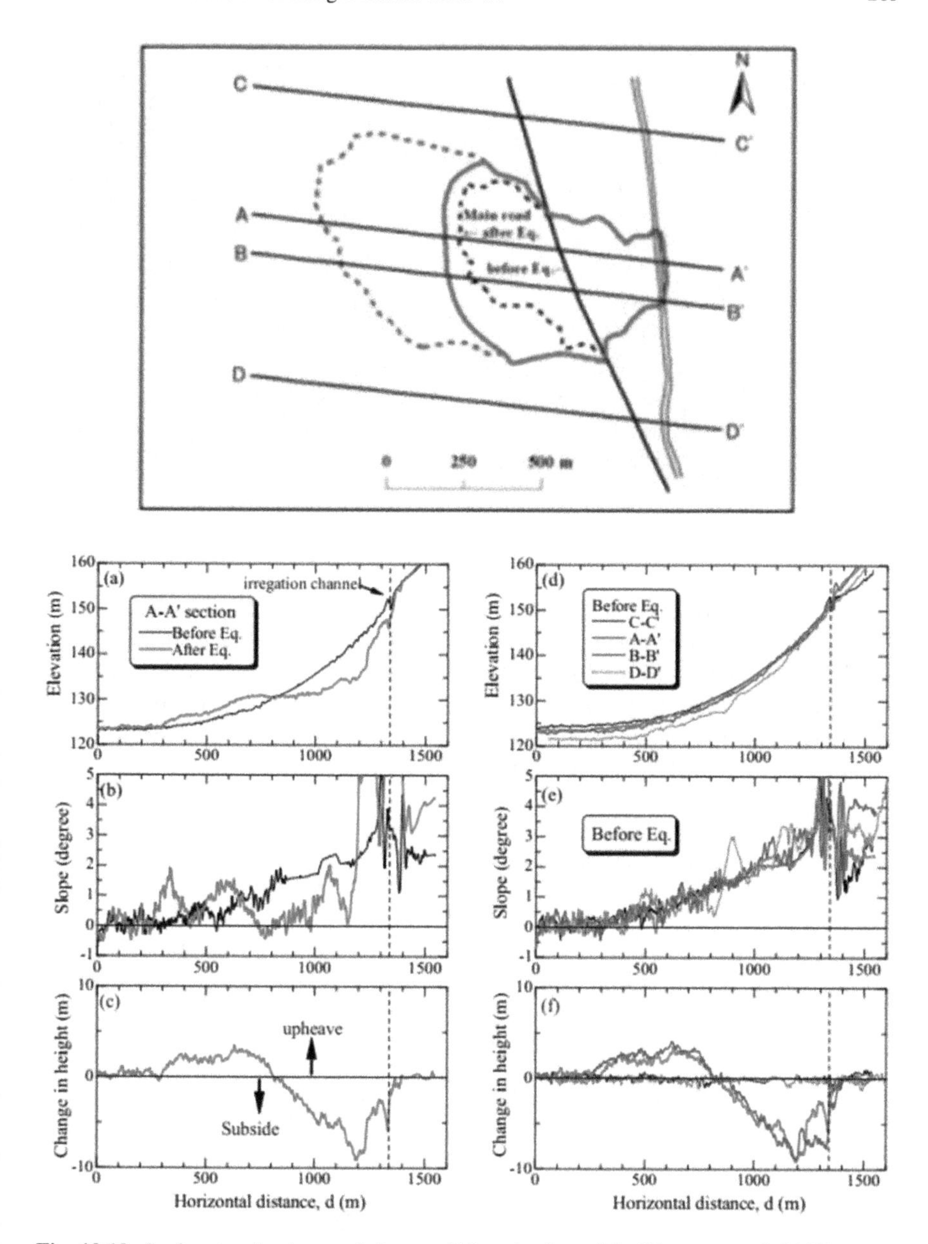

Fig. 12.19 Surface terrain characteristics parallel to the flow-slide (Okamura et al. 2020)

sandy gravel layers, is assumed to be the cause of failure by trapping the excess pore water pressure developed during the ground motion and thereby destabilizing the underlying sandy layers.

Fig. 12.20 Soil profile observed near the main scarp of failure in Sibalaya

The research group from Ehime University conducted multiple trench excavation in the flow-slide zone in Sibalaya, to study the subsurface soil profile, the locations (indicated as #1 ~ #7) of which are shown in Fig. 12.21. This subsection will discuss the findings from one particular trench referred here as Trench #1.

Figure 12.22a–d shows the images of the excavated Trench #1 as well as soil stratification. Figure 12.22a shows the aerial view of the excavated trench, while

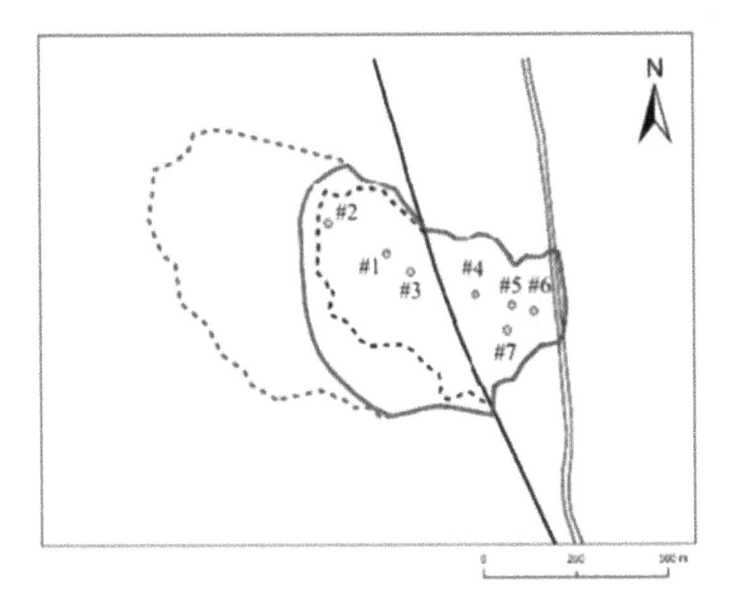

Fig. 12.21 Locations of excavated trench in Sibalaya (Okamura et al. 2020)

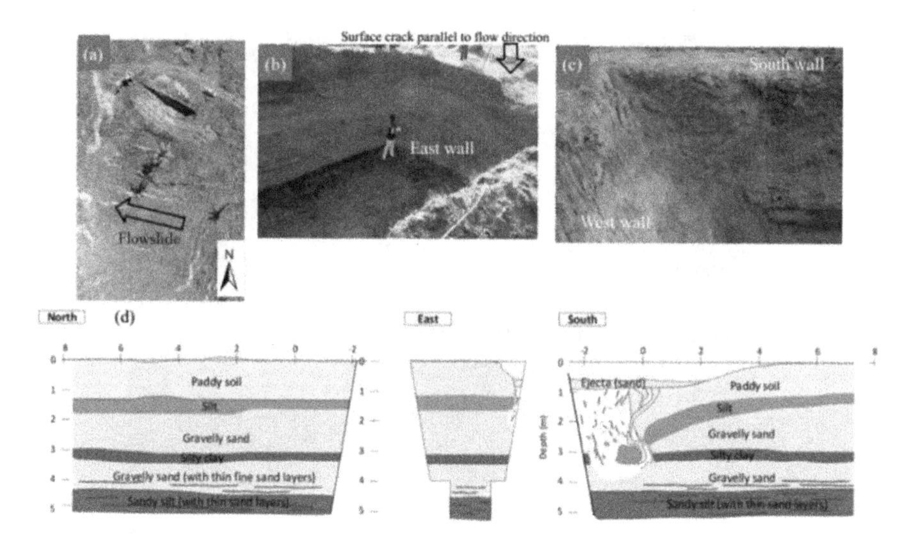

Fig. 12.22 Details of Trench #1 at Sibalaya. **a** Aerial photograph, **b** north and west walls in the trench, **c** southeast corner in the trench and **d** sketch indicating stratification of soil layer exposed on the walls (Okamura et al. 2020)

Fig. 12.22b, c shows the faces of east wall and west wall of the trench. The sketch of the north, east and west walls of the trench, depicting the soil profile, is shown in Fig. 12.22d. The trench was located upstream, about 150 m downslope of the irrigation channel. The longitudinal axis of the trench was oriented parallel to the flow-slide direction. In Fig. 12.22d, the gravelly sand layer underlying the surface layers is confined by multiple sand and sandy gravel layers of 10–30 cm thick, with some sandwiched silty sand layers, whose thicknesses are of the order of 10 cm. In the south wall, the surface of gravelly sandy layer sinking toward the southeast corner implies the potential liquefaction of this layer during the earthquake. Beneath the gravelly sand layer, there exists a thin stiff silty clay layer that separates the two gravelly sand layers, which seemed to be intact except at the corners. The underlying gravelly sand layer was mostly uniform, with only a few thin interlayers near its bottom. It has been observed that, in all the exposed walls, the typical configuration of non-liquefied gravelly sand layer was a stack of sand and sandy gravel layers in every 10–30 cm, with thin silty sand layers having thicknesses of the order of 10 cm sandwiched in between. This implies the potential liquefaction of both the upper and lower gravelly sand layers, resulting in shearing, which was observed in the lower gravelly sand layer.

Interview of eyewitnesses

Okamura et al. (2020) conducted interviews of multiple residents living near the flow-slide area, who witnessed the flow-slide event (Fig. 12.23a). All the interviewees confirmed a strong ground shaking that continued for 10–20 s, which is in accordance with the ground acceleration recorded at the observatory near Balaroa.

(a)

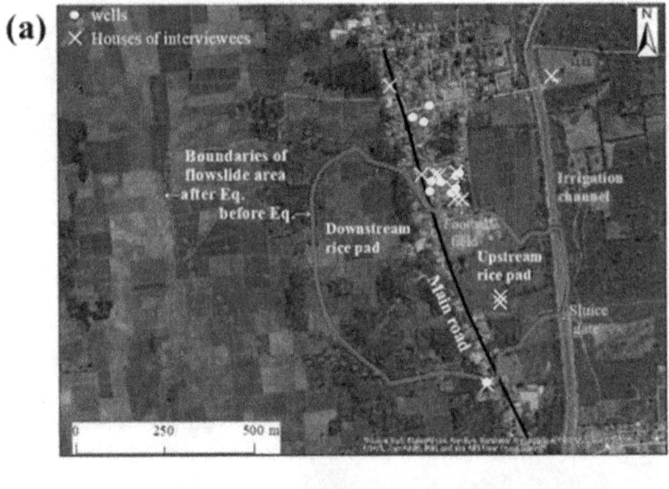

(b)

Fig. 12.23 Sibalaya flow-slide zone, **a** satellite image before the earthquake with locations of houses of interviewees, surveyed wells, main road, irrigation channel and flow-slide-affected area and **b** image and schematic diagram of well (Okamura et al. 2020)

Soon after the shaking ceased, within 10 s, almost all the interviewees felt a sudden drop in elevation. This vertical displacement, after the mainshock, was not recorded by the accelerometers near Balaroa, but was felt by many residents near the flow-slide areas on the east side of the Palu River, including residents of Sibalaya as well as Petobo and Jono Oge (Mason et al. 2019).

Thereafter, within a minute, almost all the interviewees said to have heard an extremely loud sound, similar to structural collapse or thunder. Concurrently, eight interviewees residing near the flow-slide area observed that houses and coconut trees were moving downslope toward the Palu River at a rate faster than the average walking speed, which continued for about a minute. Multiple interviewees claimed to have seen that the water from the breached irrigation channel was flowing downhill, immediately after the ground motion stopped. All these events occurred within few minutes.

Historically, the level of groundwater table is a key in earthquake damage investigation and evaluating the liquefaction potential of ground. The research team of Ehime University also interacted with the owners of the existing wells in the area. One of the owners of a well with internal diameter 1 m and tip elevation 10 m below ground level (Fig. 12.23b) confirmed that the water level in the well before the earthquake was around 5.6 m below the ground surface, which varied minimally throughout the year. Just after the earthquake, the owner observed water overflowing from the well, which lasted for two days, after which the water level gradually reduced and finally dried up in the next couple weeks.

12.3.3 Flow-Slide at Balaroa

Flow-slide extent and ground movement

This subsection constitutes the summary of outcomings from the field survey conducted by the research team of University of Tokyo. In Balaroa, the flow-slide area was smaller as compared to Jono Oge and Sibalaya. The failed area in Balaroa was spread to about 40 ha, located on the west side of Palu Valley. The watershed area of the damaged area is approximately at 4 km^2. The failure occurred at the base of alluvial fan formed near the valley mouth, where there was no presence of discharged water at the time of survey. The maximum displacement of flowed object observed in Balaroa was about 300 m. The ground inclination before the flow-slide was very gentle (3–5%), although it was steeper as compared to those in Petobo, Jono Oge and Sibalaya. Also, in this area no irrigation channel existed.

A satellite image of Balaroa flow-slide area after the earthquake is shown in Fig. 12.24a, while the ground movement direction denoted by arrows is shown in Fig. 12.24b. The soil profile at the crown was found to be coarse sand deposit of thickness 3–4 m which was covered under debris of granite boulders. Sand ejecta from the lower layers were found in the surface layer which was covered from the upstream debris. A significant amount of ground subsidence was found near the crown of the flow-slide, while most of the overlying soil and structures moved in the east direction forming large tensile cracks. The ground subsided by an average depth of 3–4 m in the upper region of the flow-slide, while in the lower region, the ground was uplifted by 5 m, which could have been due to the accumulation of soil blocks and structural debris. Unlike Jono Oge and Sibalaya, Balaroa was a densely populated area, where around 1300 units of houses were destroyed. The presence of

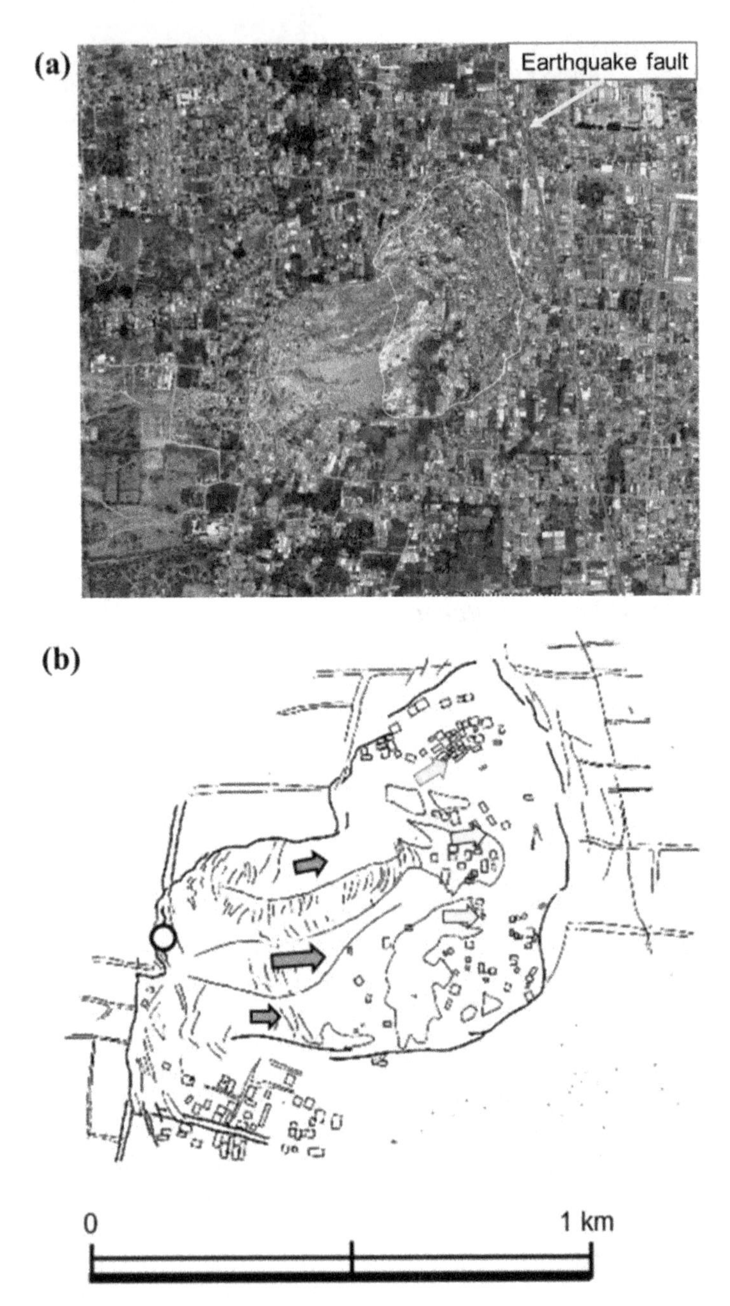

Fig. 12.24 Flow-slide in Balaroa area: **a** satellite image and **b** the direction of ground movement (Kiyota et al. 2020)

spring water inundation on the surface was found in the middle portion of the flow-slide during the survey (Fig. 12.25).

Multiple eyewitness accounts describe the movement of ground in the form of fluidized flow downhill, carrying along the houses and mosques (Fig. 12.26) with it. Also, a clear evidence of sand ejecta on the ground surface was found during the survey, and that confirmed the authors' hypothesis that liquefaction took place in the underlying soil layers (Fig. 12.27).

Furthermore, the residents of this area confirmed that the groundwater table was shallow. The presence of a confined aquifer could also be confirmed by a well digger, who observed that the groundwater sprouted to a height of 0.5–1 m, on inserting a water pipe until 6 m depth from the ground.

Fig. 12.25 Soil profile observed at the crown (Kiyota et al. 2020)

Fig. 12.26 Inundation of water after the earthquake and destroyed mosque structure due to the flow-slide (Kiyota et al. 2020)

Fig. 12.27 Sand ejecta observed in the upper part of the flow-slide (Kiyota et al. 2020)

12.3.4 Flow-Slides at Petobo

Flow-slide extent and ground movement

In this subsection, assessment of the damage, based on the survey conducted by the research team of the University of Tokyo, is described. Petobo is located on the west side of Palu Valley. It had a flow-slide area of about 180 ha. The watershed area of the failed region is approximately 14 km^2. The ground failed at the bottom of alluvial fan formed near the valley mouth. Figure 12.28a depicts an aerial image of the post-earthquake flow-slide zone in Petobo, using Google Earth along with the locations of the PDCPTs ①~④ conducted in the area. The ground displacement vectors during the flow-slide can be seen in Fig. 12.28b. The ground inclination of the failed site was very gentle (2–3%). Similar to Jono Oge and Sibalaya, Petobo also had an irrigation channel passing directly above the flow-slide area, which was damaged (Fig. 12.29). The water from irrigation channel disappeared after the earthquake. The ground movement in this area occurred mainly in two directions, as evident from Fig. 12.29.

The surface soil profile at the crown shown in Fig. 12.30 depicts the presence of alternating layer of silty sand and gravelly soil, which runs parallel to the ground surface. The soil deposits shown in Fig. 12.30 are assumed to be from the deposits caused by flow-slide. Similar to other flow-slide sites, many tensile cracks were observed in upstream slope of the flow-slide in Petobo. The ground elevation was found to be lowered by an average of 5 m.

Similar to Balaroa, Petobo was also a densely populated area, where approximately 1920 housing units were reportedly damaged due to the flow-slide. Contrastingly, the housing structures outside the boundary of flow-slide zone were

(a)

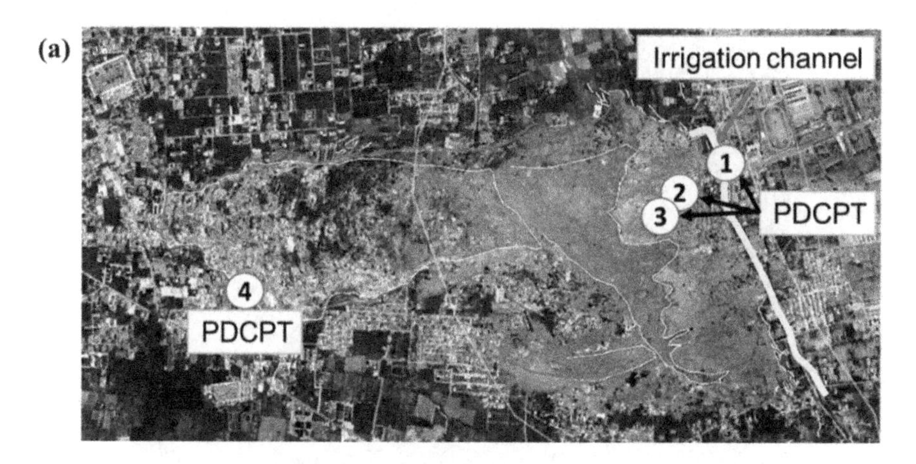

(b)

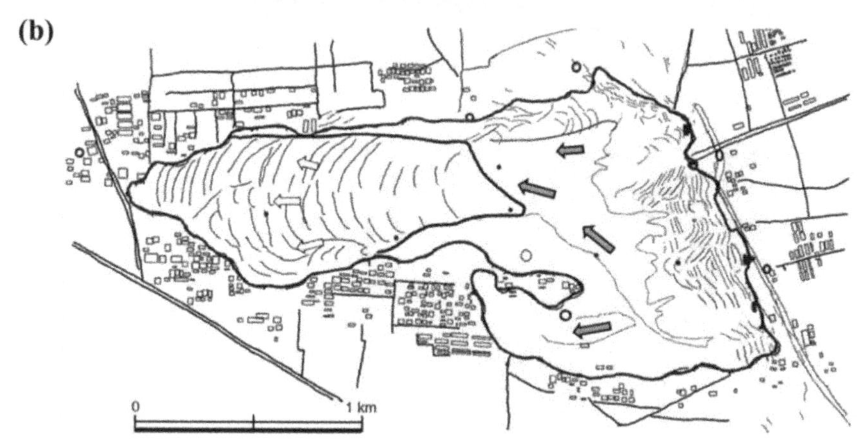

Fig. 12.28 Petobo area after the earthquake: **a** satellite imagery after the earthquake along with the location of PDCPTs ①~④ (Google Earth, 2018) and **b** ground displacement during the flow-slide shown by arrows (Kiyota et al. 2020)

not damaged, as shown in Fig. 12.31, where undamaged structure stands near the damaged structure. Housing structures which moved downstream were buried by thick soil mass and other debris (Fig. 12.32). As observed in Balaroa, the presence of surface flow of water due to formation of small springs was seen in Petobo too. Additionally, similar springs were found in the downstream sedimentation zone, which experienced about 5 m increase in its elevation, probably due to compressive strain from surrounding soil. The presence of springs at the site implies the presence of a confined aquifer (artesian water) below the ground surface. Similar to Balaroa, in Petobo too sand ejecta were observed.

Fig. 12.29 Damage to the embankment of irrigation channel (Kiyota et al. 2020)

Fig. 12.30 Soil profile observed at the main scarp of flow-slide in Petobo (Kiyota et al. 2020)

In situ tests

As described before in Fig. 12.28a, PDCPTs ①～④ were performed in and around the flow-slide zone to evaluate the subsoil strength and condition. It is to be noted that the sand ejecta were found near PDCPT nos. ②, ③ and ④. As shown in Fig. 12.33, the subsoil condition was weak with N_{SPT} values for the PDCPT that are small which implies very weak soil deposits. It can be inferred that a combination of weak soil and shallow groundwater with artisan pressure led to the occurrence of liquefaction due to the earthquake.

Fig. 12.31 Damaged and undamaged structures (Kiyota et al. 2020)

Fig. 12.32 Spring water inundation on the ground surface (Kiyota et al. 2020)

12.4 Probable Flow-Slide Mechanism

Due to complex mechanism of failure and nonidentical geological conditions at some of the flow-slide locations, the exact mechanism of the flow-slides due to the 2018 Sulawesi earthquake cannot be clearly predicted at this stage. Furthermore, owing to the total destruction and displacement of the top soil due to the flow-slide,

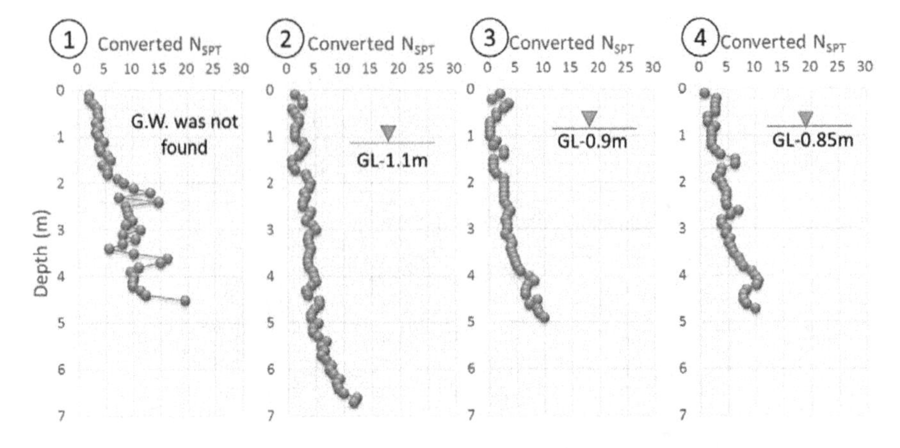

Fig. 12.33 Converted N_{SPT} values from PDCPT performed at Petobo (Kiyota et al. 2020)

added to the change in morphology, and the absence of historic geotechnical or geological data of the locations, it is difficult to arrive at the exact cause of the flow-slide at this stage and suggest a concrete hypothesis for the failure mechanism.

The flow-slide in Jono Oge initiated a minute after ceasing of the earthquake. The flow-slide carried away soil blocks with overlying houses and trees downstream toward the Palu River. The initial movement of ground occurred in Zone II, due to which the surrounding soil lost its retaining support and started moving gradually in the primary zone. The soil blocks, houses and trees were displaced by a distance of few meters (Zone I) to around 1.5 km (Zone II). Furthermore, the breach in irrigation channel caused mudflow into the already subsided area of the flow-slide. A similar case of flow failure was reported in the Lower San Fernando Dam during the 1971 San Fernando earthquake, during which the failure occurred after around 1 min of cessation of the ground motion (Bolton 1987), although the ground displacement of this huge magnitude was not observed. While, in the case of Lower San Fernando Dam, the inertia force is mainly considered as the triggering force due to the slope profile of the ground, this cannot be attributed to the flow-slides in Palu, due to gentle gradients of the areas.

Further, the subsoil lithology in Jono Oge was found to be highly stratified with fine silty and clayey sub-layers sandwiched between coarse sandy layers. These sandwiched layers acted as low-permeable capping layers trapping the liquefied sandy soils and prevent those from reaching the ground surface, thereby causing failure in the underlying soil layer. A similar case of stratified silty sub-layers was observed during the investigation aftermath the 1964 Niigata earthquake, where lateral flow up to maximum 10 meters was observed in gently sloped ground (Kishida 1966).

Low-permeable cap layer hypothesis has been studied in past through model tests by various researchers (Kokusho 1999; Kokusho and Kojima 2002; Kokusho 2015). The model tests exhibited the formation of a thin 'water film' beneath the cap layer, due to liquefaction of the underlying sand during earthquake loading. The water film is formed due to the low permeability of overlying layer which inhibits the dissipation of excess pore water pressure developed in the underlying permeable sandy layer. The water film thus formed reduces the shear strength of the underlying sandy soil layer to less than the initial static shear stress. This leads to lateral flow in a very gentle gradient under the influence of gravitational force until a static equilibrium is achieved.

In consistent with previous studies and the soil stratification observed in this forensic investigation, Hazarika et al. (2020) have proposed a hypothesis explaining the combined effects including typical geology and terrains, and water film formation under the capping layer due to liquefaction of the sandy layer, and existence of underlying artesian aquifer could be a possible reason for such long-distance flow-slide (Fig. 12.34). At this stage, no general mechanism of failure could be hypothesized for all the sites.

On the other hand, hypothesis proposed by Kiyota et al. (2020) suggested that the artisan pressure from the underlying confined aquifer triggered the flow-slide by reducing the shear strength of the overlying ground surface as shown in Fig. 12.35. Although the authors could not confirm the presence of such an aquifer yet, the JICA (2019) report confirmed the presence of a confined aquifer at a depth of 30 m through borehole tests, which is much deeper than expected. Also, the highly stratified subsoil lithology with alternating layers of sand, silt, gravel and organic soils reduced the effective permeability of the surface layers. The low permeability of surface layers inhibited the dissipation of excess water pressure, in spite of the upward pressure exceeding the critical hydraulic gradient contributed by the confined aquifer.

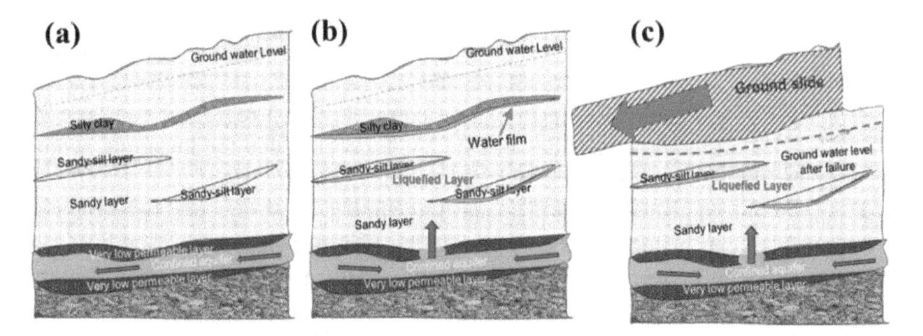

Fig. 12.34 Mechanism of flow-slide and development of water film under low-permeable layers **a** before the earthquake, **b** immediately after initiation of liquefaction and **c** after the earthquake (Hazarika et al. 2020)

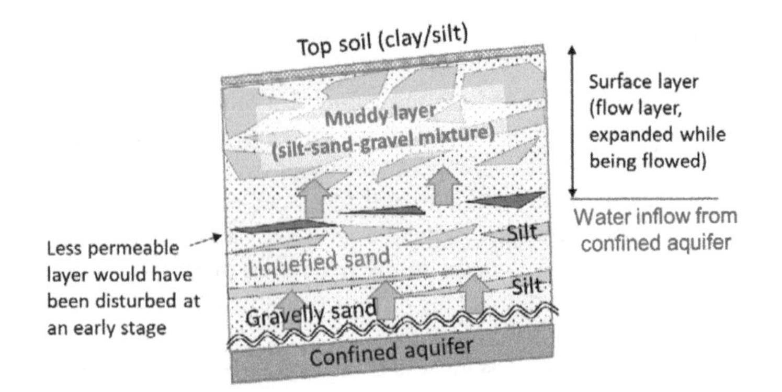

Fig. 12.35 Schematic illustration of long-distance flow-slide based on water inflow from confined aquifer (Kiyota et al. 2020)

12.5 Concluding Remarks

This paper presents a summary of the field reconnaissance conducted by the authors on the geotechnical and infrastructural damage caused by the 2018 Sulawesi earthquake. Key findings from the site investigations of all the flow-slide sites in Palu, performed by three individual research teams from Kyushu University, University of Tokyo and Ehime University, are reported in this paper. The teams collected various information using survey equipments, such as GPS device, cameras and UAV. In addition, trench surveying and in situ testing such as PDCPT and soil sampling (both disturbed and undisturbed) were conducted to investigate the ground conditions in the affected areas.

Based on the forensic analysis performed by the authors, the following conclusions could be drawn.

- The complex geology of Central Sulawesi as well as magnitude and intensity of the extreme event caused by Palu–Koro fault had significant contribution in severity and extension of geotechnical failures, such as liquefaction-induced flow-slides.
- Considering the geological features of Central Sulawesi region, it can be concluded that all major flow-slides in Palu Valley occurred at the locations, where the new alluvial fans meet the old alluvial fans.
- Complex mechanism of long-distance flow-slide at Jono Oge can be explained using interlayer water film theory due to the presence of low-permeable layers (silt and clay) over loosely deposited sandy and sandy gravel layers. Furthermore, damage in underlying artesian aquifer during the earthquake may also have contributed to the development of water film and liquefaction-induced flow-slide on the layers with very low mobilized shear resistance (nearly zero).

Further research, focusing on geophysical tests as well as laboratory tests (such as triaxial test and torsional shear test), and physical modeling are required to arrive at the actual mechanism of such flow failures and to suggest preventive measures against future geohazard risks in the areas having similar geological features.

Acknowledgements The authors would like to express their sincere gratitude to the members of the JICA Domestic Committee for technical support to the 2018 Sulawesi earthquake, Indonesia, and, in particular, Prof. Kenji Ishihara, Prof. Takaji Kokusho, Prof. Susumu Yasuda, Prof. Ikuo Towhata and Dr. Kimio Takeya, for their valuable comments and suggestions.

Prof. Towhata's special initiation to start the reconnaissance survey immediately after the earthquake is also highly appreciated. Previous research done by Prof. Kokusho in the area of fluidized flow has been a guiding principle for the authors, and his encouragement to take up this research is acknowledged with special thanks. The authors also would like to acknowledge JICA for the partial support provided in the initial phase of the investigation. Special thanks go also to Dr. Naoto Tada, JICA Indonesia, and Mr. Hisashi Furuichi, Yachiyo Engineering Co., Ltd., for their encouragement and timely support during the investigations. In addition, the authors like to acknowledge the financial support for this research under Japan Society for Promotion of Science (JSPS) KAKENHI (Grant Number 20H02244). The authors also would likely to acknowledge the partial support provided by JSPS KAKENHI (Grant Number JP19KK0108) and grants provided by Japan Society for Civil Engineers (JSCE) Committee for promoting innovation in infrastructure management. The authors further acknowledge Prof. Masyhur Irsyam (Bandung Institute of Technology, Indonesia) and Dr. Ardy Arsad (Hasanuddin University, Indonesia) for the logistical support during the field investigations. Last but not least, the overwhelming support by Ms. Ode Wa Sumartini, graduate student, Kyushu University, and students of Tadulako University, Palu, during the field investigations is gratefully acknowledged.

References

Bao H, Ampuero JP, Meng L, Fielding E, Liang C, Milliner CWD, Feng T, Huang H (2019) Early and persistent supershear rupture of the 2018 magnitude 7.5 Palu earthquake. Nature Geosci 12:200–205

Bellier O, Sebrier M, Beaudouin T, Villeneuve M, Braucher R, Bourles D, Siame L, Putranto E, Pratomo I (2001) High slip rate for a low seismicity along the Palu-koro active fault in central Sulawesi (Indonesia). Terra Nova 13(6):463–470

Bolton SH (1987) Design problems in soil liquefaction. J Geotech Eng 113(8):827–845

Hazarika H, Rohit D, Pasha SMK, Masyhur I, Arsyad A, Nurdin S (2020) Large distance flow-slide at Jono-Oge due to the 2018 Sulawesi earthquake. Soils and Foundations (Accepted for publication), Indonesia

Kadarusman A, van Leeuwen TM, Sopaheluwakan J (2011) Eclogite, peridotite, granulite, and associated high-grade rocks from The Palu region, central Sulawesi, Indonesia: an example of mantle and crust interaction in a young orogenic belt. Proceedings of JCM Makassar

Kishida H (1966) Damage to reinforced concrete buildings in Niigata city with special reference to foundation engineering. Soils Found 6(1):71–88. https://doi.org/10.3208/sandf1960.6.71

Kiyota T, Furuichi H, Hidayat RF, Tada N, Nawir H (2020) Overview of long-distance flow-slide caused by the 2018 Sulawesi earthquake, Indonesia. Soils Found. https://doi.org/10.1016/j.sandf.2020.03.015

Kokusho T (1999) Water film in liquefied sand and its effect on lateral spread. J Geotechn Geoenviron Eng 125(10):817–826. https://doi.org/10.1061/(ASCE)1090-0241(1999)125:10(817)

Kokusho T (2015) Major advances in liquefaction research by laboratory tests compared with in-situ behavior. Soil Dyn Earthq Eng 91:3–22. https://doi.org/10.1015/j.soildyn.2015.07.024

Kokusho T, Kojima T (2002) Mechanism for post liquefaction water film generation in layered sand. J Geotech Geoenviron Eng 128(2):129–137. https://doi.org/10.1061/(ASCE)1090-0241(2002)128:2(129)

Mason B, Gallant A, Hutabarat D, Montgomery J, Reed A, Wartman J, Irsyam M, Prakoso W, Djarwadi D, Harnanto D, Alatas I, Rahardjo P, Simatupang P, Kawanda A, Hanifa R (2019) Geotechnical reconnaissance: the 28 September 2018 M7.5 Palu-Donggala, Indonesia Earthquake, GEER association Report No. GEER-061. https://doi.org/10.18118/G63376

National Agency of Disaster Management (2018) Gempa Bumi Sulteng: Retrieved 12 Nov 2018. https://bnpb.go.id/geoportal.html

Okamura M, Ono K, Arsyad A, Minaka US, Nurdin S (2020) Large scale flow-side in Sibalaya caused by the 2018 Sulawesi earthquake. Soils Found 59(5):1148–1159

Socquet A, Hollingsworth J, Pathier E, Bouchon M (2019) Evidence of supershear during the 2018 magnitude 7.5 Palu earthquake from space geodesy. Nature Geosci 12:192–199

U.S. Geological Survey (2019) M 7.5–70 km N of Palu, Indonesia. Retrieved 22 March 2019, from https://earthquake.usgs.gov/earthquakes/eventpage/us1000h3p4/map

van Leeuwen TM (2005) Stratigraphy and tectonic setting of the Cretaceous and Paleogene volcanic-sedimentary successions in northwest Sulawesi, Indonesia: implications for the Cenozoic evolution of western and northern Sulawesi. J Asian Earth Sci 25:481–511

Watkinson MI (2011) Ductile flow in the metamorphic rocks of central Sulawesi, the SE Asian gateway: history and tectonics of the Australia-Asia collision 355:157–176

Chapter 13
Empirical Predictions of Fourier Amplitude and Phase Spectra Including Local Site Effects for Simulation of Design Accelerograms in Western Himalayan Region

Ishwer Datt Gupta

13.1 Introduction

For earthquake-resistant design of structures by detailed dynamic response analysis (Chopra 2007; Kausel 2017), the input ground motion is required to be specified in terms of the site-specific time histories of ground acceleration, known as design accelerograms. The response spectra with different damping ratios are also used commonly for dynamic analysis using somewhat simpler spectrum superposition approach (Gupta 2018a). However, the time histories and the response spectra of real earthquakes are not considered suitable for practical design applications due to stochastic and random uncertainties associated with the recorded accelerograms. The response spectra of real accelerograms are characterized by random peaks and valleys, the amplitudes and frequencies of which are not expected to be repeated exactly in any of the future earthquakes. This may result in unrealistic overestimation or underestimation of the structural response, depending upon the structural frequency matching with the frequency of a peak or a valley. It is thus necessary that the random fluctuations in the recorded ground motion are smoothed out to define the design ground motion for practical applications.

The smoothed design response spectra for desired combination of earthquake and site condition parameters are obtained by using empirical prediction equations for the median spectral amplitudes and the associated standard deviation at different natural periods developed using recorded strong motion data in a region (Douglas 2017; Gupta and Trifunac 2018). As it is not feasible to develop the prediction equations directly for the accelerograms, the design accelerograms for practical design applications are generated synthetically to be compatible with a smooth design response spectrum or by modification of suitably selected real accelerograms

I. D. Gupta (✉)
Row House 04, Suncity, Anandnagar, Pune 411051, India

T. G. Sitharam et al. (eds.), *Latest Developments in Geotechnical Earthquake Engineering and Soil Dynamics*, Springer Transactions in Civil and Environmental Engineering, https://doi.org/10.1007/978-981-16-1468-2_13

(Al Atik and Abrahamson 2010; Gupta and Joshi 1993). The synthetic accelerograms are unable to represent the non-stationary frequency characteristics, which are important for the response analysis of nonlinear structures. Also, due to limited number of recorded accelerograms available, it is generally difficult to find out suitable real accelerograms in every case. Further, the generation of accelerograms from a given response spectrum is not a mathematically accurate process, because the response spectrum is not a direct representation of the ground motion.

An exact and equivalent representation of the ground acceleration time history in the frequency domain is provided by its complex Fourier transform, from which it is possible to get back the original time history by inverse Fourier transformation. Thus, in place of the response spectrum, the use of the Fourier amplitude spectrum would be a scientifically better approach for generation of compatible accelerograms. Similar to that for the response spectrum amplitudes, it is possible to develop the empirical prediction equations for the Fourier spectrum amplitudes also, but not many such relations exist in the literature (Gupta and Trifunac 2017; Trifunac 1989). However, the phase spectra of real accelerograms are seen to be highly random with no apparent and easily identifiable trends to develop the predictive relationships. Thus, the Fourier approach is being presently used with the phase spectrum of a limited segment of Gaussian white noise process along with the Fourier amplitude spectra based on the seismological source model approach (Boore 2003a; Motazedian and Atkinson 2005). Thus, not only the phases are unrealistic, but the estimated Fourier amplitude spectra are also unrealistic due to various simplifying idealizations and generalizations made in defining the source models. To get acceptable results, it is necessary to use the empirical prediction equations for the Fourier spectrum amplitudes or to develop an adequately calibrated source model using recorded strong motion data.

This paper has developed an empirical framework to generate realistic design accelerograms by inverse Fourier transform technique using a database of 217 (35 analog and 182 digital) three components of accelerograms recorded in western Himalayan region of India during 1986–2014 (Kumar et al. 2012; Gupta 2018b). This database has been used to develop the empirical prediction equations for the mean Fourier spectrum amplitudes at different wave periods in terms of earthquake magnitude, an effective source-to-site distance including the effect of the source rupture dimension and indicator variables for component of motion (0 for horizontal and 1 for vertical), local geological condition (0 for thick sediments, 1 for intermediate sediments and 2 for basement rock) and the site soil condition (0 for rock soil, 1 for stiff soil and 2 for soft soil). Statistics of the residuals between the Fourier amplitudes of the recorded accelerograms and the mean predictions has been also developed to predict the Fourier amplitude spectrum with any desired confidence level. The predicted Fourier spectra are able to portray the magnitude and distance saturation effects and the dependences on local geologic and soil conditions in a physically realistic manner and are found to show very good agreement with the spectra of several real accelerograms with widely differing amplitude and frequency characteristics. Thus, the Fourier amplitude spectra obtained from the proposed empirical prediction model can be considered to provide very good representation of the true spectra in western Himalayan region.

The horizontal components of only 182 digital accelerograms have been used to develop a prediction model for the Fourier phase spectrum, because the frequency contents in the analog records are not represented very accurately. Each of these records is used to compute the group delay times at all the Fourier frequencies (Kumari et al. 2018; Liao and Jin 1995), which in turn are used along with the estimated travel time of the first *P*-waves to define the equivalent group velocity values for the various waves at each frequency. These group velocity values are used to develop empirical prediction relations for the frequency-dependent mean group velocity and the frequency-independent standard deviation as a function of hypocentral distance alone. The group velocity values for a specified hypocentral distance are then proposed to be simulated by generating a series of Gaussian random numbers with zero mean and predicted standard deviation value and adding those to the predicted mean group velocity values. The simulated group velocity values are next used to compute the phase velocity values, which are finally used to generate the unwrapped phase spectrum by estimating the relative travel times of the waves at all the Fourier frequencies.

The empirically predicted phase and Fourier amplitude spectra are used to simulate the site-specific design accelerograms for western Himalayan region by inverse Fourier transformation. Quite realistic non-stationary characteristics in both amplitudes and frequency contents are attained in these accelerograms automatically due to the use of realistic phase spectra based on empirically predicted group velocity values. The predicted Fourier amplitude spectra are also of site-specific nature due to the comprehensive dependence on the earthquake and the site condition parameters developed using strong motion data recorded in the western Himalayan region. The response spectra for different damping ratios computed from the simulated accelerograms can thus be considered to provide the site-specific design response spectra. Several example results are computed to illustrate the application of the method developed in this paper, which establishes its suitability for generation of site-specific design accelerograms and response spectra in the western Himalayan region of India.

13.2 Study Region and Strong Motion Database

The region of western Himalaya selected for the present study is characterized by very high seismicity with potential to generate earthquakes with magnitude 8.0(+). The great Kangra earthquake of April 4, 1905 is known to be the largest earthquake in the region, which has been assigned magnitude value between 7.5 and 8.6 by different investigators (Ambraseys and Bilham 2000). This earthquake is included among the four great contemporary earthquakes in India. Major tectonic features in the region of study are shown in Fig. 13.1 (Dasgupta et al. 2000), the most

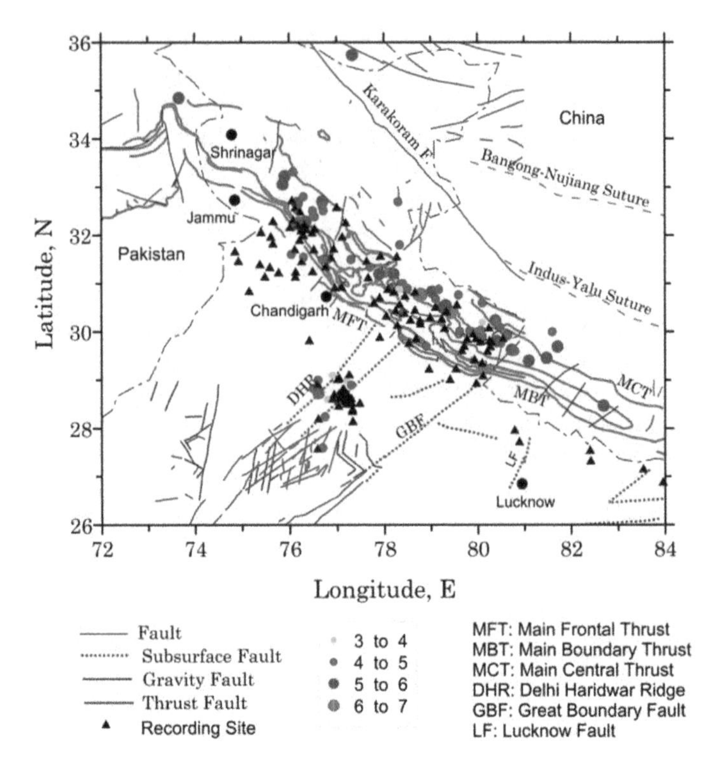

Fig. 13.1 Map of the study region showing major tectonic features, recording sites for strong motion data and the epicenters of the earthquakes contributing the data for the present study

prominent among which from north to south are the main central thrust (MCT), the main boundary thrust (MBT) and the main frontal thrust (MFT). The region between MFT and MBT is known as the Sub-Himalayan range or the Siwaliks, which comprises Miocene to Pleistocene molasses; that between the MBT and the MCT is known as the Lesser Himalaya, which comprises metamorphosed Precambrian Paleozoic sediments; and that beyond MCT encompassing the zone with the highest topography is known as the Higher Himalaya, which comprises meta-sedimentary rocks with granitic intrusions in the south and metamorphosed sedimentary rocks in the north (Gansser 1964; Gupta 2006; Ni and Barazangi 1984).

In spite of very high level of seismicity, the recording of strong motion data in the region of study was taken up only in 1983 by installing a set of 50 analog type SMA-1 recorders in Kangra area (Gupta 2018b). These instruments recorded the first set of nine accelerograms from an earthquake of magnitude 5.5 on April 26, 1986. Another set of 40 analog recorders was installed in the region much later in 1990 in Garhwal area of Uttrakhand state (then Uttar Pradesh state), which recorded the Uttarkashi earthquake of October 20, 1991 with magnitude of 6.9 and the Chamoli earthquake of March 29, 1999 with magnitude of 6.6. A total of 35 analog accelerograms was recorded by the above two sets of instruments from five different earthquakes during 1986–1999. The next phase of strong motion recording in India started in 2005 with the installation of nearly 270 digital accelerographs by IIT Roorkee, covering entire Himalaya, adjoining Gangetic plains, and the northeast India (Kumar et al. 2012). These instruments added 182 digital accelerograms from 66 earthquakes in the region during 2005–2014, which are available on the PESMOS Web site (http://www.pesmos.in).

A total of 217 three components of strong motion accelerograms recorded at 65 sites from 71 different earthquakes is thus available for the present study. Figure 13.1 also shows the locations of the recording stations along with the epicenters of the contributing earthquakes. The metadata on the details of the earthquakes (date, location, magnitude, focal depth) and the recording sites (location and geology) are available in Gupta (2018b). The magnitude values used in this study are referred to as published magnitude M_p which is assigned directly from the M_L, M_S, m_b and M_W values of the contributing earthquakes available in the published literature. For magnitudes greater than or equal to 6.5, the M_p is taken in order of priority equal to M_W or M_S, whereas it is taken equal to m_b, M_L or M_S in this order of priority for smaller earthquakes. The distribution of the available 217 records with respect to the magnitude and hypocentral distance is shown in Fig. 13.2. The data are seen to be spread over the magnitude range from 3.0 to 6.9 and the distance range from 8.6 to 492.8 km without any prominent gaps.

The amplitudes of strong motion at a site are affected strongly by both the local geology and site soil condition (Trifunac 2016). A quantitative characterization of

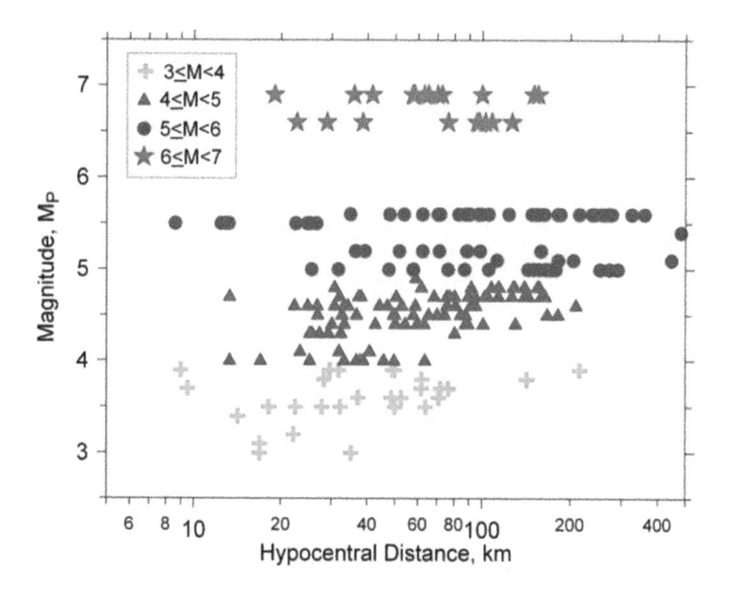

Fig. 13.2 Distribution of the 217 strong motion records available and used in this study as a function of the earthquake magnitude and the hypocentral distance

the site soil condition is commonly made in terms of the average shear wave velocity in top 30 m of ground (V_{s30}), whereas the local geology is characterized by depth of the sedimentary deposits over the bedrock. Lacking such details for the recording stations in the western Himalayan region, qualitative characterizations in terms of indicator variables have been adopted in the present study. The classification for local geology is adopted from Trifunac and Brady (1975), who have used a parameter s taking values of 0, 1 or 2 for sites located on thick sediments, intermediate type of rocks or complex geological environment to permit clear classification and the basement rocks, respectively. The site soil condition in this study is defined using the indicator variable s_L as per the site classifications due to Seed et al. (1976) and used in Trifunac (1989). Three different site soil classes used are: rock sites with $s_L = 0$, stiff soil sites with $s_L = 1$ and deep soil sites with $s_L = 2$.

The indicator variables for local geological and the site soil condition have been assigned to all the 65 recording stations of the database used. The local geology parameter s to each of the 65 recording sites is assigned using the contours for the basement-depth given in the Tectonic Atlas of India (Dasgupta et al. 2000), published crustal velocity model for the region (Parvez et al. 2003), and the receiver function modeling of the region (Borah et al. 2015). On the other hand, the parameter s_L characterizing the site soil condition at each recording site is decided subjectively from knowledge of the surface geology given in various maps by Geological Survey of India or using the predominant frequency and associated amplitude of the H/V spectral ratios of the accelerograms recorded at the site (Zhao et al. 2006).

The distributions of all the 217 strong motion records among various local geologic and site soil condition parameters are given in Table 13.1 for four different magnitude intervals of 3.0–3.9, 4.0–4.9, 5.0–5.9 and 6.0–6.9, respectively. The data are not seen to be distributed uniformly among different magnitude bins with respect to the geological and the soil conditions at the various recording sites. About 72% of the accelerograms are recorded on basement rock-type geology ($s = 2$) and 20.0% on deep sedimentary deposits ($s = 0$), while only 8% are recorded on intermediate type of geology ($s = 1$). On the other hand, the distribution with respect to the local site soil condition indicates recording of 54.4%, 29.0% and 16.6% of the data on the deep soil sites ($s_L = 2$), stiff soil sites ($s_L = 1$)

Table 13.1 Distribution of the 217 strong motion records available in the western Himalaya among different combinations of local geology and site soil condition with respect to four different magnitude bins

Magnitude range	$s = 0$ (44)			$s = 1$ (17)			$s = 2$ (156)		
	$s_L = 0$	$s_L = 1$	$s_L = 2$	$s_L = 0$	$s_L = 1$	$s_L = 2$	$s_L = 0$	$s_L = 1$	$s_L = 2$
3.0–3.9	0	0	1	0	0	2	8	6	11
4.0–4.9	0	1	18	0	0	9	18	23	26
5.0–5.9	0	0	24	0	0	6	9	17	15
6.0–6.9	0	0	0	0	0	0	1	16	6

and rock-type soils (s_L = 0), respectively. To eliminate the possible bias that may result in the scaling relations developed due to this non-uniform distribution of data, a decimation scheme due to Trifunac and Anderson (1977) is adopted before use in the regression analysis.

13.3 Prediction Relations for Fourier Amplitude Spectra

The available strong motion database has been used to develop empirical prediction relations for the Fourier spectrum amplitudes FS*(T)* at different wave periods *T* as a function of earthquake magnitude *M*, a representative source-to-site distance Δ, local geologic condition parameter *s* and the site soil condition parameter s_L. The functional form of the regression model considered is (Gupta and Trifunac 2017; Trifunac 1989)

$$\begin{aligned}\log_{10}\mathrm{FS}(T) &= M + C_0(T)\log_{10}\Delta + C_1(T) + C_2(T)M + C_3(T)M^2 \\ &\quad + C_4(T)v + C_5(T)s + C_6^0(T)S_L^0 + C_6^1(T)S_L^1 + C_6^2(T)S_L^2\end{aligned} \tag{13.1}$$

Here, *v* is an index variable taking values of 0 and 1 for horizontal and vertical components of motion, respectively, and S_L^0, S_L^1 and S_L^2 are the variables for the site soil condition having a value of 1 for the parameter s_L = 0, 1 and 2 corresponding to rock site, stiff soil site and deep soil site, respectively, and zero otherwise. The coefficients $C_0(T)$, $C_1(T)$, $C_2(T)$, $C_3(T)$, $C_4(T)$, $C_5(T)$, $C_6^0(T)$, $C_6^1(T)$ and $C_6^2(T)$ are the scaling functions to be determined at each period by performing the regression analysis on the recorded data.

The representative source-to-site distance Δ in Eq. (13.1) is defined as (Gupta and Trifunac 2017; Trifunac 1989)

$$\Delta = S\left(\ln\frac{R^2+H^2+S^2}{R^2+H^2+S_o^2}\right)^{-\frac{1}{2}} \tag{13.2}$$

in which, S_0 is the correlation radius for the source, *S* is the fault size for a given magnitude, *R* is the epicentral distance and *H* is the focal depth. Following Trifunac (1989), the correlation radius is taken as

$$S_0 = \min\left(\frac{\beta T}{2}, \frac{S}{2}\right) \tag{13.3}$$

where β is the shear wave velocity at the seismic source, which for the western Himalayan region is taken equal to 3.3 km/s. The size of the fault size *S* in km is estimated from the following empirical relationship (Gupta and Trifunac 2017)

$$S = \begin{cases} 0.2 \text{ for } M \leq 3.0 \\ -13.557 + 4.586M \text{ for } 3.0 < M \leq 6.0 \\ 13.959 \text{ for } M > 6.0 \end{cases} \quad (13.4)$$

13.3.1 Estimation of Regression Coefficients

The various coefficients involved in the prediction relationship of Eq. (13.1) are estimated by regression analysis based on the available database of 651 (217 × 3) components of accelerograms. All the accelerograms were processed to obtain the Fourier amplitude spectra free from the high-frequency and baseline errors (Gupta 2018b). Further, the Fourier amplitudes at every period are screened to minimize any possible bias that may arise due to uneven distribution of data over different magnitude ranges, local geology parameter values and site soil condition parameters for each of the component orientations (Trifunac and Anderson 1977). For this purpose, the 651 Fourier amplitudes available at each period are first divided into four magnitude groups as 3–3.9, 4–4.9, 5–5.9 and 6–6.9. The Fourier amplitudes in each magnitude group are next subdivided into horizontal ($v = 0$) and vertical ($v = 1$) components of motion. Data for each magnitude bin and component of motion are further divided into three subgroups as per the local geological condition parameter ($s = 0$, 1 and 2), and the resulting data in each subgroups are finally divided into three more subgroups as per the site soil condition parameter ($s_L = 0$, 1 and 2). The data at each wave period are thus divided into a total of 72 subgroups. If the number of data points in any of these subgroup is less than or equal to 33, all the data are used in the regression analysis. But, if the data points are more, then only 33 values corresponding to 3rd, 6th, 9th, …, 96th and 99th percentile positions, when all the $FS(T)$ values are arranged in increasing amplitude, are retained. This helps to utilize the data over the complete range in a uniform way without any undue weightage to the subgroups with more number of data points.

The regression coefficients in Eq. (13.1) are estimated in a sequential manner to eliminate the effects of any possible interaction among the various predicting variables. To start with, the coefficient $C_0(T)$, representing the attenuation with distance, is estimated using the reduced data as above by least squares regression analysis on the following equation

$$\log_{10} \mathrm{FS}(T) - M = C_0(T) \log \Delta + C_4(T)v + C_5(T)s + \sum_{j=1}^{J} B_j(T)e_j \quad (13.5)$$

In this equation, e_j is a dummy variable, which is assigned a value of 1.0 when the data point belongs to the jth earthquake and value of 0.0 otherwise. Due to random fluctuations in the Fourier amplitudes of the recorded accelerograms, the coefficient $C_0(T)$ is also characterized by a fluctuating nature. This coefficient is therefore smoothed to have statistically stable mean trend, and the smoothed coefficient is designated as the attenuation function $A_0(T)$.

The remaining scaling coefficients are estimated using a two-step weighted regression method, in which the reduced data are used to fit the following equation by the least squares method in the first step

$$\log_{10} \mathrm{FS}(T) - M - A_0(T) \log \Delta = C_4(T)v + C_5(T)s + \sum_{j=1}^{J} B_j(T)e_j \quad (13.6)$$

In the second step, the weighted least squares regression method is used to estimate the magnitude dependence by fitting the following equation to the coefficients $B_j(T)$ obtained in the above

$$B_j(T) = C_1(T) + C_2(T)M_j + C_3(T)M_j^2 \quad (13.7)$$

The weights required for fitting of Eq. (13.7) in the second step are defined by (Gupta and Trifunac 2017)

$$w_j = (\sigma_1^2/n_j + \sigma_2^2)^{-1} \quad (13.8)$$

where σ_1^2 and σ_2^2 are the variances of the fitting of Eqs. (13.6) and (13.7), respectively. As the variance σ_2^2 is unknown, a search method is used to find its value that leads to convergence, indicated by weighted variance of the fitting of Eq. (13.7) close to 1.0.

To have statistically stable estimates, all the regression coefficients obtained from fitting of Eqs. (13.6) and (13.7) are also smoothed. Representing the smoothed coefficients by a hat over them as $\hat{C}_1(T)$, $\hat{C}_2(T)$, $\hat{C}_3(T)$, $\hat{C}_4(T)$ and $\hat{C}_5(T)$, the dependence on the site soil condition parameter is developed by fitting of the following equation using all the 651 Fourier amplitudes at each period:

$$\begin{aligned} \log_{10} \mathrm{FS}(T) - M - A_0(T) \log \Delta - \hat{C}_1(T) - \hat{C}_2(T)M - \hat{C}_3(T)M^2 - \hat{C}_4(T)v - \hat{C}_5(T)s \\ = C_6^0(T)S_L^0 + C_6^1(T)S_L^1 + C_6^2(T)S_L^2 \end{aligned} \quad (13.9)$$

The above equation is not defined with the reduced data, because of very limited data points for the rock soil ($s_L = 0$) and stiff soil ($s_L = 1$) conditions. The additional regression coefficients obtained from Eq. (13.9) have been also smoothed to obtain the coefficients $\hat{C}_6^0(T)$, $\hat{C}_6^1(T)$ and $\hat{C}_6^2(T)$ for use in the final prediction model.

13.3.2 Prediction Model and Statistics of Residues

The empirical prediction model for the expected Fourier spectrum amplitudes $\overline{\text{FS}}(T)$ is defined using the smoothed regression coefficients as follows

$$\log_{10}\overline{\text{FS}}(T) = M + A_0(T)\log\Delta + \hat{C}_1(T) + \hat{C}_2(T)M + \hat{C}_3(T)M^2 + \hat{C}_4(T)v + \hat{C}_5(T)s + C_6^0(T)S_L^0 + C_6^1(T)S_L^1 + C_6^2(T)S_L^2 \tag{13.10}$$

The value of $\log_{10}\overline{\text{FS}}(T)$ defined by this relationship for specified values of variables s(local geology parameter), s_L (site soil condition parameter), v(component parameter) and Δ (representative source-to-site distance) has a parabolic dependence on magnitude M. As the coefficient of M^2 in Eq. (13.10) has negative values, the parabola has a maximum at $M = M_{\max}(T)$

$$M_{\max}(T) = -(1 + \hat{C}_2(T))/2\hat{C}_3(T) \tag{13.11}$$

Following Trifunac (1989), it is assumed that the model of Eq. (13.10) applies only in the range of $M_{\min}(T) \leq M \leq M_{\max}(T)$, with

$$M_{\min}(T) = -\hat{C}_2(T)/2\hat{C}_3(T) \tag{13.12}$$

For magnitudes below $M_{\min}(T)$, M is used only in the first term of Eq. (13.10) and $M_{\min}$ is used with $\hat{C}_1(T)$ and $\hat{C}_2(T)$. For magnitudes greater than $M_{\max}(T)$, $M_{\max}$ is used in all the terms for M.

To describe the distribution of the observed Fourier spectral amplitudes $\text{FS}(T)$ about the expected amplitudes $\overline{\text{FS}}(T)$ predicted by Eq. (13.10), we analyze the statistics of the residues

$$\in(T) = \log_{10}\text{FS}(T) - \log_{10}\overline{\text{FS}}(T) \tag{13.13}$$

The residues $\in(T)$ at each period are found to follow a Gaussian distribution with mean $\mu(T)$ and standard deviation $\sigma(T)$. However, similar to the regression coefficients in the mean prediction equation, to have statistically stable predictions, the $\mu(T)$ and $\sigma(T)$ are also described by their smoothed values $\hat{\mu}(T)$ and $\hat{\sigma}(T)$, respectively. The attenuation function $A_0(T)$, values of the smoothed scaling functions $\hat{C}_1(T)$, $\hat{C}_2(T)$, $\hat{C}_3(T)$, $\hat{C}_4(T)$, $\hat{C}_5(T)$, $\hat{C}_6^0(T)$, $\hat{C}_6^1(T)$ and $\hat{C}_6^2(T)$ in Eq. (13.10) and the smoothed statistical parameters $\hat{\mu}(T)$ and $\hat{\sigma}(T)$ are listed in Table 13.2 for 13 selected wave periods between 0.04 and 3.0 s. These parameters can be used to predict the Fourier amplitude spectrum specific to a site in western Himalayan region with any specified confidence level (probability of not exceeding).

Table 13.2 Smoothed values of the various scaling functions involved in the mean prediction model of Eq. (13.10) and statistical parameters used to define the probability distribution of residues

T (s)	$A_0(T)$	$\hat{C}_1(T)$	$\hat{C}_2(T)$	$\hat{C}_3(T)$	$\hat{C}_4(T)$	$\hat{C}_5(T)$	$\hat{C}_6^0(T)$	$\hat{C}_6^1(T)$	$\hat{C}_6^2(T)$	$\hat{\mu}(T)$	$\hat{\sigma}(T)$
0.04	−1.63821	−1.90659	0.09972	−0.07540	0.25250	0.05910	0.15436	0.00745	0.06486	0.00997	0.46446
0.06	−1.48400	−1.18018	−0.12090	−0.05409	0.13010	0.08140	0.14405	0.07168	0.08100	0.02262	0.45072
0.08	−1.37414	−1.15341	−0.12809	−0.05029	0.04829	0.08157	0.11256	0.09377	0.08528	0.03061	0.44164
0.10	−1.29028	−1.42046	−0.04499	−0.05442	−0.01573	0.07316	0.07532	0.09760	0.08583	0.03555	0.43645
0.15	−1.14894	−2.27304	0.20810	−0.07014	−0.13254	0.04357	−0.00444	0.07859	0.08402	0.03902	0.43269
0.20	−1.06385	−2.87850	0.36398	−0.07874	−0.20509	0.01493	−0.05373	0.04893	0.08037	0.03623	0.43363
0.40	−0.92924	−3.78863	0.46806	−0.07369	−0.29025	−0.05802	−0.10200	−0.02311	0.05841	0.02604	0.44116
0.60	−0.89051	−4.33198	0.50445	−0.06795	−0.28505	−0.08983	−0.10278	−0.03636	0.04404	0.02657	0.44675
0.80	−0.87250	−4.84001	0.57249	−0.06785	−0.26725	−0.10544	−0.10609	−0.04082	0.03599	0.02638	0.44973
1.00	−0.86043	−5.26158	0.63797	−0.06934	−0.24791	−0.11307	−0.11202	−0.04913	0.02950	0.02324	0.44995
1.50	−0.83735	−5.93637	0.73436	−0.07149	−0.20374	−0.11389	−0.12539	−0.08323	0.01115	0.00909	0.44048
2.00	−0.81835	−6.27544	0.75841	−0.06999	−0.16860	−0.10303	−0.13141	−0.11984	−0.00973	−0.00556	0.42315
3.00	−0.78843	−6.59706	0.73795	−0.06350	−0.11835	−0.07660	−0.13268	−0.17863	−0.04749	−0.02821	0.38696

13.3.3 Examples of Predicted Fourier Spectra

We now show that the prediction relationship developed in this study is able to model the dependence on the earthquake magnitude and the source-to-site distance in a physically realistic manner. For dependence on the magnitude, left-side panel in Fig. 13.3 gives the horizontal median FS(T) for M values of 3.5, 4.5, 5.5, 6.5 and 7.5 for epicentral distance $R = 5$ km and focal depth $H = 10$ km. Similarly, for dependence on distance, right-side panel gives the spectra for R values of 0, 10, 25, 50, 100 and 200 km for $M = 6.5$ and $H = 10$ km. All the results are computed for local geological parameter $s = 2$ (basement rock) and site soil parameter $s_L = 0$ (rock).

In Fig. 13.3, the Fourier amplitudes are seen to grow at slower rate with increase in the magnitude, which illustrates the magnitude saturation effect. This effect is exhibited more significant for lower wave periods. Similarly, the distance saturation effect is also illustrated by slower increase in the Fourier amplitudes with decrease in the distance, which is seen to decrease due to the distance saturation effect.

To illustrate the dependence on local geologic condition, Fig. 13.4 compares the median estimates of horizontal FS(T) for the three values of the local geological condition parameter $s = 0$ (thick sediments), 1 (intermediate geology or complex rock type) and 2 (geological basement rock), and for $M = 6.5$, $R = 25$ km and $H = 10$ km. The left, middle and right panels in Fig. 13.4 are for site soil condition parameter $s_L = 0$ (rock soil), 1 (stiff soil) and 2 (deep soil), respectively. The Fourier amplitudes up to about 0.24 s periods are seen to be lower on sites with thick sedimentary deposits for all the site soil conditions, but the trend is reversed

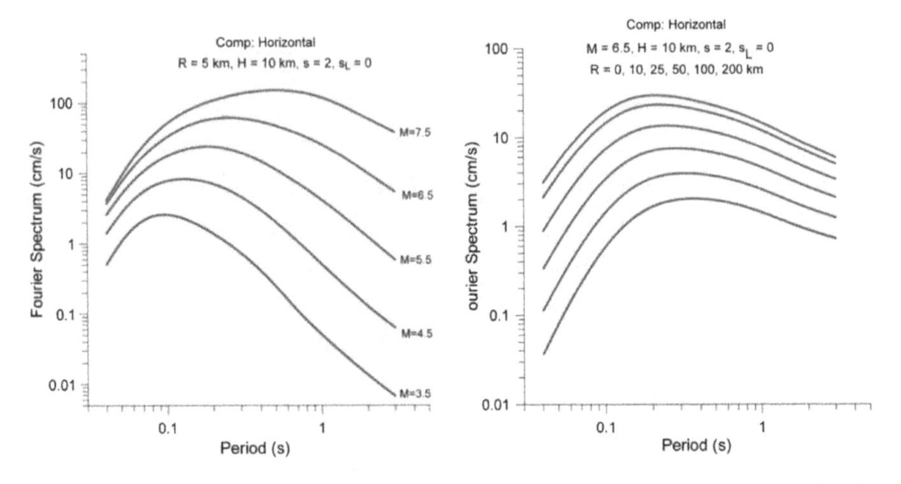

Fig. 13.3 Illustration of the dependence of predicted Fourier amplitude spectra on earthquake magnitude (left-side plot) and epicentral distance (right-side plot)

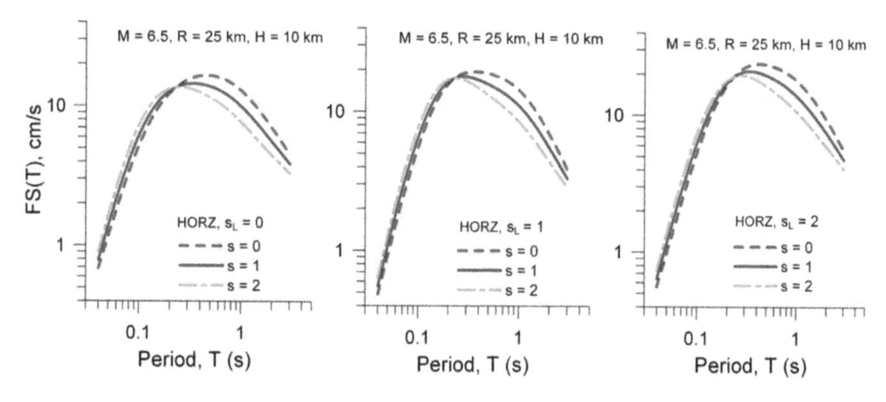

Fig. 13.4 Illustration of the dependence of predicted median Fourier amplitude spectra on local geologic condition for each of the three types of site soil conditions

for longer periods. This is due to much higher amplification of longer period waves compared to the de-amplification of shorter period waves on sites with thick sedimentary deposits.

To illustrate the dependence on the soil condition at the site, Fig. 13.5 compares the median estimates of horizontal FS(T) for three values of the soil condition parameter $s_L = 0$ (rock soil), 1 (stiff soil) and 2 (deep soil) and for $M = 6.5$, $R = 25$ km and $H = 10$ km. The left, middle and right panels are for local geological condition parameter $s = 0$ (thick sediments), 1 (intermediate geology or complex rock type) and 2 (geological basement rock), respectively. The Fourier amplitudes up to about 0.1 s periods are seen to be slightly higher on rock soil ($s_L = 0$) compared to that on deep soil ($s_L = 2$), but the trend is reversed at longer periods. The Fourier amplitudes in the intermediate period range of about 0.16–2.0 s are seen to lie in between on the stiff soil ($s_L = 1$), whereas they are seen to have the lowest values near both the short- and long-period ends. They are closer to the Fourier amplitudes on deep soil at the short-period end and closer to those on rock sites near the long-period end.

13.3.4 Comparisons Between Predicted and Real Fourier Spectra

The empirical prediction equations for Fourier amplitudes developed in this study are not only able to predict physically realistic dependence on various predicting variables, but are also able to predict the realistic values of the spectral amplitudes. In this regard, Fig. 13.5 shows two typical examples of the comparisons between the Fourier spectra of the recorded accelerograms and the predictions from the model developed (Fig. 13.6).

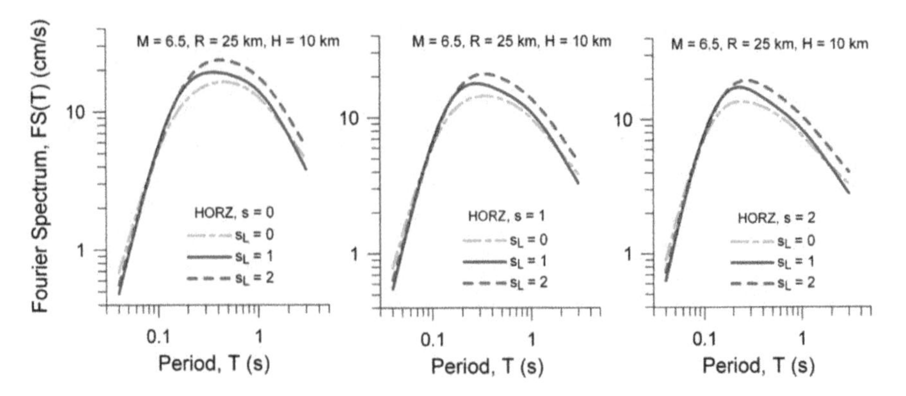

Fig. 13.5 Illustration of the dependence of predicted median Fourier amplitude spectra on varying site soil conditions for each of the three types of local geologic condition

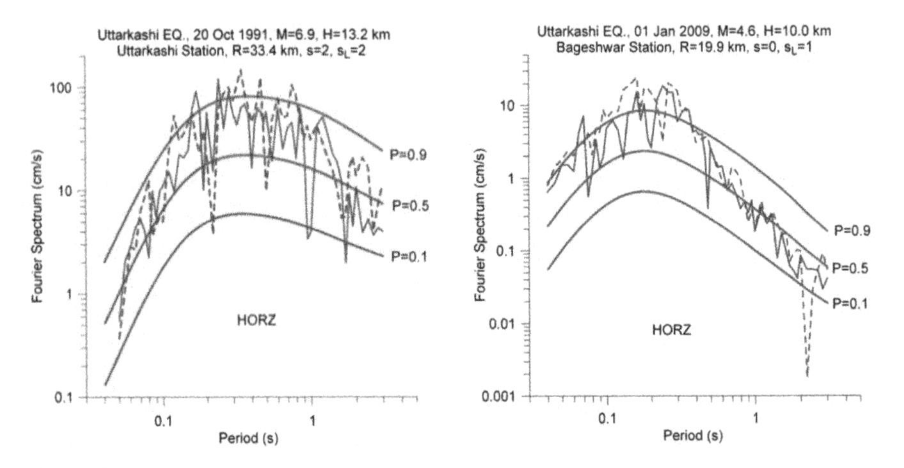

Fig. 13.6 Typical comparisons between actual and estimated Fourier spectra for two different accelerograms with earthquake and recording site details given in the headings

The Fourier spectra for both the horizontal components of the recorded accelerograms are plotted along with the predicted spectra computed with confidence levels of $p = 0.1$, 0.5 and 0.9 for the earthquake and site parameters of the real accelerograms, details of which are given in the headers of Fig. 13.5. The spectra of real accelerograms are seen to lie mostly between the predicted spectra for $p = 0.1$ and 0.9, indicating good agreement between the recorded and the estimated Fourier spectrum amplitudes. The overall shapes of the actual and the predicted spectra are also seen to be in good agreement.

13.4 Prediction Methodology for Fourier Phase Spectra

A simple methodology is also developed to predict the Fourier phase spectra by developing empirical prediction relations for equivalent group velocity dispersion values at all the Fourier frequencies, such that the value at each frequency gives an aggregate representation of the various wave types at that frequency. The database of 364 horizontal components of digital accelerograms only has been used for this purpose, because the frequency contents are not represented very accurately in the analog records. The predicted group velocity values are used to compute the phase velocity values, which in turn are used to simulate the Fourier phase spectrum for generation of non-stationary ground acceleration time histories from the predicted Fourier amplitude spectra.

13.4.1 Prediction of Group Velocity Dispersion Curves

Assuming that the ground motion in a narrow frequency band, $\omega_n \pm \Delta\omega_n$, can be approximated by a constant Fourier amplitude, A_n, and that the phase angles over this frequency band vary linearly with slope t_n and central value ψ_n, the corresponding time history is given by inverse Fourier transform as (Kimura and Izumi 1989).

$$a_n(t) = \frac{2A_n}{\pi} \frac{\sin \Delta\omega_n(t - t_n)}{(t - t_n)} \cos(\omega_n t + \psi_n) \tag{13.14}$$

The complete accelerogram can be generated by superposition of $a_n(t)$ for all the frequency bands. The time t_n in the above corresponds to the maximum amplitude of the wave group $a_n(t)$, and it is thus also termed as the group delay time. Mathematically, t_n represents the slope of the unwrapped phase spectrum $\phi(\omega)$

$$t_n = \left.\frac{d\phi(\omega)}{d\omega}\right|_{\omega=\omega_n} = \frac{\phi(\omega_n + \Delta\omega_n) - \phi(\omega_n - \Delta\omega_n)}{2\Delta\omega_n} \tag{13.15}$$

To avoid the imprecise process of phase unwrapping, following Boore (2003b), we have computed the group delay times for an accelerogram using the following expression in terms of the Fourier transform $A(\omega)$ of an accelerogram $a(t)$ and the Fourier transform $B(\omega)$ of $ta(t)$

$$t_n = \left.\frac{d\phi(\omega)}{d\omega}\right|_{\omega=\omega_n} = \frac{\mathrm{Re}A(\omega_n)\mathrm{Re}B(\omega_n) + \mathrm{Im}A(\omega_n)\mathrm{Im}B(\omega_n)}{|A(\omega_n)|^2} \tag{13.16}$$

The group delay times, t_n, computed as above can be used to estimate of the equivalent group velocity values at the central frequencies, f_n, as (Kumari et al. 2018; Liao and Jin 1995)

$$U(f_n) = \frac{D}{(t_n + t_0)} \tag{13.17}$$

Here, D is the hypocentral distance and t_0 is the travel time of the first arriving P-waves. The t_0 is obtained by dividing the hypocentral distance D with the average P-wave velocity, taken as 6.0 km/s for the western Himalayan region (Mahesh et al. 2013). Plots in Fig. 13.7 show two typical examples of the acceleration records (top panels), computed time delays (middle panels) and the group velocity values obtained (bottom panels).

The mean trend of the group velocity values $U(f_n)$ plotted in Fig. 13.7 can be approximated well by a quadratic function of frequency as

$$\bar{U}(f) = a_0 + a_1 \log_{10} f + a_2 (\log_{10} f)^2 \tag{13.18}$$

The least squares fitting of Eq. (13.18) is also shown in Fig. 13.7. The dispersion of the observed group velocity values around the mean curves can be characterized by a frequency-independent standard deviation value defined by

$$\mathrm{SD} = \sqrt{\frac{1}{N-1} \sum_{n=1}^{N} (U(f_n) - \bar{U}(f_n))^2} \tag{13.19}$$

Here, N is the number of discrete frequencies used to estimate the group velocity values.

The values of the coefficients a_0, a_1 and a_2 obtained by fitting of Eq. (13.18) and the standard deviation SD obtained from Eq. (13.19) for all the 364 digital

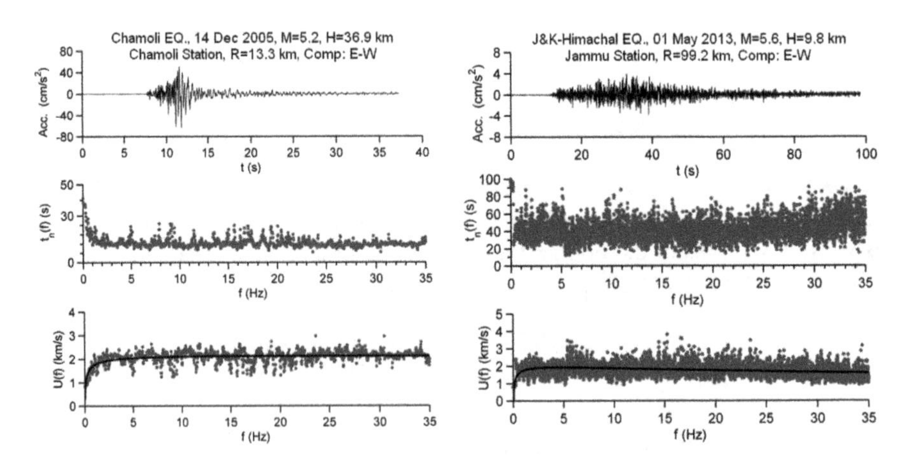

Fig. 13.7 Typical examples of equivalent group velocity values computed from recorded accelerograms with details given in the headings and the mean quadratic curves fitted by least squares fitting

accelerograms are now used to fit the following forms of predicting relations by least squares regression analysis

$$a_i = c_1 + c_2 \log R + c_3 (\log R)^2; \quad i = 0,\ 1,\ 2 \tag{13.20}$$

$$\mathrm{SD} = c_1 + c_2 \log R + c_3 (\log R)^2 \tag{13.21}$$

These functional forms in terms of the hypocentral distance alone are found to be adequate after trying several alternatives in terms of both magnitude and distance. The values of the regression coefficients c_1, c_2 and c_3 obtained for a_i $(i = 0,\ 1,\ 2)$ and SD are listed in Table 13.3. The values of the coefficients a_0, a_1, a_2 and the standard deviation SD obtained by using the least squares from Table 13.3 are designated as $\hat{a}_0$, $\hat{a}_1$, $\hat{a}_2$ and $\widehat{\mathrm{SD}}$, respectively. Also, the group velocity values obtained by using the coefficients $\hat{a}_0$, $\hat{a}_1$ and $\hat{a}_2$ into Eq. (13.18) are designated as $\hat{U}(f_n)$.

13.4.2 Simulation of Fourier Phase Spectra

The empirically predicted mean values $\hat{U}(f_n)$ and the standard deviation $\widehat{\mathrm{SD}}$ of the group velocity for a specified hypocentral distance are proposed to be used to simulate the group velocity values as

$$U(f_n) = \hat{U}(f_n) + \xi_n \widehat{\mathrm{SD}} \tag{13.22}$$

Here, ξ_n is a series of random numbers simulated from the standard normal distribution with zero mean and unit standard deviation. The group velocity values thus obtained can be used to compute the phase velocity values $C(f_n)$ using (Kumari et al. 2018; Liao and Jin 1995)

$$C(f_n) = \frac{f_n}{\int_0^{f_n} \frac{1}{U(f)} \mathrm{d}f} \tag{13.23}$$

Table 13.3 Least squares estimates of the regression coefficients in Eqs. (13.20) and (13.21)

Predicted parameter	Regression coefficients		
	c_1	c_2	c_3
a_0	0.92362	−0.66292	0.63449
a_1	−1.35974	1.77437	−0.40108
a_2	0.93758	−1.04353	0.21054
SD	−0.15041	0.350605	−0.05369

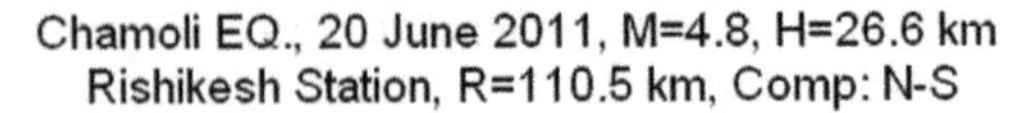

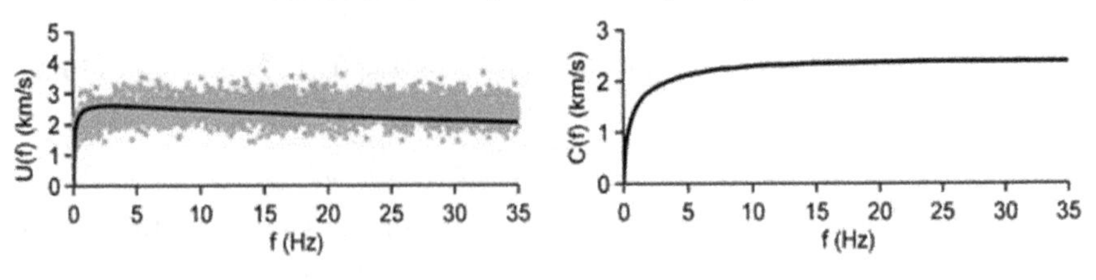

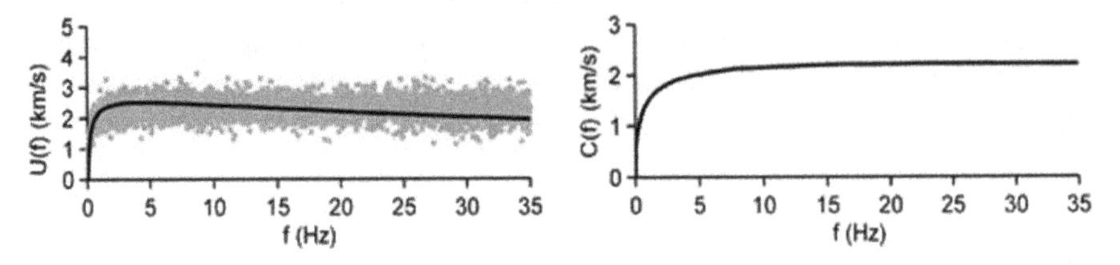

Fig. 13.8 Typical comparisons between the simulated group velocity values and the mean quadratic curves for recorded accelerograms (left-side panels) and the examples of the phase velocity dispersion curves computed from the simulated group velocity values

The left-side panels in Fig. 13.8 present typical examples of the comparison between the mean group velocity dispersion curves $\bar{U}(f)$ of two real accelerograms obtained by fitting Eq. (13.18) to the values obtained by Eq. (13.17) and the group velocity values $U(f_n)$ simulated using Eq. (13.22). The mean group velocity dispersion curves of real accelerograms are, in general, seen to lie within the range of the empirically simulated group velocity estimates. Thus, the phase velocity dispersion curves computed from the simulated group velocity data and shown in the right-side panels of Fig. 13.8 can be considered to be very good representation of the phase velocity curves of the real accelerograms. The phase velocity values are finally used to define the unwrapped phase spectrum at central frequency f_n as (Kumari et al. 2018; Liao and Jin 1995)

$$\phi(f_n) = -2\pi f_n \cdot \left[\frac{D}{C(f_n)} - t_0\right] \tag{13.24}$$

The quantity within the brackets on right-hand side represents the relative travel time of the wave group with central frequency f_n.

13.5 Generation of Design Accelerograms

The empirically predicted Fourier amplitude and phase spectra, using the relationships developed in the preceding two sections, can be used to define the complex Fourier transform $A(f)$ of the ground acceleration $a(t)$ as

$$A(f) = \mathrm{FS}(f)\mathrm{e}^{i\phi(f)} \quad (13.25)$$

Here, $\mathrm{FS}(f)$ is the Fourier amplitude spectrum predicted using the empirical relations developed in this study and $\phi(f)$ is Fourier phase spectrum obtained from Eq. (13.24) using the empirically estimated phase velocity values. The $\mathrm{FS}(f)$ can be computed with any desired confidence level for specified magnitude M, epicentral distance R, focal depth H, local geological condition parameter s and the site soil condition parameter s_L. The phase spectrum $\phi(f)$, on the other hand, depends only on the hypocentral distance, because the group velocity dispersion curves of real accelerograms were not found to have significant magnitude dependence similar to some other studies (e.g., Sato et al. 2002). The design accelerogram $a(t)$ is obtained by computing the inverse Fourier transform of the complex Fourier spectrum $A(f)$ using fast Fourier transformation algorithm

$$a(t) = \int_{-\infty}^{\infty} A(\omega)\mathrm{e}^{2\pi ift}\mathrm{d}f \quad (13.26)$$

Similar to the real accelerograms, the accelerograms generated from Eq. (13.26) also suffer from baseline errors, leading to physically unrealistic velocity and displacement time histories by integration of the accelerograms. The baseline error in a simulated accelerogram may be associated mainly with the incorrect initial conditions. To illustrate the effect of the baseline error, the left-hand side plots in Fig. 13.9 show an original simulated accelerogram without baseline correction (top plot) along with the computed velocity (middle plot) and the displacement (bottom plot) records corresponding to the median Fourier amplitude spectrum of horizontal motion for magnitude $M = 7.0$, epicentral distance $R = 30$ km, focal depth $H = 10$ km, local geological condition as thick sediments ($s = 0$) and site soil condition as deep soil ($s_L = 2$). The velocity and displacement records are seen to be completely unrealistic with large baseline distortions. It is thus necessary that baseline correction is applied to the simulated accelerograms to get the design accelerograms for practical applications. The method proposed in Gupta (2018b) has been used to apply the baseline correction in the present study, which combines the merits of two different methods due to Trifunac (1971) and Converse and Brady (1992). This method used is based on the high-pass filtering by non-causal Ormsby filter that does not attenuate the frequencies in the pass band and needs much less number of leading zero pads to absorb the filter transients. The right-hand side plots in Fig. 13.9 show the baseline corrected accelerogram (top plot) and the computed

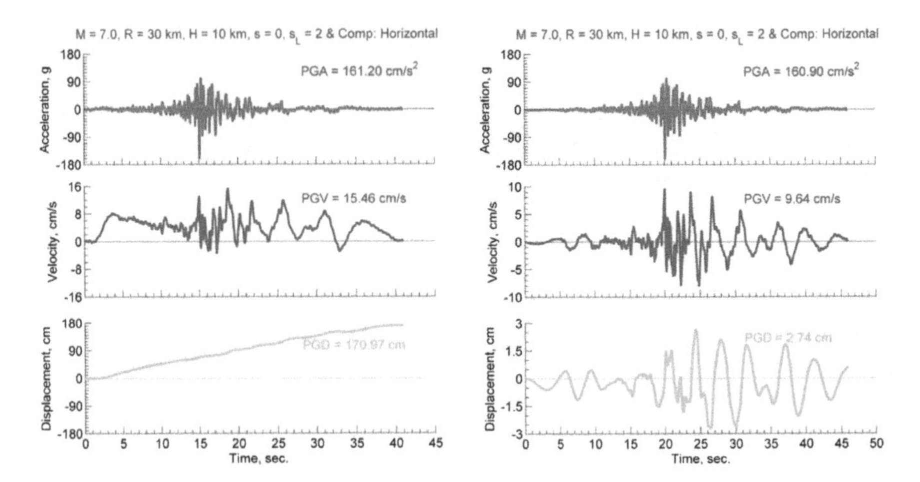

Fig. 13.9 A typical illustration of the uncorrected (left-side plots) and baseline corrected (right-side plots) simulated accelerogram and the corresponding velocity and displacement records obtained by integration

velocity (middle plot) and the displacement (bottom plot) records. The velocity and displacement time histories of the corrected accelerogram are now seen to be quite realistic with no baseline distortions.

The accelerograms obtained by using the simulated phase spectrum are able to automatically attain the realistic non-stationary characteristics in both amplitude and frequency contents. These are thus suitable for design of important structures by detailed dynamic response analysis considering linear as well as nonlinear behavior. To use the simplified response spectrum superposition method of response analysis, the design accelerograms can also be used to compute the design response spectra with different damping ratios. A response spectrum represents the maximum response amplitudes of single degree of freedom oscillators with natural periods covering the complete range of engineering interest and a fixed damping ratio. Three different types of response spectra can be defined, depending upon the type of response as the relative displacement, relative velocity or absolute acceleration response (Hudson 1979). In common engineering applications, only the displacement spectrum (SD) is computed and the velocity and acceleration spectra are approximated by the pseudo-spectral velocity (PSV) and pseudo-spectral acceleration (PSA), respectively. The design spectra are most commonly specified in terms of the PSA amplitudes (Reston 2010; BIS 1893).

Figure 13.10 presents three different sets of four plots each to illustrate the methodology developed in this study for generation of site-specific design accelerograms and the PSA spectra. The three sets correspond to three different combinations of earthquake magnitude M, epicentral distance R and focal depth H with fixed local geologic condition as thick sediments ($s = 0$), site soil condition as deep soil ($s_L = 0$) and component of motion as horizontal ($v = 0$). In each set of plots, the top left plot gives the empirically simulated phase spectrum $\phi(f)$, the top

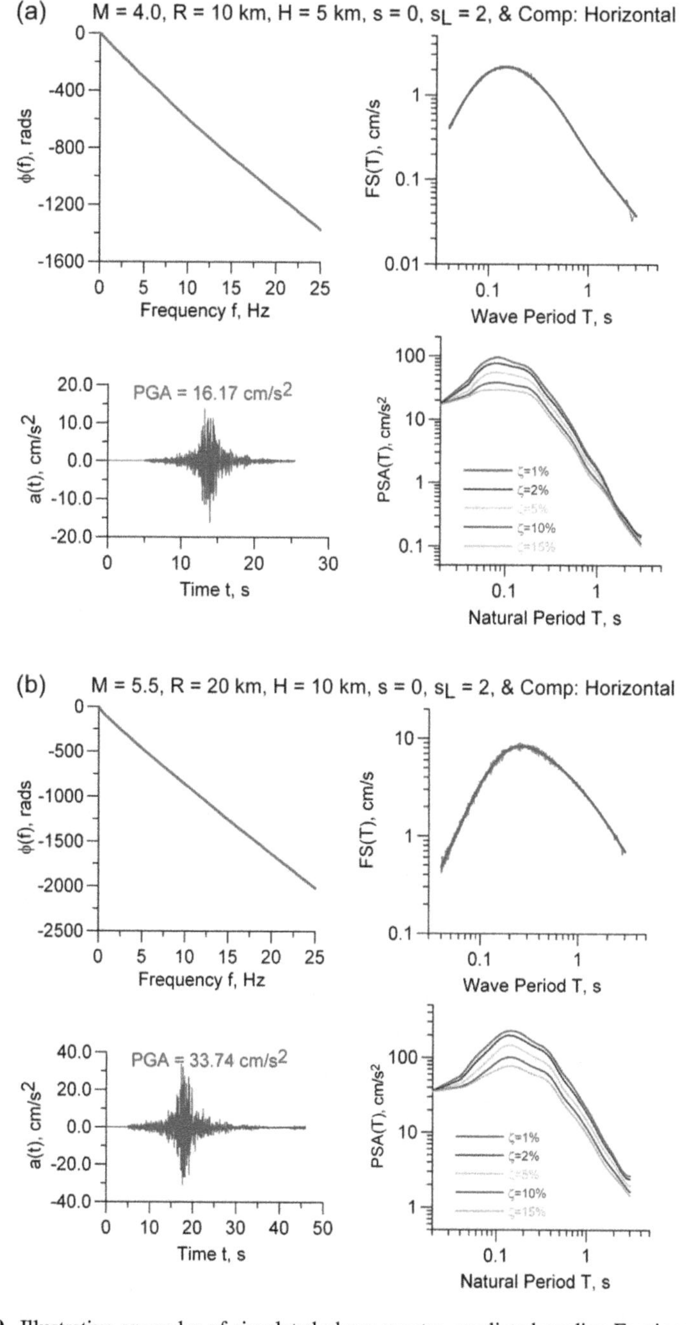

Fig. 13.10 Illustrative examples of simulated phase spectra, predicted median Fourier amplitude spectra of horizontal motion, generated design accelerograms and the computed design response spectra with damping ratios of 1, 2, 5, 10 and 15%. The results are for local geology as thick sediments ($s = 0$) and site soil as deep soil ($s_L = 2$) and for **a** $M = 4.0$, $R = 10$ km and $H = 5$ km, **b** $M = 5.5$, $R = 20$ km and $H = 10$ km and **c** $M = 7.0$, $R = 30$ km and $H = 10$ km

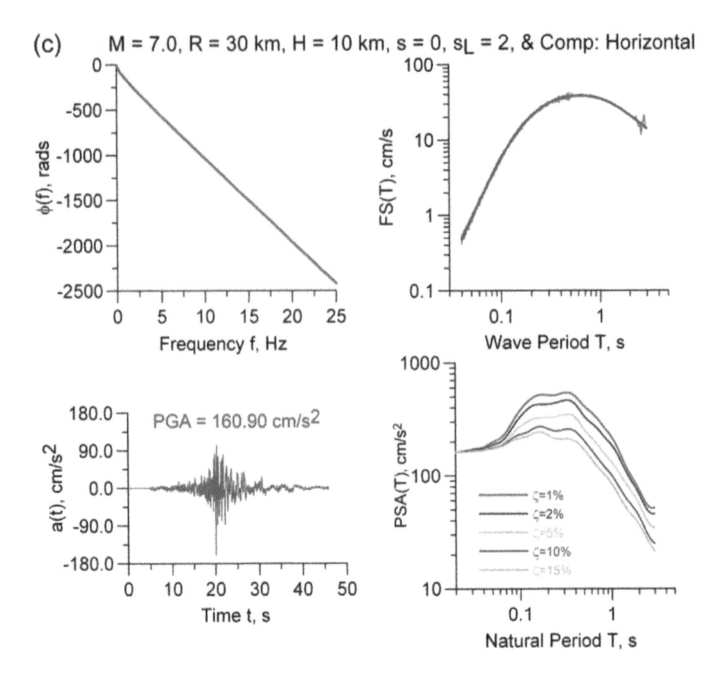

Fig. 13.10 (continued)

right plot gives the predicted median Fourier amplitude spectrum FS(T), the bottom left plot gives the simulated design accelerogram $a(t)$ and the bottom right plot gives the computed PSA spectra for five damping ratios of ς = 1, 2, 5, 10 and 15%.

To show the compatibility of the simulated design accelerograms with the empirically predicted Fourier amplitude spectra, the Fourier spectra of the accelerograms are also shown by zig-zag gray lines in the top right panels of the three sets of plots in Fig. 13.10. Very close matching between the two Fourier spectra indicates that all the site-specific characteristics of the amplitude spectra have been carried forward into the simulated accelerograms. Further, the simulated accelerograms are able to represent realistic phase characteristics also due to the use of the phase spectra generated from the empirically predicted group velocity values matching closely with the group velocity values of real accelerograms.

13.6 Discussion and Conclusions

The most commonly used approach for generation of design accelerograms is to artificially synthesize the accelerograms compatible with 5% damped response spectrum, termed as target response spectrum. As the synthetic accelerograms are not able to represent the non-stationarity in frequency contents, suitably selected real accelerograms modified to match a target spectrum are used instead, particularly when nonlinear structural behavior is of importance. However, due to limited database on recorded accelerograms, it is normally difficult to get the real accelerograms matching closely the desired earthquake and site condition parameters.

To define the design accelerograms, the target response spectra are generally obtained using empirical prediction equations. To account for strong regional dependence, such equations are required to be developed using recorded strong motion data in a region of interest. Lacking the region-specific prediction equations, weighted average of several equations from other regions with subjective weights is used commonly in practical applications, the reliability and validity of which remains unknown. Alternatively, the prediction equations developed using the data simulated by seismological source model approach or developed using a combination of simulated and limited recorded data are also used in many applications. The reliability of this approach also remains highly questionable due to several simplifying idealizations made and inadequate calibration of the source model due to lack of or inadequacy of recorded strong motion data.

To overcome the limitations and difficulties associated with the generation of response spectrum compatible accelerograms, this paper has developed a methodology based on empirical prediction of Fourier amplitude and phase spectra for generation of realistic site-specific design accelerograms for the western Himalayan region. A comprehensive prediction model is developed for the Fourier spectrum amplitudes at various wave periods using a database of 217 strong motion accelerograms recorded in the region. The predicted Fourier amplitude spectra are shown to have physically realistic dependence on earthquake magnitude, representative source-to-site distance, local geological condition and the site soil condition, and also to match closely with the trends and amplitudes of the spectra of the real accelerograms.

A subset of 182 digital accelerograms is used to develop a prediction model for the equivalent group velocity values in terms of the hypocentral distance only. The phase spectra are simulated via computation of the phase velocity values from the group velocity values. The predicted group velocity values are shown to match very well with the group velocity values of the real accelerograms, implying the realistic nature of the simulated phase spectra. The design accelerograms are generated by inverse Fourier transformation of the complex Fourier spectra defined in terms of the predicted amplitude and phase spectra. In view of the realistic nature of the phase spectra, the design accelerograms obtained are able to represent realistic non-stationary characteristics in both amplitudes and frequencies.

It may thus be concluded that the predictive models developed in this paper provide a scientifically sound basis for generation of site-specific design accelerograms for any site in the western Himalayan region. For simplified dynamic response analysis using spectrum superposition approach, the design accelerograms can also be used to compute the site-specific design response spectra for different damping ratios.

References

Al Atik L, Abrahamson N (2010) An improved method for nonstationary spectral matching. Earthq Spectra 26(3):601–617

Ambraseys NN, Bilham R (2000) A note on the Kangra Ms = 7.8 earthquake of 4 April 1905. Curr Sci 79(1):101–106

ASCE (2010) Minimum design loads for buildings, ASCE/SEI 7–10. American Society for Civil Engineering (ASCE), Reston, VA

BIS (2016) Criteria for earthquake resistant design of structures part 1 general provisions and buildings. IS 1893, Bureau of Indian Standards (BIS), New Delhi

Boore DM (2003a) Simulation of ground motion using the stochastic method. Pure Appl Geophys 160:635–676

Boore DM (2003b) Phase derivatives and simulation of strong ground motions. Bull Seismol Soc Am 93:1132–1143

Borah K, Kanna N, Rai SS, Prakasam KS (2015) Sediment thickness beneath the Indo-Gangetic Plain and Siwalik Himalaya inferred from receiver function modeling. J Asian Earth Sci 99:41–56

Chopra AK (2007) Dynamics of structures: theory and applications to earthquake engineering, 7th edn. Pearson/Prentice Hall, Upper Saddle River

Converse AM, Brady AG (1992) BAP—basic strong-motion accelerogram processing software; version 1.0. Open-File Report 92–296A, United States Geological Survey, 174 pp

Dasgupta S, Pande P, Ganguly D, Iqbal Z, Sanyal K, Venkatraman NV, Dasgupta S, Sural B, Harendranath L, Mazumdar K, Sanyal S, Roy A, Das LK, Misra PS, Gupta H (2000) Seismotectonic Atlas of India and its environs. In: Narula PL, Acharyya SK, Banerjee J (eds) Special publication no. 59, Geological Survey of India, 87 pp

Douglas J (2017) Ground motion prediction equations 1964–2016. Report, Department of Civil and Environmental Engineering, University of Stratchlyde, Glasgow, UK. www.gmpe.org.wk/gmpereport2014.pdf

Gansser A (1964) Geology of the Himalayas. Wiley Interscience, New York

Gupta ID (2006) Delineation of probable seismic sources in India and neighbourhood by a comprehensive analysis of seismotectonic characteristics of the region. Soil Dyn Earthq Eng 76:766–790

Gupta ID (2018a) An overview of response spectrum superposition methods for MDOF structures. In: Sharma ML, Shrikhande M, Wason HR (eds) Advances in Indian earthquake engineering and seismology, chapter 16. Springer, Switzerland, pp 335–364

Gupta ID (2018b) Uniformly processed strong motion database for Himalaya and northeast region of India. Pure Appl Geophys 175(3):829–863. https://doi.org/10.1007/s00024-017-1703-y

Gupta ID, Joshi RG (1993) On synthesizing response spectrum compatible accelerograms. Eur Earthq Eng 7(2):25–33

Gupta ID, Trifunac MD (2017) Scaling of Fourier spectra of strong earthquake ground motion in western Himalaya and northeastern India. Soil Dyn Earthq Eng 102:137–159

Gupta ID, Trifunac MD (2018) Empirical scaling relations for pseudo relative velocity Spectra in western Himalaya and northeastern India. Soil Dyn Earthq Eng 106:70–89

Hudson DE (1979) Reading and interpreting strong motion accelerograms. Monograph MNO-1, Earthquake Engineering Research Institute, Oakland, CA, pp 94612–1934

Kausel E (2017) Advanced structural dynamics. Cambridge University Press, Cambridge

Kimura M, Izumi M (1989) A method of artificial generation of earthquake ground motion. Earthquake Eng Struct Dyn 18:867–874

Kumar A, Mittal H, Sachdeva R, Kumar A (2012) Indian strong motion instrumentation network. Seismol Res Lett 83(1):59–66

Kumari N, Gupta ID, Sharma ML (2018) Synthesizing nonstationary earthquake ground motion via empirically simulated equivalent group velocity dispersion curves for Western Himalayan region. Bull Seismol Soc Am 108(6):3469–3487. https://doi.org/10.1785/0120170387

Liao ZP, Jin X (1995) A stochastic model of the Fourier phase of strong ground motion. Acta Seismol Sin 8:435–446

Mahesh P, Rai SS, Sivaram K, Paul A, Gupta S, Sarma R, Gaur VK (2013) One-dimensional reference velocity model and precise locations of earthquake hypocenters in the Kumaon–Garhwal Himalaya. Bull Seismol Soc Am 103(1):328–339. https://doi.org/10.1785/0120110328

Motazedian D, Atkinson GM (2005) Stochastic finite-fault modeling based on a dynamic corner frequency. Bull Seismol Soc Am 95(3):995–1010

Ni J, Barazangi M (1984) Seismotectonics of the Himalayan collision zone: geometry of the underthrusting Indian plate beneath the Himalaya. J Geophys Res 89(B2):1147–1163

Parvez IA, Vaccari F, Panza GF (2003) A deterministic seismic hazard map of India and adjacent areas. Geophys J Int 155(2):489–508

Sato T, Murono Y, Nishimura A (2002) Phase spectrum modeling to simulate design earthquake motion. J Nat Dis Sci 24:91–100

Scherbaum F, Cotton F, Staedtke H (2006) The estimation of minimum-misfit stochastic models from empirical ground-motion prediction equations. Bull Seismol Soc Am 96(2):427–445

Seed HB, Ugas C, Lysmer J (1976) Site-dependent spectra for earthquake-resistant design. Bull Seismol Soc Am 66(1):221–243

Trifunac MD (1971) Zero baseline correction of strong-motion accelerograms. Bull Seismol Soc Am 61(5):1201–1211

Trifunac MD (1989) Dependence of Fourier spectrum amplitudes of recorded strong earthquake accelerations on magnitude, local soil conditions and on depth of sediments. Earthq Eng Struct Dyn 18(7):999–1016

Trifunac MD (2016) Site conditions and earthquake ground motion—a review. Soil Dyn Earthq Eng 90:88–100

Trifunac MD, Anderson JG (1977) Preliminary empirical models for scaling absolute acceleration spectra. Report CE 77-03, University of Southern California, Los Angeles, USA

Trifunac MD, Brady AG (1975) On the correlation of seismic intensity scales with the peaks of recorded strong ground motion. Bull Seismol Soc Am 65(1):139–162

Zhao JX, Irikura K, Zhang J, Fukushima Y, Somerville PG, Asano A, Ohno Y, Oouchi T, Takahashi T, Ogawa H (2006) An empirical site-classification method for strong-motion stations in Japan using H/V response spectral ratio. Bull Seismol Soc Am 914–925. https://doi.org/10.1785/0120050124

Chapter 14
Regional–Local Hybrid Seismic Hazard and Disaster Modeling of the Five Tectonic Province Ensemble Consisting of Westcentral Himalaya to Northeast India

Sankar Kumar Nath, Chitralekha Ghatak, Arnab Sengupta, Arpita Biswas, Jyothula Madan, and Anand Srivastava

14.1 Introduction

Earthquakes are most threatening to mankind, killing thousands of people and damaging properties worth billions of dollars every year in various parts of the globe. Seismic loss depends not only on the hazard caused by moderate to large earthquakes, but also on the socioeconomic/structural exposure and its vulnerability. Indian subcontinent resonates as one of the most earthquake-prone regions in the world. The seismotectonic regimes across India are significantly diverse. The Indian plate boundary encompasses transverse fault system of the Chaman, the Ornach Nal, and the Sulaiman-Kirthar ranges to the northwest, the Himalayan arc to the north, the subduction zones of the Hindu Kush Pamir to the northwest, the Indo-Myanmar arc to the northeast, and the Andaman–Nicobar–Sumatra tracts to the southeast. The Indian subcontinent consists of pre-cambrian cratons of Archean age and rift zones filled with proterozoic and phanerozoic sediments. The Indian subcontinent can be divided into three main sub-regions based on the geologic and tectonic regime, viz. (a) the Himalayan frontal arc in the north extending from north-west to the Arakan-Yoma Mountain ranges covering a distance of 2,500 km as a result of Mesozoic subduction and the collision between the Indian and the Eurasian plates; (b) the Indo-Gangetic plains located between the abruptly rising Himalaya in the north and the Indian Peninsula in the south formed by the vast alluvial plains in the north along the basin of the River Ganges and Sindhu (Indus) extending from east to west; and (c) the Indian Peninsula in the south, which

S. K. Nath (✉) · C. Ghatak · A. Sengupta · A. Biswas · J. Madan · A. Srivastava
Department of Geology and Geophysics, Indian Institute of Technology Kharagpur, Kharagpur 721302, India
e-mail: nath@gg.iitkgp.ac.in

T. G. Sitharam et al. (eds.), *Latest Developments in Geotechnical Earthquake Engineering and Soil Dynamics*, Springer Transactions in Civil and Environmental Engineering, https://doi.org/10.1007/978-981-16-1468-2_14

comprises the Indian Shield with the Deccan traps and the Dharwar cratons. Three seismic gaps along the Himalaya are located as: 'Assam seismic gap' between the occurrences of 1950 Assam earthquake and the 1934 Nepal-Bihar earthquake, 'Central (Himalaya) seismic gap' between the epicenters of 1934 Nepal-Bihar earthquake and the 1905 Kangra earthquake, and 'Kashmir seismic gap' to the west of the 1905 Kangra earthquake occurrence.

Records of earthquake occurrences in the form of a catalogue constitute an important database for seismotectonic and seismic hazard studies. Nath et al. (2017) compiled the homogeneous declustered seismicity of Southeast Asia covering the period 1900–2014 comprising of 58,256 Mainshock events which further updated upto 2018 in the present study. This complete and homogeneous earthquake catalogue prepared for India and the adjoining region has been used for probabilistic seismic hazard assessment of the region in terms of seismogenic source zonation, seismicity analysis, smoothened seismicity modeling, and in framing the seismic hazard assessment protocol.

Based on the underlying tectonic setup and the past seismic activities in the Indian Subcontinent, we identified eleven Tectonic Provinces, viz. (i) the Bengal Basin, (ii) the Indo-Gangetic Foredeep, (iii) the Central India, (iv) the Kutch Region, (v) the Koyna-Warna Region, (vi) the Western Ghats Region, (vii) the Eastern Ghats Region, (viii) the Kashmir Himalaya, (ix) the Westcentral Himalaya, (x) the Darjeeling-Sikkim Himalaya and (xi) the Northeast India which have the potential of generating moderate to large magnitude earthquakes as shown in Fig. 14.1.

Probabilistic seismic hazard maps have been adopted in several countries for national seismic provisions of building codes, insurance rate of structures, risk assessments and other public policies. Such maps depict ground motion values, namely peak ground acceleration (PGA), and spectral amplitude at different periods, and describe the frequency of exceeding a set of ground motions. The hazard analysis involves integration of different seismic hazard components such as the seismogenic source models, the ground motion prediction equations, and the seismic site conditions. In the probabilistic approach, the incorporation of uncertainties associated with the hazard components that are of both aleatory and epistemic nature is explicit in the analysis. Nath and Thingbaijam (2012) carried out probabilistic seismic hazard analysis (PSHA) of entire India. Fundamental studies have been carried out to deliver the hazard components, including seismogenic source zonation and seismicity modeling in the Indian Subcontinent, assessment of site conditions across the country (Nath et al. 2013), and a suitability test for the ground motion prediction equations in the regional context (Nath and Thingbaijam 2011). They formulated a layered seismogenic source zonation corresponding to four hypocentral depth ranges (in km): 0–25, 25–70, 70–180 and 180–300. They adopted 16 ground motion prediction equations which were selected according to the assessment criteria carried out by Nath and Thingbaijam (2011). These components were integrated to deliver a preliminary model consisting of spatial distributions of PGA and 5%-damped pseudo-spectral acceleration (PSA). The results indicate that the hazard distribution in the country is significantly higher than those

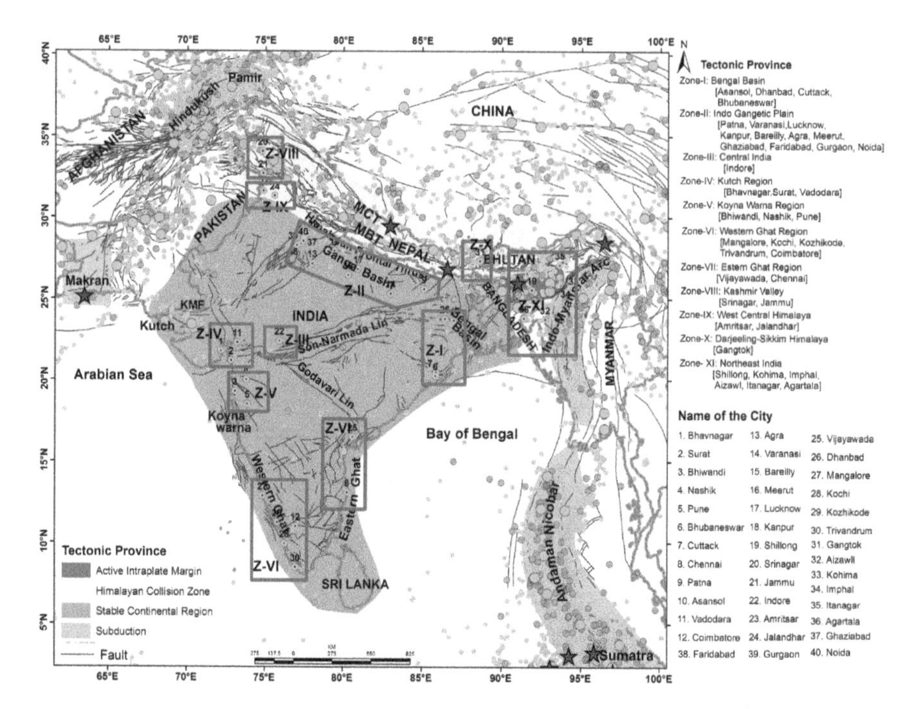

Fig. 14.1 Seismotectonic map of the Indian Subcontinent with eleven Tectonic Provinces identified as: **(i)** Bengal Basin, **(ii)** Indo-Gangetic Foredeep, **(iii)** Central India, **(iv)** Kutch Region, **(v)** Koyna-Warna Region, **(vi)** Western Ghats Region, **(vii)** Eastern Ghats Region, **(viii)** Kashmir Himalaya, **(ix)** Westcentral Himalaya, **(x)** Darjeeling-Sikkim Himalaya and **(xi)** Northeast India region

specified previously by Global Seismic Hazard Assessment Program (GSHAP) and BIS (2002). Figure 14.2a depicts the seismic hazard distribution of the Indian subcontinent in terms of PGA for firm rock site conditions (Nath and Thingbaijam 2012). The PGA is seen to vary from 0.02 to 0.94 g on the Indian territory while that generated at the surface using site amplification factor from IBC (2006) in compliance with the site classes in the Indian subcontinent as shown in Fig. 14.2b which depicts a variation of 0.048–1.392 g in the subcontinent.

The maximum estimated surface PGA of 1.392 g is associated with Northwestern Himalaya and Northeast India. The cities of Srinagar, Jammu, Kangra, Shimla, Dehradun, Chamoli, Lakimpur, Lucknow, Muzaffarpur, Kishanganj, Darjeeling and Gangtok exhibit relatively higher hazard level. So is the case in the Northeast India region at the state capitals of Itanagar, Tezpur, Dispur, Guwahati, Kohima, Imphal, Aizwal, Manipur and Nagaland being exposed to very high hazard level. The cities of Bhuj, Kacchh, Rajkot, Pune and Ratnagiri exhibit high hazard level. The cities of Delhi, Udaipur, Jaisalmer, Patna, Jabalpur and Latur are seen to experience moderate hazard to the tune of 0.44–0.60 g. In order to understand the implications of this probabilistic seismic hazard, preliminary

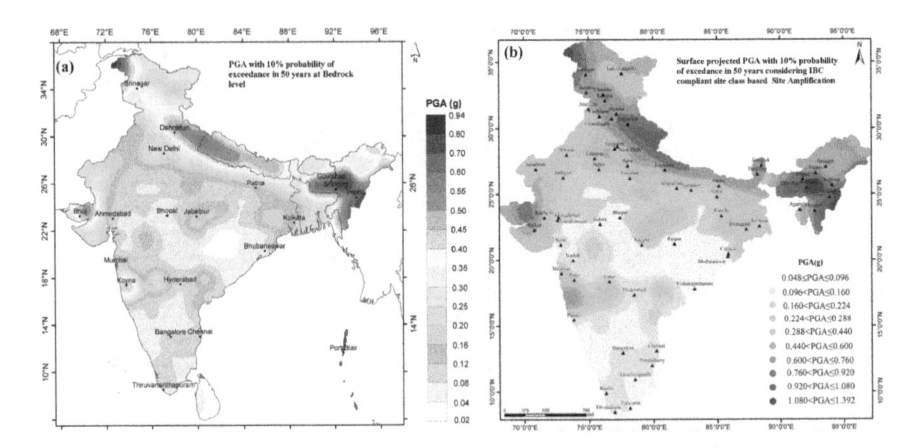

Fig. 14.2 Spatial distribution of seismic hazard in terms of peak ground acceleration (PGA in g) in India and its adjoining region for **a** 10% probability of exceedance in 50 years with a return period of 475 years at engineering bedrock [after Nath and Thingbaijam (2012)], and **b** at the surface considering IBC (2006) compliant site class-based site amplification

regional seismic risk is computed for the subcontinent on the vulnerability components, viz. population density (PD) and landuse and landcover (LULC), building rural wall (BRW), building rural roof (BRR), building urban wall (BUW), building urban roof (BUR), building density (BD) and integrated through a fuzzy protocol and presented in Fig. 14.3.

Regionally the risk level has been classified into five categories, such as severe, high, moderately high, moderate and low. The entire Himalayan belt and Indo-Gangetic plains fall under severe risk zone while few patches of high-risk zones have also been spotted in this region. The cities situated in the Northern region of the Indian territory such as Srinagar, Jammu, Amristar, Kangra, Shimla, Dehradun, Chandigarh, Gurgaon, Chamoli, Delhi and Lucknow fall in the severe risk zone. The entire Northeastern region has also been identified as severe risk zone encompassing Gangtok, Darjeeling, Itanagar, Dispur, Kohima, Shillong, Imphal, Manipur, Nagaland, Agartala and Aizwal. Likewise western parts of Kutch, Rajkot, Jaisalmer, Pune, and Ratnagiri fall in the severe risk zone. Although located in the stable region of the Indian subcontinent a few cities like Latur, Jabalpur, Vijayawada and Kolkata fall under severe risk zone. High seismic risk zone is identified along the Indo-Gangetic plain, in the central, western and south-eastern parts of Indian territory. The cities such as Udaipur, Jaipur, Allahabad, Varanasi, Ahmadabad, Surat, Hyderabad, Chennai and Bangalore are located in this zone. The Northwestern region, part of the Southern Peninsular region and a few patches in the central part of the Indian territory is demarcated under moderately high seismic risk zone. The moderate seismic risk is exhibited in most parts of the northwestern region and the central part of the Peninsular Shield.

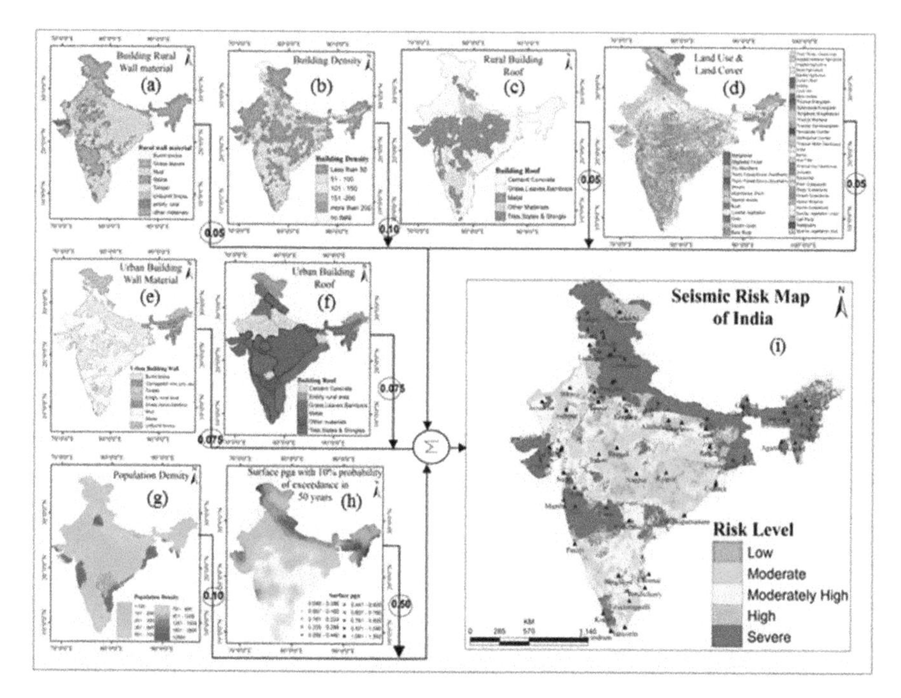

Fig. 14.3 Seismic risk assessment for the Indian subcontinent by integrating **a** building rural wall, **b** building density, **c** building rural roof, **d** LULC, **e** building urban wall, **f** building urban roof, **g** population density, **h** spatial distribution of PGA (*g*) with 10% probability of exceedance in 50 years at IBC (2006) compliant to surface level and **i** regional seismic risk map of the subcontinent. Weight assigned to each of these themes is also indicated in the integration protocol

14.2 Second-Order Seismic Hazard Assessment

Fault-based source consideration has never been used in most of the previous studies by Bhatia et al. (1999), and Nath and Thingbaijam (2012) for probabilistic seismic hazard analysis. In the present investigation active tectonic features, viz. faults and lineaments extracted from the Seismotectonic Atlas of India (Dasgupta et al. 2000), National Geomorphological Lineament Mapping on 1:50,000 scale (http://www.portal.gsi.gov.in/portal/page; http://bhuvan.nrsc.gov.in/gis/thematic/index.php) and additional features by image processing of Landsat TM, SRTM, ASTER and LISS IV data have been considered as shown in Fig. 14.4a in addition to the layer-wise polygonal sources identified in the 0–25, 25–70, 70–180 and 180–300 depth ranges, which expectedly have the potential of generating earthquakes of M_w 3.5 and greater in the seismogenic source formulation for the five seismotectonic blocks as shown in Fig. 14.4b starting from Westcentral Himalaya (WCH) all the way up to Northeast India (NEI) including Indo-Gangetic Foredeep (IGF), Bengal Basin (BB) and Darjeeling-Sikkim Himalaya (DSH) while postulating the PSHA regime for these blocks on regional/local-specific site condition and its

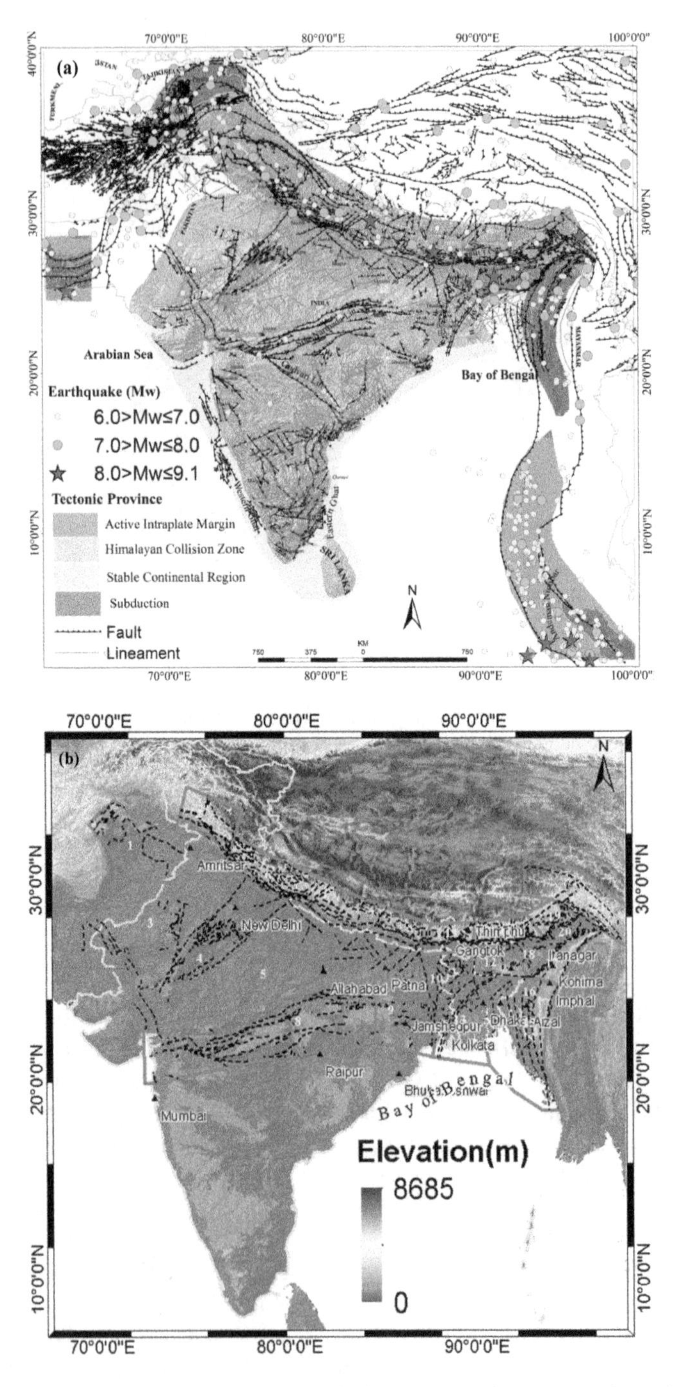

Fig. 14.4 **a** Major fault and lineaments constituting the tectonic source regime of the Indian Subcontinent considered for its seismic hazard assessment. **b** Layered polygonal seismogenic source definition for the tectonic study region starting from Westcentral Himalaya all the way up to Northeast India overseeing Indo-Gangetic Foredeep, Bengal Basin and Darjeeling-Sikkim Himalaya at 0–25 km hypocentral depth range as adopted from Nath and Thingbaijam (2012)

secondary hazard and urban impact on the city of Amritsar located in the Westcentral Himalaya, Agra in the Indo-Gangetic Foredeep, Kolkata and Dhaka in Bengal Basin, Gangtok in Darjeeling-Sikkim Himalaya and Guwahati in the Northeastern Indian Seismic Province.

14.2.1 Smoothened Gridded Seismicity Model

The contribution of background events in the hazard perspective is calculated using smoothened gridded seismicity models. It allows modeling of discrete earthquake distributions into spatially continuous probability distributions. The technique given by Frankel (1995) is employed here for seismicity smoothening. The technique was previously employed by several workers (Nath and Thingbaijam 2012; Frankel et al. 2002). In the present analysis, the study region, i.e., the tectonic ensemble of Westcentral Himalaya all the way up to Northeast India including Bhutan, Indo-Gangetic Foredeep, Bengal Basin and Darjeeling-Sikkim Himalaya, is gridded at a regular interval of 0.1°; each grid point encompassing a cell of 0.1° × 0.1°. The smoothened function is given by,

$$N(m_r) = \frac{\sum_j n_j(m_r) \mathrm{e}^{-\left(d_{ij}/c\right)^2}}{\sum_j \mathrm{e}^{-\left(d_{ij}/c\right)^2}} \tag{14.1}$$

where $n_j(m_r)$ is the number of events with magnitude $\geq m_r$, d_{ij} is the distance between *i*th and *j*th cells, and *c* denotes the correlation distance. The annual activity rate λ_{m_r} is computed each time as $N(m_r)/T$ where *T* is the (sub)catalogue period. The smoothened seismicity analysis for M_w 3.5 at four hypocentral depth ranges of 0–25, 25–70, 70–180 and 180–300 km are shown in Fig. 14.5a–d. From the smoothened seismicity analysis, the probable asperity zones could be identified. It is observed that at the threshold magnitude of M_w 3.5 patches of well clustered activity rate indicating possible stress concentration are seen within the Northeast India which has been the source of the deadly 1950 Assam earthquake of M_w 8.6, 1897 Shillong earthquake of M_w 8.1, 1869 Cachar earthquake of M_w 7.6 and so on.

In the present study, seismic activity rates are also calculated for each active tectonic source using three different threshold magnitudes of M_w 3.5, M_w 4.5 and M_w 5.5 depending upon the focal depth ranges of 0–25, 25–70, 70–180 and 180–300 km. The tectonic-based seismic activity has also been estimated by fault degradation technique following the methodology of Iyengar and Ghosh (2004). The number of earthquake occurrences per year with $m > m_o$ in a given source zone consisting of *n* number of faults is denoted as $N(m_o)$. According to the fault degradation technique, $N(m_o)$ should be equal to the sum of the number of earthquakes $N_s(m_o)$, that is, possible to occur at different faults available in the source

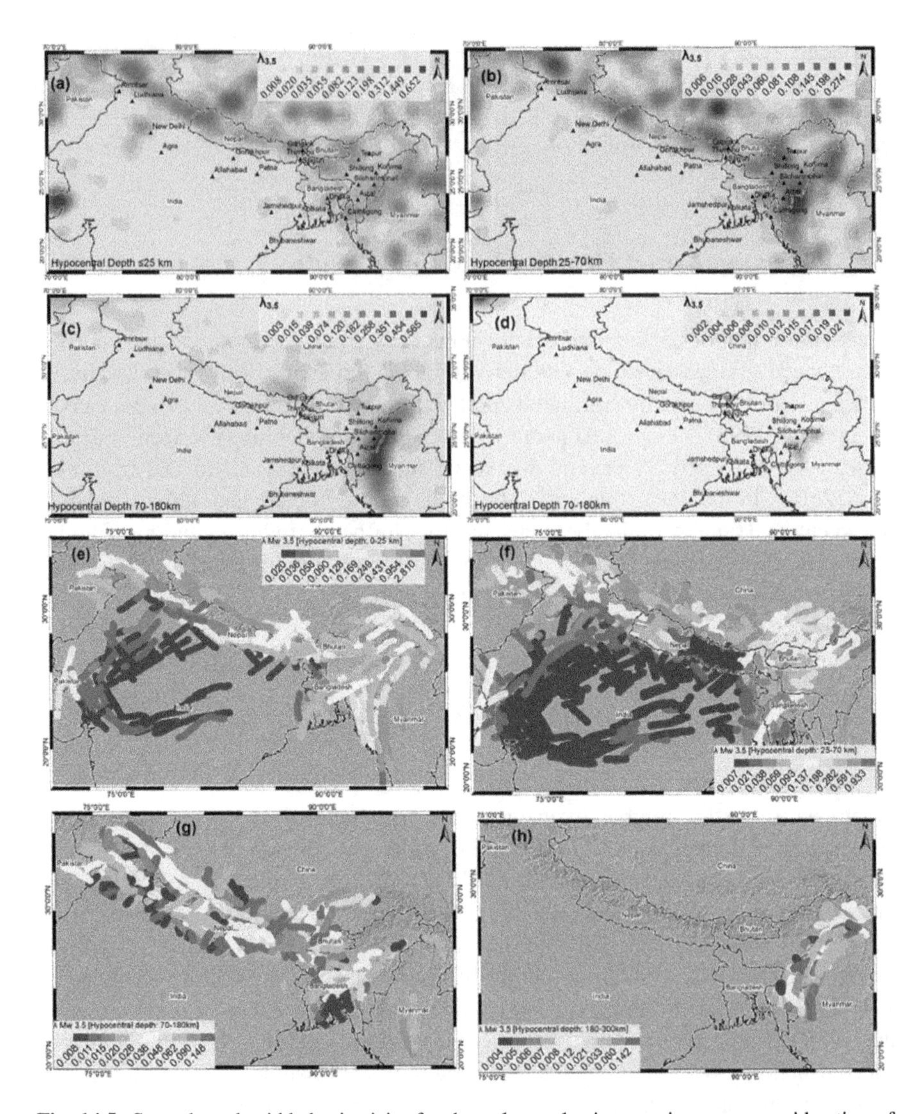

Fig. 14.5 Smoothened gridded seismicity for the polygonal seismogenic source consideration of the tectonic ensemble from Westcentral Himalaya all the way up to Northeast India including Bhutan, Indo-Gangetic Foredeep, Bengal Basin and Darjeeling-Sikkim Himalaya for the threshold magnitude M_w 3.5 at the hypocentral depth range of **a** 0–25 km, **b** 25–70 km, **c** 70–180 km and **d** 180–300 km. Fault/lineament activity rates computed for the threshold magnitude M_w 3.5 are also distributed at the hypocentral depth range of **e** 0–25 km, **f** 25–70 km, **g** 70–180 km and **h** 180–300 km

zone, i.e., $N(m_o) = \sum N_s(m_o)$, where $N_s(m_o)$ represents the annual frequency of occurrence of an event on *s*th subfault (s = 1,2….n) and m_o of 3.5, 4.5 and 5.5 have been used based on the threshold magnitudes. Thus, the two parameters, viz. length of the fault (L_s) and the number of past earthquakes (n_s) of magnitude m_o associated with the *s*th fault, have been used as weights for estimating $N_s(m_o)$. If N_t is the total number of events occurred within the source zone, the weighting factor is estimated as,

$$\alpha_s = L_s/\Sigma L_s \quad \text{and} \quad \delta_s = n_s/N_t \tag{14.2}$$

Taking the mean of the above two weight factors as indicating the seismic activity of the *s*th fault in the zone, we get

$$N_s(m_o) = 0.5(\alpha_s + \delta_s) * N(m_o) \tag{14.3}$$

The annual activity rate of each tectonic feature has been computed using the above expression where the regional recurrence is degraded into individual faults/lineaments. Figure 14.5e–h depicts the tectonic annual activity rate for all the faults and lineaments with the occurrence of threshold magnitude of M_w 3.5 at four hypocentral depth ranges. These diagrams also depict stress regime of each fault/lineament triggering earthquakes of magnitude M_w 3.5 and above.

14.2.2 Probabilistic Seismic Hazard Analysis

Seismic hazard at a particular locality in an earthquake county is usually quantified in terms of the level of ground shaking observed in the region. The methodology for probabilistic seismic hazard analysis (PSHA) incorporates how often the annual rate of ground motion exceeds a specific value for different return periods of the hazard level at a particular site of interest. In the hazard computation, all the relevant sources and possible earthquake events are considered. A synoptic probabilistic seismic hazard model is generated at engineering bedrock based on the protocol given by Nath and Thingbaijam (2012). The basic methodology of PSHA involves computation of ground motion thresholds that are exceeded with a mean return period of say 475 years/2,475 years of return period at a particular site. The effects of all the earthquakes of different sizes occurring at various locations for all the seismogenic sources at various probabilities of occurrences are integrated into one curve that shows the probability of exceeding different levels of a ground motion parameter at the site during a specified time period. The computational formulation as developed by Cornell (1968), Esteva (1970), McGuire (1976) is given by,

$$\nu(a > A) = \sum_i \lambda_i \int_m \int_r \int_\delta P(a > A|m, r, \delta) \; f_m(m) f_r(r) f_\Delta(\delta) \mathrm{d}m \, \mathrm{d}r \, \mathrm{d}\delta \tag{14.4}$$

where ν $(a > A)$ is the annual frequency of exceedance of ground motion amplitude A, λ_i is the annual activity rate for the *i*th seismogenic source for a threshold magnitude, function P yields probability of the ground motion parameter a exceeding A for a given magnitude m at source-to-site distance r. The corresponding probability density functions are represented by $f_m(m)$, $f_r(r)$ and $f_\Delta(\delta)$. The probability density function for the magnitudes is generally derived from the GR relation (Gutenberg and Richter 1944). In practice, this relationship is truncated at some lower and upper magnitude values which are defined as the truncation parameters related to the minimum (m_{min}) and maximum (M_{max}) values of magnitude, obtained by various methods. The computational protocol of PSHA is presented in Fig. 14.6.

The fuzzy protocol/the logic tree framework shown in Fig. 14.7 is employed for the computation of PSHA at each site of the five tectonic blocks, viz. Westcentral Himalaya, Indo-Gangetic Foredeep region (Nath et al. 2019), Bengal basin (Nath et al. 2014), Darjeeling-Sikkim Himalaya and Northeast India, to incorporate multiple models in the source considerations, GMPEs, and seismicity parameter definitions. In the present study, the seismogenic sources, i.e., tectonic and layered polygonal sources, are assigned weights equal to 0.60 and 0.40, respectively. The

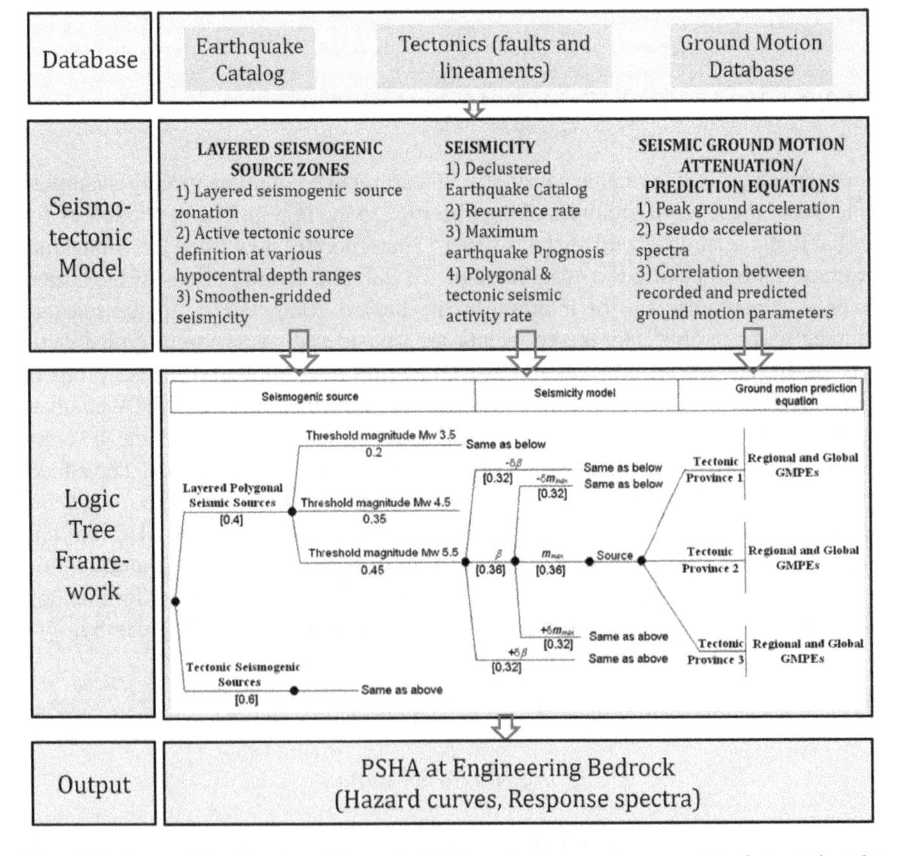

Fig. 14.6 Computational protocol for probabilistic seismic hazard assessment of an earthquake province

three derivatives for the threshold magnitude of M_w 3.5, 4.5 and 5.5 are assigned weights equal to 0.20, 0.35 and 0.45, respectively. The seismicity model parameters, namely the annual rate of earthquakes $\lambda(m)$ and β pair, are assigned weights of 0.36 while the respective $\pm$ 1 standard deviation gets weight equal to 0.32. Similar weight allotment is assigned for $M_{\max}$. The weights are allocated following the statistical rationale suggested by Grünthal and Wahlström (2006). In order to define appropriate weights, the percentage of probability mass in a normal distribution for the mean value and ± 1 standard deviation is considered corresponding to the center of two equal halves.

The hazard distribution is estimated for the source zones at all the four hypocentral depth ranges separately and thereupon integrated to obtain a pragmatic total value. The seismic hazard maps of the study region corresponding to the spatial distribution of peak ground acceleration (PGA) for 10% probability of exceedance in 50 years with a return period of 475 years are shown in Fig. 14.8 depicting an overall PGA variation of 0.02–0.58 g in the study region encompassing five tectonic blocks of WCH, IGF, BB, DSH and NEI. The cities situated in the Northern region of the Indian territory, viz. Shillong, Imphal, Itanagar and Tezpur, are identified with higher hazard level, while moderate hazard level is associated with the upper part of Indo-Gangetic Foredeep region and some part of Bengal Basin adjoining the Chota Nagpur Plateau. Low hazard level is observed in the southern part of Odisha, Punjab and Varanasi. The seismic hazard curves show the probability of exceeding different ground motion parameters at a particular location. Figure 14.8 depicts representative seismic hazard curves for the cities of Amritsar, Agra, Kolkata, Gangtok, and Imphal corresponding to PGA, pseudo-spectral acceleration (PSA) at 0.2 and 1.0 s, respectively, at engineering bedrock. Both 2% and 10% probability of exceedance in 50 years have been demarcated by dotted lines in the diagram presenting both 475 and 2,475 years of return period scenarios at firm rock condition (NEHRP site class BC:V_s 760 ms^{-1}).

14.3 Site Classification

The effective shear wave velocity (V_s^{30}) is a proxy for defining a site class since it is a good indicator of shallow subsurface condition and is directly related to sediment stiffness; thus, its spatial distribution helps in delineating the presence of soft to hard sediments in the terrain based on sharp acoustic impedance contrast. National Earthquake Hazards Reduction Program (NEHRP) and Uniform Building Code (UBC) based on V_s^{30} with similar site response have recommended five site classes or ground profiles (usually for soil types), viz. site class A and site class B with $V_s^{30} \geq$ 1,500 m/s and 1,500 $\geq V_s^{30} >$ 760 m/s representing hard rock and rock site conditions, respectively, while site class C with 760 $> V_s^{30} \geq$ 360 m/s corresponds to soft rock, hard or very stiff soils or gravels, whereas stiff soils with 360 $> V_s^{30} \geq$ 180 m/s designate site class D (Nath and Thingbaijam 2011). On the other hand, Sun et al. (2014) proposed subclasses in site classes C and D,

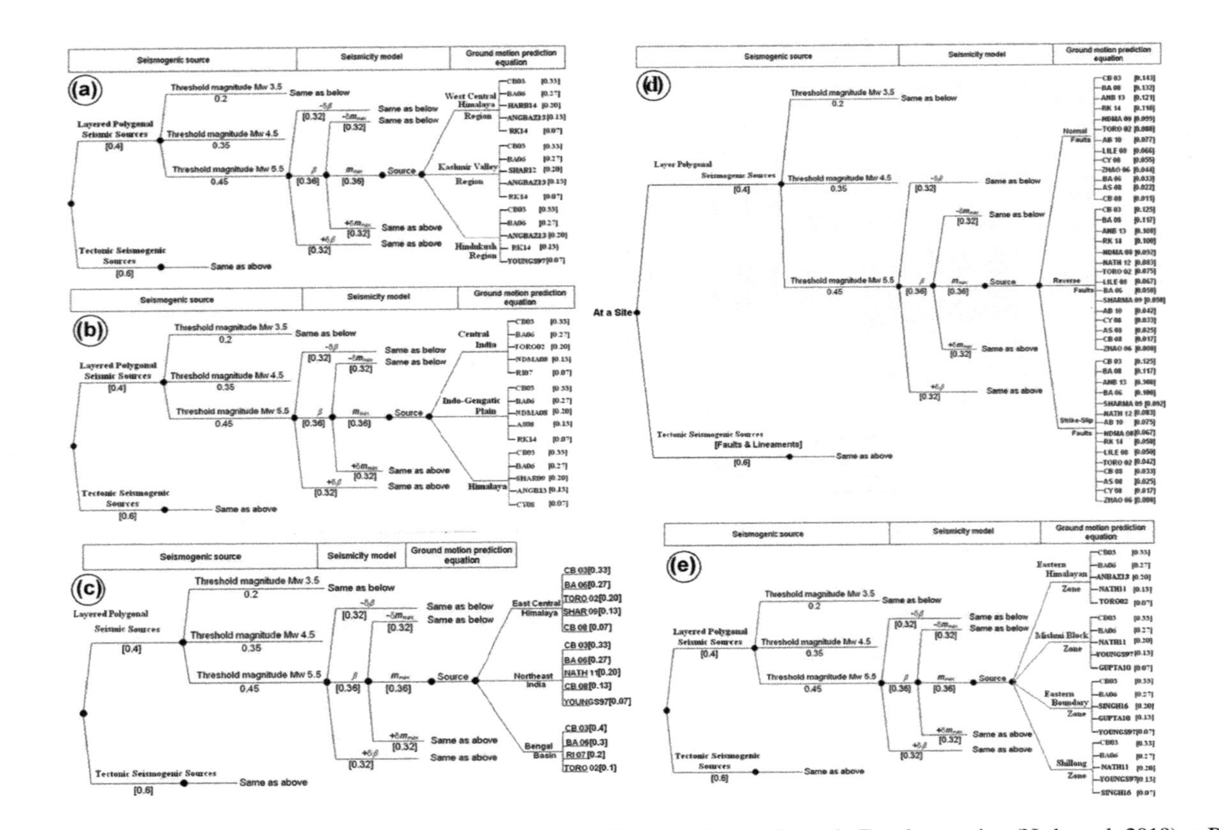

Fig. 14.7 Fuzzy protocol/logic tree framework for PSHA in **a** Westcentral Himalaya, **b** Indo-Gangetic Foredeep region (Nath et al. 2019), **c** Bengal Basin tectonic province (Nath et al. 2014), **d** Darjeeling-Sikkim Himalaya and **e** Northeast India tectonic province

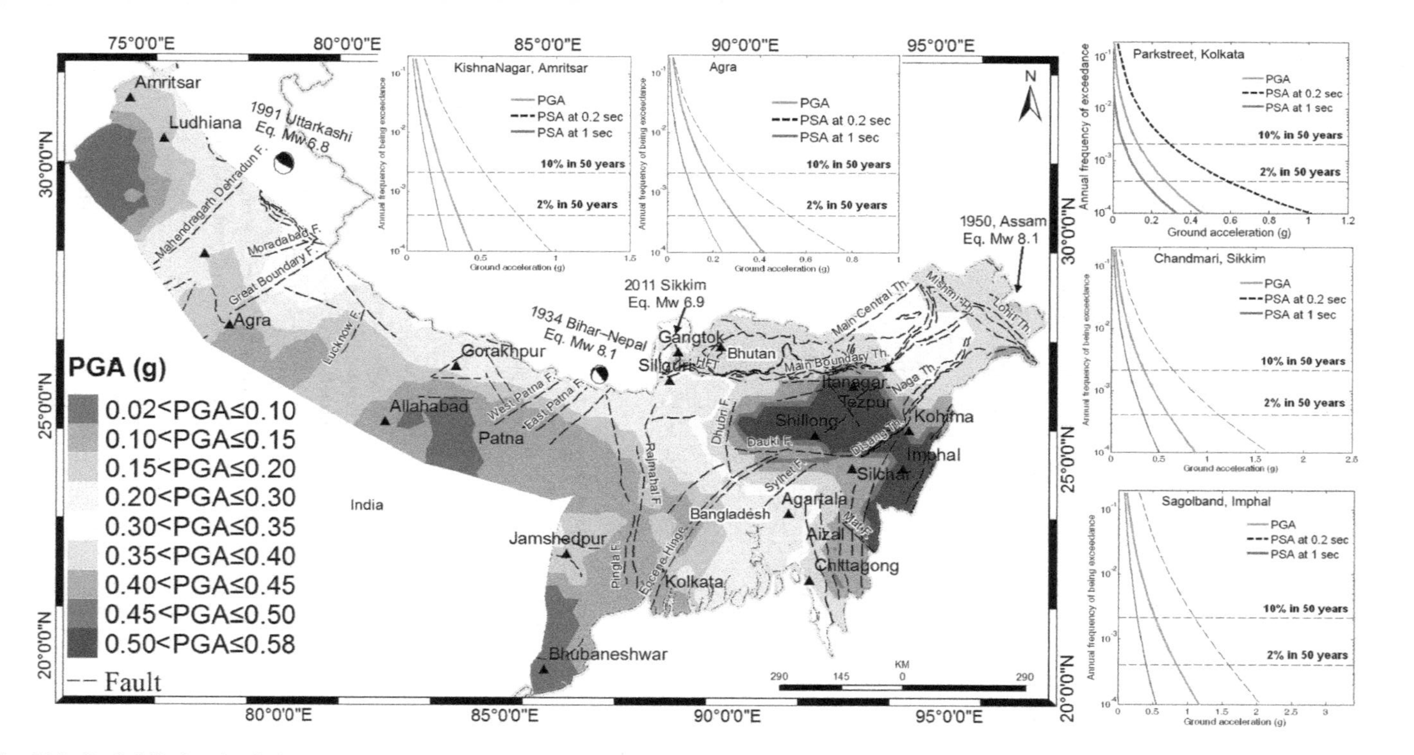

Fig. 14.8 Probabilistic seismic hazard at the engineering bedrock at firm rock condition with NEHRP site class BC for the tectonic study region starting from Westcentral Himalaya all the way upto Northeast India overseeing Indo-Gangetic Foredeep, Bengal Basin and Darjeeling-Sikkim Himalaya in terms of PGA (g) distribution for 10% probability of exceedance in 50 years. The annual frequency of exceedence versus ground acceleration plots usually termed as seismic hazard curves for the selected cities of Amritsar in WCH, Agra in IGF, Kolkata in BB, Chandmari in DSH, and Imphal in NEI for PGA and PSA at 0.2 and 1.0 s for uniform firm rock site condition have also been depicted with 10% and 2% probabilities of exceedance in 50 years' levels being demarketeed by horizontal dotted lines in each of these seismic hazard plots

subdividing site classes C and D into four subcategories as shown in Table 14.1 as: C1 (V_s^{30}: 620–760 m/s), C2 (V_s^{30}: 520–620 m/s), C3 (V_s^{30}: 440–520 m/s), C4 (V_s^{30}: 360–440 m/s), D1 (V_s^{30}: 320–360 m/s), D2 (V_s^{30}: 280–320 m/s), D3 (V_s^{30}: 240–280 m/s) and D4 (V_s^{30}: 180–240 m/s), respectively. In both the classifications, soft clay defined as soil with plasticity index (PI) > 20, moisture content (w) $\geq$ 40% and average undrained shear strength (Su) < 25 kPa is categorized in site class E, while similar liquefiable soft clays are classified under site class F.

Table 14.1 Seismic site classification based on effective shear wave velocity (V_s^{30}) of 30 m soil column (after NEHRP, UBC 1997; Sun et al. 2014)

V_s^{30}(m/s)			Description
NEHRP	UBC (1997)	Sun et al. (2014)	
A (>1,500)	S_A (>1,500)		Hard rock
B (760–1,500)	S_B (760–1,500)	>760	Rock site
C (360–760)	S_C (360–760)	C1 (>620)	Soft rock, hard or very stiff soils or gravels
		C2 (>520)	
		C3 (>440)	
		C4 (>360)	
D (180–360)	S_D (180–360)	D1 (>320)	Stiff soils
		D2 (>280)	
		D3 (>240)	
		D4 (>180)	
E (<180)	S_E (<180)	$\leq$ 180	More than 3 m of soft clay defined as soil with plasticity index (PI) > 20, moisture content (w) $\geq$ 40%, and average undrained shear strength (Su) < 25 kPa
F (requires site-specific evaluation of the soil)	S_F (special soils)		Either of the following four categories of soil are considered: (i) Soils vulnerable to potential failure or collapse under seismic loading such as liquefiable soils, quick and highly sensitive clays, and collapsible weakly cemented soils, (ii) Peats and/or highly organic clays (soil thickness >3 m) of peat and/or highly organic clay, (iii) Very high plasticity clays (soil thickness >8 m) with PI > 75, and, (iv) Very thick soft/medium stiff clays (soil thickness >36 m)

14.3.1 Regional Site Classification

Shear wave velocity is more often connected to the geological attributes, geomorphological units, elevations, slope gradients, and distances from mountains or hills. Therefore, terrain information, such as surface geology, geomorphology, landform and topography, are used in a power law relationship to predict the effective shear wave velocity of a region. In the present study, a regional site classification map for the Northeast India tectonic province encompassing Bhutan has been prepared based on these four terrain attributes in a nonlinearly regressed relationship.

14.3.1.1 Geology

Northeast India is geologically very complex since it represents the dynamic frontal part of the Indian plate on the one hand and the relatively stable central (Chinese) plate on the other. Geologically, the region presents a stratigraphic sequence ranging from pre-Cambrian to Quaternary (Nandy et al. 1975). The oldest geological formation is represented by the pre-Cambrian gneissic complex of Meghalaya Plateau, a craton, and the Karbi Anglong Plateau. The Himalayas occupying the northern border of the region consists of Protozoic to early Palaeozoic age formations. These consist of low-grade metamorphic rocks in the southern section to high-grade schists toward the crest of the mountains. The foothill zone of the Himalaya is formed by the Tertiary rocks, and the rest of the region is formed by Tertiary rocks with different marine facies, ranging from Eocene to Pliocene age (Dikshit and Dikshit 2014).

14.3.1.2 Geomorphology

The northeastern region of India presents a unique mosaic of landforms with great diversity reflecting a complex geotectonic setup. In a regional scale, the most striking morphological expression is given by the Arunachal Himalaya toward north, the Naga-Patkai range of hills toward south, the fold belts of Cachar-Tripura and Mizoram, the Shillong Plateau and the Brahmaputra Valley. The Brahmaputra within Assam represents the general westward gradient while its numerous north and south bank tributaries represent the other two gradient trends. The major geomorphic units can be broadly delineated as, dissected structural hills of the upper catchment, the piedmont zone, Glacial Valleys, and the vast Brahmaputra alluvial plain occupying the most segment of the study area.

14.3.1.3 Landform

Shear wave velocity is directly connected with the landform attributes like stream, drainage, valleys and ridges. Therefore, landform map for the Northeast India is prepared based on Topographic Position Index (TPI). TPI is an algorithm for determining an object or point's relative topographic position to a landform (Weiss 2001). TPI works by comparing the value of each cell in a DEM with the mean value of its neighborhood. Positive values indicate areas of relatively high elevation, like ridges and negative values indicate areas of relative lows, like valleys. Values close to zero indicate areas of constant slope. Landform classification maps are generated based on the computed TPI values for the terrain.

Once the geology, geomorphology, slope and landform maps are prepared for the Northeast India region including Bhutan as shown in Fig. 14.9a–e, respectively; those are nonlinearly regressed with 80% supervision using already available or sparsely acquired in-situ shear wave data to obtain a power polynomial relationship like the 5th degree polynomial given in Eq. (14.5) for the Northeast India region. Considering this 5th degree nonlinear polynomial equation the terrain attributes of Northeast India are combined converted into shear wave velocity (V_s) which is effectively used to generate the regional site classification map of Northeast India as given in Fig. 14.9f following NEHRP, FEMA (2000) and Sun et al. (2018) site classification nomenclature.

$$V_s^{30} = A * \ln(\mathrm{LF}) + B * (\mathrm{GGM})^5 + C * (\mathrm{GGM})^4 + D * (\mathrm{GGM})^3 + E * (\mathrm{GGM})^2 + F * (\mathrm{GGM}) + G * \ln(\mathrm{SLP}) + H \tag{14.5}$$

where $A = -28.887$, $B = 0.174$, $C = -4.881$, $D = 48.668$, $E = -201.934$, $F = 347.873$, $G = 173.646$, $H = -34.572$, LF = landform, GGM = geology and geomorphology and SLP = slope.

In order to check the reliability of this regional classification map, 614 geotechnical borehole data converted (both in-situ shear wave velocity and SPT-N value corrected shear wave velocity) shear wave velocity have been considered for the correlation study with both the topography-based site classification map of Nath et al. (2013) and regional site classification map just produced and depicted in Fig. 14.9f. From this correlation between 614 geotechnical and regional dataset, a good clustering is observed along 1:1 correspondence line as shown in Fig. 14.10a depicting a 70–75% confidence bound in the regional site class estimation through above-mentioned nonlinear regression protocol. Even the residual plot between these two dataset as shown in Fig. 14.10b exhibits a strong clustering along zero residual line. In contrast correlation between topography gradient-based site class maps of Nath et al. (2013) has predicted an underestimated shear wave velocity in comparison with the geotechnical data depicting a large negative scatter with respect to 1:1 correspondence line as shown in Fig. 14.10b. This, therefore, establishes the authenticity and reliability of the shear wave velocity estimated through a hybrid combination of

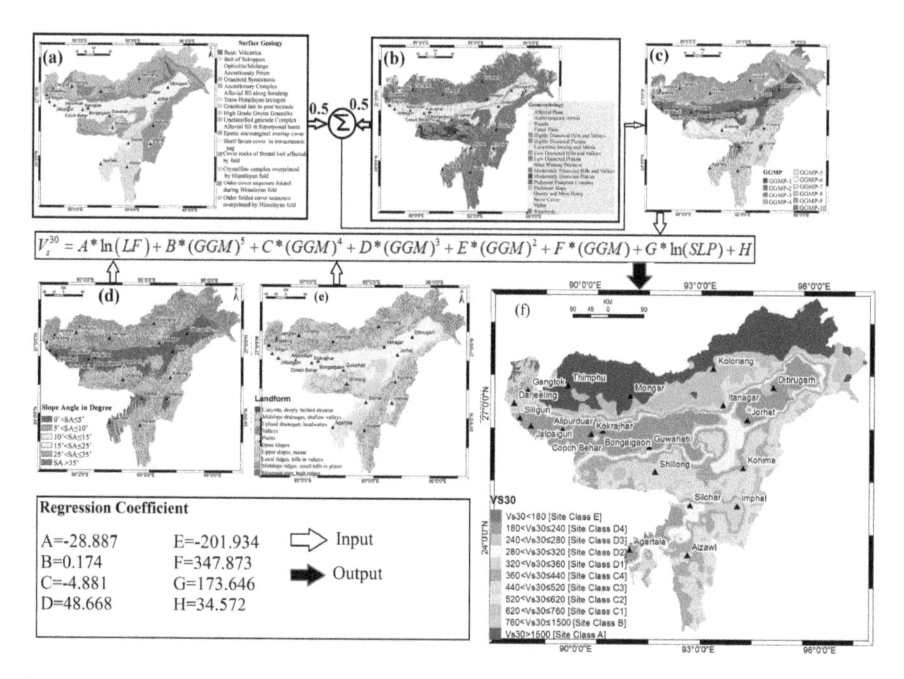

Fig. 14.9 Regional level seismic site classification of Northeast India is performed through a nonlinearly regressed 5th degree polynomial Eq. (14.5) combining shear wave velocity with: **a** surface geology, **b** geomorphology, **c** integrated geology and geomorphology, **d** slope angle and **e** landform and thus evolved the **f** regional site map of this tectonic ensemble

geology, geomorphology topography and landform in a power law relationship of the terrain which can be used for site classification modeling in the absence of both the detail in-situ and surface measurements.

However, this cannot act as an alternative to the invasive and non-invasive investigations generally carried out in a seismic province for site-specific study of seismic hazard at the surface, seismic vulnerability, risk and damage potential considered integral part of urbanization of an earthquake-prone terrain which is why in-situ and surface measurements are carried out for the entire study region for its detailed local-specific site class mapping, its characterization, surface level hazard assessments, etc., as enunciated in the following sections.

14.4 Site Characterization

14.4.1 In-Situ Measurements

In order to understand the nature, sequence and thickness of subsurface strata and to determine the associated engineering properties which quantifies the soil strength,

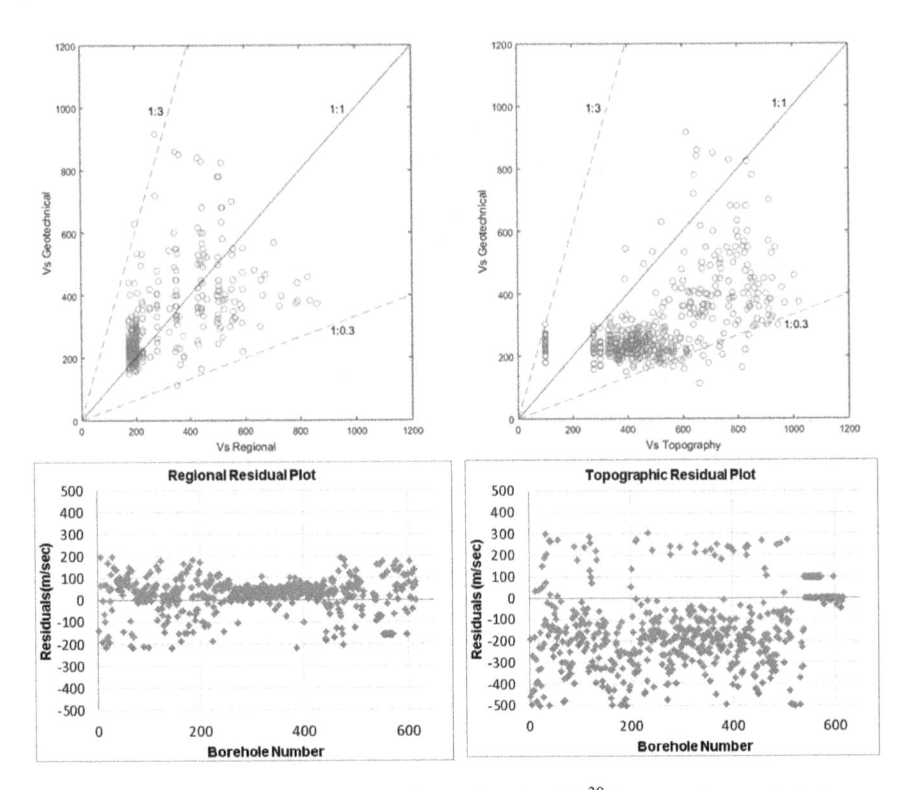

Fig. 14.10 **a** Correlation between geotechnical and regional V_s^{30} dataset and geotechnical versus topographical gradient-based V_s^{30} dataset with 70% confidence bound following the 1:1 correspondence line. **b** Residual plot obtained from geotechnical versus regional V_s^{30} dataset and geotechnical versus topographical gradient-based V_s^{30} dataset in which most of the datasets are having negative residuals and also seen to drift significantly away from the zero residue-line

composition, water content, density and other soil parameters, intended for site classification and its characterization for the present vast seismogenic tectonic study region, extensive surface measurements have been carried out in the form of ambient noise survey, multi-channel analysis of surface wave (MASW) survey, spectral analysis of surface wave (SASW) survey, joint microtremor and MASW data acquisition and processing and in-situ measurements, viz. downhole seismic survey, standard penetration test and geotechnical investigation involving bulk density, unit weight, moisture test, fine content, Atterberg limits test (PL, LL), grain size analysis, etc. Effective shear wave velocity (V_s^{30}) is a good indicator of soil stiffness and acoustic impedance contrast across sediment stratum; thus, its spatial distribution can help delineating the presence of various sediment strata in this tectonic study region from Westcentral Himalaya upto Northeast India including Bhutan, Indo-Gangetic Foredeep, Bengal Basin and Darjeeling-Sikkim Himalaya, thus rendering it an effective proxy for site classification. This vast seismotectonic province stretching over five tectonic blocks comprises of both alluvial and rugged

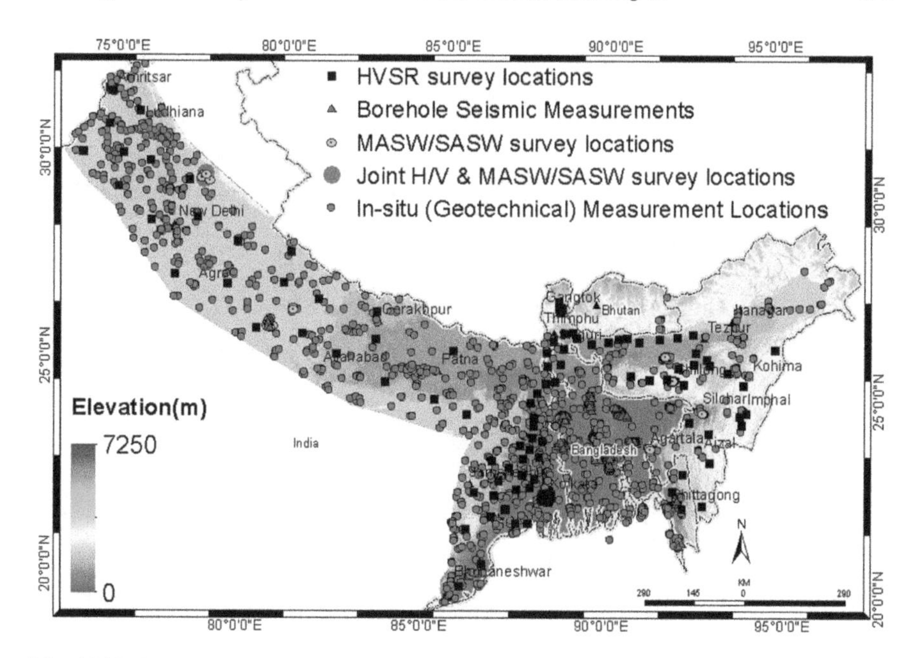

Fig. 14.11 Data coverage map depicting about 10,000 measurement points of HVSR, MASW/SASW, Borehole Seismics, Joint H/V & MASW/SASW and geotechnical SPT-N and other investigations for direct/empirical estimation of shear wave velocity at each of those locations considering an average of 30 m sediment cover bringing in the data density to 01 observation per 7.5 km^2 in the five tectonic block ensemble consisting of Westcentral Himalaya to Northeast India including Bhutan, Indo-Gangetic Foredeep, Bengal Basin and Darjeeling-Sikkim Himalaya

hilly terrain, thus necessitating a dense areal coverage of both surface and in-situ measurements. Figure 14.11 presents a glimpse of the total data coverage of about 10,000 measurements in the form of HVSR, MASW/SASW, borehole seismics, Joint H/V & MASW/SASW and Geotechnical SPT-N and others for direct or empirical estimation of shear wave velocity at each of those locations considering an average of 30 m sediment cover bringing in the data density to 01 observation per 7.5 km^2.

Some representative SPT datasets with allied downhole lithology, physical and shear parameters from the study region encompassing five tectonic blocks have been presented in Fig. 14.12 for the cities of Chandigarh and Amritsar from Westcentral Himalaya; Agra and Noida from Indo-Gangetic Foredeep region; Kolkata and Dhaka from Bengal Basin; 6th mile Gangtok from Darjeeling-Sikkim Himalaya; and Guwahati from Northeast India tectonic province.

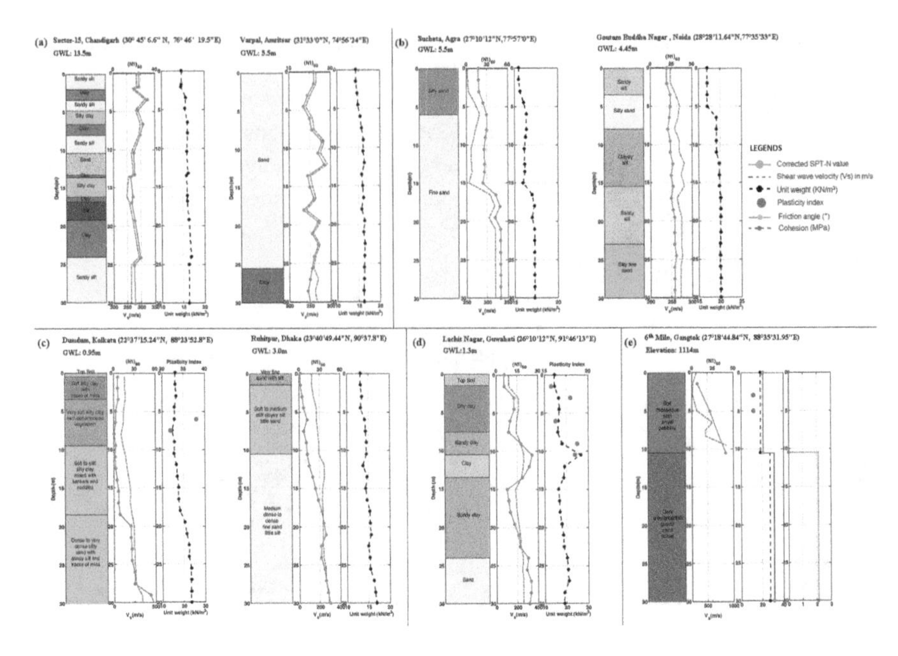

Fig. 14.12 Representative geotechnical borehole dataset with depth-wise lithology, corrected SPT-N, Shear wave velocity, unit weight, plasticity index for the cities of **a** Chandigarh and Amritsar from Westcentral Himalaya, **b** Agra and Noida from Indo-Gangetic Foredeep region, **c** Kolkata and Dhaka from Bengal Basin, **d** Guwahati from Northeast India and **e** Gangtok from Darjeeling-Sikkim Himalaya

14.4.2 Surface Measurements

Surface measurements comprising of various geophysical investigations are used to ascertain high-resolution shallow subsurface S-wave velocity imaging using surface geophysical measurements that comprise of microtremor survey which is capable of estimating predominant frequency of a site through inversion of H/V or HVSR curve obtained from ambient noise recording at a location and Multi-channel Analysis of Surface Waves (MASW) Survey to generate a near-surface 2D shear wave velocity tomogram along a profile of induced energy. Even an advanced joint microtremor and MASW/SASW survey is used in array pattern for guided inversion of phase velocity curve vis-à-vis HVSR curve for generating the most reliable near-surface shear wave velocity tomogram of the subsurface. A representative field setup and processing protocol for microtremor and MASW survey and joint fit of H/V and dispersion curves for estimating 2D S-wave velocity model of the subsurface are shown in Fig. 14.13.

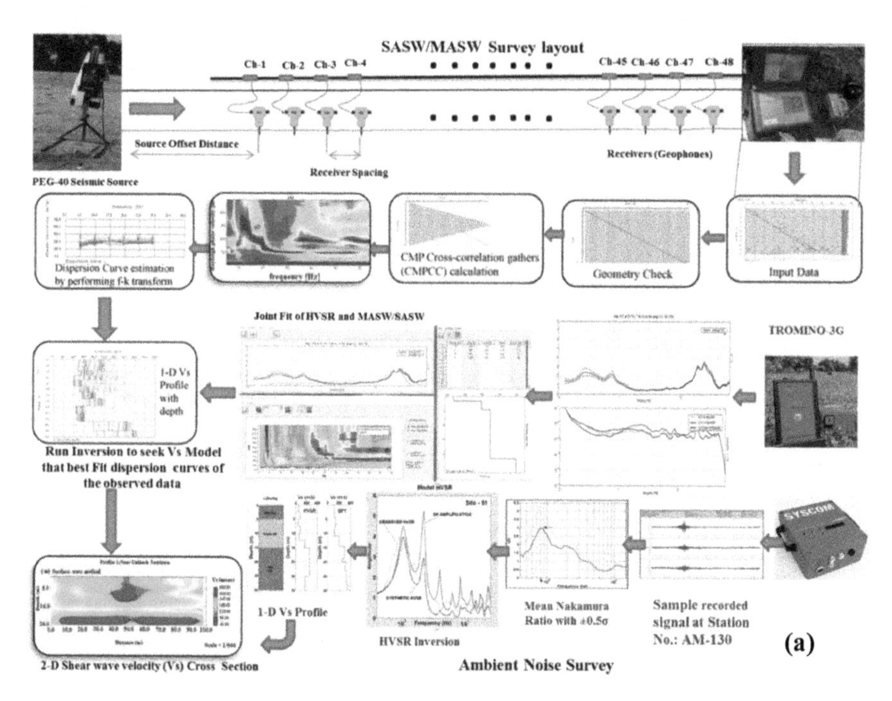

Fig. 14.13 **a** Representative field setup and data processing protocol for joint microtremor field survey and MASW data acquisition, processing and joint analysis of H/V and dispersion curves for estimating 2D S-wave velocity model of the subsurface and **b** Representative joint analysis of H/V and dispersion curves generated from microtremor and MASW/SASW survey, respectively, as depicted for the cities of **(i)** Medinipur in Bengal Basin, and **(ii)** Roorkee in the Indo-Gangetic Foredeep region

14.4.3 Generation of Site- and Lithology-Specific, Depth-Dependent Empirical Relations Between SPT-N and V_s

An attempt has been made in the present investigation to develop site and lithology-specific and depth-dependent empirical relations between SPT-N and V_s for the alluvial plain region in which lithological units have been classified into sixteen categories according to their grain size, plasticity index and presence or absence of decomposed wood, etc., as: (i) Top Soil, (ii) Sand, (iii) Sandy Silt (iv) Silty Clay with Decomposed Wood, (v) Silty Clay with Mica, Sand and/or Kankar, (vi) Clay with Decomposed Wood, (vii) Silty Sand with Mica and/or Clay (vii) Silty Clay with rusty Silty Spots, (ix) Sand with Silt and Clay, (x) Silty Sand with Mica and Kankar, (xi) Bluish/Yellowish gray Silt, (xii) Silt, (xiii) Sand and (xiv) Fine Sand with Gravel (xv) Clayey Silt and (xvi) All Soils. A total of regressed equations for sixteen lithological units identified in the Bengal Basin, Indo-Gangetic Foredeep region and Northeast India are presented in Table 14.2.

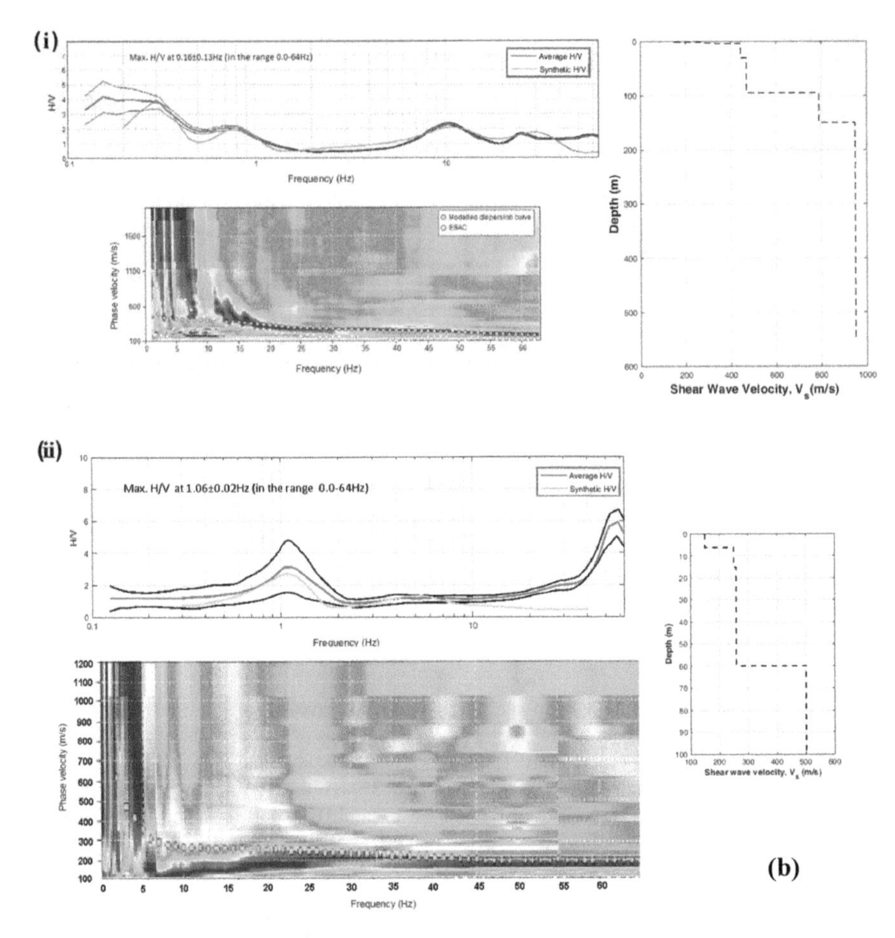

Fig. 14.13 (continued)

Apart from these region-specific original equations developed here, some equations reported from other parts of India given in Table 14.3 and some global empirical relations between SPT-N and V_s as presented in Table 14.4 have also been used in the SPT-N to V_s conversion with appropriate weight assignments.

The final site classification map of the study region starting from Westcentral Himalaya all the way up to Northeast India including Bhutan, Indo-Gangetic Foredeep, Bengal Basin and Darjeeling-Sikkim Himalaya is generated on GIS platform as presented in Fig. 14.14 which exhibits the presence of site classes, E, D4, D3, D2, D1, C4, C3, C2, C1, B and A with the dominance of site classes D4, D3, D2 and D1 in the sediment filled areas of Bengal Basin, part of Westcentral Himalaya, Indo-Gangetic Foredeep region, some part of Siliguri, the Assam valley pertaining to low shear wave velocity. These areas are predominantly underlain by silty clay/clayey silt and silty sandy clay. In contrast, patches of site classes C4 and

Table 14.2 Empirically derived nonlinear regression equations between the corrected SPT-N and in-situ downhole shear wave velocities for different lithological units at various depths of Bengal Basin (BB), Indo-Gangetic Foredeep (IGF) region and Northeast India (NEI)

Lithology	Depth range (m)	Relation established
Top fill (BB + NEI)	0–1.95	$V_S = 74.9 * N^{0.2679}$
Sand (BB)	0–4.5	$V_{sS} = 64.306 * N^{0.4897}$
Sandy silt (BB)	0–9.0	$V_S = 90.732 * N^{0.2301}$
Silty clay with mica, sand and kankar (BB + NEI)	1.5–4.95	$V_S = 78.55N^{0.3116}$
Silty clay with decomposed wood (BB + NEI)	1.5–9.96	$V_S = 85.18 * N^{0.3196}$
Fine sand with silt and clay (BB + NEI)	1.5–4.97	$V_S = 81.14 * N^{0.2448}$
Clay with decomposed wood (BB + NEI)	4.5–10.1	$V_S = 102.2 * N^{0.286}$
Silty clay with mica, sand and kankar (BB + NEI)	4.5–10.95	$V_S = 71.52 * N^{0.3894}$
Sand (BB)	9.0–15.0	$V_S = 89.65 * N^{2954}$
Silty sand with mica and kankar (BB + NEI)	6.0–15.5	$V_S = 68.18 * N^{0.39929}$
Silty clay with decomposed wood (BB + NEI)	10.5–18.45	$V_S = 85.6 * N^{0.2149}$
Silty clay with kankar and silty spots (BB + NEI)	12–18.95	$V_S = 115.0 * N^{0.2914}$
Silt (BB + NEI)	14.9–17.2	$V_S = 117.3 * N^{0.246}$
Sand (BB)	15.0–24.0	$V_S = 122.42 * N^{0.2082}$
Sandy silt (BB)	15.0–24.0	$V_S = 120.25 * N^{0.2814}$
Silty sand with mica and clay (BB + NEI)	18.25–26.95	$V_S = 57.57 * N^{0.4568}$
Silty clay/clayey silt with micaceous sand (BB + NEI)	16.5–22.5	$V_S = 72.13 * N^{0.4802}$
Silty clay with mica, sand and kankar (BB + NEI)	23.95–34.45	$V_S = 63.33 * N^{0.4497}$
Sandy silt (BB)	24.0–30.0	$V_S = 227.66 * N^{0.128}$
Fine sand with gravel (BB + NEI)	24.6–34.6	$V_S = 69.73 * N^{0.4053}$
Silty clay/clayey silt with micaceous fine sand (BB + NEI)	34.45–45.45	$V_S = 67.47 * N^{0.5101}$
Silty sand with mica and clay (BB)	39.45–50.45	$V_S = 139.2 * N^{0.3182}$
Silty clay/clayey silt with mica (BB)	40.5–54.5	$V_S = 87.68 * N^{0.4772}$
All soil (BB + NEI)	0–30.0	$Vs = 116.00 * N^{0.27}$
All soil (IGF)	0–30.0	$V_S = 95.926N^{0.3183}$
Silt (IGF)	0–30.0	$V_S = 83.392N^{0.3995}$
Clay (IGF)	0–30.0	$Vs = 5.975N^{0.421}$
Sand (IGF)	0–30.0	$Vs = 92.126N^{0.3234}$
Clayey silt (IGF)	0–30.0	$Vs = 88.326N^{0.3417}$
Sandy silt (IGF)	0–30.0	$Vs = 113.24N^{0.2627}$

C3 have been found in some part of Odisha and Jharkhand in Bengal Basin, small patches in Indo-Gangetic Foredeep region, and also in Darjeeling-Sikkim Himalaya and Northeast India including Bhutan followed by site class C2, and site class C1 which comprise of very stiff to very dense soil and soft rock, such as boulders,

Table 14.3 Relations between SPT-N value and shear wave velocity of different soil types reported for various regions of India which have also been used in the present study region by assigning appropriate weights (adopted from Nath 2016; Nath 2011)

Lithology	V_s	Region
Sand	$V_s = 79.0N^{0.434}$	Delhi
	$V_s = 57\ (N1)_{60},^{0.44}$	Bangalore, India
	$V_s = 100.53N^{0.265}$	Chennai, India
Clay	$V_s = 80(N)^{0.33}$	Bangalore, India
	$V_s = 89.31N^{0.358}$	Chennai, India
Silt	$V_s = 86N^{0.420}$	Delhi, India
All soils	$V_s = 95.64N^{0.301}$	Chennai, India
	$V_s = 116N^{0.27}$	Agartala, India

Table 14.4 Relations between SPT-N value and shear wave velocity of various soil types reported for various parts of the globe which have been used in the present study region by assigning appropriate weights (adopted from Nath 2016; Nath 2011)

Lithology	V_s	Region
Sand	$V_s = 57.4N^{0.49}$	USA
	$V_s = 56.4N^{0.50}$	USA
	$V_s = 162.0N^{0.17}$	Greece
	$V_s = 100.0N^{0.24}$	Greece
	$V_s = 123.4N^{0.29}$	Greece
	$V_s = 80.6N^{0.331}$	Japan
	$V_s = 90.8N^{0.319}$	Turkey
	$V_s = 79.0N^{0.434}$	Delhi
	$V_s = 32N^{0.5}$	Japan
	$V_s = 87N^{0.36}$	Japan
	$V_s = 145\ (N_{60})^{0.178}$	Greece
	$V_s = 5.1N + 152$	USA
	$V_s = 88.0N^{0.34}$	Japan
	$V_s = 73N^{0.33}$	Western Taiwan
	$V_s = 80.0N^{0.33}$	Japan
Clay	$V_s = 5.3N + 134$	USA
	$V_s = 114.4N^{0.31}$	USA
	$V_s = 165.7N^{0.19}$	Greece
	$V_s = 105.7N^{0.33}$	Greece
	$V_s = 184.2N^{0.17}$	Greece
	$V_s = 27.0N^{0.73}$	South of Tehran
	$V_s = 80.2N^{0.292}$	Japan
	$V_s = 97.9N^{0.269}$	Turkey
	$V_s = 132(N60)^{0.178}$	Greece
	$V_s = 44N^{0.48}$	Western Taiwan
	$V_s = 100.0N^{0.33}$	Japan

(continued)

Table 14.4 (continued)

Lithology	V_s	Region
Silt	$V_s = 105.6N^{0.32}$	Taiwan
	$V_s = 104(N + 1)^{0.334}$	Taiwan
	$V_s = 145.6N^{0.178}$	Greece
	$V_s = 22N^{0.770}$	Tehran
	$V_s = 60N^{0.360}$	Turkey
	$V_s = 88.8(N_{60})^{0.370}$	Greece
All soils	$V_s = 91.0N^{0.34}$	Japan
	$V_s = 85.35N^{0.348}$	Japan
	$V_s = 121.0N^{0.27}$	Japan
	$V_s = 61.0N^{0.50}$	USA
	$V_s = 107.6N^{0.36}$	Greece
	$V_s = 22.0N^{0.85}$	Iran
	$V_s = 116.1(N + 0.3185)^{0.202}$	China
	$V_s = 51.5N^{0.516}$	Turkey
	$V_s = 97.0N^{0.314}$	Japan
	$V_s = 82.0N^{0.39}$	Japan
	$V_s = 92.1N^{0.337}$	Japan
	$V_s = 90.0N^{0.309}$	Turkey
	$V_s = 92N^{0.329}$	Japan
	$V_s = 32.8N^{0.51}$	Turkey
	$V_s = 68.3N^{0.292}$	Turkey
	$V_s = 58N^{0.39}$	Western Taiwan
	$V_s = 76.2N^{0.24}$	Greece

cobbles or near-surface fractured rocks. The most upper part of Northeast India including Bhutan and Darjeeling-Sikkim Himalaya is attributed to the presence of high S-velocity layer associated with site class A and site class B comprising of hard Crystalline rocks. Site class E in the region is attributed to the presence of a low velocity layer of high plastic silty clay. Site class E marks its presence in lower parts of Bengal Basin and in some part of Indo-Gangetic Foredeep region and in the state of Haryana.

14.5 Site Amplification

Evidences from past earthquakes clearly exhibit that the damages due to an earthquake and its severity are controlled mainly by earthquake source and path characteristics along with the local site conditions. Therefore, the assessment of site amplification factor for a region located over soft sediments is crucial to account for

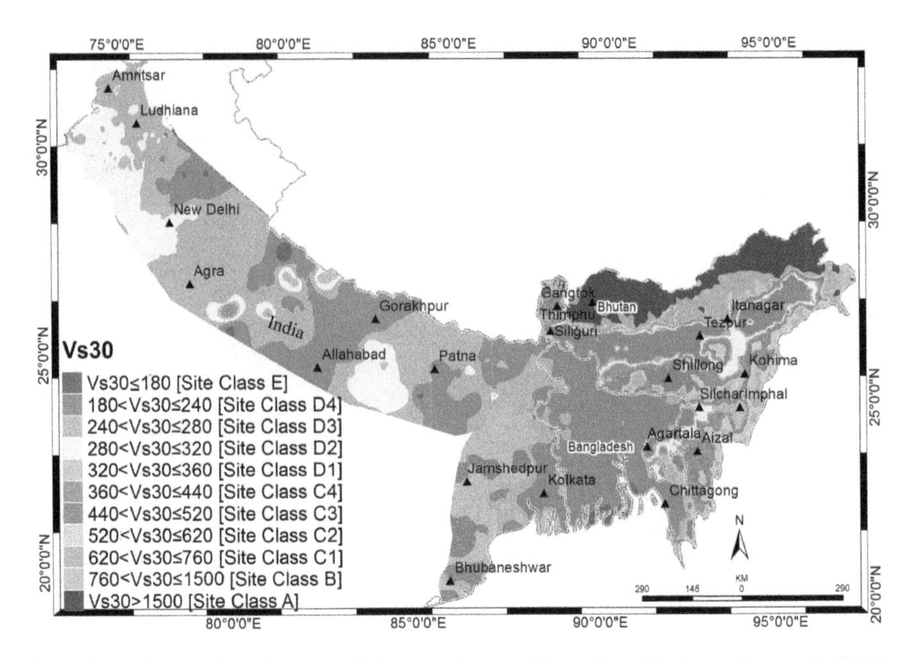

Fig. 14.14 Site classification map of the tectonic ensemble region adhering to Sun et al. (2014, 2018) and displaying the presence of Site Class E, D4, D3, D2, D1, C4. C3, C2, C1, B and A in the terrain with the dominance of site class D4 followed by Site Class D3 and site class D2 in the alluvial filled region, while site class A followed by site class B are seen predominantly present in the hilly terrain

the changes in ground motion at the surface as compared to that at the engineering bedrock due to nonlinear behavior of the in-between sediment column and also from site to site on the surface of the earth. Site characterization study is needed to understand the soil behavior during seismic shaking and also to understand site response of any earthquake-prone county. In the present study, our motivation is to perform a holistic site characterization of the entire tectonic study region from Westcentral Himalaya to Northeast India including Bhutan, IGF, BB and DSH in order to provide information for surface consistent seismic hazard and risk assessment and to identify potential sources that may be subject to future rupturing.

14.5.1 Ground Motion Simulation

Strong ground motion of an earthquake is of main significance from both seismological and engineering point of view and is needed to comprehend the rupture process, the complexity of the fault system and the near-field propagation of seismic waves. The stochastic method of estimating strong ground motion is based on the fact that near-field high-frequency ground motion due to an earthquakes can be determined by a finite duration white Gaussian noise band restricted by the corner frequency f_o and the peak frequency $f_{\max}$ (Hanks and McGuire 1981). In accordance with the stochastic method (Boore 1983), the synthesized acceleration spectrum $A(\omega)$ of shear waves at a distance R from the fault rupture with the seismic moment M_o can be expressed as,

$$A(\omega) = \left[\frac{C.M_0.S\ (\omega).P\ (\omega).\exp\left(\frac{-\omega.R}{2.Q.\beta}\right)}{R}\right] \quad C = \left[\frac{R_{\theta\phi}.FS.M_0}{4.\pi.\rho.\beta^3}\right]$$
$$S(\omega) = \left[\frac{\omega^2}{1+\left(\frac{\omega}{\omega c}\right)^2}\right], \tag{14.6}$$

where ω is the angular frequency, $S(\omega)$ is the source spectrum, $P(\omega)$ is a high cut filter, Q is the quality factor, β is the shear wave velocity and C is a constant. Beresnev and Atkinson (1997) extended this stochastic method to major faults. The rupture plane of the finite fault is subdivided into sub-faults, each of which is then regarded as point source (Boore 1983), and then the contribution of each fault is summed up. The source functions for earthquake simulation using EXSIM software package have been obtained from GCMT catalogue and various published literatures. 1D crustal velocity model for the Westcentral Himalaya is adopted from Gilligan et al. (2015), for Indo-Gangetic Foredeep the same has been taken from Monsalve et al. (2006), for Bengal Basin the crustal model is adopted from Mitra et al. (2008), for the Darjeeling-Sikkim Himalaya it is taken from Acton et al. (2011) and for Northeast India, it has been taken from Mitra et al. (2005). The 1D velocity model is used for the estimation of crustal amplification of each tectonic region using the quarter wavelength approximation of Boore and Joyner (1997). The representative simulated acceleration time history for the 1905 Kangra Earthquake of M_w 7.8 at Mansa City of Punjab, the 1934 Bihar-Nepal earthquake of M_w 8.1 at Bhagalpur, the 1930 Dhubri earthquake of M_w 7.1 at Sylhet city, the 1918 Srimangal earthquake of M_w 7.6 at Rajshahi City of Bangladesh, the 1897 Shillong earthquake of M_w 8.1 at Guwahati and the 2009 Bhutan earthquake of M_w 6.1 at Trashigang City of Bhutan are presented in Fig. 14.15.

14.5.2 Site Response

Earthquake ground motion is seen to vary with local geological conditions in site-specific terms where soil or sediment layers can cause amplification and resonance as seismic waves propagate up to the ground surface because of significant acoustic impedance contrast (Nath and Thingbaijam 2009). In the hilly terrain, apart from incident waves, scattered and diffracted waves constructively interfere amplifying ground motion manifolds (Nath et al. 2008). Therefore, the assessment of site amplification factor for a region located over alluvium is crucial to account

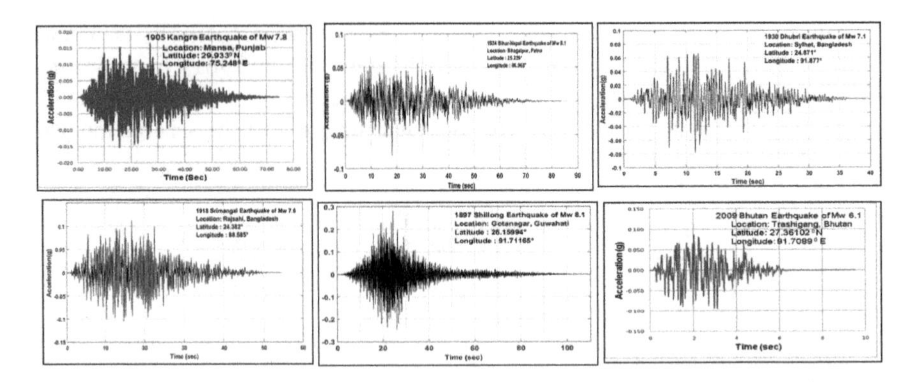

Fig. 14.15 Representative simulated acceleration time history for the 1905 Kangra Earthquake of M_w 7.8 at Mansa City of Punjab, the 1934 Bihar-Nepal earthquake of M_w 8.1 at Bhagalpur, the 1930 Dhubri earthquake of M_w 7.1 at Sylhet City, the 1918 Srimangal earthquake of M_w 7.6 at Rajshahi City of Bangladesh, the 1897 Shillong earthquake of M_w 8.1 at Guwahati and the 2009 Bhutan earthquake of M_w 6.1 at Trashigang City of Bhutan are presented

for the changes in ground motion at the surface due to impedance contrast, resonance, trapping, of energy, focusing, basin edge scattering and damping. Different strategies have been proposed by various researchers to account for soil nonlinearity behavior using numerical techniques. This study adopts the nonlinear analysis, initially proposed by Idriss and Seed (1968) through DEEPSOIL software module proposed by Hashash et al. (2011) following the schematic illustration of wave propagation through seismic bedrock to engineering bedrock and through soil profile to the surface as shown in Fig. 14.16a and the computational workflow shown in Fig. 14.16b. DEEPSOIL uses the geotechnical parameters viz. soil type, thickness of the layer, unit weight of the material, and shear wave velocity of the material along with the acceleration time history at the engineering bedrock level as inputs. The nonlinear effect of soil/sediment is approximated by modifying the linear elastic properties of the soil based on the induced strain level. Further, the strain compatible shear modulus and damping ratio values are iteratively calculated to generate the transfer function for each soil layer. This transfer function is convolved with the Fourier series of the input (bedrock) motion to generate the Fourier series of the output motion at ground surface, thus delivering peak ground acceleration, acceleration time history, stress–strain time history, response spectra and amplification spectra at each site of interest.

In the present investigation, stochastically simulated synthetic ground motions of the earthquakes which have affected the study region most, viz. the 1897 Shillong earthquake of M_w 8.1, 1869 Cacher earthquake of M_w 7.6, 1950 Assam earthquake of M_w 8.6, 2016 Manipur earthquake of M_w 6.7, 1930 Dhubri earthquake M_w 7.6, 2009 Bhutan earthquake of M_w 6.1 for Northeast India; 1918 Srimangal earthquake of M_w 7.6, the 1930 Dhubri earthquake of M_w 7.1, 1885 Bengal earthquake of M_w

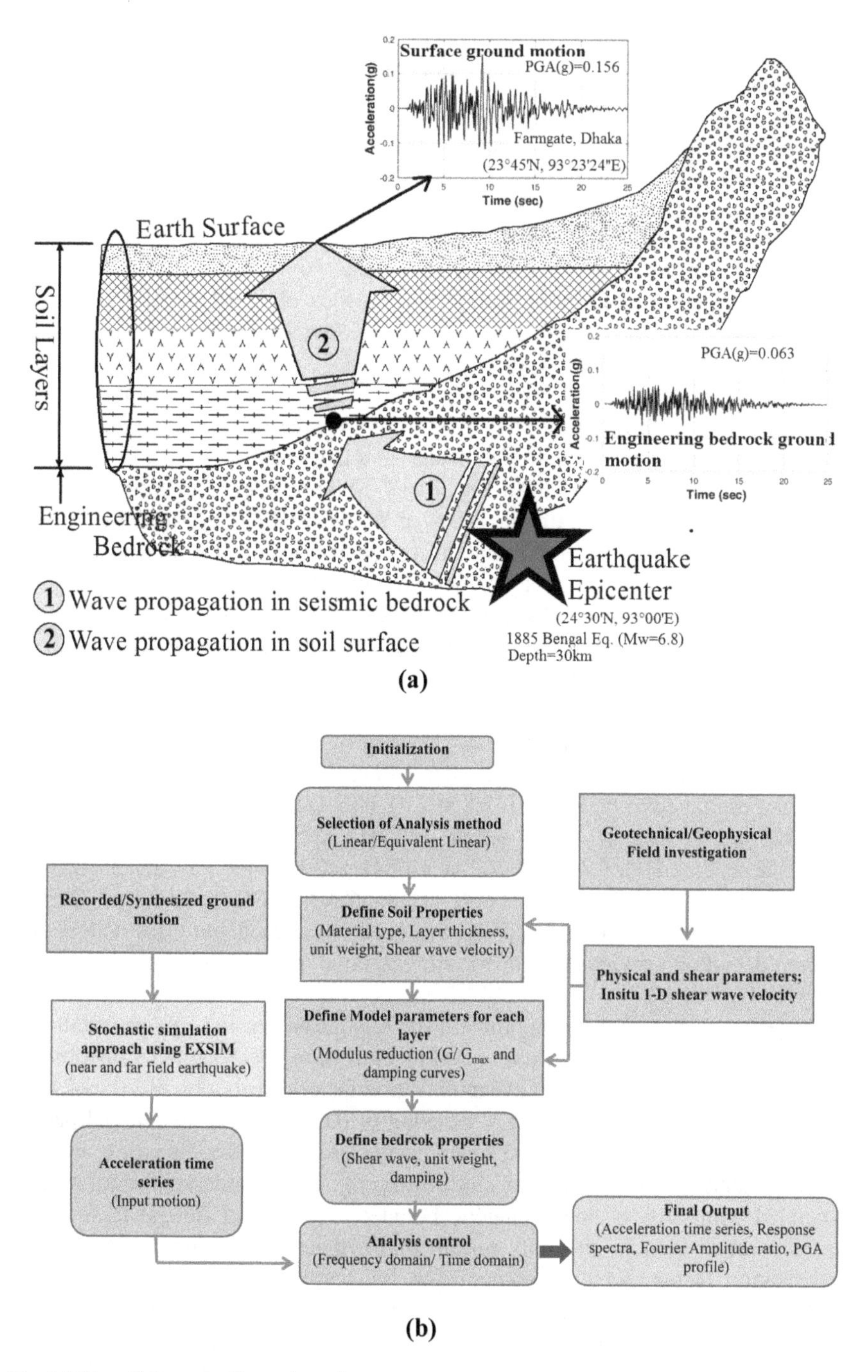

Fig. 14.16 **a** Schematic illustration of wave propagation through seismic bedrock to engineering bedrock and through soil profile to the surface with variation in PGA (*g*) for the same. **b** Flowchart for 1D nonlinear site response analysis through DEEPSOIL software (after Nath 2016; Hashash et al. 2011)

6.8, 1934 Bihar-Nepal earthquake of M_w 8.1 for Bengal Basin region; 1934 Bihar-Nepal earthquake of M_w 8.1, 1988 Bihar earthquake of M_w 6.8, 1991 Uttakashi earthquake of M_w 6.8, 1999 Chamoli earthquake of M_w 6.6, 1905 Kangra earthquake of M_w 7.8 for Indo-Gangetic Foredeep region; 1945 Chamba earthquake of M_w 6.4, 1905 Kangra earthquake of M_w 7.8, 1991 Uttarkashi earthquake of M_w 6.8 for Westcentral Himalaya have been used by inheriting the earthquake parameters from different published literatures with an initial estimate of 5% damping for all soil types to perform nonlinear dynamic analysis using the DEEPSOIL package. The amplification spectra thus obtained at each of around 10,000 locations in the study region for at least four earthquakes at each site to obtain the mean amplification spectrum and also the peak ground acceleration (PGA) amplification factor at zero time periods each of those locations in the study region. The spatial distribution of PGA site amplification factor thus estimated for the entire study region at around 10,000 sites and mapped on GIS platform depict a variation of 1.0–3.58 as shown in Fig. 14.17. Maximum ground motion amplification is associated with the alluvial plains of Westcentral Himalaya, Bengal Basin, Indo-Gangetic Foredeep region and Bramhaputra valley of Northeast India. High PGA amplification is directly associated with low shear wave velocity regions of site class E, followed by D4 to D1 gradually. The firm rock condition of site classes A, B and C has low amplification factor that ranges between 1 and 1.75 mostly in the upper part of Indo-Gangetic region and Northeast India including Bhutan and also in the Darjeeling-Sikkim Himalaya. The spectral site amplification factor (SAF) at a predominant frequency derived by considering both the near- and far-field aforesaid earthquakes for the sites falling in each of the eleven site classes have been estimated to be E: (SAF-5.8 at 1.41 Hz), D4: (SAF-4.8 at 1.66 Hz), D3: (SAF-4.2 at 1.81 Hz), D2: (SAF-3.9 at 1.86 Hz), D1: (SAF-3.3 at 1.95 Hz), C4: (SAF-2.58 at 2.22 Hz), C3: (SAF-2.2 at 3.17 Hz), C2: (SAF-1.87 at 4.1 Hz), C1: (SAF-1.81 at 4.6 Hz), B: (SAF-1.3 at 6.7 Hz) and A: (SAF-1.0 at 8.5 Hz). The site amplification spectra amalgamated and averaged out at each site class with $\pm$ one standard deviation provides the generic site response spectrum for each site class as shown in Fig. 14.17.

Thereafter, surface consistent probabilistic seismic hazard for 10% probability of exceedance in 50 years has been assessed by convolving bedrock PGA distribution of Fig. 14.8 with site-specific PGA amplification factor distribution of Fig. 14.17 as displayed in Fig. 14.18 exhibiting a variation of 0.04–1.23 g. The alluvial part of Northeast India is seen to adhere to higher hazard level encompassing areas of Shillong, Imphal, Itanagar, Tezpur and Chittagong, while moderate hazard level is associated with parts of Indo-Gangetic Foredeep region and Bengal Basin. Low hazard level is seen in the hilly terrain of Northeast India due to its low site amplification factor and high predominant frequency. Similarly, pseudo-spectral acceleration (PSA) at the surface is generated by convolving those at the bedrock level with the spectral site amplification. The 5% damped design response spectra have been generated using surface consistent PSA at 1.0 and 0.2 s with 10% probability of exceedance in 50 years following the International Building Code (IBC 2009) for some of the selected cities of the tectonic study region, viz.

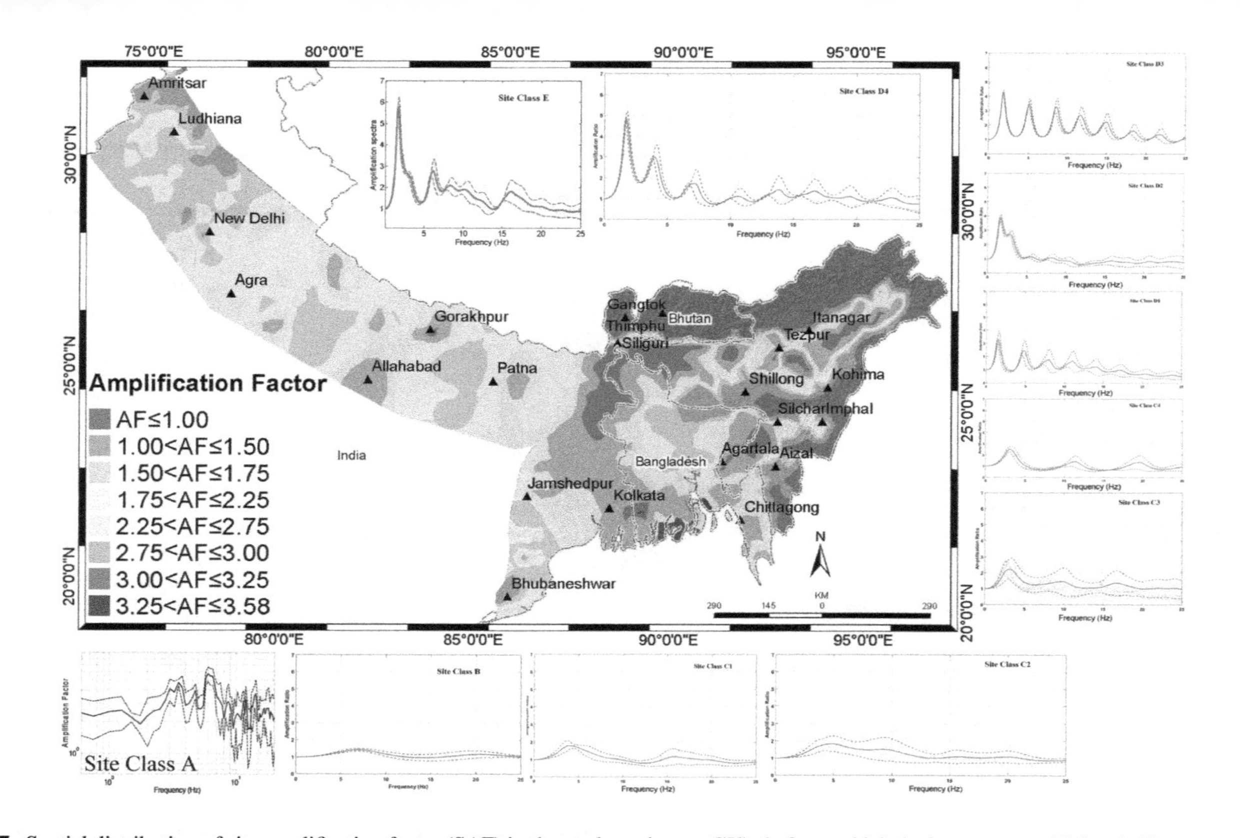

Fig. 14.17 Spatial distribution of site amplification factor (SAF) in the study region on GIS platform which depicts a range of 1.0 to 3.58 encompassing both the alluvial and hilly terrain of Westcentral Himalaya all the way up to Northeast India together with Indo-Gangetic Foredeep, Bengal Basin and Darjeeling-Sikkim Himalaya also depicted are the generic site amplification spectra with $\pm$ one standard deviation for each of the eleven site classes in the region

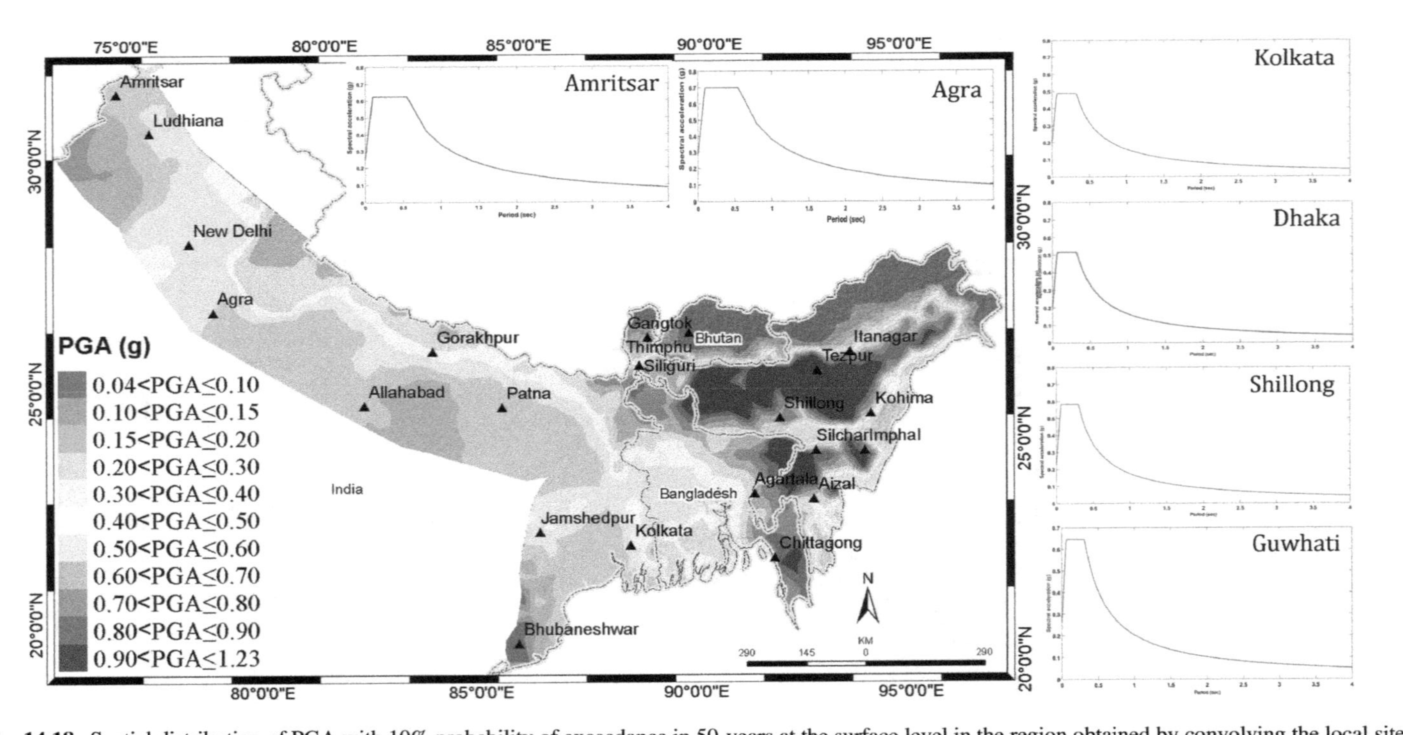

Fig. 14.18 Spatial distribution of PGA with 10% probability of exceedance in 50 years at the surface level in the region obtained by convolving the local site effect with PGA at the bedrock level along with the surface consistent design response spectra for the city of Amritsar, Agra, Kolkata, Dhaka, Shillong and Guwahati

Amritsar, Agra, Kolkata, Dhaka, Shillong and Guwahati as presented in Fig. 14.18 itself exhibiting an increase in the design values as compared to BIS (2002) implying a probable escalation in the urbanization cost if designed with the updated building code.

14.6 Induced Hazards

14.6.1 Liquefaction

Soil liquefaction is a secondary phenomenon triggered by a large earthquake in an alluvium filled terrain like Kolkata (Nath et al. 2018), Dhaka, Amritsar and Guwahati, which causes increase in pore water pressure resulting in the reduction of shear strength of soil when monotonic, cyclic or shock loading is applied. Recent sediments especially fluvial and aeolian deposits, water table information along with physical soil characteristics such as type of soil, degree of water saturation, grain size, and plasticity are the key inputs for the liquefaction hazard assessment. The fine grain criteria for sands [e.g., Seed and Idriss (1982), Bray and Sancio (2006)] allows quick assessment while the standard geotechnical evaluation prospects into the mechanical properties of the soil. A widely used technique for liquefaction hazard assessment is the simplified procedure of Seed and Idriss (1982) and its upgraded versions. The factor of safety against liquefaction is evaluated as 'Cyclic Stress Ratio'/'Cyclic Resistance Ratio,' i.e., the earthquake induced loading divided by the liquefaction resistance of the soil.

14.6.1.1 Factor of Safety Assessment

This study follows the methodology developed by Youd et al. (2001) and Idriss and Boulanger (2004, 2010) and is based on the borehole geotechnical database carrying information about SPT-N values and other index properties, viz. unit weight, Atterberg limits, percent of fine content, etc., at each location. Semiempirical field-based procedure is used to quantify liquefaction in terms of factor of safety (FOS) against liquefaction which is defined as the ratio of in-situ soil resistance expressed in terms of cyclic resistance ratio (CRR) and earthquake induced loading in terms of cyclic stress ratio (CSR) (Youd et al. 2001) given as,

$$\mathrm{FOS} = \left(\frac{\mathrm{CRR}}{\mathrm{CSR}}\right) \tag{14.7}$$

If the FOS value is less than 1, the site is considered to be liquefiable and if it is greater than 1, the site is considered to be non-liquefiable, thus classifying the soil layer as safe or unsafe (Seed and Idriss 1982).

14.6.1.2 Cyclic Resistance Ratio (CRR)

The capacity of an in-situ soil layer to resist liquefaction due to the applied seismic demand has been assessed by using SPT-N values in terms of cyclic resistance ratio or CRR. The presence of fine content (FC) affects the soil resistance, higher the FC percentage in the sediment more resistive it will be toward liquefaction. Therefore, fine content correction has been applied to $(N_1)_{60}$ in order to convert it into an equivalent clean sand value (Idriss and Boulanger 2004),

$$(N_1)_{60cs} = (N_1)_{60} + \Delta(N_1)_{60} \tag{14.8}$$

where

$$\Delta(N_1)_{60} = \exp\left[1.63 + \frac{9.7}{\mathrm{FC}+0.01} - \left(\frac{15.7}{\mathrm{FC}+0.01}\right)^2\right] \tag{14.9}$$

These $(N_1)_{60cs}$ values have further been used to compute cyclic resistance ratio by using the following formulation of Idriss & Boulanger (2004),

$$\mathrm{CRR} = \exp\left\{\frac{(N_1)_{60cs}}{14.1} + \left(\frac{(N_1)_{60cs}}{126}\right)^2 - \left(\frac{(N_1)_{60cs}}{23.6}\right)^3 + \left(\frac{(N_1)_{60cs}}{25.4}\right)^4 - 2.8\right\} \tag{14.10}$$

14.6.1.3 Cyclic Stress Ratio (CSR)

The seismic demand induced by an earthquake can be quantified in terms of cyclic stress ratio (CSR) by using peak ground acceleration (PGA) and the causing earthquake magnitude for a particular site. Idriss & Boulanger (2010) and Youd et al. (2001) have modified CSR previously defined by Seed & Idriss (1982) by introducing magnitude scaling factor (MSF) and overburden correction factor (K_σ) in the expression as given below:

$$\mathrm{CSR} = 0.65\left(\frac{\sigma_v}{\sigma'_v}\right)\left(\frac{a_{\max}}{g}\right)(r_d)\left(\frac{1}{\mathrm{MSF}}\right)\left(\frac{1}{K_\sigma}\right) \tag{14.11}$$

where σ_v, σ'_v = total and effective overburden stresses, respectively, g = acceleration of gravity, $a_{\max}$ = peak ground acceleration (PGA), r_d = stress reduction coefficient, MSF = magnitude scaling factor, K_σ = effective overburden correction. The factor of 0.65 has been introduced by Seed et al. (1985) to convert the peak cyclic shear stress ratio into a cyclic stress ratio that is representative of the most significant cycles over the full duration of loading (Idriss and Boulanger 2010).

14.6.1.4 Liquefaction Potential Index (LPI)

Iwasaki et al. (1982) introduced Liquefaction Potential Index (LPI) in order to estimate the severity of liquefaction for the upper 20 m soil column of a specific borehole location. As surface effects resulting from liquefaction are seldom reported beyond the depth of 20 m, the computation of LPI has been restricted upto this depth.

$$\mathrm{LPI} = \sum_{i=1}^{n} w_i S_i H_i \tag{14.12}$$

where 'n' is the number of layers present in the upper 20 m of the soil deposit, 'w' is the weighting function introduced to account for liquefaction extent with respect to depth and is expressed as,

$$w(z) = 10 - 0.5z \tag{14.13}$$

'z' is the depth of layers in meters, 'H' is the thickness of each layer, and 'S' is estimated by,

$$S = \left\{ \frac{0 \quad \text{for} \quad \mathrm{FOS} > 1}{1 - \mathrm{FOS} \quad \text{for} \quad \mathrm{FOS} < 1} \right\} \tag{14.14}$$

Thus to understand the liquefaction potential of the underlying alluvium below Kolkata and Dhaka, these two capital cities in Bengal basin and the city of Guwahati in Northeast India, the impact of the 1897 Great Shillong Earthquake is considered that reportedly caused sporadic damage in these cities presumably due to the effect of soil liquefaction triggered by the intensifying ground motion coupled with shallow groundwater table and soft unconsolidated and semi-consolidated sediments in both the Bengal basin and the Brahmaputra valley over which these cities are located. A detailed liquefaction potential analysis of the city of Kolkata has already been reported by Nath et al. (2018) considering the existing geotechnical practices by adopting geological and geomorphological criteria, compositional and grain-size criteria. Factor of safety (FOS) and probability of liquefaction (P_L) have been assessed to categorize each shallow litho-stratum as safe or unsafe. To consider the severity of liquefaction for the entire 20–30-m-thick alluvial column at each investigation site, Liquefaction Potential Index (LPI) has also been estimated. A conservative deterministic liquefaction hazard potential has also been predicted for all the three Cities for 475 years of return period considering the surface-consistent probabilistic seismic hazard model of Kolkata, Dhaka and Guwahati with 10% probability of exceedance in 50 years.

The liquefaction susceptibility map of all the cities is prepared by categorizing LPI in four subclasses, viz. 'Low (LPI = 0),' 'Moderate (0 < LPI $\leq$ 5),' 'High (5 < LPI $\leq$ 15)' and 'Severe Susceptibility (LPI > 15)' as shown in Figs. 14.19 and 14.20. The spatial distribution of Liquefaction Potential Index (LPI) provides the liquefaction susceptibility map of Amritsar considering 1905 Kangra Earthquake of M_w 7.8 as shown in Fig. 14.19a. While majority of the regions of Amritsar, viz. Gandiwind, Majitha, Bhikiwind, etc., are placed in the High Liquefaction Susceptible Zone ($5 < LPI \leq 15$); the Golden Temple, Tarshika, Ajanala Chogwan, Verka, Ajnala and most parts of Northwest Amritsar are associated with Low Susceptibility Zone ($LPI \leq 5$) for the Kangra Earthquake Scenario. The 1897 Shillong Earthquake of M_w 8.1 is comparatively less impactful in Kolkata due to its location from this earthquake epicenter. Only a few small patches of moderate to high LPI are seen in Saltlake and Dhapa Region of the city of Kolkata as shown in Fig. 14.19b. But the 1897 Shillong Earthquake induced moderate to severe liquefaction hazard condition in the Dhaka City, seen mostly clustered in Shimuliya, Hazatpur and Basundhara as shown in Fig. 14.19c while the same earthquake induced liquefaction in Dharapur, Guwahati University area, and Chandmari in the city of Guwahati in Northeast India as shown in Fig. 14.19d.

The deterministic liquefaction potential for the city of Amritsar has been estimated as shown in Fig. 14.20a which places most part of the City under 'Low to

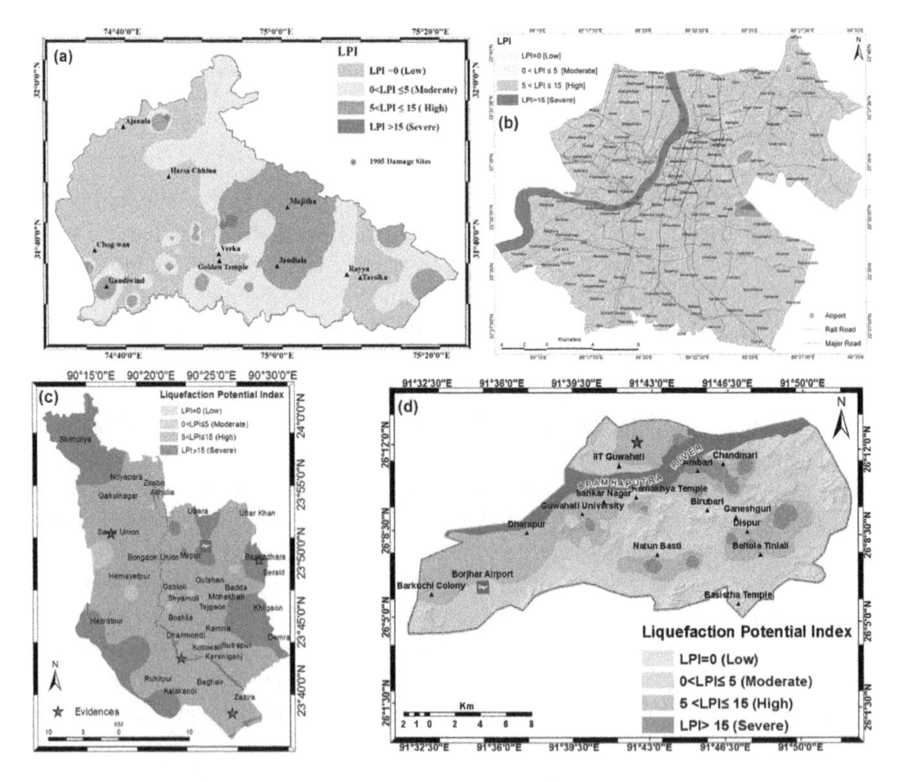

Fig. 14.19 Liquefaction Potential Index distribution on GIS platform as induced by the 1905 Kangra earthquake of M_w 7.8 at **a** the city of Amritsar in WCH. The 1897 Shillong Earthquake of M_w 8.1 induced liquefaction scenario at the city of **b** Kolkata in BB (Nath et al. 2018), **c** Dhaka in BB and **d** Guwahati in the Northeast India

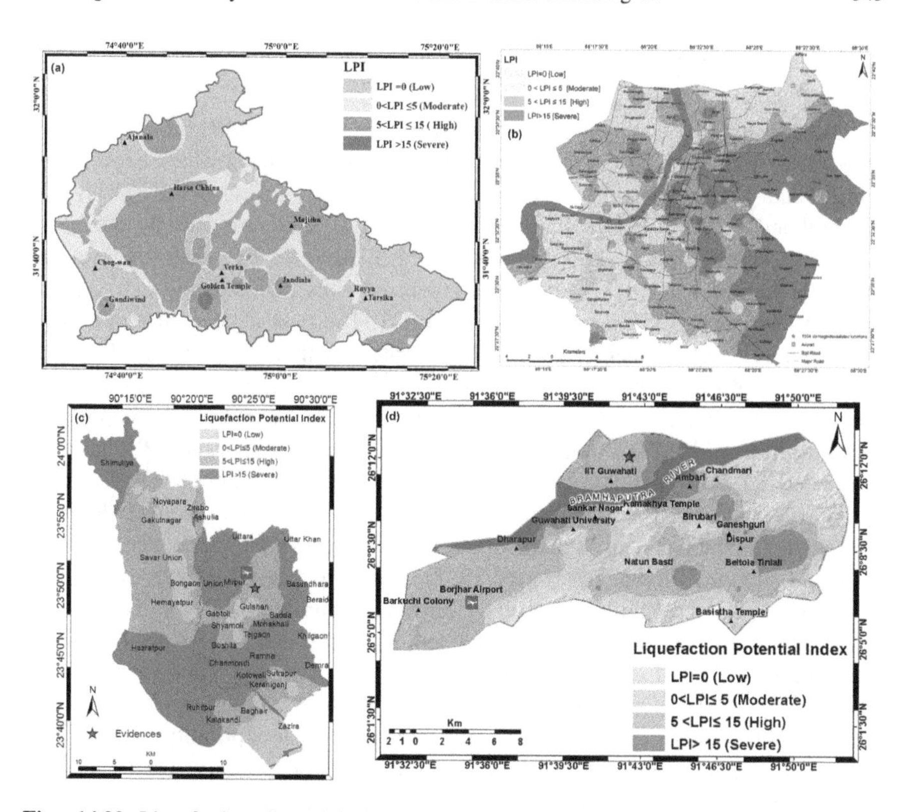

Fig. 14.20 Liquefaction Potential Index spatial distributions on GIS platform for the Deterministic scenario compliant with 10% probability of exceedance in 50 years for the city of **a** Amritsar in WCH, **b** Kolkata in BB (Nath et al. 2018), **c** Dhaka in BB, and d Guwahati in NEI

Moderate' Liquefaction Potential for the surface PGA at 10% of probability exceedance in 50 years. The regions of Harsa Chhina, Jandiala, Majitha, etc., located in the city are placed in the 'High' Liquefaction Susceptible Zone ($5 < \mathrm{LPI} \leq 15$), while Golden Temple Chogwan, Verka, Ajnala and most parts of Northwest Amritsar are associated with 'Low' Susceptibility Zone ($\mathrm{LPI} \leq 5$) for the futuristic scenario of the city.

Deterministic scenario for the city of Kolkata shown in Fig. 14.20b suggests the liquefaction vulnerability of the northeastern region of the city encompassing Saltlake, Rajarhat and Central Kolkata encompassing Park street region, while the southwestern Alipore region is seen to be comparatively safer. The deterministic liquefaction potential of Dhaka as shown in Fig. 14.20c places almost the entire city under 'High to Severe' Liquefaction regime. In case of Guwahati City, the fluvial marine sediments of the Brahmaputra Basin exhibit moderate to severe liquefaction potential under a conservative deterministic scenario for 475 years of return period. IIT Guwahati, Guwahati University, Dispur and Ambari are placed in the 'High' Liquefiable Zone with LPI varying between 5 and 15 as shown in Fig. 14.20d.

14.6.2 Landslides

Landslide is another earthquake induced hazard generally experienced in the rugged topography like the hilly tracts of Darjeeling-Sikkim Himalaya which experienced widespread land sliding in various parts of the city of Gangtok during the September 18, 2011 Sikkim Earthquake of M_w 6,9. Here we initially attempted Landslide Hazard Index (LHI) mapping in 1:5000 scale by integrating various landslide hazard contributory thematic layers on GIS platform through analytic hierarchy process (AHP) as shown in Fig. 14.21. The rank for each thematic layer attribute is assigned using the frequency ratio method on training landslide inventory (70% used here) database. Individual thematic layers are in-direct or indirect relation with landslides in this region. In order to determine the importance of occurrences of landslides on each thematic layer, the predictor rate (PR) has been calculated for each thematic layer based on its spatial association with landslide inventory. In AHP, the consistency ratio (CR) implies an acceptable level of consistency in the pairwise comparison matrix to recognize the factor weights in the landslide susceptibility model (Saaty 1977). Thereupon, the LHI values are classified into five susceptibility zones, viz. (i) low susceptibility (LHI < 0.20) covering around 20.2 km^2 with important destinations like lower part of Gangtok, (ii) moderate susceptibility ($0.20 <$ LSI ≤ 0.40) covering 10km^2, (iii) high susceptibility ($0.40 <$ LSI ≤ 0.60) covering around 8.6 km^2 with important destinations like Tadong, Chandmari and Bhurtuk, (iv) very high susceptibility ($0.60 <$ LSI ≤ 0.80) covering around 7.9 km^2 and (v) severe susceptibility ($0.80 <$ LSI $\leq$ 1.00) covering around 7.8 km^2 as shown in Fig. 14.22.

14.6.2.1 Slope Stability Analysis

We also attempt to understand the landsliding mechanism by performing slope stability analysis by considering a Newmark displacement model for the synthesis of the critical slip surface and the estimation of the factor of safety (FoS) in 'Severe' Landslide Susceptible Zones (LSZ) in the Greater Gangtok region. The model geometry is drawn by creating a 2D cross-sectional profile along the slope from Google Earth imagery and the elevation extracted from ALOS PALSAR DEM.

Analysis of 6th Mile Landslide

6th Mile sinking zone is located at 27°19′4.05″N–88°35′42.63″E along a distance of 240 m with an elevation of 70 m. It is tranquiled with steep to moderate slope characterized by debris slide under the metamorphic terrain within the lower grade, and it is composed of quartzitic phyllites with mica schist, quartz phyllites interbedded with quartz-sericite and quartz chlorite sericite phyllites. The lithological strata shows that the region comprises of three layers, viz. organic soil, red

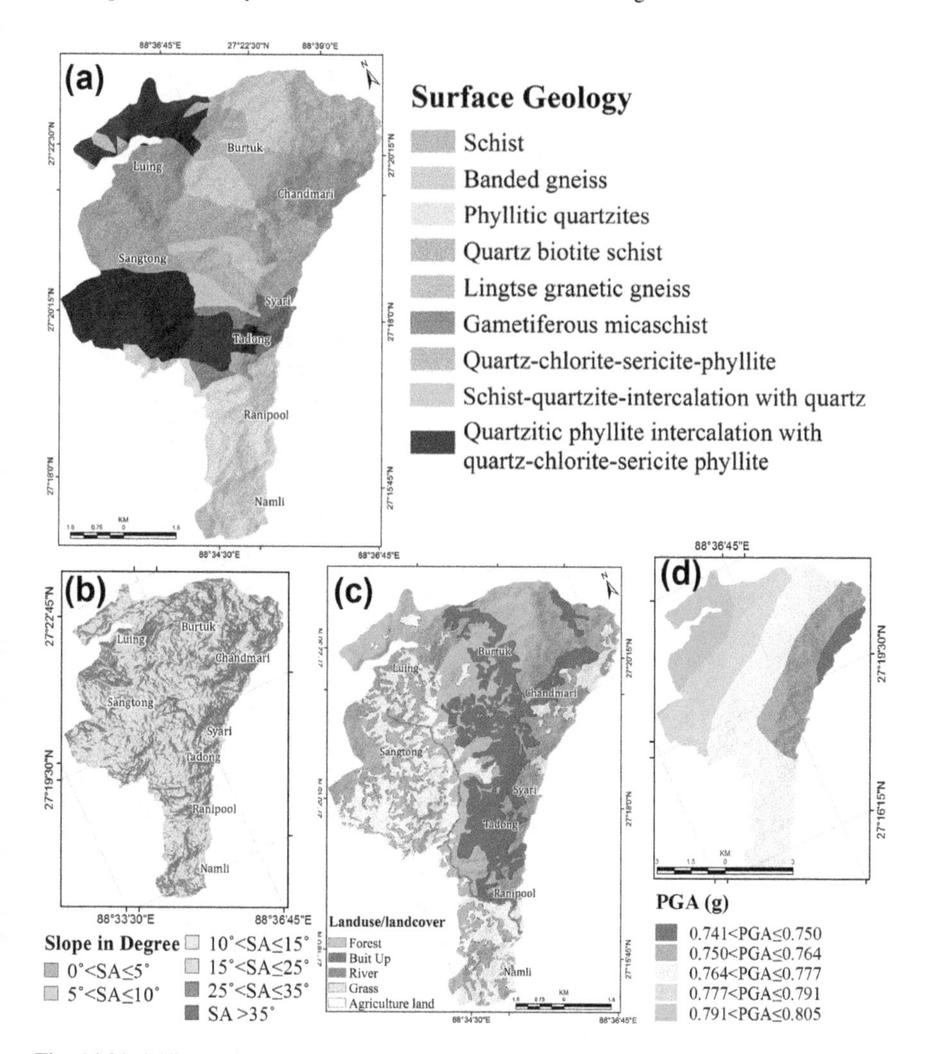

Fig. 14.21 Different themes used for landslide susceptibility zonation in Gangtok

soil and highly weathered rock, in which the weathered rock extends from 4.5 m to 23 m depth overlain by organic and red soil. The material properties have been incorporated for the slope geometry from the shear strength parameters using Mohr–Coulomb equation, and the geometrical configuration of the slope is defined by 15 nodal points. The initial in-situ stress distribution ranges from <0 to >1,000 kPa as shown in Fig. 14.23a. FoS calculated under the static condition is 1.281 as presented in Fig. 14.23c. The velocity vector and the displacement of the slope after applying seismic loading due to Sikkim Earthquake is shown in Fig. 14.23b with FoS 0.949 indicating an unstable slope due to gravity sliding/ water seepage and/or seismic shaking. The radius and depth of the sliding block

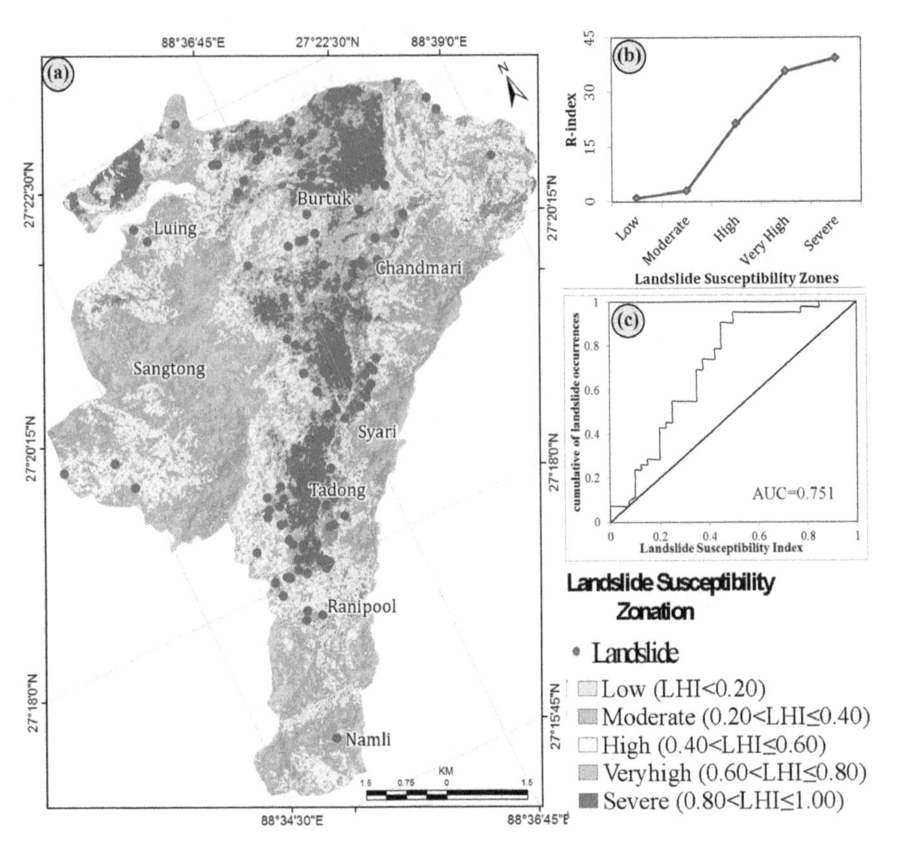

Fig. 14.22 a Analytic hierarchy process (AHP)-based landslide susceptibility Map of Greater Gangtok. Five susceptible zones have been identifying, viz. 'low' (LHI < 0.20), 'moderate' (0.20 < LHI ≤ 0.40), 'high' (0.40 < LHI ≤ 0.60), 'very high' (0.60 < LHI ≤ 0.80) and 'severe' (0.80 < LHI ≤ 1.00). **b** ROC curve for LHI derived from susceptibility index, graphs show the cumulative landslide occurrences versus landslide susceptibility index. The large value of the area confirms the accuracy of the classification under the ROC curves with AUC = 0.751, and **c** relative landslide density index (R-index) of the AHP derived susceptibility map

critical slip circle is found to be 101.03 and 22.53 m, respectively, as shown in Fig. 14.23c, d under static and dynamic loading, respectively. The yield acceleration and the maximum deformation for the critical slip surface are 0.10579 m/s^2 and 0.00001837 m, respectively, as shown in Fig. 14.23e, f.

14.7 Urban Seismic Hazard Impact Assessment

At the onslaught of a destructive earthquake in a region, the pre-disaster preparedness and post-disaster relief, rescue and rehabilitation are worked out using any of the tools such as, HAZUS (Hazard-US), RADIUS (Risk Assessment Tools

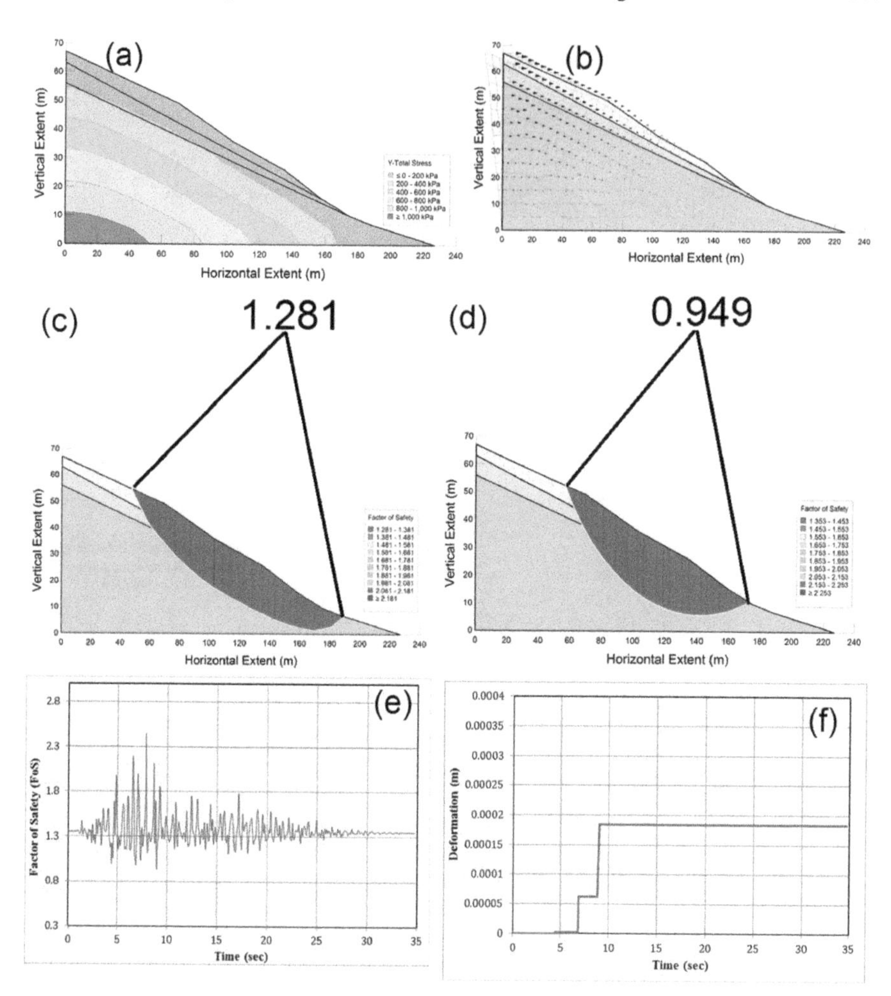

Fig. 14.23 Slope stability modeling of 6th Mile: **a** initial in-situ stress condition, **b** displacement with velocity vector, **c** critical slip surface and FOS distribution under Static loading condition, **d** critical slip surface and FOS distribution on the impact of 2011 Sikkim Earthquake of M_w 6.9, **e** the factor of safety versus time from Newmark deformation for the sliding block and **f** cumulative deformation over time caused by acceleration greater than the yield acceleration

for Diagnosis of Urban areas against Seismic Disasters), ELER (Earthquake Loss Estimation Routine), EPEDAT (The Early Post-Earthquake Damage Assessment Tool), SELENA (SEismic Loss EstimatioN using a logic tree Approach) either individually or in unison. In the present study the probability of damage to the built environment of the cities in the study region encompassing five tectonic blocks has been estimated by using SELENA, based on capacity spectrum method (CSM). The basic methodology of SELENA is shown in Fig. 14.24.

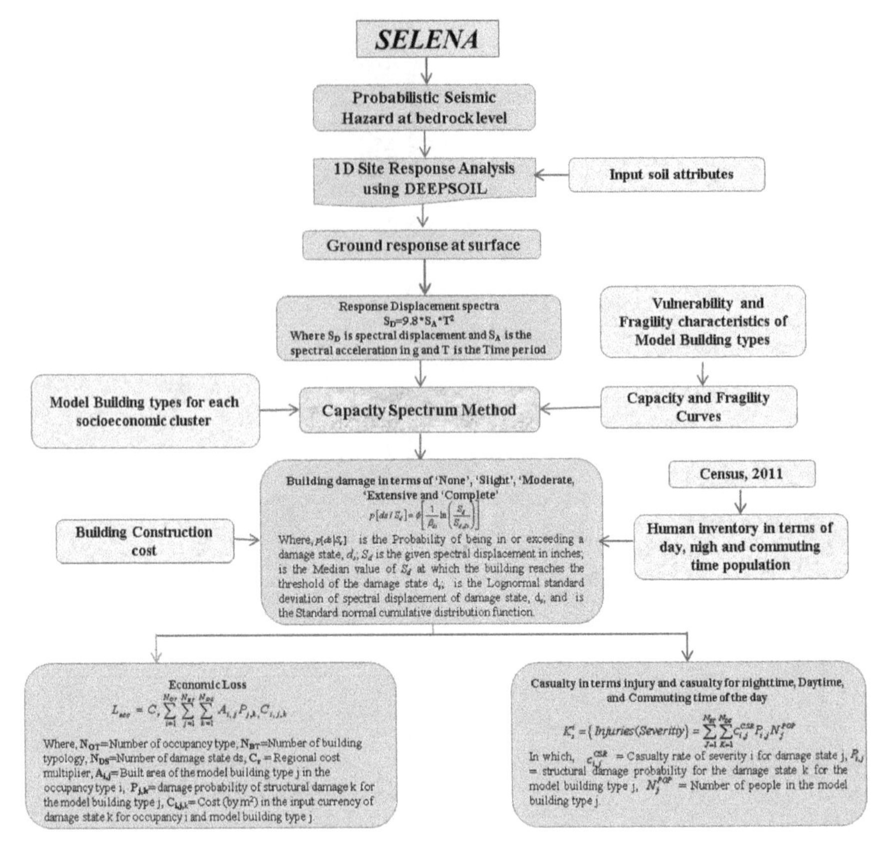

Fig. 14.24 Computational framework of SELENA for damage, casualty and loss assessment (after Nath 2016; Ghatak et al. 2017)

The capacity spectrum method is a nonlinear static analysis, which compares the capacity curve of a structure in terms of force and displacement with seismic response spectrum (Freeman 1978). To calculate the cumulative damage probabilities of the entire model building types, demand spectrum curve as a function of spectral displacement, at the period 0.3 and 1.0 s, has been considered. The building capacity curve has three control points: design, yield and ultimate capacity as presented in Fig. 14.25a. It is assumed that building capacity curve behave elastically linear upto the yield point, the curve changes from elastic to plastic state from yield point to ultimate point and the curve behaves totally plastically as it crosses the ultimate point. The performance point (d_p) is identified from the intersection between the seismic demand and building capacity curve. The probability of damage in each point has been calculated in relationship with the provided ground motion (Freeman 1978; ATC 1996). It consists of steps like generation of capacity spectrum, computation of demand spectrum and determination of performance point as shown in Fig. 14.25a. Structural capacity is represented by a force–

displacement curve. A pushover analysis is performed for a structure with increasing lateral forces, representing the inertial forces of the structure under seismic demand. The process is continued till the structure becomes unstable. The seismic demand curve is represented by the response spectrum curve in the spectral displacement–spectral acceleration space. For the computation of damage probabilities, vulnerability curves or fragility curves for four damage states, viz. 'Slight,' 'Moderate,' 'Extensive' and 'Complete' are essential, which are developed as lognormal probability distribution of damage from the capacity curve as shown in Fig. 14.25b. For an expected displacement, cumulative probabilities are defined to obtain discrete probabilities of being in each of the five different damage states as shown in Fig. 14.25c.

In the present study, on the basis of building typology and height along with the stipulated building nomenclature given in WHE-PAGER (2008), FEMA (2000) twelve model building types have been identified in the study region encompassing the five tectonic provinces WCH, IGF, BB, DSH and NEI of which building types A1, URML, URMM, C1L, C1M, C1H, C3L, C3M, C3H and HER have been identified in the city of Amritsar in Westcentral Himalaya; building types A1, RM2L, URML, C1L, C1M, C1H, C3L, C3M, C3H and HER have been recognized in the city of Agra in Indo-Gangetic Foredeep region; building types A1, RS2,

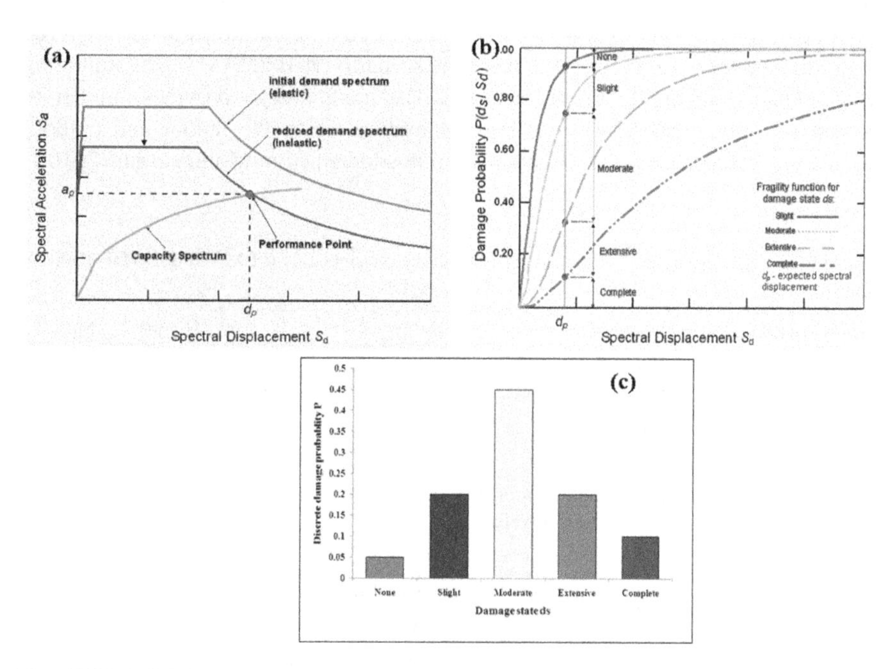

Fig. 14.25 SELENA computation steps: **a** building specific capacity spectrum intersected by the demand spectrum representing the performance point, **b** fragility curves showing extent of different damage states (ds) and **c** the discrete probabilities of different damage states, ds (Nath 2016; Molina et al. 2010)

URML, URMM, C1L, C1M, C1H, C3L, C3M, C3H and HER have been recognized in the city of Kolkata while building types A1, URML, URMH, C1L, C1M, C1H, C3L, C3M, C3H have been found in Dhaka in Bengal Basin; building types A1, URML, URMM, C1L, C1M, C1H, C3L, C3M and C3H have been identified in Guwahati in Northeast India while in the city of Gangtok in Darjeeling-Sikkim Himalaya model building types W1, URML, URMH, C1L, C1M, C1H, C3L, C3M and C3H have been recognized as presented in Table 14.5. The most predominant building types in all the six cities are C1 and C3, while Adobe building A1 and URM building types are also common in all the five tectonic blocks. The building inventories of all the cities have various occupancy classes such as residential, commercial, educational, government and religious. It considers assessment at the level of a tiny area termed as 'Geounit' in which the damage probabilities for the number of buildings of the model building types have been computed in terms of 'None,' 'Slight,' 'Moderate', 'Extensive' and 'Complete' damage. The damage probability in the geounit has been calculated from the provided ground motion relationship, which consists of capacity spectrum generation, demand spectrum computation and performance point determination (Freeman 1978; ATC 1996). The respective capacity curves for all the building types have been collected from NIBS (2002). On the other hand, human casualty has been estimated by considering the probabilistic seismic hazard condition and demographic distribution of the cities as per Census 2011. In order to consider the occupancy cases, the number of casualties has been computed in terms of total injury at three different times of the day, viz. night time scenario (at 02:00AM), day time scenario (at 10:00AM) and commuting time scenario (at 05:00PM), respectively. The methodology provides number of human casualties caused by building collapse in which the indoor and outdoor population percentage for a particular time is adopted from Molina et al. (2010).

Table 14.5 Different model building types used in the present study (FEMA 2000; WHE-PAGER 2008)

Model building type	Description	Height	Stories
HER	Heritage building		
C1L	Ductile reinforced concrete frame with or without infill	Low-rise	1–3
C1M		Mid-rise	4–6
C1H		High-rise	7+
C3L	Non-ductile reinforced concrete frame with masonry infill walls	Low-rise	1–3
C3M		Mid-rise	4–6
C3H		High-rise	7+
A1	Adobe block, mud mortar, wood roof and floors	Low-rise	1–2
W1	Adobe block, wood roof and floors	Low-rise	1–2
RS2	Rubble stone masonry walls with timber frame and roof	Low-rise	1–2
URML	Unreinforced masonry bearing wall	Low-rise	2–3
URMM		Mid-rise	3–4

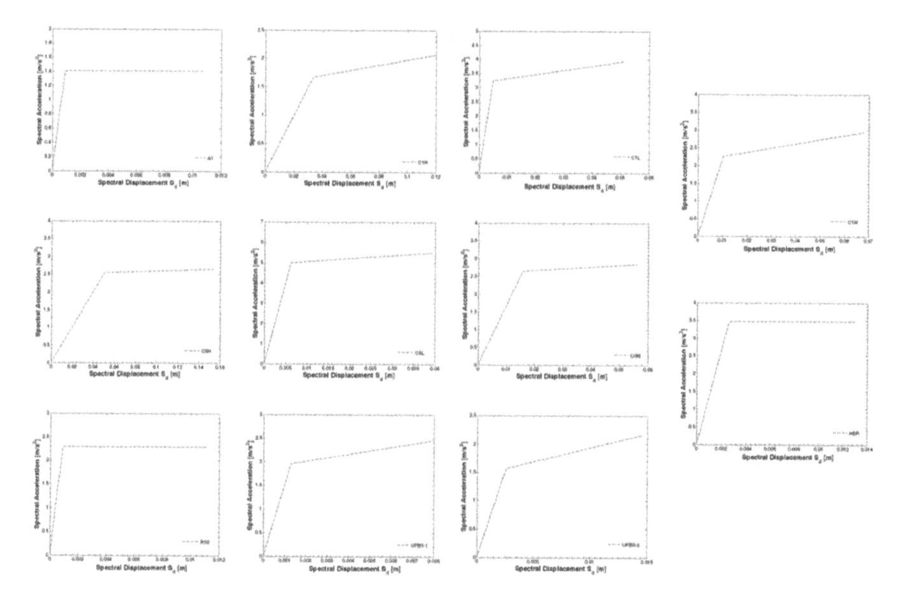

Fig. 14.26 Capacity curves for eleven model building types (NIBS 2002) in the study region

14.7.1 Capacity and Fragility Curves

The capacity curves for eleven model building types for the cities of Amritsar, Agra, Kolkata, Dhaka, Guwahati and Gangtok in the study region are given in Fig. 14.26. For the computation of damage probabilities, vulnerability curves or fragility curves for four damage states are essential, which are developed as log-normal probability distribution of damage from the capacity curve. Figure 14.27 depicts the fragility curves for eleven model building types for the cities of Amritsar, Agra, Kolkata, Dhaka, Guwahati and Gangtok in the study region. Both the capacity and fragility curve parameters have been adopted from NIBS (2002).

14.8 Structural Damage and Casualty Scenario for the City of Amristar, Agra, Kolkata, Dhaka, Gangtok and Guwahati

14.8.1 Damage Scenario

Damage is computed based on model building types, as the structural parameters are directly related to building performance under seismic loading. In the present study, a total of 30,000 buildings have been analyzed for the city of Amritsar picked

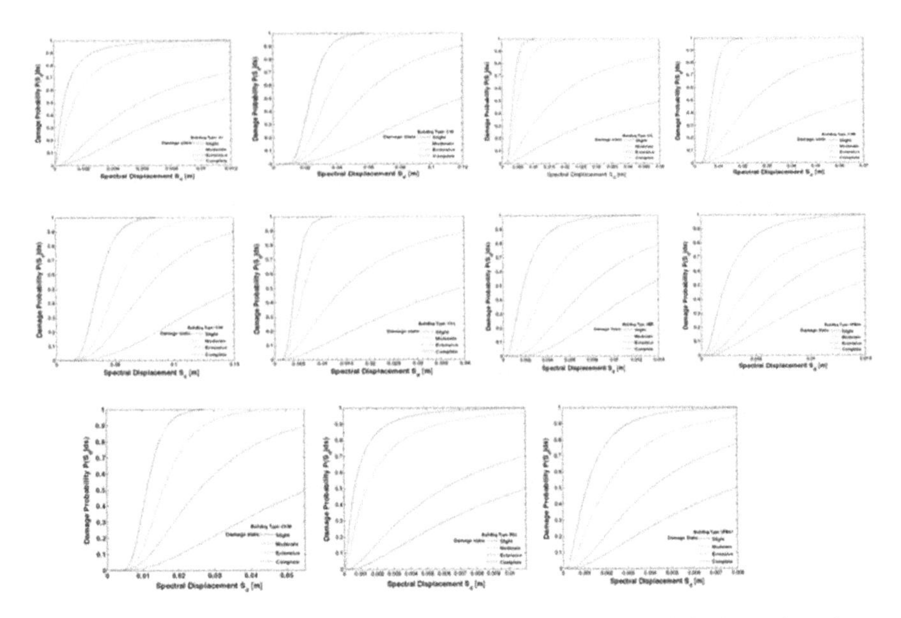

Fig. 14.27 Fragility curves for eleven model building types (NIBS 2002) in the study region

from Google Earth imagery and verified by rapid visual screening. It has been estimated from the protocol that under the prevailing surface consistent probabilistic seismic hazard scenario with 475 years of return period out of 30,000 buildings in Amritsar approximately 10% will suffer from 'Complete' damage followed by ~20% 'Extensive,' ~35% 'Moderate,' ~25% 'Slight' damage while ~10% buildings will be seismic resistant as shown in Fig. 14.28a. Unreinforced masonry buildings are the most seismically vulnerable ones and, therefore, will face the maximum brunt of the seismic impact. The same has been estimated for the city of Agra, in which approximately 40% of mostly A1 and URM buildings are expected to suffer 'Complete' damage followed by ~20% 'Extensive,' ~25% 'Moderate,' and ~5% will face 'Slight' damage as shown in Fig. 14.28b.

Out of 554,907 buildings of Kolkata approximately 34% is expected to suffer from 'Moderate' damage followed by ~26% 'Complete,' ~18% 'Extensive,' and ~15% 'Slight' damage. Approximately, 7% buildings are seismic resistant in the city as shown in Fig. 14.28c. For the city of Dhaka in Bangladesh as shown in Fig. 14.28d, maximum damage will be expected for URM and C1 model building types at about 60–80% 'Extensive' to 'Complete' damage. Out of 23,093 buildings of Gangtok approximately 32% is expected to suffer from 'Moderate' damage followed by ~15% 'Complete,' 22% 'Extensive' and ~17% 'Slight' damage. Approximately, 14% buildings are not vulnerable in the city as shown in Fig. 14.28e.

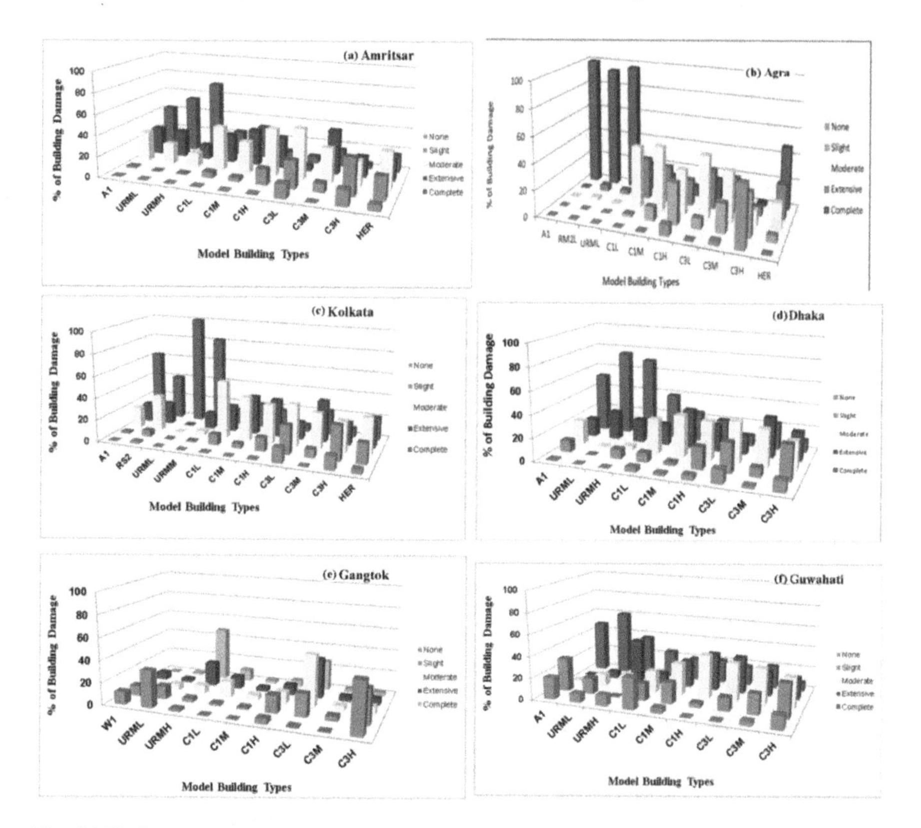

Fig. 14.28 Damage probability computed in terms of 'None,' 'Slight,' 'Moderate,' 'Extensive' and 'Complete' for the model building types in the city of **a** Amritsar, **b** Agra, **c** Kolkata, **d** Dhaka, **e** Gangtok and **f** Guwahati from the entire tectonic ensemble study region

Same has been estimated for Guwahati for the surface hazard scenario, and it has been observed that more or less all the model building types of this city will undergo various degree of 'Slight' to 'Complete' damage as shown in Fig. 14.28f.

14.8.2 Human Casualty Scenario

The building damage is the direct consequence of the impact of a large earthquake; casualty is, however, the indirect effect. By considering the severity state of injury from Coburn and Spence (2002) protocol and the Molina et al. (2010) formulation for the computation of the casualty as shown in Fig. 14.25, the state of injury level may range from 'Low,' 'Medium,' 'Heavy' to even succumbing to 'Death.' The total number of injured persons at various stages for the city of Amritsar, Agra, Kolkata, and Gangtok is shown in Fig. 14.29. In the proposed protocol, it has been

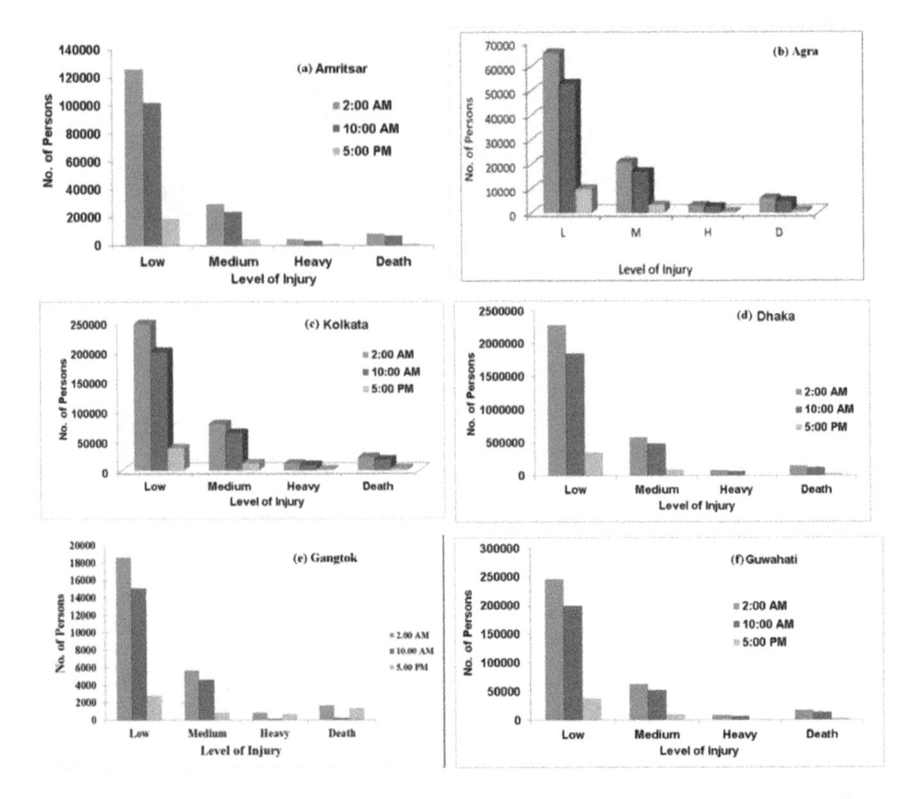

Fig. 14.29 Number of human casualties at various levels of injury at the three distinctly different times of the day for the city of **a** Amritsar, **b** Agra, **c** Kolkata, **d** Dhaka, **e** Gangtok and **f** Guwahati belonging to the present study region encompassing five tectonic blocks

considered that about 98%, 90% and 36% of the population resides at home or indoors in the night, day and commuting time, respectively. The estimated level of injury for the city of Amritsar depict that, respectively, ~167,592 persons, ~135,886 persons and 26,044 persons will suffer from various stages of injury at night time, day time and commuting time scenario as shown in Fig. 14.29a. The same has been estimated for the city of Agra which reveals that approximately 95,188 persons, 76,919 persons and 14,422 persons will suffer from various levels of injury, viz. 'Low,' 'Medium,' 'Heavy' and even succumbing to 'Death' in the night time, day time and commuting time scenario, respectively, as shown in Fig. 14.29b.

For the city of Kolkata, approximately 356,374 persons, 287,978 persons and 53,995 persons will suffer from various levels of injury in the night time, in the day time and at the commuting time scenario as shown in Fig. 14.29c. Casualty in terms of injury level has also been estimated for the city of Dhaka in Bangladesh, Gangtok in Darjeeling-Sikkim Himalaya and Guwahati in Northeast India as presented in Fig. 14.29d–f, respectively. The results of this analysis reveal that approximately 60% of the population of these cities will suffer from 'Low' to 'Heavy' injury states.

14.9 Conclusions

Seismic hazard analysis has emerged as an important issue in high-risk urban centers across the globe and is considered an integral part of earthquake induced disaster mitigation practices. Site conditions significantly influence seismic hazard potential of a region, and therefore, its characterization is of utmost importance in the quantification of site-specific surface consistent seismic hazard of any earthquake-prone province anywhere in the globe. The present study appraises the site characteristics vis-à-vis the site-specific seismic hazard potential of the ensemble of five tectonic province Westcentral Himalaya to Northeast India including Bhutan, Indo-Gangetic Foredeep, Bengal Basin and Darjeeling-Sikkim Himalaya based on an enriched surface geological, geomorphological, topography, landform and in-situ downhole geotechnical and geophysical database of the region with a new regional fifth degree nonlinear power law polynomial combining shear wave velocity with geology, geomorphology, landform and topography gradient, a set of new lithology-based depth-dependent SPT-N value derived/in-situ downhole seismic measurement yielded shear wave velocity, DEEPSOIL-based nonlinear soil–structure interaction, earthquake induced liquefaction and landslide potential and SELENA-HAZUS-based historic/maximum earthquake and surface-level probabilistic seismic hazard triggered urban impact assessment on some Capital–Spiritual–Commercial cities like Amritsar, Agra, Kolkata, Dhaka, Guwahati and Gangtok belonging to these five tectonic province ensemble, thus bringing in an unique text book like regional–local hybrid seismic hazard–disaster model for pre-disaster preparedness in the form of updated urban By-laws and post-disaster rescue, relief and rehabilitation.

Acknowledgments This work is supported partially by the Geosciences/ Seismology Division of the Ministry of Earth Sciences, Government of India through the Projects: CS/EHRA/5/2013 and MoES/P.O. (Seismo)/1(60)/2009. The critical review and constructive suggestions of the anonymous reviewers greatly helped in bringing the manuscript to its present shape with enhanced scientific and technical exposition.

References

Acton CE, Priestley K, Mitra S, Gaur VK (2011) Crustal structure of the Darjeeling-Sikkim Himalaya and southern Tibet. Geophys J Int 184(2):829–852

ATC: 40 (1996) Seismic evaluation and retrofit of concrete buildings. Applied technology council, report ATC-40, Redwood City

Beresnev IA, Atkinson GM (1997) Modeling finite-fault radiation from the ω^n spectrum. Bull Seismol Soc Am 87(1):67–84
Bhatia SC, Kumar MR, Gupta HK (1999) A probabilistic seismic hazard map of India and adjoining regions. Ann Geofis 42(6):1153–1166
BIS: IS1893–2002 (2002) (Part 1): Indian standard criteria for earthquake resistant design of structure part 1—resistant provisions and buildings. Bureau of Indian Standards, New Delhi
Boore DM (1983) Stochastic simulation of high-frequency ground motions based on seismological models of the radiated spectra. Bull Seismol Soc Am 73(6A):1865–1894
Boore DM, Joyner WB (1997) Site amplifications for generic rock sites. Bull Seismol Soc Am 87:327–341
Bray JD, Sancio RB (2006) Assessment of the liquefaction susceptibility of fine-grained soils. J Geotech Geoenviron Eng 132(9):1165–1177
Coburn A, Spence R (2002) Earthquake protection, 2nd edn. Wiley
Cornell CA (1968) Engineering seismic risk analysis. Bull Seismol Soc Am 58(5):1583–1606
Dasgupta S, Pande P, Ganguly D, Iqbal Z, Sanyal K, Venaktraman NV, Dasgupta S, Sural B, Harendranath L, Mazumdar K, Sanyal S, Roy A, Das LK, Misra PS, Gupta H (2000) Seismotectonic atlas of India and its environs. Special publication 59, 87. Geological Survey of India, Calcutta, India
Dikshit KR, Dikshit JK (2014) Relief features of north-east India. In: North-East India: land, people and economy. Springer, Dordrecht, pp 91–125
Esteva L (1970) Seismic risk and seismic design decisions. In: Seismic design for nuclear power plants. Massachusetts Institute of Technology Press, Cambridge, pp 142–82
FEMA (2000) Prestandard and commentary for the seismic rehabilitation of buildings. Federal Emergency Management Agency 356, Washington DC
Frankel A (1995) Mapping seismic hazard in the central and eastern United States. Seismol Res Lett 66(4):8–21
Frankel AD, Petersen MD, Mueller CS, Haller KM, Wheeler RL, Leyendecker EV, Wesson RL, Harmsen SC, Cramer CH, Perkins DM, Rukstales KS (2002) Documentation for the 2002 update of national seismic hazards maps. U.S. Geological Survey Open-File Report, vol 2(420)
Freeman SA (1978) Prediction of response of concrete buildings to severe earthquake motion. Special Publication 55, American Concrete Institute, Detroit, pp 589–605
Ghatak C, Nath SK, Devaraj N (2017) Earthquake induced deterministic damage and economic loss estimation for Kolkata, India. J Rehabil Civ Eng 5(2):1–24
Gilligan A, Priestley KF, Roecker SW, Levin V, Rai SS (2015) The crustal structure of the western Himalayas and Tibet. J Geophys Res Solid Earth 120(5):3946–3964
Grünthal G, Wahlström R (2006) New generation of probabilistic seismic hazard assessment for the area Cologne/Aachen considering the uncertainties of the input data. Nat Hazards 38:159–176
Gutenberg B, Richter CF (1944) Frequency of earthquakes in California. Bull Seismol Soc Am 34(4):185–188
Hanks T, McGuire R (1981) The character of high-frequency strong ground motion. Bull Seismol Soc Am 1897–1919
Hashash YMA, Groholski DR, Phillips CA, Park D, Musgrove M (2011) DEEPSOIL 5.0, user manual and tutorial. University of Illinois, Urbana, IL, USA
IBC International Building Code (2006) International Code Council, Inc. Country Club Hills, Illinois
IBC: International Building Code (2009) International Code Council, Inc. Country Club Hills, Illinois
Idriss IM, Boulanger RW (2004) Semi-empirical procedures for evaluating liquefaction potential during earthquakes. In: Proceedings of 11th international conference on soil dynamics and earthquake engineering, and 3rd international conference on earthquake geotechnical engineering, Stallion Press vol 1, pp 32–56

Idriss IM, Boulanger RW (2010) SPT-based liquefaction triggering procedures. Rep. UCD/CGM-10, vol 2, pp 4–13
Idriss IM, Seed HB (1968) Seismic response of horizontal soil layers. J Soil Mech Found Div ASCE 94(SM4):1003–1031
Iwasaki T, Tokida KI, Tatsuoka F, Watanabe S, Yasuda S, Sato H (1982) Microzonation for soil liquefaction potential using simplified methods. In: Proceedings of the 3rd international conference on microzonation 1982, June, Seattle, vol 3, pp 1310–1330
Iyengar RN, Ghosh S (2004) Microzonation of earthquake hazard in greater Delhi area. Curr Sci 87(9):1193–1202
McGuire RK (1976) FORTRAN computer program for seismic risk analysis. US Geol Survey 76–67
Mitra S, Priestley K, Bhattacharyya AK, Gaur VK (2005) Crustal structure and earthquake focal depths beneath northeastern India and southern Tibet. Geophys J Int 160(1):227–248
Mitra S, Bhattacharya SN, Nath SK (2008) Crustal structure of the Western Bengal basin from joint analysis of teleseismic receiver functions and rayleigh wave dispersion. Bull Seismol Soc Am 98:2715–2723
Molina S, Lang DH, Lindholm CD (2010) SELENA-an open-source tool for seismic risk and loss assessment using a logic tree computation procedure. Comput Geosci 36:257–269
Monsalve G, Sheehan A, Schulte-Pelkum V, Rajaure S, Pandey MR, Wu F (2006) Seismicity and one-dimensional velocity structure of the Himalayan collision zone: earthquakes in the crust and upper mantle. J Geophys Res Solid Earth 111(B10)
Nandy DR, Mullick BB, Chowdhurys B, Murthy MVN (1975) Geology of the NEFA Himalayas: recent geological studies in the Himalayas. Geol Surv India, Misc Publ 24(1):91–114
Nath SK (2011) Seismic microzonation manual and handbook. Geoscience Division, Ministry of Earth Sciences, Govt. of India, New Delhi (2011)
Nath SK (2016) Seismic hazard vulnerability and risk microzonation atlas of Kolkata. Open file report. Geoscience Division, Ministry of Earth Sciences, Government of India
Nath SK, Thingbaijam KKS (2009) Seismic hazard assessment-a holistic microzonation approach. Nat Hazard 9(4):1445
Nath SK, Thingbaijam KKS (2011) Peak ground motion predictions in India: an appraisal for rock sites. J Seismol 15(2):295–315
Nath SK, Thingbaijam KKS (2012) Probabilistic seismic hazard assessment of India. Seismol Res Lett 83(1):135–149
Nath SK, Thingbaijam KKS, Raj A (2008) Earthquake hazard in northeast India—a seismic microzonation approach with typical case studies from Sikkim Himalaya and Guwahati city. J Earth Syst Sci 117(2):809–831
Nath SK, Thingbaijam KKS, Adhikari MD, Nayak A, Devaraj N, Ghosh SK, Mahajan AK (2013) Topographic gradient based site characterization in India complemented by strong ground-motion spectral attributes. Soil Dyn Earthq Eng 55:233–246
Nath SK, Adhikari MD, Maiti SK, Devaraj N, Srivastava N, Mohapatra LD (2014) Earthquake scenario in West Bengal with emphasis on seismic hazard microzonation of the city of Kolkata, India. Nat Hazards Earth Syst Sci 14(9):2549
Nath SK, Mandal S, Adhikari MD, Maiti SK (2017) A unified earthquake catalogue for South Asia covering the period 1900–2014. Nat Hazards 85(3):1787–1810
Nath SK, Srivastava N, Ghatak C, Adhikari MD, Ghosh A, Sinha Ray SP (2018) Earthquake induced liquefaction hazard, probability and risk assessment in the city of Kolkata, India: its historical perspective and deterministic scenario. J Seismol 22(1):35–68
Nath SK, Adhikari MD, Maiti SK, Ghatak C (2019) Earthquake hazard potential of Indo-Gangetic Foredeep: its seismotectonism, hazard, and damage modeling for the cities of Patna, Lucknow, and Varanasi. J Seismol 23(4):725–769
NIBS (2002) HAZUS99—earthquake loss estimation methodology, technical manual. In: Technical manual. FEMA, Federal Emergency Management Agency, National Institute of Building Sciences (NIBS), Washington DC, pp 325

Saaty TL (1977) A scaling method for priorities in hierarchical structures. J Math Psychol 15(3): 234–281
Seed HB, Idriss IM (1982) Ground motions and soil liquefaction during earthquakes. Earthquake engineering research institute 5
Seed HB, Tokimatsu K, Harder LF, Chung RM (1985) Influence of SPT procedures in soil liquefaction resistance evaluations. J Geotech Eng 111(12):1425–1445
Sun CG, Kim HS, Chung CK, Chi HC (2014) Spatial zonations for regional assessment of seismic site effects in the Seoul metropolitan area. Soil Dyn Earthq Eng 56:44–56
Sun CG, Kim HS, Cho HI (2018) Geo-proxy-based site classification for regional zonation of seismic site effects in South Korea. Appl Sci 8(2):314
UBC: Uniform building code (1997) International Conference of Building Officials. Whittier, CA
Weiss A (2001) Topographic position and landforms analysis. In: Poster presentation, ESRI user conference, San Diego, CA, vol 200
WHE-PAGER: WHE-PAGER Phase 2 (2008) Development of analytical seismic vulnerability functions. EERI-WHE-US Geological Survey
Youd TL, Idriss IM, Andrus RD, Arango I, Castro G, Christian JT, Dobry R, Finn WDL, Harder LF Jr, Hynes ME, Ishihara K, Koester JP, Liao SSC, Marcuson-III WF, Martin GR, Mitchell JK, Moriwaki Y, Power MS, Robertson PK, Seed RB, Stokoe-II KH (2001) Liquefaction resistance of soils: summary report from the 1996 NCEER and 1998 NCEER/NSF workshops on evaluation of liquefaction resistance of soils. ASCE J Geotech Geoenviron Eng 127:817–833

Chapter 15
Geosynthetics in Retaining Walls Subjected to Seismic Shaking

G. Madhavi Latha, A. Murali Krishna, G. S. Manju, and P. Santhana Kumar

15.1 Introduction

Usage of synthetic materials has gained a high level of confidence with the civil engineering community, after years of research and successful installations during the past half century. Constrained infrastructure budgets and rise in environmental concerns about the depleting natural construction materials have necessitated and popularized the use of synthetics in lieu of traditional construction materials in various geotechnical structures like foundations, pavements, retaining walls, embankments and engineered slopes. Special polymeric products called geosynthetics have been developed specifically to serve various functions like separation, filtration, drainage, reinforcement, containment and erosion control in variety of civil engineering applications. Reinforcing the soils with geosynthetics to improve its strength and mechanical properties has gained wide acceptance worldwide, and this technique is adopted to increase the bearing capacity of foundations, improving the stability of slopes and embankments and build vertical walls of greater heights (Koerner 2012; Chen et al. 2007; Li and Rowe 2008). Common types of geosynthetics used for soil reinforcement in retaining walls are—geotextiles, geogrids and geocells.

Under static loading conditions, role of tensile reinforcement in geotechnical structures is to provide additional confinement to the soil, which can be realized as apparent cohesion, providing additional stability against failures. During earthquake, the soil element under constant overburden pressure is subjected to cyclic simple shear stresses with alternating positive and negative values in addition to the vertical and horizontal normal stresses. These earthquake-induced shear stresses increase the difference in principal stresses, thus enlarging the Mohr circle to bring the soil element close to a failure state (Ling et al. 2009). If the induced shear

G. Madhavi Latha (✉) · A. Murali Krishna · G. S. Manju · P. Santhana Kumar
Indian Institute of Science, Bangalore 560012, India
e-mail: madhavi@iisc.ac.in

T. G. Sitharam et al. (eds.), *Latest Developments in Geotechnical Earthquake Engineering and Soil Dynamics*, Springer Transactions in Civil and Environmental Engineering, https://doi.org/10.1007/978-981-16-1468-2_15

stresses are very high, the minor principal stress can be negative, thus inducing tension in soil. Since structures like retaining walls are usually built with granular soils, which cannot sustain any tension, ground surface cracks develop. Cyclic rotation of principal stress directions also occurs under seismic loading conditions, which can significantly reduce the shear strength of soils, causing further instabilities. The tensile reinforcement offers restraint to the shear deformations in soil induced by seismic events. The use of tensile reinforcement for improving the seismic stability of retaining walls has gained considerable attention in recent times and has become a very common practice.

Seismic vibrations can induce instability in otherwise stable slopes. Seismic vibrations break the contacts between soil grains and also impose additional driving forces on the slopes, thus triggering failure in cases where the shear strength of the soils is exceeded by these driving forces. Reinforced soil slopes provide better resistance to the seismic forces and possess higher yield accelerations compared to unreinforced slopes. Experiences from recent earthquake records all over the world suggest that reinforced slopes perform better during earthquakes and in many cases the unreinforced failed slopes are rebuilt using reinforcement. The confinement effect generated by the layers of reinforcement in a reinforced slope prevents the vibrations to easily transmit through soil layers unlike in the case of unreinforced slope and hence improves the stability to a great extent. Generally, geosynthetic-reinforced soil slopes are more ductile and flexible and hence more tolerant to seismic loading conditions (Leshchinsky et al. 2009).

Shaking table model studies are widely popular among geotechnical engineers to study the seismic performance of soil structures. Several earlier researchers have used shaking tables to study the seismic performance of soil structures. Shaking table tests are *1-g* model tests with the limitations of stress dissimilarities between the model and the prototype. The larger size of *1-g* shaking table tests makes it easier to use a larger number and wider range of instruments to record deformations and accelerations. Major disadvantages of shaking table modeling are the low confining pressure applied to stress-dependent backfill and reinforcement layers and the requirement to properly scale the reinforcement stiffness (El-Emam and Bathurst 2004). Rigid shaking table model containers can have undesirable influences of reflecting energy boundaries (Coe et al. 1985), which can be overcome by using laminar or shear boxes that reduce these boundary effects (Wood et al. 2002; Turan et al. 2009; Krishna and Latha 2009).

From early 1980s, many studies were reported in the literature on the use of shaking tables to understand the seismic response of reinforced retaining walls (Sakaguchi 1996; Bathurst and Hatami 1998; Perez and Holtz 2004; El-Emam and Bathurst 2007; Lo Grasso et al. 2005; Krishna and Latha 2007; Panah et al. 2015; Latha and Santhanakumar 2015; Wang et al. 2015). Height of the models was 1 m or less in majority of these studies. Most of these studies brought out the advantages of reinforced retaining walls over unreinforced walls under earthquake conditions. Since shaking table studies are *1-g* model tests, direct extrapolation of the test results to the field structures is not possible. Several researchers in the past attempted to develop similitude laws for shaking table test results in order to apply the results to field situations (Wood et al. 2002; Clough and Pirtz 1956; Iai 1989; Sugimoto et al. 1994; Telekes et al. 1994).

Beneficial effects of soil reinforcement using geosynthetics for various types of retaining walls under different earthquake loading conditions are demonstrated in this study through shaking table model studies. Models of wrap-faced, rigid-faced and segmental retaining walls were built in a laminar box mounted on a uniaxial shaking table. These models were tested at sinusoidal earthquake loading conditions of different acceleration amplitudes and frequencies. Results from the model tests are analyzed to understand the effects of reinforcement on acceleration amplifications and deformations of the soil structures.

15.2 Shaking Table and Instrumentation

A shaking table with uniaxial degree of freedom was used for simulating earthquake motion in this study. The table is square in plan with 1 m sides and can carry a load of 10 kN. The table can be operated at an acceleration range of 0.05–2 g and frequency range of 1–50 Hz. Natural frequency of the shaking table is 100 Hz, which is much higher than the operating frequency, to make sure that the table is not subjected to resonance during model testing. A rectangular laminar box with clear inner dimensions of 1 m length, 0.5 m width and 0.8 m height was fabricated using a steel outer frame and 15 hollow aluminum panels stacked on each other with frictionless roller bearings. These bearings allow the aluminum panels to move independently during the movement of model, thus reducing the boundary effects on models. During testing, accelerations were measured using piezo-electric-type accelerometers and non-contact ultrasonic displacement transducers (USDT). Sensitivity of accelerometers is 0.001 g within the bandwidth of 1 Hz–2 kHz. The displacement sensors have response time of 30 ms and sensitivity of 0.01 mm.

15.3 Model Studies on Retaining Walls

Four different types of retaining walls, namely wrap-faced, rigid-faced, segmental and geocell walls were built inside the laminar box. Results from shaking table studies on these models are discussed in following subsections. All retaining wall models were built using poorly graded dry sand. The maximum and minimum dry unit weights of sand are 18 and 14 kN/m^3, respectively. Friction angle measured from direct shear test at 65% relative density was 45°. Sand pluviation technique was used for backfilling, achieving a relative density of 65%.

15.3.1 Wrap-Faced Walls

Wrap-faced retaining walls are the retaining walls where the geotextile reinforcement itself is extended as the facing for the wall. A polypropylene multifilament-type woven geotextile was used to build these walls. The geotextile has ultimate tensile strength of 55 kN/m at peak strain of 38%, measured in wide-width tension test conducted at 10% strain rate. At low strain level (2%), the geotextile has mobilized a tensile strength of 3 kN/m, exhibiting a secant modulus of 152 kN/m. Models of wrap-faced retaining walls were of 0.75 m length oriented in the direction of horizontal shaking, 0.5 m width perpendicular to the direction of shaking and 0.6 m height. Walls were constructed in 4 lifts, each of 0.15 m thickness. Models were instrumented with accelerometers (A) and pressure transducers (P) at different locations, and three displacement transducers (U) were fixed to a rigid steel bracket connected to the laminar box frame at a distance from the retaining wall facing, as shown in Fig. 15.1. A surcharge pressure of 0.5 kPa is applied over the constructed retaining wall.

Retaining wall models were built with different lengths of reinforcement (L_{rein}) and were subjected to 20 cycles of horizontal shaking of acceleration amplitude 0.2 g and frequency 3 Hz. Reinforcement lengths of 420, 300 and 600 mm were adopted in the tests WT17, WT19 and WT20, respectively, with the corresponding normalized reinforcement length (L_{rein}/H), where H is the height of the wall, as 0.7, 0.5 and 1.0. Results from these three tests are analyzed to understand the effect of reinforcement length on facing deformations and acceleration amplifications of the wall at different elevations. Acceleration amplifications are converted to root mean square acceleration (RMSA) amplification factors so that the entire acceleration time history for each accelerometer device is integrated over the entire duration with positive and negative acceleration measurements squared and averaged over the duration to get one absolute value.

Figure 15.2 shows the displaced profiles for the test walls WT17, WT19 and WT20 with different reinforcement lengths (L_{rein}). Displacements reduced considerably with the increase in the length of reinforcement. With L_{rein}/H of 0.5 in test WT19, a maximum displacement of 95 mm was observed at the top of the wall and the displacement reduced to 85 mm when L_{rein}/H was increased to 0.7, and it further reduced to 28 mm when the L_{rein}/H was 1.0. Reduction in displacements from L_{rein}/H of 0.5 to 0.7 is marginal. However, displacements for L_{rein}/H of 1.0 are very low, about 33% lesser compared to displacements observed when L_{rein}/H was 0.7, highlighting the importance of maintaining the minimum length of reinforcement in a reinforced retaining wall. In case of geosynthetic-reinforced retaining walls subjected to static loading conditions, a minimum recommended length of reinforcement is about $0.6H$, so that the reinforcement can extend beyond the failure wedge to contribute to the resistance against pullout failure and reduce face deformations (Wu 2019). However, under seismic shaking conditions, the minimum length of reinforcement required is more than $0.7H$ as per the present study because increasing the length of reinforcement beyond $0.7H$ has resulted in

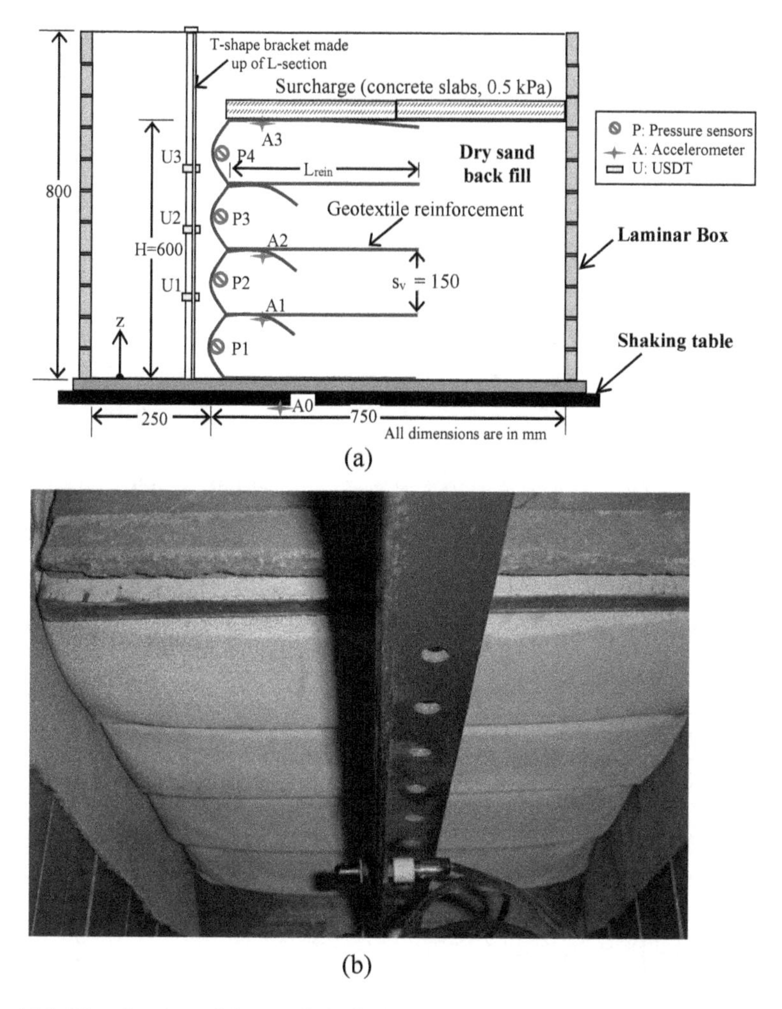

Fig. 15.1 Wrap-faced retaining wall built inside a laminar box. **a** Schematic diagram. **b** Photograph

considerable decrease in lateral deformations. Seismic loads cause additional pullout of reinforcement layers, which needs to be resisted through additional length of reinforcement.

Figure 15.3 shows the effect of reinforcement length on RMSA amplification factors for the tests WT17, WT19 and WT20. Accelerations were amplified by about 1.6 times at the top of the wall for all models, and the difference in acceleration amplifications is not significant for different models. All accelerometers are placed close to the facing, within the reinforced soil zone. Since the relative density of soil is same in all models, soil stiffness and the shear wave velocity within the soil remain same at different elevations. Hence, rate of acceleration amplification

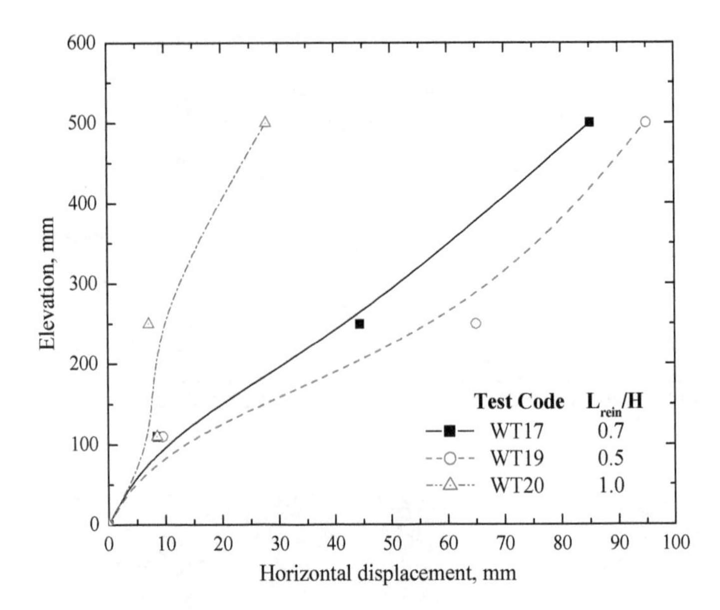

Fig. 15.2 Displacements of wrap-faced walls with change in length of reinforcement

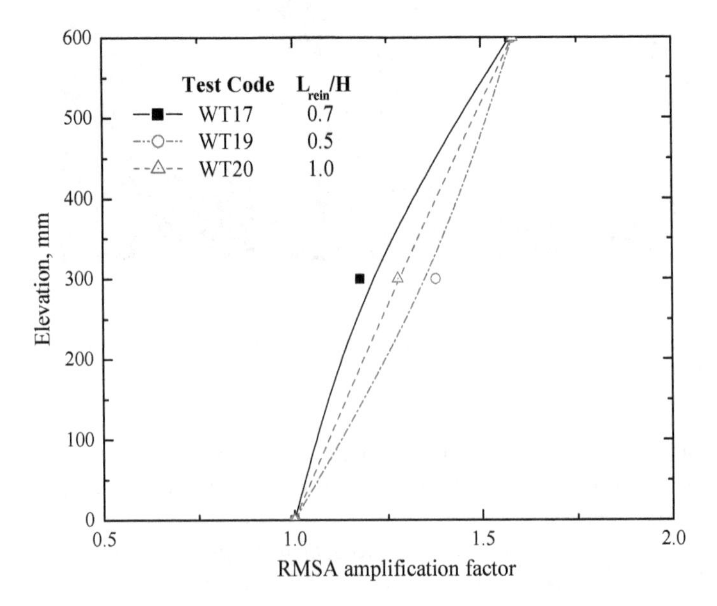

Fig. 15.3 RMSA amplification factors for wrap-faced walls with change in length of reinforcement

does not change with elevation with change in reinforcement length. Similar negligible effect of reinforcement design parameters on amplification factors was reported by El-Emam and Bathurst (2004).

15.3.2 Rigid-Faced Retaining Walls

Rigid-faced retaining walls were constructed to same dimensions as the wrap-faced walls. However, facing in these wall models was created using 12 rectangular hollow steel sections connected through three vertical steel rods running through them, which are rigidly fixed to the base of the laminar box to create a rigid steel panel facing of 600 mm height and 25 mm thickness. Both unreinforced and geogrid reinforced retaining wall models were tested. Biaxial geogrids made of polypropylene with ultimate tensile strength of 26 kN/m at peak strain of 16.5%, and low strain (2%) secant modulus of 219 kN/m were used as reinforcement in these models. Geogrids were run through the steel rods connected with the rigid facing. Dimensional and construction details of rigid faced walls along with the instrumentation are presented in Fig. 15.4.

Length of reinforcement was 420 mm in all reinforced walls. The number of reinforcing geogrid layers was varied in different model tests. Test UT4 represents unreinforced model, while tests RT6, RT7, RT8 and RT9 represent models with 4, 3, 2 and single-layer geogrid models, respectively. Geogrid layers were equally spaced in models. A surcharge pressure of 0.5 kPa is applied over the retaining wall. Shaking table tests were conducted for 20 cycles of sinusoidal dynamic motion of 0.1 g acceleration and 1 Hz frequency.

Figure 15.5 shows the wall deformations for unreinforced and rigid-faced retaining walls at 1 and 2 Hz frequencies. These figures clearly demonstrate the beneficial effects of geosynthetic reinforcement in reducing the deformations of rigid faced walls. Unreinforced retaining wall deformed to a maximum of 1.6 mm, and the deformations were reduced to 0.25 mm with four layers of geogrid reinforcement. Acceleration amplifications were not significant in rigid-faced retaining walls because of the confinement effect created by the rigid wall facing, which imparts additional stiffness to the soil, allowing the shear waves to pass quickly through the soil from bottom to top, thus causing lesser differential accelerations at the bottom and top of the wall. However, difference in acceleration amplifications with increase in the number of reinforcing layers is not significant.

15.3.3 Segmental Retaining Walls

Segmental retaining wall models were built to a length of 0.7 m in the direction of shaking, 0.5 m in the horizontal direction perpendicular to it and 0.6 m height inside the laminar box. The facing for these walls was made using concrete blocks

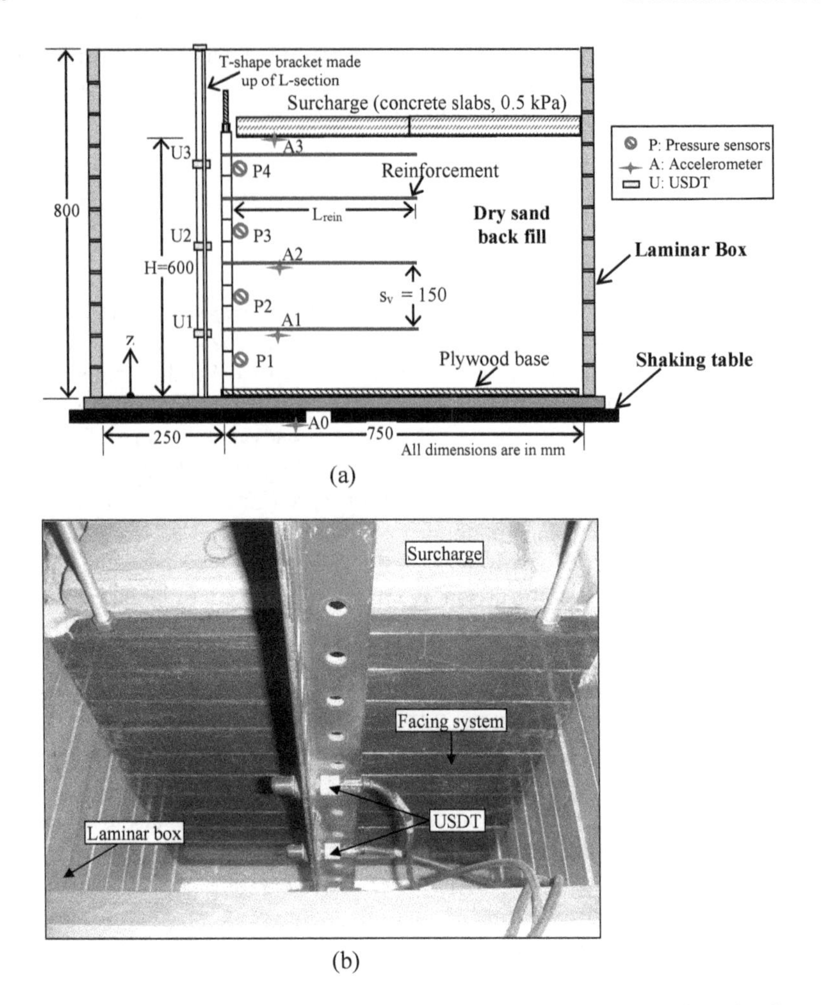

Fig. 15.4 Rigid faced retaining wall built inside a laminar box. **a** Schematic diagram. **b** Photograph

of dimensions 125 mm × 100 mm × 150 mm with an interlocking lip, forming an inward batter of 7.2°. Backfill sand was placed in equal lifts of 150 mm thickness, a layer of geogrid placed at the between two modular facing blocks, as shown in Fig. 15.6. Biaxial geogrids made of polypropylene with ultimate tensile strength of 26 kN/m at peak strain of 16.5% and low strain (2%) secant modulus of 219 kN/m were used as reinforcement in these models. The number of reinforcement layers was 2 and 3 in different tests with equal spacing between the layers. Length of reinforcement was kept as 420 mm in all reinforced walls. A surcharge pressure of 0.5 kPa is applied over the retaining wall. Shaking table tests were conducted for 20 cycles of 0.3 g acceleration at 2 Hz frequency.

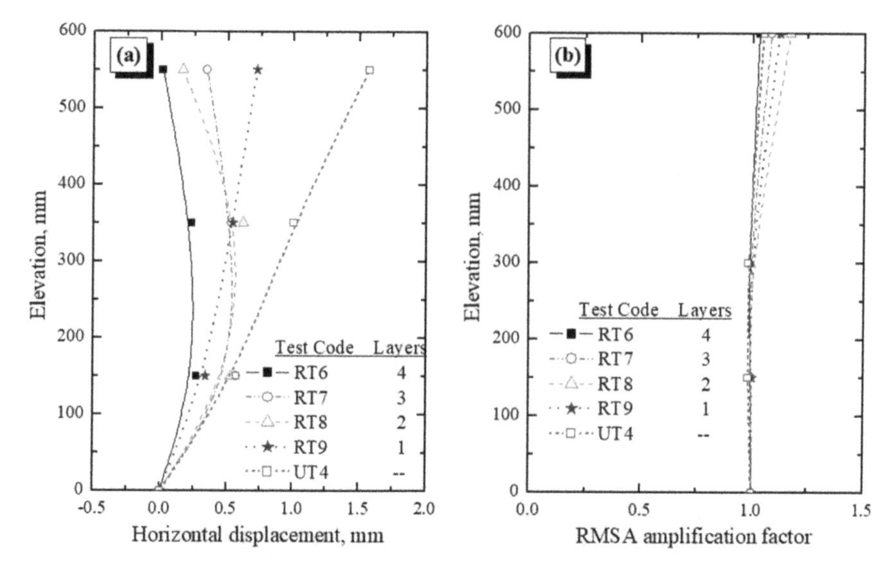

Fig. 15.5 Performance of rigid-faced retaining walls under base shaking conditions. **a** Displacement response. **b** Acceleration amplification response

Figure 15.7 shows the variation of wall deformations and RMSA amplification factors for unreinforced and reinforced modular block walls. Maximum displacement of 6 mm observed in case of unreinforced segmental retaining walls was reduced to less than 1 mm with three layers of geogrid reinforcement. This high reduction in deformations can be related to the increased connectivity between the wall facing and backfill in reinforced walls and additional confinement provided by the reinforcement layers, which restricts the movement of the backfill and thus controls the wall deformations. RMSA amplification factors estimated for the reinforced retaining walls in the present study ranged between 1.1 and 1.3, which is same for unreinforced walls. Appreciable difference is not observed in acceleration amplifications of unreinforced and reinforced segmental retaining walls. Since the stiffness of reinforcement is very low compared with the stiffness of the segmental blocks at the facing and the inward batter of the walls with larger base provides a stable configuration of the wall even without the reinforcement, change in the velocity of wave passing vertically from the base to the top of the wall is not significant with the inclusion of reinforcement.

15.3.4 Geocell Retaining Walls

Geocell retaining walls have backfill retained by the three-dimensional polymeric cells filled with granular fill acting like a facing. In this study, geocell retaining walls of 0.6 m height with and without reinforcement layers in backfill are

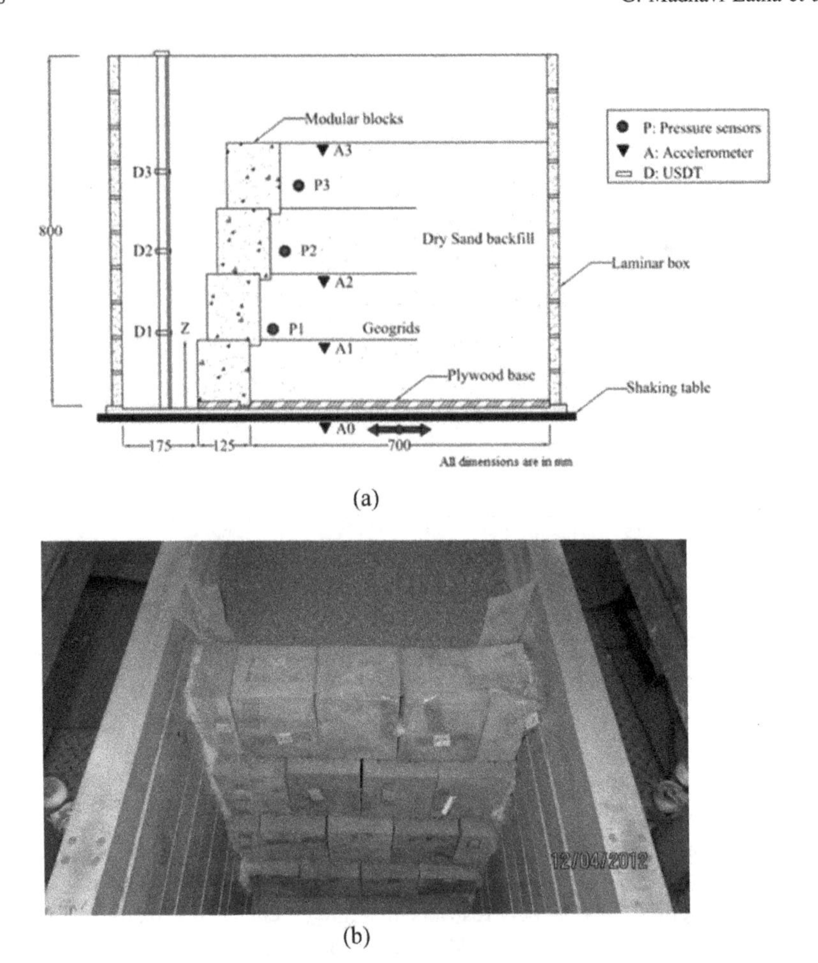

Fig. 15.6 Segmental retaining wall built inside a laminar box. a Schematic diagram. b Photograph

constructed inside the laminar box to base dimensions of 0.8 m in the direction of shaking and 0.5 m in the orthogonal direction to shaking in plan. The walls are bettered toward the backfill, with the crest dimensions coming to 0.55 m in the direction of shaking and 0.5 m in the direction perpendicular to it in plan. Backfill was placed using dry pluviation technique, in six layers of 0.1 m height each, which is equal to the height of geocell layer. Geocells were manufactured using polyvinylchloride (PVC) sheets to dimensions of 100 mm × 100 mm × 100 mm, with laminated joints. The PVC sheet has ultimate tensile strength of 0.6 kN/m at a peak strain of 24% and secant modulus of 5 kN/m with seam strength of 0.45 kN/ m. Geocell layers were sequentially laid and filled with aggregate of 12 mm average size and compacted to a relative density of 70%. Backfill sand was poured to the height of 0.1 m to the specific relative density before the second layer of geocell is stacked. An offset is maintained to get the retaining wall constructed to

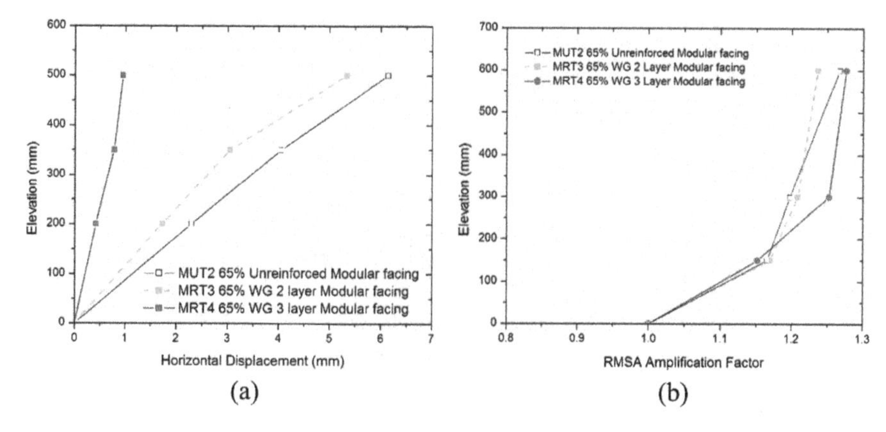

Fig. 15.7 Performance of unreinforced and reinforced segmental retaining walls under base shaking conditions. **a** Displacement response. **b** Acceleration amplification response

with a batter of about 65°. In case of reinforced backfill, a layer of geonet of 0.42 m length with ultimate tensile strength of 11 kN/m is placed between the geocell layers, extending into backfill. Figure 15.8 gives the dimensional details of geocell wall models along with instrumentation.

Geocell retaining walls were subjected to horizontal shaking of 0.3 g and 7 Hz for 20 cycles. Test code S7A3F7 represents the test without geonet in backfill, and the test code S7BA3F7 represents the test with geonet layers in backfill. Difference in facing deformations and acceleration amplifications of the walls without geonet reinforcement in backfill and walls with geonet reinforcement are compared in Fig. 15.9.

Results showed that reinforcement in backfill resulted in about 60% reduction in wall deformations. When the geocell walls have no reinforcement in backfill, lateral pressures on the wall facing are more, resulting in more wall deformations. Layers of geonet provided extra confinement effect to the backfill because of which the backfill exerts lesser lateral pressures on wall facing, causing the walls move relatively lesser during shaking. Also, the internal movement of backfill is restricted due to the separation effect of geonet layers, which resulted in better friction mobilization during shaking and hence acted against the movement of wall.

Considerable reduction in RMSA amplification factors was observed with the geonet reinforcement. The reduction was from a value of 2.2 to 1.5 at the top of the wall where the amplifications were maximum. These huge reductions in RMSA amplification factors can be attributed to the overall increase in wall stiffness due to the inclusion of geonets. In case of models without a basal geonet, the connection between the wall facing and the backfill is established only through the stepped geometry of the geocell walls. Hence, the wall components (facing and backfill) might lose connection during seismic shaking, which results in higher deformations and amplification factors. In case of walls geonet, since the geonet is connecting the wall facing with the backfill more effectively, the wall moves as a whole, and

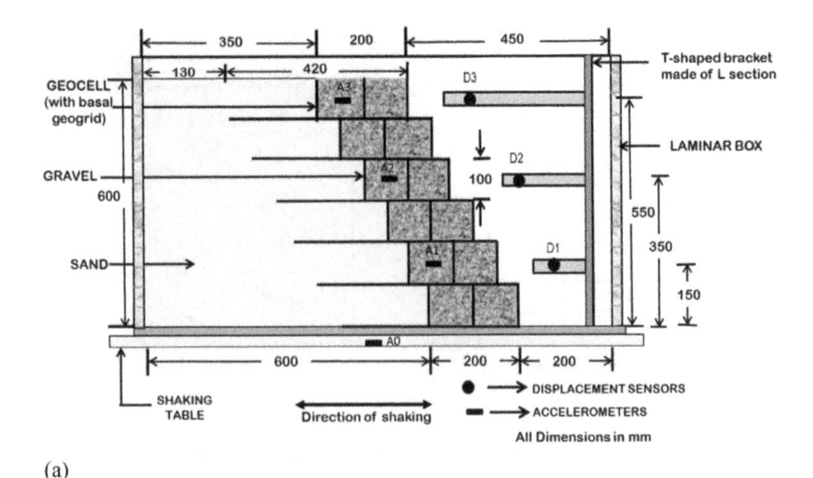

(a)

(b)

Fig. 15.8 Geocell retaining wall built inside a laminar box. **a** Schematic diagram. **b** Photograph

hence, even stronger ground shaking cannot separate them, and hence, the deformations and acceleration amplifications are lesser. The geonet used in this study has very low tensile strength, and in prototype models, it matches with the tensile strength of commercial geosynthetics with higher tensile strength.

15.4 Conclusions

Through shaking table studies on models of wrap-faced, rigid-faced, segmental and geocell retaining walls, it is demonstrated that geosynthetic reinforcement provides appreciable benefits for the seismic performance of these walls. These benefits are highly significant in reducing the deformations of the walls. Even with weaker geosynthetic reinforcement, deformations were reduced by a minimum of 30% and

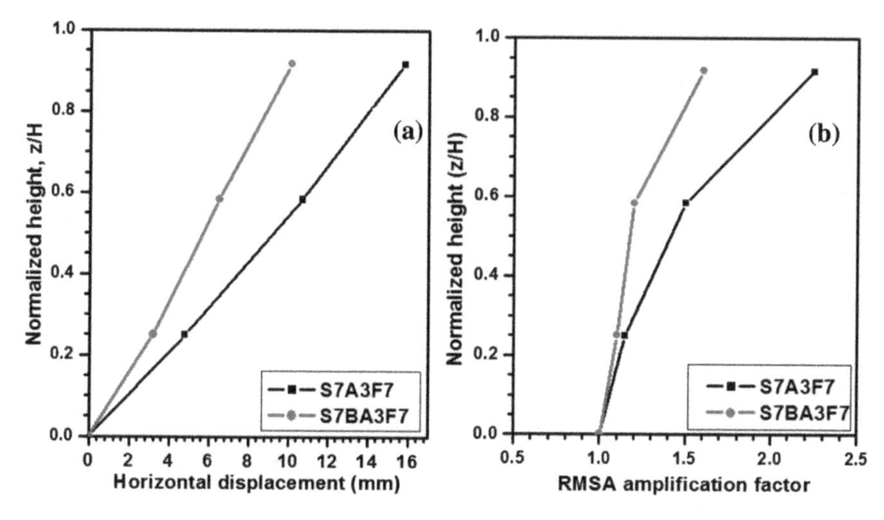

Fig. 15.9 Performance of geocell walls with and without backfill reinforcement

in some cases more than 80%. Reinforcement was found to be an effective means to confine the backfill soil and to establish stable connectivity between wall facing and the backfill soil so that the lateral pressures on the facing from backfill can be effectively controlled. However, effect of reinforcement on acceleration amplifications in retaining walls is not significant because the improvement in stiffness within the model scale walls is not good enough to alter the velocity of shear waves traveling from wall base to the top of the wall. Studies reported in this paper are carried out using 1-g model tests, and hence, the results are subjected to scaling effects and stress discrepancies between the models and prototype walls. However, qualitatively, the benefits of using geosynthetic reinforcement in retaining walls for their enhanced seismic performance are clearly established from this study.

References

Bathurst RJ, Hatami K (1998) Seismic response analysis of a geosynthetic-reinforced soil retaining wall. Geosynth Int 5(1–2):127–166

Chen HT, Hung WY, Chang CC, Chen YJ, Lee CJ (2007) Centrifuge modeling test of a geotextile-reinforced wall with a very wet clayey backfill. Geotext Geomembr 25(6):346–359

Clough RW, Pirtz D (1956) Earthquake resistance of rockfill dams. J Soil Mech Found Div 82 (2):1–26

Coe CJ, Prevost JH, Scanlan RH (1985) Dynamic stress wave reflection/attenuation: earthquake simulation in centrifuge soil models. Earthquake Eng Soil Dyn 13:109–128

El-Emam M, Bathurst RJ (2004) Experimental design, instrumentation and interpretation of reinforced soil wall response using a shaking table. Int J Phys Model Geotech 4:13–32

El-Emam MM, Bathurst RJ (2007) Influence of reinforcement parameters on the seismic response of reduced-scale reinforced soil retaining walls. Geotext Geomembr 25(1):33–49

Iai S (1989) Similitude for shaking table tests on soil-structure fluid model in 1-g gravitational field. Soils Found 29(1):105–118

Koerner RM (2012) Designing with geosynthetics, 4th edn. Prentice Hall Inc., New Jersey

Krishna AM, Latha GM (2007) Seismic response of wrap-faced reinforced soil retaining wall models using shaking table tests. Geosynth Int 14(6):355–364

Krishna AM, Latha GM (2009) Container boundary effects in shaking table tests on reinforced soil wall models. Int J Phys Model Geotech 9(4):1–14

Latha GM, Santhanakumar P (2015) Seismic response of reduced-scale modular block and rigid faced reinforced walls through shaking table tests. Geotext Geomembr 43(4):307–316

Leshchinsky D, Ling HI, Wang J-P, Rosen A, Mohri Y (2009) Equivalent seismic coefficient in geocell retention systems. Geotext Geomembr 27(1):9–18

Li AL, Rowe RK (2008) Effects of viscous behaviour of geosynthetic reinforcement and foundation soils on the performance of reinforced embankments. Geotext Geomembr 26: 317–334

Ling HI, Leshchinsky D, Wang J, Mohri Y, Rosen A (2009) Seismic response of geocell retaining walls, experimental studies. J Geotech Geo-Environ Eng ASCE 135(4):515–524

Lo Grasso AS, Maugeri M, Recalcati P (2005) Seismic behaviour of geosynthetic-reinforced slopes with overload by shaking table tests. In: Slopes and retaining structures under seismic and static conditions. ASCE Geotechnical Special Publication No. 140, CDROM

Panah AK, Yazdi M, Ghalandarzadeh A (2015) Shaking table tests on soil retaining walls reinforced by polymeric strips. Geotext Geomembr 43:148–161

Perez A, Holtz RD (2004) Seismic response of reinforced steep soil slopes: results of a shaking table study. In: Geotechnical engineering for transportation projects, vol 126. ASCE Geotechnical Special Publication, pp 1664–1672

Sakaguchi M (1996) A study of the seismic behavior of geosynthetic reinforced walls in Japan. Geosynth Int 3(1):13–30

Sugimoto M, Ogawa S, Moriyama M (1994) Dynamic characteristics of reinforced embankments with steep slope by shaking model tests. In: Recent case histories of permanent geosynthetic-reinforced soil walls, Seiken symposium, Tokyo, Japan, pp 271–275

Telekes G, Sugimoto M, Agawa S (1994) Shaking table tests on reinforced embankment models. In: Proceedings of 13th international conference on soil mechanics and foundation engineering, New Delhi, India, vol 2, pp 649–654

Turan A, Hinchberger SD, El Naggar H (2009) Design and commissioning of a laminar soil container for use on small shaking tables. Soil Dyn Earthq Eng 29:404–414

Wang L, Chen G, Chen S (2015) Experimental study on seismic response of geogrid reinforced rigid retaining walls with saturated backfill sand. Geotext Geomembr 43(1):35–45

Wood DM, Crewe A, Taylor C (2002) Shaking table testing of geotechnical models. Int J Phys Model Geotech 1:1–13

Wu JTH (2019) Geosynthetic reinforced soil (GRS) walls. Wiley, Hoboken, NJ

Chapter 16
Studies on Modeling of Dynamic Compaction in a Geocentrifuge

B. V. S. Viswanadham and Saptarshi Kundu

16.1 Introduction

The engineering properties of various problematic soils, fills and waste materials existing in the field need to be improved prior to their use as construction sites and their exposure to various loading conditions. The problematic sites encountered in the field involve weak compressible soils, collapsible soils, expansive soils, fills, MSW materials, fly ash/coal ash deposits, etc. The selection of a particular improvement method depends on the type and degree of improvement required and the soil type. The standard ground improvement methods adopted in the field can be broadly classified as reinforcement techniques, densification techniques, grouting/mixing techniques and drainage techniques. Dynamic compaction (DC) is one of the most widely adopted densification techniques for geomaterials in view of its simplicity, cost-effectiveness and ease of implementation (Menard and Broise 1975; Leonards et al. 1980; Mayne et al. 1984; Lukas 1986; Rollins and Rogers 1994). The technique, also referred to as impact densification and heavy tamping, has evolved as a routine method of site improvement for treating poor soils in situ. Densification by DC is performed by dropping a heavy tamper of steel or concrete in a grid pattern from heights of 5–30 m. Liquefaction is initiated locally beneath the drop point making it easier for the sand grains to densify. When the excess pore water pressure from dynamic loading dissipates, additional densification occurs. The process is usually repeated in several passes until the required post-treatment density is achieved.

B. V. S. Viswanadham (✉) · S. Kundu
Department of Civil Engineering, Indian Institute of Technology Bombay, Mumbai 400076, India
e-mail: viswam@civil.iitb.ac.in

T. G. Sitharam et al. (eds.), *Latest Developments in Geotechnical Earthquake Engineering and Soil Dynamics*, Springer Transactions in Civil and Environmental Engineering, https://doi.org/10.1007/978-981-16-1468-2_16

Numerous studies have been reported in the literature on DC being applied on diverse geomaterials (Mayne and Jones 1983; Lutenegger 1986; Zou et al. 2005; Bo et al. 2009; Feng et al. 2011; Zekkos et al. 2013; Kundu and Viswanadham 2018, 2020). The effectiveness of DC in the context of remediation of loose soil deposits is hereby illustrated in Fig.16.1a–d. Figure 16.1a depicts the problems encountered in loose subsoils, wherein superstructure load from adjacent infrastructural facilities results in large differential settlements of the foundation and subsequent instability of the building. Figure 16.1b, c presents two alternatives in this regard, in the form of installation of conventional pile foundations and adoption of DC for soil densification, respectively. The resultant improvement achieved using DC is presented in Fig. 16.1d, wherein reduced settlements and enhanced stability to superstructure loads can be observed in the soil stratum densified by DC.

DC possesses a number of advantages compared to other ground improvement methods. DC has been successfully applied over a range of soil types, including loose granular deposits, saturated clayey soils, MSW, rockfills and mine spoils. As per the compilation of Yee and Ooi (2010), DC has the third least CO_2 emission as a ground improvement method after vacuum consolidation and vertical drain installations. Furthermore, as per the database of Geotechnical Engineering Circular No. 1 prepared by Lukas (1995) (Table 16.1), the cost for executing DC in the field is significantly less compared to other ground remediation methodologies.

During DC, the blow energy is applied in single or multiple passes in a grid pattern over the entire area. In the first stage, the blows are spaced at a distance dictated by the depth of the compressible layer, the depth of existing groundwater table and the grain size distribution of soil. Initial grid spacing is usually at least equal to the thickness of the compressible layer, and 6–50 tamper drops are imparted at each point. This first phase of the treatment with widely spaced blows is designed to improve the deeper layers. In saturated fine-grained soils, a sufficient time interval is planned between succeeding passes to allow the excess pore water pressure to dissipate. After each pass, backfilling is done periodically with surrounding materials available at the site. The initial passes are also called 'high-energy pass,' as the tamper energy is higher than subsequent passes. The second pass is generally made at the centroid points of the first pass and consists of several tamper drops at the same point, which lead to closure of the voids for achieving minimum void ratio. Finally, an 'ironing' pass with a low-energy blow and reduced drop height is performed to compact shallow soil layers. The field procedure of DC as discussed is shown in Fig. 16.2.

16.2 Scaling Considerations of DC

Physical modeling of geotechnical engineering problems can be executed at full-scale level or as a reduced scale model to replicate various field situations. However, a limitation of small-scale laboratory models under normal gravity is that the stress levels are much smaller than in prototype structures. Only full-scale

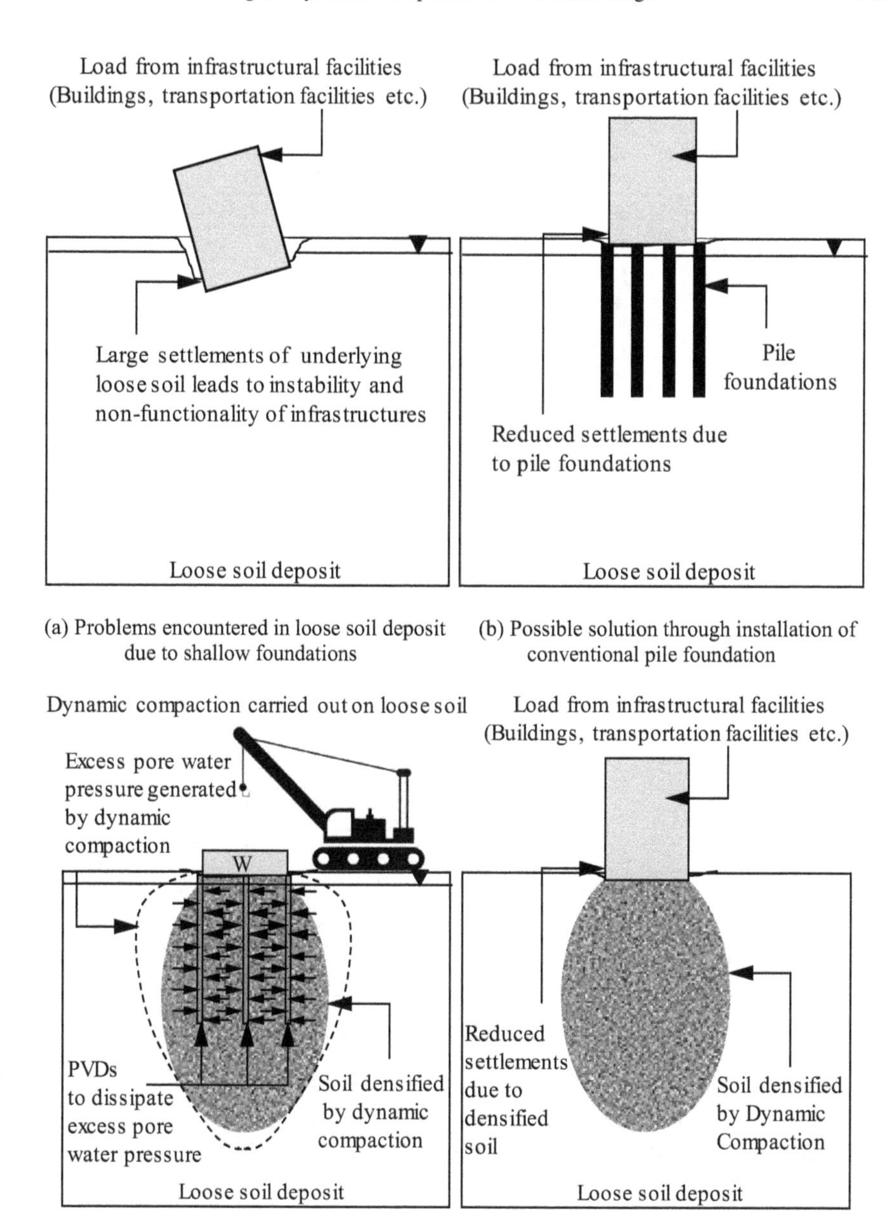

(a) Problems encountered in loose soil deposit due to shallow foundations

(b) Possible solution through installation of conventional pile foundation

(c) Alternative solution attained through densification of loose soils by DC

(d) Effective ground remediation induced by DC

Fig. 16.1 Effectiveness of DC as a ground remediation technique for geomaterials

Table 16.1 Comparative cost of different ground improvement techniques [based on the compilation of Lukas (1995)]

Treatment method	Basis of cost calculation		
	Volume of treated soil (US$/m^3)	Surface (US$/m^2)	Length (US$/m)
[a]Dynamic compaction	0.7–3	4.3–22	–
Vibro-replacement	4–12	–	30–52
Vibro-compaction	1–7	–	16–39
Excavate–replace	10–20	–	–
Slurry grouting	40–80	–	–
Chemical grouting	160–525	–	–
Compaction grouting	30–200	–	–
Jet grouting	100–400	–	82–325
[b]Freezing	275–650	110–160	–

[a]Estimate is prepared based on projects undertaken during 1985–1993
[b]Plus $2–$10.75 per sq-m/week for maintaining frozen zones

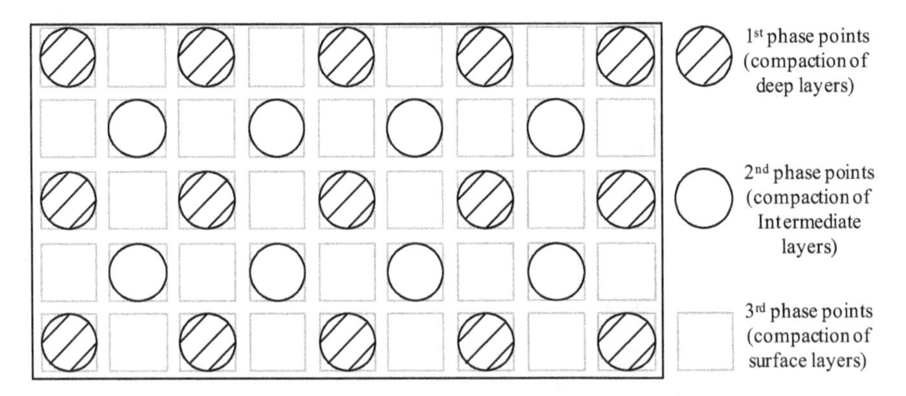

Fig. 16.2 Typical phases involved in the field during DC process

physical models can include all these complexities, but they are expensive, time-consuming and difficult to replicate with dynamic loadings similar to earthquakes, DC, blasting, etc. In such situations, geotechnical centrifuge modeling can be used as an effective tool to replicate identical stress–strain response in the model as that of the prototype (Schofield 1980; Taylor 1995; Madabhushi 2014). This enables study of the behavior of engineered earth structures in a controlled environment by deriving standard scaling laws linking the model behavior to that of corresponding prototype. The main governing parameters for modeling DC in a geocentrifuge include scaling of the mass of tamper, height of fall and frequency of tamper drops. Considering a soil deposit of thickness d_p in the field and corresponding centrifuge model (Ng model) of thickness d_m, the relevant scaling law is presented in Eq. (16.1):

$$d_m = \frac{d_p}{N} \tag{16.1}$$

where the notations p and m are used to denote the prototype and model, respectively, and N indicates the gravity level or scale factor. The above scaling factor is also applicable for modeling the tamper radius (r), drop height (H), improvement depth (d_i), crater depth (d_c) and radial distance from tamper center (x) (Eq. 16.2a–e).

$$r_m = \frac{r_p}{N} \tag{16.2a}$$

$$H_m = \frac{H_p}{N} \tag{16.2b}$$

$$(d_i)_m = \frac{(d_i)_p}{N} \tag{16.2c}$$

$$(d_c)_m = \frac{(d_c)_p}{N} \tag{16.2d}$$

$$x_m = \frac{x_p}{N} \tag{16.2e}$$

The base area of tamper (A) ($A = \pi r^2$) used in inducing DC in centrifuge is reduced by N^2 times that of the tamper used in the prototype [Eq. (16.3)], while the volume of crater (V_c) induced in model surface is scaled by a factor of N^3 [Eq. (16.4)]. In addition, the tamper velocity [$v = (2gH)^{0.5}$] at the moment of blow (v) is related to its height of fall (H) and gravitational acceleration (g), and is scaled using Eq. (16.5):

$$\frac{A_m}{A_p} = \frac{\pi r_m^2}{\pi r_p^2} = \frac{1}{N^2} \tag{16.3}$$

$$\frac{(V_c)_m}{(V_c)_p} = \frac{1}{N^3} \tag{16.4}$$

$$\frac{v_m}{v_p} = \sqrt{\frac{2(Ng)H_m}{2gH_p}} = 1 \tag{16.5}$$

The work done by tamper in inducing crater at the soil surface is derived from its kinetic energy (KE) at the onset of tamper blow. By definition, work done is the product of force and displacement, and is scaled by a factor of $1/N^3$ [Eq. (16.6)].

Based on scaling of the tamper velocity and kinetic energy, scale factor for tamper mass (m) can be further derived [Eq. (16.7)].

$$\frac{\mathrm{KE_m}}{\mathrm{KE_p}} = \frac{F_\mathrm{m}(d_\mathrm{c})_\mathrm{m}}{F_\mathrm{p}(d_\mathrm{c})_\mathrm{p}} = \frac{1}{N^3} \tag{16.6}$$

$$\frac{m_\mathrm{m}}{m_\mathrm{p}} = \left[\frac{2(\mathrm{KE})_\mathrm{m}}{v_\mathrm{m}^2}\right]\left[\frac{v_\mathrm{p}^2}{2(\mathrm{KE})_\mathrm{p}}\right] = \frac{1}{N^3} \tag{16.7}$$

As the tamper strikes the soil surface, Rayleigh waves are generated, which spread radially on the ground. The peak ground velocity (PGV) of the soil surface is used for quantifying these waves, which is scaled as in Eq. (16.8) similar to tamper velocity.

$$\frac{(\mathrm{PGV})_\mathrm{m}}{(\mathrm{PGV})_\mathrm{p}} = \frac{v_\mathrm{m}}{v_\mathrm{p}} = 1 \tag{16.8}$$

The relevant scaling laws for modeling DC in centrifuge are summarized in Table 16.2.

Table 16.2 Scaling laws applicable for modeling DC in a geotechnical centrifuge

Parameters	Prototype	Model scale
Cohesion (c) (kPa)	1	1
Angle of internal friction (ϕ) (°)	1	1
Unit weight of the soil (γ) (kN/m^3)	1	[a]N
Relative density of soil (RD) (%)	1	1
Pore water pressure in soil (u) (kPa)	1	1
Seepage time (t_s) (s)	1	$1/N^2$
Coefficient of permeability (k) (m/s)	1	N
Tamper mass (m) (t)	1	$1/N^3$
Tamper radius (r) (m)	1	$1/N$
Tamper drop height (H) (m)	1	$1/N$
Time interval between successive blows (t_i) (min)	1	$1/N$
Time for generation of pore water pressure (t_g) (s)	1	$1/N$
Frequency of blows (f_b) (min^{-1})	1	N
Velocity of tamper (v) (m/s)	1	1
Peak ground acceleration (PGA) (m/s^2)	1	N
Peak ground velocity (PGV) (mm/s)	1	1
Crater depth (d_c) (m)	1	$1/N$
Crater volume (V_c) (m^3)	1	$1/N^3$
Depth of improvement (d_i) (m)	1	$1/N$
Kinetic/potential energy of tamper blow (E) (t-m)	1	$1/N^3$
Momentum of tamper (M) (t-m/s)	1	$1/N^3$

N Gravity level or scale factor
[a]For example, $\gamma_\mathrm{m}/\gamma_\mathrm{p} = N$; m: model; p: prototype

16.3 Design Details of Actuator

An actuator was custom-designed and developed in the present study based on the scaling laws presented in Table 16.2. The model test package and actuator assembly are shown in Fig. 16.3. The developed actuator consists of primarily four different components, referred to as the support system, impactor assembly, tamper—hook

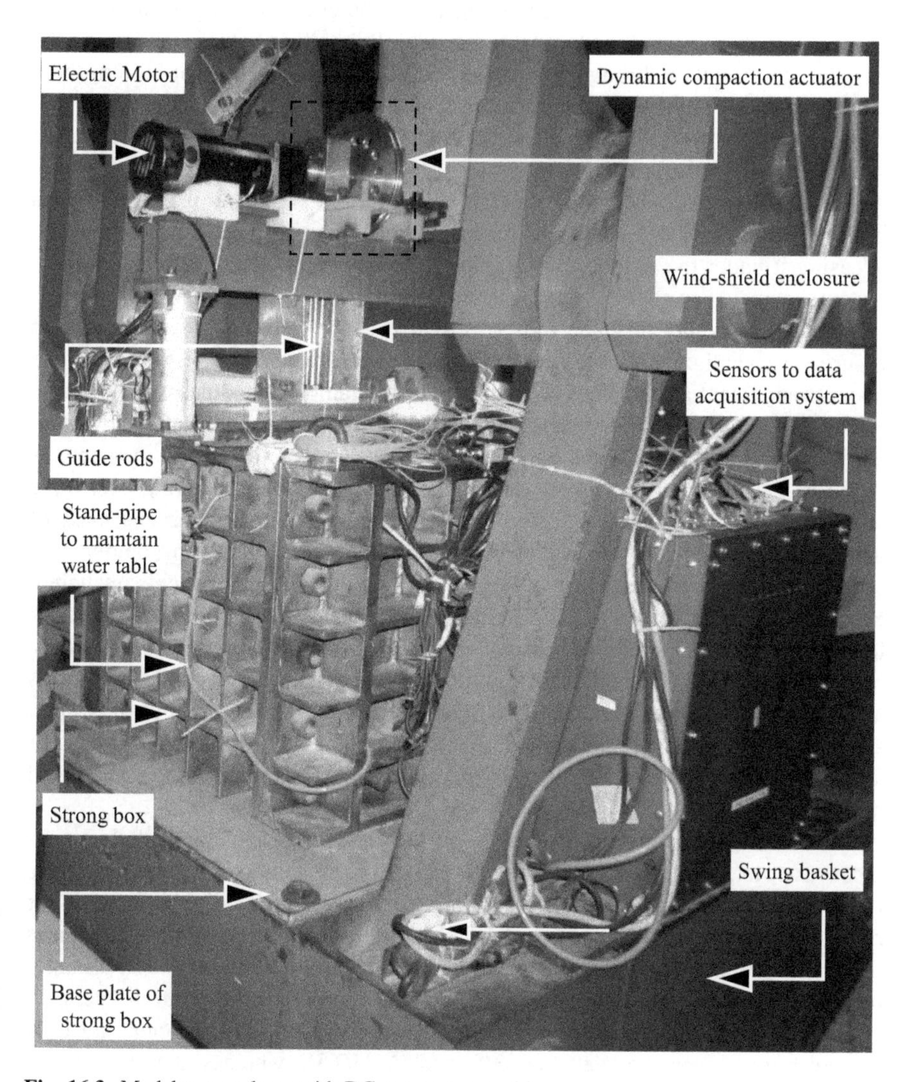

Fig. 16.3 Model test package with DC actuator assembly

arrangement and guiding rod assembly. The support system of the developed actuator consists of three distinctive components, referred to as supporting beam, supporting columns and associated flanges with slot arrangement. The beam has been custom-designed to support the loading transferred by the entire actuator and subsequently transfers the structural load to the supporting columns by a suitably designed nut–screw arrangement. In addition, the presence of supporting columns on either side of the beam component ensures elevation of tamper to a considerable height above the model soil surface in centrifuge, thereby increasing the drop height to the range that is usually adopted in the field. Supporting columns distribute the loads to the strongbox through two properly designed flanges made of 10-mm-thick mild steel plates. Slots provided along the centerline of the flanges enable horizontal shift of the DC assembly, which is advantageous while modeling the effects of variable tamper radii in centrifuge.

The impactor assembly is the fundamental lifting component of the developed in-flight actuator. It is responsible for elevating tamper to a specified height and releasing it under the influence of gravity with the help of four fundamental components, namely drive shaft, sheave, steel hoist rope and a pair of bearing housings. Besides, an assembly of guide rods is provided at the impactor assembly base to ensure vertical alignment of tamper during all stages of centrifuge test. Tamper and hook arrangement are the core components of the tamping module of the in-flight actuator. Three distinctive components, namely base plate, guide shafts and collar, are integrated to constitute the tamper. The base plate is half-circular in shape, as only one-half of circular tamper used in the field was modeled during centrifuge tests. In order to model the effects of variable tamper radii and variable mass of tamper in centrifuge, the base plate is proportioned accordingly to comply with scaling requirements, while keeping guide shafts and collar components unaltered. Hook arrangement consists of two primary components, the hook itself and hook holding block. The guiding rod assembly comprises two individual sets of guiding rods, referred herein as central rods and peripheral rods. The central guiding rods are made to pass through hollow guide shafts and enable linear guidance of tamper. The peripheral guide rods are compactly secured to the bottom surface of impactor assembly. Their primary function is to counter Coriolis acceleration generated during flight.

Additional components associated with the developed actuator include standpipes for maintaining groundwater level, electric motor for providing the mechanical power necessary for rotating the drive shaft, digital camera for recording proceedings of experiment, windshield enclosure to prevent disturbances from artificial air currents above the model soil surface and illumination arrangement in the form of thin strips of LED light. The instrumentation included accelerometers and pore water pressure transducers. The accelerometers are DJB piezoelectric sensors (models: A/23/S and A/23/TS), whereas the pore water pressure transducers are Druck PDCR81 type miniature PPTs (GE make, UK). These miniature Druck PDCR 81 PPTs have been used extensively by researchers to monitor pore water pressure variations in centrifuge (Muraleetharan and Granger 1999; Ghayoomi et al. 2011; Bhattacherjee and Viswanadham 2018). The data from PPTs were acquired at sampling rate of 10,000 data per sec during DC and at a normal rate of 1 datum per sec before and after inducing DC. The accelerometers were positioned on the soil surface at 0.15, 0.30 and 0.45 m distances from tamper center to measure vertical vibrations (A1_V, A3_V and A5_V) and horizontal vibrations (A2_H, A4_H and A6_H).

16.4 Salient Features of Developed Actuator

The in-flight actuator exhibits the following advantages over existing simulators:

- The actuator is robust and versatile and can be controlled remotely in-flight for replicating DC on geomaterials with variable drop heights, tamper shapes, tamper radii and tamper mass. Thus, the actuator can model the effects of both low-energy and high-energy DC processes adopted in the field within the controlled conditions of a geotechnical centrifuge.
- Adequate measures are taken by providing an assembly of guide rods to counter Coriolis effects in centrifuge and to prevent lateral shift of tamper.
- Another notable advantage of the actuator is its capability to model the time interval between successive drops during DC by regulating the drop frequency through a remotely operated motor. This facilitates monitoring of pore water pressure response in saturated soils subjected to DC.

16.5 Test Procedure and Model Materials

Centrifuge model tests were conducted at 30 gravities using the 4.5-m radius large beam centrifuge available at IIT Bombay, India. Details of the centrifuge facility are summarized briefly in Table 16.3. During discussion, model values have been referred, with corresponding prototype values within parenthesis. Model tests were conducted on loose granular soil deposits of 330 mm (9.9 m) depth subjected to DC under dry and saturated conditions. The model soil used is poorly graded Goa sand (SP) of specific gravity of 2.654, having a permeability of 1.85×10^{-4} m/s at a relative density (R.D.) of 35%. Detailed properties of sand are presented in Table 16.4.

A rigid container with internal dimensions of 720 mm × 450 mm × 410 mm provided with a front transparent perspex plate was used for model preparation and testing (Fig. 16.3). Permanent markers made of thin transparency sheets were pasted at fixed intervals to serve as reference points for subsequent GeoPIV analysis. The sand bed was prepared in a loose dry state at 35% R.D. by adopting the air pluviation technique. Details of model preparation are outlined in Kundu and Viswanadham (2021). Among the two tests conducted (Table 16.5), Model TC1

Table 16.3 Details of the large beam centrifuge facility at IIT Bombay

Sr. no.	Parameters	Details
1	Configuration type	Beam centrifuge
2	Radius	4.5 m (measured up to top surface of the basket from center of the shaft)
3	Radial acceleration range	10–200 g
4	Maximum pay load (at 100 g)	2.5 tons
5	Capacity	250 g-tons
6	Run-up time from 1 to 200 g	6 min
7	Model area	1.00 m × 1.2 m (up to 0.66 m height) 0.76 m × 1.2 m (up to 1.20 m height)

Table 16.4 Summary of properties for Goa sand

Properties	Unit	Soil A (sand)
Specific gravity (G_s)	[a]–	2.654
Sand (0.075–4.75 mm)	%	100
Effective particle size (D_{10})	mm	0.101
Average particle size (D_{50})	mm	0.191
Coefficient of uniformity (C_u)	[a]–	2.065
Coefficient of curvature (C_c)	[a]–	1.117
Maximum void ratio (e_{max})	[a]–	0.94
Minimum void ratio (e_{min})	[a]–	0.63
[b]Cohesion (c')	kPa	0
[b]Friction angle (ϕ')	°	32
[d]Coefficient of permeability at 35% relative density	m/s	[c]1.85×10^{-4}

[a]Not relevant
[b]CU test [35% R.D.]
[c]Average of three tests
[d]Constant head permeability test

Table 16.5 Summary of centrifuge tests conducted in the present study

Test legend	[a]Parameters varied	[a]Constant parameters
TC1–TC2	Depth of groundwater table d_w = Nil, 1.5 m	$N = 30$, $\gamma_{d,i} = 14.22$ kN/m^3, $m = 20.79$ t, $r = 1.2$ m, $H = 10$ m, $E = 208$ t-m, *soil type*: sand

N Gravity level; d_w depth of GWT from soil surface; *m* mass of tamper; *r* tamper radius; *H* drop height; *E* tamper energy in each blow corresponded to a dry model, whereas an initial water table was simulated prior to DC for Model TC2. The groundwater table was established in Model TC2 at normal gravity (1 g) by employing the bottom to top flow method. Standpipes connected to the model container ensured maintenance of desired water head at a depth of 50 mm (1.5 m at 30 g) from the model surface. The mass of semicylindrical tamper used in centrifuge tests was 0.385 kg (10.395 t). Due to axial symmetry, only half of the tamper was modeled, thereby replicating a cylindrical tamper of 0.77 kg (20.79 t) mass. An energy level of 7.7 kg-m (208 t-m) was simulated, thereby replicating high-energy DC process. A total of 16 drops were delivered on the soil surface in each test, in view of the marginal increase in improvement depth observed beyond 16 blows, as per the numerical simulations of Gu and Lee (2002). The pore water pressure generation during DC and corresponding vibration levels induced at the ground surface were analyzed based on data recorded by PPTs and accelerometers, whereas the ground improvement induced by DC was ascertained through contours of displacements and volumetric strains.

16.6 Results and Discussion

The interpretation of centrifuge model tests through instrumentation data and GeoPIV analysis on in-flight images are discussed in this section and summarized in Table 16.6.

16.6.1 Crater Profiles Induced by DC

The crater profiles for dry sand (Model TC1) and saturated sand (Model TC2) with increasing distance from tamper center are presented in Fig. 16.4 corresponding to the 1st, 4th, 8th and 16th blows. The maximum crater depth is observed to be about 38 mm (1.15 m) in Model TC1 and 53 mm (1.6 m) in Model TC2. In addition, the

Table 16.6 Summary of centrifuge test results

Test legend	d_w (m)	$d_{i,e}$ (m)	$(d_c)_{max}$ (m)	[b]PGA (g)		[b]PGV (mm/s)	
				Radial	Vertical	Radial	Vertical
TC1	[a]–s	5.59	1.16	0.191	0.284	32.8	18.2
TC2	1.5	5.03	1.58	0.204	0.233	27.0	17.1

All tests were conducted at gravity level of 30 g (N = 30)
Note All values are reported in prototype scale
[a]Dry soil sample
[b]Reported after 16th blow at 13.5 m from tamper center
d_w Depth of water table from soil surface; $d_{i,e}$ effective depth of improvement; $(d_c)_{max}$ maximum crater depth; PGA peak ground acceleration; PGV peak ground velocity

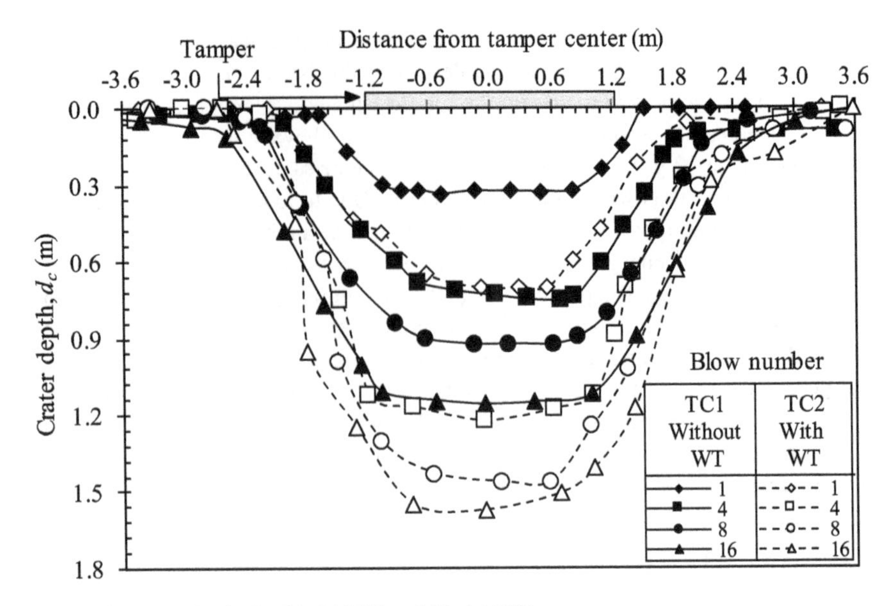

Fig. 16.4 Crater depths for Model TC1 and Model TC2

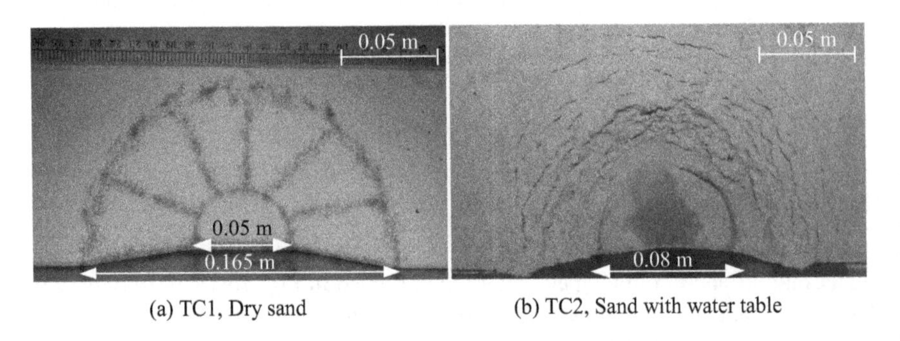

(a) TC1, Dry sand (b) TC2, Sand with water table

Fig. 16.5 Top view of crater induced by DC on various soil types

top view of crater observed during posttest investigations after completion of DC is shown in Fig. 16.5a, b. In general, well-defined crater surfaces are observed in dry soil (Model TC1) (Fig. 16.5a) with considerable collapse of crater boundaries. In comparison, the presence of moisture due to groundwater table in Model TC2 prevented collapse of crater boundaries (Fig. 16.5b), resulting in higher crater depths for the same blow number, as evident from Fig. 16.4.

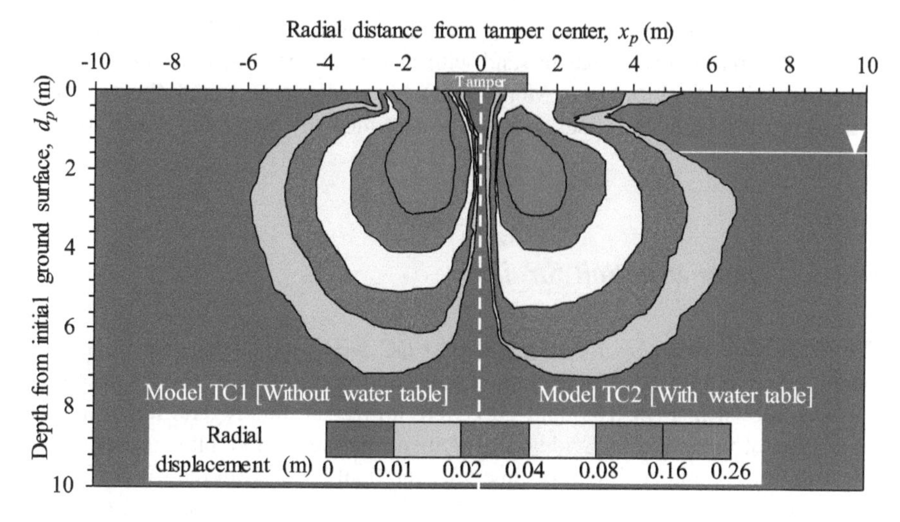

Fig. 16.6 Radial displacement contours for Models TC1 and TC2

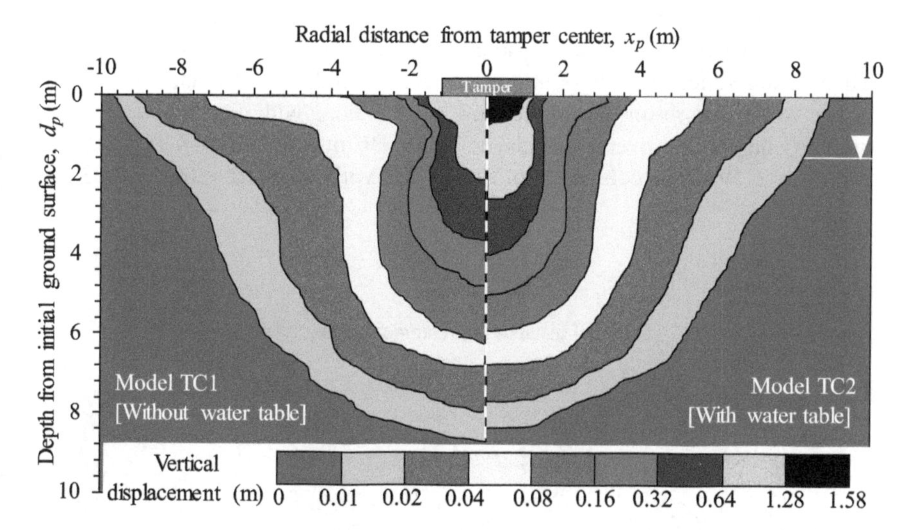

Fig. 16.7 Vertical displacement contours for Models TC1 and TC2

16.6.2 Displacement Contours

The radial and vertical displacement contours of Models TC1–TC2 at the end of 16 blows are presented in Figs. 16.6 and 16.7, respectively. The contours were derived from displacement vectors using GeoPIV software as per the procedure outlined in White et al. (2003). The displacement contours indicated considerable soil movement and associated disturbance in the vicinity of tamper, which reduced with depth and radial distance from the tamper center. The extent of the disturbed zone was found to be marginally higher in sand with water table (6.6 m in Model TC2) as compared to dry sand (6.0 m in Model TC1), especially in the radial direction. The reason is attributed to lesser resistance provided by saturated sand to soil displacement.

16.6.3 Volumetric Soil Strains

The volumetric strains (ε_v) within soil post-DC were studied for each centrifuge model test as an indication of the extent of ground improvement. The displacement contours shown in Figs. 16.6 and 16.7 were utilized in this regard, together with calculation of volume of individual soil elements before the first blow occurs on the soil surface (V_i), and at a point of time after the final blow (16th blow) is delivered (V_f). Using the above information, R.D. of soil after every blow was ascertained. In the present study, the depth of improvement (d_i) was considered as the thickness of soil strata measured from initial ground surface to a depth below which ΔRD is less than 10% (which corresponds to ε_v = 1.7%). Additionally, an effective depth of improvement ($d_{i,e}$) was defined measured from the base of crater. Numerically, $d_{i,e}$ is equal to difference of depth of improvement and depth of crater, and equal to (d_i–d_c). The volumetric strains (ε_v) plotted in Fig. 16.8 along with d_c, d_i and $d_{i,e}$ indicate marginally higher improvements induced during DC in case of Model TC1 with dry sand ($d_{i,e}$ = 5.59 m) as compared to Model TC2 with water table ($d_{i,e}$ = 5.03 m).

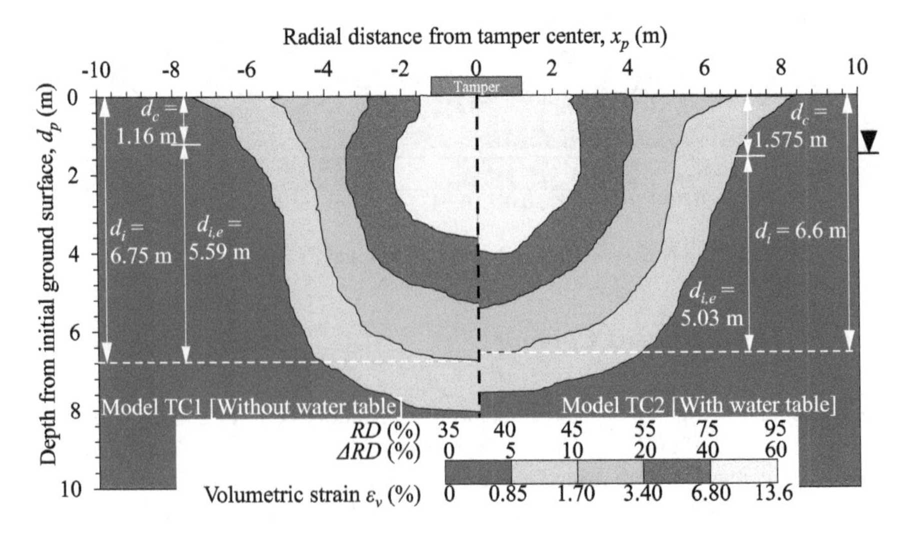

Fig. 16.8 Volumetric strain contours for Models TC1 and TC2

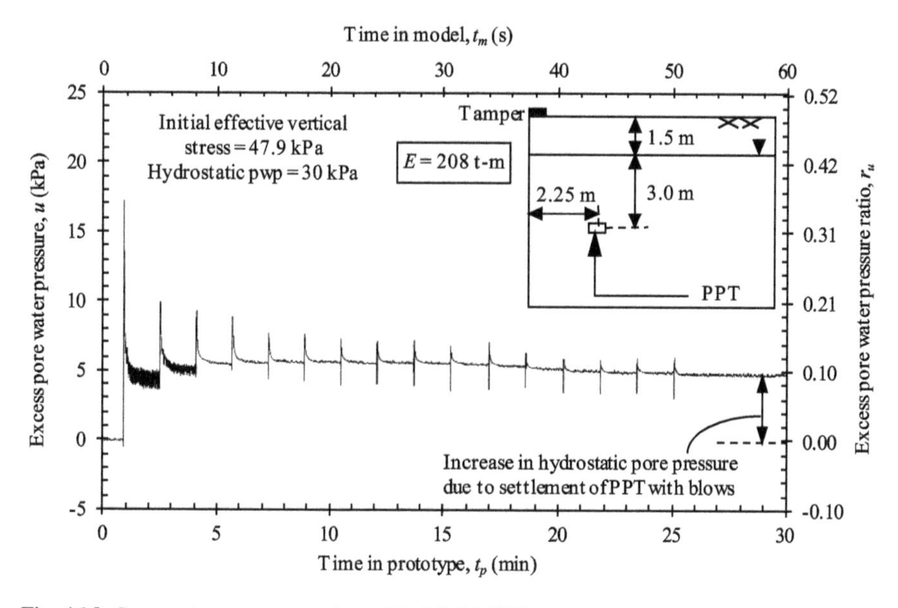

Fig. 16.9 Pore water pressures registered in Model TC2

16.6.4 Pore Water Pressure Developments

The pore water pressures developed in Model TC2 with successive blows are presented in Fig. 16.9 corresponding to a typical PPT placed at 100 mm (3.0 m) below the water table and at a radial distance of 75 mm (2.25 m) from tamper. The peak-induced pore water pressure was observed to be maximum after the 1st blow, which was in the magnitude of 17 kPa. In addition, the excess pore water pressure ratio (r_u) defined as the ratio between excess pore pressure and the total overburden pressure is presented in Fig. 16.9 as a measure of liquefaction potential of the soil. The peak r_u value was about 0.40 in Model TC2. Higher magnitudes of r_u beyond the observed limits could not be captured during centrifuge tests to prevent damage of PPTs placed close to the point of tamper drop.

16.6.5 Ground Vibrations Associated with DC

The peak ground accelerations (PGAs) and peak ground velocities (PGVs) induced during DC were investigated in the study based on data recorded by accelerometers.

The accelerometer data at the end of 4th blow and 16th blow are presented in prototype dimensions in Fig. 16.10a to analyze the PGA induced during tamper blows. The magnitude of PGA after 16th blow in dry sand [Model TC1] was 45 g (1.5 g) [radial] and 37 g (1.23 g) [vertical], respectively, and that in case of

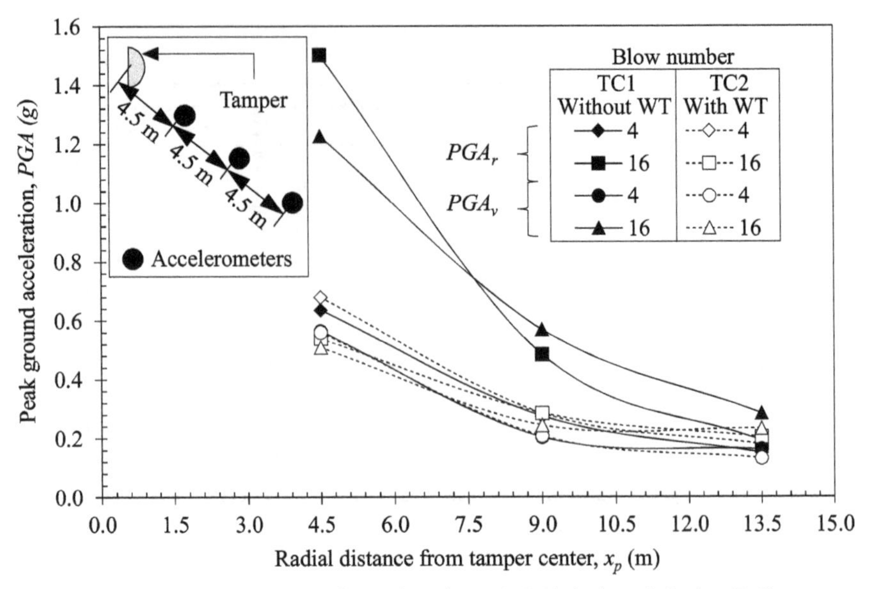

(a) Peak ground accelerations (PGA) induced during DC

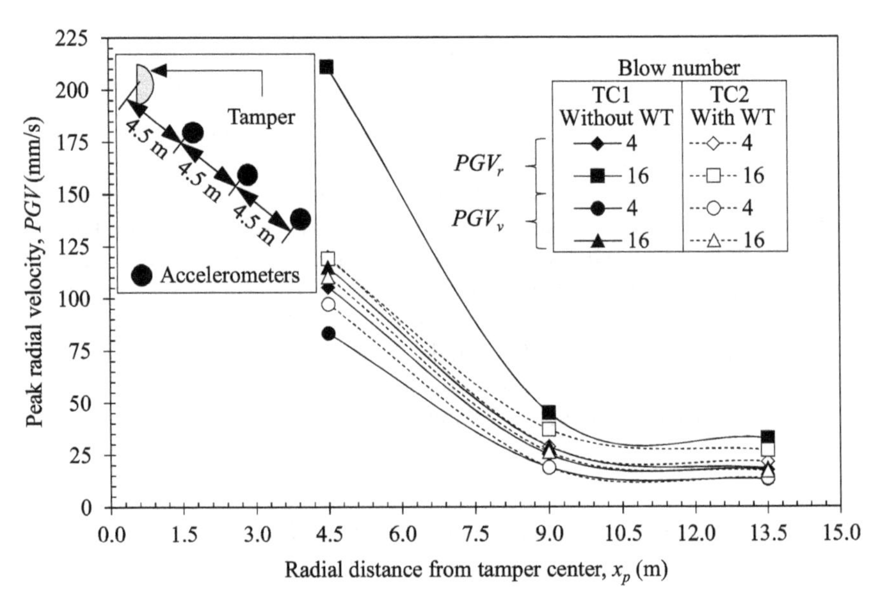

(b) Peak ground velocities (PGV) induced during DC

Fig. 16.10 Ground vibrations associated with DC

saturated sand [Model TC2] was about 20.4 g (0.68 g) [radial] and 16.8 g (0.56 g) [vertical]. In the next step, the ground velocities induced during DC were evaluated by integrating the area under the acceleration–time plots with the progress of DC. The resultant PGV values after 4th blow and 16th blow are presented in Fig. 16.10b. The peak radial velocity near tamper after 16th blow (PGV_r) was about 211 and 120 mm/s, respectively, in Model TC1 (dry) and Model TC2 (saturated). The corresponding peak vertical velocity (PGV_v) was about 112 mm/s in Model TC1 (dry) and Model TC2 (saturated). Thus, it can be observed that the radial component of vibrations was generally higher than corresponding vertical ones induced by DC. Further, the PGA and PGV induced during DC decreased in the presence of water table owing to damping effects in saturated soil.

16.7 Conclusions

The present paper discusses the development and key features of an actuator for simulating DC on geomaterials within the high-gravity environment prevailing in a geotechnical centrifuge. The governing laws, components of the actuator, model materials, model test package and instrumentation details are discussed explicitly. The actuator was employed to replicate high-energy DC process on sand with and without water table using the 4.5-m radius beam centrifuge at IIT Bombay, India. The primary findings are summarized herein:

- Based on GeoPIV analysis of selected images captured in-flight during experimentation, crater profiles, contours of soil displacement and volumetric strains were plotted to quantify the ground improvement induced by DC. The improvement depth was found to be comparable for both models, with marginally higher values in dry sand.
- The pore pressure magnitudes in sand with groundwater table peaked after 1st blow in tamper vicinity, which reduced gradually with successive blows.
- The corresponding ground vibrations induced by DC interpreted in terms of PGA and PGV values indicated that the radial component was higher than corresponding vertical ones. Further, the PGA and PGV induced during DC increased with successive blows and decreased in the presence of water table.

Based on the above, it can be inferred that the actuator can effectively model DC in dry and saturated soils within a geotechnical centrifuge. In addition, the developed actuator can provide an insight into the response of diverse field deposits subjected to DC, including dumped fills, hydraulically deposited fills, peats, collapsible soils, municipal solid waste landfills, dredged soils, reclaimed fills and so on.

References

Bhattacherjee D, Viswanadham BVS (2018) Design and performance of an inflight rainfall simulator in a geotechnical centrifuge. Geotech Test J ASTM 41(1):72–91

Bo MW, Na YM, Arulrajah A, Chang MF (2009) Densification of granular soil by dynamic compaction. Ground Improv 162(3):121–132

Feng SJ, Shui WH, Tan K, Gao LY, He LJ (2011) Field evaluation of dynamic compaction on granular deposits. J Performance Constr Facilities ASCE 25(3):241–249

Ghayoomi M, McCartney J, Ko HY (2011) Centrifuge test to assess the seismic compression of partially saturated sand layers. Geotech Test J ASTM 34(4):1–11

Gu Q, Lee FH (2002) Ground response to dynamic compaction of dry sand. Géotechnique 52 (7):481–493

Kundu S, Viswanadham BVS (2018) Numerical studies on the effectiveness of dynamic compaction using shear wave velocity profiling. Indian Geotech J Springer 48(2):305–315

Kundu S, Viswanadham BVS (2020) Numerical modelling of dynamic compaction induced settlement of MSW landfills. Int J Geomech ASCE 20(8):04020125–1:12

Kundu S, Viswanadham BVS (2021) Design and development of an in-flight actuator for modelling dynamic compaction in a geotechnical centrifuge. Geotech Test J ASTM 44(4):28 (published ahead of print)

Leonards GA, Cutter WA, Holtz RD (1980) Dynamic compaction of granular soil. J Geotech Eng ASCE 106(1):35–44

Lukas RG (1986) Dynamic compaction for highway construction, vol. 1: design and construction guidelines. Federal Highway Administration, Report No. RD-86/133, Washington D.C., pp 204–219

Lukas RG (1995) Geotechnical engineering circular no. 1—dynamic compaction. U. S. Federal Highway Administration, Report No.- FHWA-SA-95–037, Washington D.C.

Lutenegger AJ (1986) Dynamic compaction in friable loess. J Geotech Geoenviron Eng ASCE 112 (6):663–667

Madabhushi G (2014) Centrifuge modeling for civil engineers. CRC Press, USA

Mayne PW, Jones JS (1983) Impact stress during dynamic compaction. J Geotech Eng ASCE 109 (10):1342–1347

Mayne PW, Jones JS, E'Dinas DC (1984) Ground response to dynamic compaction. J Geotech Eng ASCE 110(6):757–774

Menard L, Broise Y (1975) Theoretical and practical aspects of dynamic consolidation. Géotechnique 25(1):3–18

Muraleetharan K, Granger K (1999) The use of miniature pore pressure transducers in measuring matric suction in unsaturated soils. Geotech Test J ASTM 22(3):226–234

Rollins KM, Rogers GW (1994) Mitigation measures for small structures on collapsible soils. J Geotech Eng ASCE 120(9):1533–1553

Schofield AN (1980) Cambridge geotechnical centrifuge operations. Géotechnique 30(3):227–268

Taylor RN (1995) Centrifuges in modelling: principles and scale effects. In: Taylor RN (ed) Geotechnical centrifuge technology. Blackie Academic and Professional (pubs.), Glasgow, UK

White DJ, Take WA, Bolton MD (2003) Soil deformation measurement using particle image velocimetry (PIV) and photogrammetry. Géotechnique 53(7):619–631

Yee K, Ooi TA (2010) Ground Improvement—a green technology towards a sustainable housing, infrastructure and utilities developments in Malaysia. Geotech Eng J SEAGS AGSSEA 41 (3):1–20

Zekkos D, Kabalan M, Flanagan M (2013) Lessons learned from case histories of dynamic compaction at municipal solid waste sites. J Geotech Geoenviron Eng ASCE 39(5):738–751

Zou WL, Wang Z, Yao ZF (2005) Effect of dynamic compaction on placement of high-road embankment. J Performance Constr Facilities ASCE 19(4):316–323

Chapter 17
A State of Art: Seismic Soil–Structure Interaction for Nuclear Power Plants

B. K. Maheshwari and Mohd. Firoj

17.1 Introduction

Safety of lifelines structures when subjected to strong excitation is utmost important as their failure may lead to disasters. Soil–structure interaction (SSI) is a design issue which cannot be neglected for the structures founded on the soft or loose soils and subjected to the strong ground motion. Failures of many structures occurred during the 1994 Northridge, California, the 1995 Kobe, Japan, the 2001 Bhuj and 2011 Fukushima, Japan earthquakes due to SSI or a related issue. The safety of nuclear power plants (NPPs) is very important during earthquakes as the failure of these structures may cause disaster.

Due to the strong ground motion shaking, the nonlinearity of soil plays a big role and further if loose sand, it may be subjected to liquefaction. The seismic loading on foundation is resisted by the foundation–structure interaction, thus, the response depends on the dynamic properties of soil, foundation material, rigidity of the structure, type of loading, etc. Numerical modeling is a feasible solution to deal with this problem which is normally carried out using finite element method. However, soil–structure interaction plays a big role in such analyses and design.

In this paper, a state of art on soil–structure interaction during earthquakes for nuclear power plants is presented. A critical review on the available literature is made and key issues are discussed. The background literature also includes those published by first author and his research group at IIT Roorkee. Recent advances in SSI for NPPs have been discussed.

B. K. Maheshwari (✉) · Mohd.Firoj
Department of Earthquake Engineering, IIT Roorkee, Roorkee, India
e-mail: bk.maheshwari@eq.iitr.ac.in

T. G. Sitharam et al. (eds.), *Latest Developments in Geotechnical Earthquake Engineering and Soil Dynamics*, Springer Transactions in Civil and Environmental Engineering, https://doi.org/10.1007/978-981-16-1468-2_17

17.2 Background of Study

There are several examples (Seed et al. 1989, 1992), which have clearly proved that buildings, bridges, dams, NPPs, offshore structures, pile foundations have been damaged in previous earthquakes due to unawareness of soil–structure interaction. The role of SSI was significant on the failure of Hanshin Expressway in 1995 Kobe Earthquake, Hanshin expressway having length of 630 m was collapsed and overturned (Maheshwari 2014). During 2001 Bhuj earthquake, the pier cap of a bridge was damaged due to SSI ignorance. During recent 2011 Fukushima, Japan earthquake, there was failure to nuclear power plants (NPPs) which were shut down. From these and other examples on damages in past earthquakes, a need to rationally incorporate soil–structure interaction in the design of structures was realized. Though research on SSI is in quite advanced stage, however, the importance of simplified analysis is not yet diminished as demonstrated by (Maheshwari 1997; Maheshwari and Watanabe 2006, 2009).

Seismic soil–structure interaction is very important for NPP like massive structures as compared to the building structures. Several researchers studied the numerical method for the soil-structure interaction problem (Spyrakos and Beskos 1986; Gazetas 1991; Wolf 1991; Kumar et al. 2015). The effects of SSI could be neglected at the hard rock condition having the average shear wave velocity greater than 1500 m/s (ASCE 2017). By considering the effect of soil–structure interaction, overall deformation demand of the structure may increase with the spectral displacement demand (Kwon and Einashai 2006). When the structure is embedded in a weak soil stratum, to reduce the structural deformation, the pile foundations are adopted to transfer the load from the structure to soil (Chore et al. 2012; Conte et al. 2013). In most of the earlier studies of SSI, a linear spring–dashpot model based on the rigid foundation on the half rigid space or 2D finite element method (plain strain) was considered (Firoj and Maheshwari 2018).

For the modeling of the unbounded soil domain, different boundary conditions are proposed by several researchers. In the direct method of SSI analysis, some of these boundaries are perfectly matched layers (Berenger 1994) and frequency-dependent Kelvin elements (Maheshwari 2003; Maheshwari et al. 2005; Maheshwari and Sarkar 2011; Sarkar and Maheshwari 2012a). In the substructure method of analysis, boundary used is consistent infinitesimal finite element cell method (Emani and Maheshwari 2009) and coupled FEM-SBFEM (Syed and Maheshwari 2014) for 3D SSI in time domain.

Wolf et al. (1981) studied the vertical and horizontal wave propagation seismic effect on pile of NPP structure. Wolf et al. (1983) studied the effect of horizontally propagating wave on the NPP structure resting on the very large basement of hard rock. Kumar (2013) studied the effect of embedment of NPP structure in the soil with three boundary conditions, i.e., elementary boundary, viscous boundary and Kelvin elements. Varma et al. (2015) studied the linear and nonlinear soil–structure interaction effect on the Fukushima Daichii nuclear power plant. They evaluate the source of nonlinearity and consider the gapping and sliding. Other nonlinear parameters were ignored for the simplicity. Kumar et al. (2015) studied the non-linear soil–structure interaction behavior of NPP structure by applying the bidirectional ground motion. Wang et al. (2017) studied the NPP building of finite element model with transmitting boundary and subjected to vertically incident seismic forces. They consider the effect of dimension of soil domain and types of artificial boundary. They also investigate the 10 MW (HTR-10) reactor building to compare the floor response spectrum with the fixed base reactor building (Firoj and Maheshwari 2018).

17.3 Objectives, Approach and Effects of SSI

For dynamic soil–structure interaction, there are various issues which need to be incorporated accurately. These are listed as follows:

(a) Modeling of truncated soil mass so as to absorb the reflecting wave.
(b) For dynamic loading, the frequency of excitation affects the soil behavior, therefore it needs to be taken care. The analysis shall be performed in such a way that the effect of interaction forces between foundation sub-system and structure sub-system is fully accounted.
(c) Material nonlinearity of soil using advanced constitute model.
(d) Liquefaction modeling in SSI problem.

For SSI analyses, various approaches can be grouped in three categories:

(i) Continuum models: Correctly represent the geometrical damping as well as inertia effects. However, soil is assumed elastic, therefore, nonlinearity cannot be considered.
(ii) Discrete models: Lumped mass, spring and dashpot models. Geometrical damping is difficult to consider but nonlinearity can be considered.
(iii) Finite element method: Overcome the limitation of the above two models. However, the computation cost is high.

In this paper, both simplified and rigorous models to deal with SSI are considered. Soil is a semi-infinite half-space, and a major problem in dynamic SSI is the modeling of the boundary condition of truncated soil mass. The fundamental objective of the SSI analysis is that dynamic response of both the structure and soil is to be calculated, taking into the effect of material damping and radiation damping (Wolf 1985). The presence of soil will modify the control motion leading to free-field motion (FFM). Effect of foundation (neglecting its inertia) leads to kinematic interaction and resulting motion is different than FFM. Finally considering inertia of the superstructure will lead to inertial effects (Maheshwari 2014).

The stepwise procedure for performing the SSI includes control motion, free-field motion, kinematic interaction and inertial interaction. SSI effect will increase with the increase in flexibility of soil and stiffness of the structure.

The American seismic code for nuclear structures (ASCE 4–1998) indicates that fixed base conditions can be assumed to apply when $V_s > 1100$ m/s. This condition is generally satisfied in weak rocks. Dowrick (1987) indicated that the fixed base conditions can be assumed for structures when $V_s > 20\,fh$, where h and f are the height and fundamental fixed-based frequency of the structure, respectively.

17.4 Geological Background of Nuclear Power Plants in India

The first series of power reactors in India were constructed at Tarapur and in Rajasthan. The Tarapur reactor is on a basaltic formation while the Rajasthan one is on the quartzitic sandstone. The raft foundations were provided under the reactor buildings. Settlement of the foundation was not considered if resting on the rock. The Kudankulam NPP structure site at Kalpakkam in Tamil Nadu has about 8 m of sandy soils overlying rocky strata. Being on the sea coast, the ground table is also at reasonably high level at all the time. Lightly loaded structures were supported on soil on spread footings while the heavy structures were supported on the raft resting on the competent rock. The turbine and service building were supported on the bored piles. At the Kakrapar site, the basaltic rock is encountered.

The situation of Narora Atomic Power plant is challenging as it is founded on 'very poor ground' condition in a region of high seismic zone. Settlement under the static and dynamic loads was considered in the choice of the type of foundation. An examination of the liquefaction potential of soil was also carried out in view of the high seismic condition, a fairly high water table and cohesionless deposits. Some of the upcoming NPPs in India (e.g., in Gorakhpur village in Haryana) will be founded on soft or alluvium soils making SSI analysis necessary.

17.5 Review of Numerical Modeling of NPPs

As very few experimental data are available on the soil-structure interaction problem for the nuclear power plants, many researchers worked for numerical modeling of NPPs, some of these are briefly discussed here.

Wolf et al. (1981) studied the NPP structure at Angra dos Reis in Brazil (supported on pile foundation). One quarter of the foundation was considered for the analysis due to the symmetry. The pile was modeled as beam element and surrounding soil is modeled with the springs as Winkler-type foundation. The stiffness of spring along the pile length was distributed as per the dynamic shear modulus of the soil layer. The superstructure was modeled as lumped mass. It was found that the response spectrum at the top of the layer was amplified by 5 times in horizontal seismic motion while 3 times in vertical seismic motion. They studied the shear force, bending moment and axial force along the pile length and found inertial interaction forces were predominant as compared to the kinematic interaction. Further, it was reported by Wolf et al. (1981) that the dynamic stiffness of the group pile cannot be determined by a single group pile. For the vertical earthquake motion, it was found that boundary pile exhibits the more axial force (approximately 1.5 times) as compared to the central pile while the distribution of horizontal displacement, bending moment and shear force along the length of pile was hardly effected by the location of pile.

Xu and Samaddar (2009) studied the effects of SSI and incoherency on seismic response analyses of NPP structures. The SSI model includes the superstructure, represented by lumped mass and beams and foundation, represented by brick elements, same as Wolf et al. (1981). For this, software SASSI (2010) was used for the study considering a surface-founded structure.

Maheshwari (2011) reviewed the advances for soil–pile interaction with reference to nuclear structures. The effect of embedment of foundation into the soil of NPP structure considering the effect of the slip and separation is studied by Saxena and Paul (2012).

Kumar (2013) studied the dynamic behavior of NPP structure in 2D coordinate system assuming behavior of soil as linear as well as nonlinear. The elastic properties of soil were considered for the linear behavior while the nonlinear behavior is modeled using most commonly used constitutive Mohr–coulomb model. Author used three types of boundaries for the unbounded soil domain, i.e., elementary, viscous and Kelvin element boundaries. The structure was modeled using the shell element in ABAQUS (2011). It was reported that the response of the NPP structure is increased by 67% when the rock was replaced by the soil at the base of the structure. The effect of soil–structure interaction was decreased by a margin of 45% when the nonlinearity of the soil is considered. The peak acceleration at the top of the structure was found more in case of elementary boundary condition as compared to viscous and Kelvin element boundary condition.

Desai and Choudhury (2015) studied the site-specific analysis of the nuclear power plants and ports in Mumbai at the four locations, i.e., JNPT, Mumbai Port, BARC and TAPS. Compatible site-specific input acceleration time histories were developed from a wavelet-based target spectra matching technique.

Varma et al. (2015) considered solid element for the modeling of SSI for NPP. The boundary used at soil domain was transmitting boundary in LS-DYNA. To simulate the free-field condition, first seismic motion was applied at the base of the soil at the selected node without considering the structure effect. Then, the resulted motion is applied at the base of the structure to consider the SSI effect on the structure. A comparison of the linear and nonlinear soil–structure interaction was made using same soil domain, boundary condition, structure and loading condition. It was observed that the maximum acceleration in the model in which the structure was surrounded by the nonlinear soil is reduced up to 49.8% due to nonlinearity.

Kumar et al. (2015) studied the Kudankulam NPP structure (supported on raft foundation) located at north of Kanyakumari in the Tamil Nadu state. The structure was modeled using the 4-nodded quadrilateral shell element having 6-degree-of-freedom system at each node. The modeling of base slab and raft foundation was carried out using shell element. The nonlinearity of the soil is considered by modeling the soil–foundation interface using the spring–dashpot at the bottom and vertical side of the raft foundation. It was reported that the fundamental period of vibration was increased by 10.4% considering SSI effect on the structure while the base shear in longitudinal and lateral direction was reduced by 21.7 and 24%, respectively.

The seismically induced uplift effects on nuclear power plants were studied by Sextos et al. (2017). It was concluded that in the presence of soft soil formations, nonlinear soil-foundation–structure interaction and associated geometric effects are possible. Wang et al. (2017) also carried out analysis for seismic SSI for NPPs.

A simplified soil–structure interaction model of embedded foundation is developed by many researchers (Wolf 1985; Gazetas 1991; AERB 2005). All these approaches assume that the soil is an elastic semi-infinite medium. The effect of soil is represented by dashpots, effective masses and lumped springs, and foundation is assumed massless and rigid. The values of soil spring static stiffness coefficients for a rigid plate on a semi-infinite homogeneous elastic half-space as per Gazetas (1991) and AERB (2005) are given in Table 17.1.

Where G, ρ, v are shear modulus, density, Poisson's ratio; A_b—Area of base, B, L and R—half-width, half-length of the circumscribed rectangle and radius of circular basement; I_{bx}, I_{by} and I_{bz} are the moment of inertia about x-direction, y-direction and z-direction, respectively.

After the FE analysis of model, it was concluded by many researchers that there is no significant effect on the results of refinement of mesh after a certain limit. Researchers suggested the boundary should be approximately at a distance more than 3 times the length and width of the structure used for the earthquake excitation. The limitation of FE model was not considering wider range of earthquake excitation such as high PGA. In case of smaller dimension (viscous boundary) of the

Table 17.1 Stiffness for a rigid plate on a semi-infinite homogeneous elastic half-space

Force system	Static stiffness, Gazetas (1991)	Spring constant for circular base, AERB (2005)
Vertical (z)	$K_z = \frac{2GL}{1-\upsilon}\left(0.73 + 1.54\left(\frac{A_b}{4L^2}\right)^{0.75}\right)$	$K_z = \frac{4GR}{(1-v)}$
Horizontal (y) lateral	$K_y = \frac{2GL}{1-\upsilon}\left(2.00 + 2.50\left(\frac{A_b}{4L^2}\right)^{0.85}\right)$	–
Horizontal (x) longitudinal	$K_x = K_y - \frac{0.2GL}{0.75-\upsilon}\left(1 - \frac{B}{L}\right)$	$K_x = \frac{32(1-v)GR}{7-8v}$
Rocking (r_x) about x-axis	$K_{rx} = \frac{GI_{bx}^{0.75}}{1-\upsilon}\left(\frac{L}{B}\right)^{0.25}\left(2.4 + 05\,\frac{B}{L}\right)$	$K_{rx} = \frac{8GR^3}{3(1-v)}$
Rocking (r_y) about y-axis	$K_{ry} = \frac{3GI_{bx}^{0.75}}{1-\upsilon}\left(\frac{L}{B}\right)^{0.15}$	$K_{ry} = \frac{8GR^3}{3(1-v)}$
Torsion	$K_t = 3.5GI_{bz}^{0.75}\left(\frac{B}{L}\right)^{0.4}\left(\frac{I_{bz}}{B^4}\right)^{0.2}$	$K_t = \frac{16GR^3}{3}$

soil domain, the peak value of acceleration is about 52% less than the larger dimension (free boundary) of the soil domain. If the dimension of the soil domain (viscous boundary) is increased up to a certain level, then this model shows no difference in peak value as compared to extended free boundary model, Firoj and Maheshwari (2018).

17.6 Recent Advances in SSI

In the last three decades, there has been significant progress in SSI studies. With the advancement in computer technology, it is now possible to model and analyze large SSI problems such as NPPs and dams more rigorously. Most of these advancement are made on the two important issues. First, accurate modeling of the unbounded soil is a complex issue. Second, the soil behavior is highly nonlinear during the strong ground motion shaking and in saturated loose sand there may be a chance of liquefaction. The numerical modeling of these two issues further complicates the SSI problem. Recent studies have been carried out or going on these SSI problem. The first author and his Ph.D. students (Emani, Sarkar, Syed and Firoj) worked broadly in the area of SSI.

Maheshwari (2003) used Kelvin elements (combination of spring and dashpot) at the boundary of unbounded soil domain and HiSS soil model for the nonlinearity of soil. Emani (2008) carried out nonlinear dynamic SSI analysis using CIFECM for boundary and hybrid methods for computation. Sarkar (2009) considered liquefaction for SSI analysis. Syed (2014) used scaled boundary finite element method (SBFEM) to model the boundary for nonlinear seismic SSI. Firoj (2018) is working to study nonlinear SSI for NPPs. A number of research publications from these works are listed in references and further discussed in following sections.

17.6.1 Modeling of Boundary

There are many ways to model the unbounded soil domain which can be grouped into following two categories.

Approximate Boundaries. The approximate boundaries are limited in space and time. Besides modeling the soil's stiffness up to infinity, reflection of upcoming wave to the boundary is to be avoided. Elementary boundaries are usually applicable in static analysis in which stresses and displacement are considered zero at the boundaries of the calculation domain (Ghosh and Wilson 1969). Local boundaries or viscous boundary (Lysmer and Kuhlemeyer 1969) are used to prevent the back-propagation of waves into a calculating domain. Kelvin element boundary absorbs the outgoing waves and prevents them to reflect back to structure. In this boundary, outer node is attached with spring–dashpot, and this boundary is used by Maheshwari et al. (2005).

Liao and Wong (1984) proposed transmitting boundary for the numerical simulation elastic wave propagation. This boundary is applicable to 3D linear SSI problem with a time-stepping algorithm and convex artificial boundary. The accuracy of this boundary can be improved by decreasing the length of time steps. For higher transmitting orders, a local transmitting boundary was presented in a compact form, which can be directly incorporated into finite element analysis (Liao and Liu 1992). In addition, the deviation problem on high-order transmission has been removed.

Bettess (1977) gives the infinite element approach to incorporate the shape functions analogues to Lagrange polynomial including the exponential decay term. This infinite element can be used in both explicit and implicit analyses. Some researchers used to couple the finite element with the infinite element to solve the SSI problems (Godbole et al. 1990; Noorzaei 1991), subjected to static loading. Wolf and Song (1996) combine the advantage of doubly asymptotic and multidirectional formulation, and it is highly accurate for plane wave at the intermediate frequency. Even with these approximate boundary conditions, accurate result can be obtained if the soil domain size is selected large enough. But, these boundaries may lead to inaccurate results for the inclined waves. Wolf (1994) described various formulations for modeling the foundation vibrations using simple physical models.

Rigorous Boundaries. The rigorous boundaries are global in space and time. These boundaries can simulate the far field effect, and therefore the unbounded soil domain is chosen such that the near-field effects are just-enclosed. In the frequency domain, the rigorous boundaries or consistent boundaries are formulated in the form of dynamic stiffness matrices.

Boundary Element Methods: In the early stage of SSI, rigorous boundary element method is extensively used by the various researcher (Brebbia et al. 1984; Beskos 1987; Banerjee 1994). The two main advantages of BEM, namely the reduction of soil domain size and high accuracy of the method, are more pronounced in linear electrodynamic, especially when the domain is semi-infinite. Domain type of BEM, such as FEM and FDM, requires the discretization of surface and interior of domain.

Consistent Boundaries: The consistent boundaries, also called thin layer method (Lysmer and Wass 1972; Waas 1972; Kausel and Roesset 1975; Kausel et al. 1975), are developed for the analysis of footing on the layered soil mass. This limit uses precise displacement functions in the horizontal direction that satisfy the radiation conditions and are an extension in the vertical direction used for the finite element method. This consistent boundary is based on finite element techniques and thus does not need a separate fundamental explanation. This method is well suitable to analyze the horizontal layers with varying material properties in the vertical direction.

Consistent Infinitesimal Finite Element Cell Method (CIFECM): As an alternative to the BEM, which applies analytic solutions to incorporate radiation damping, Emani and Maheshwari (2009) used the CIFECM boundary (Fig. 17.1) for the SSI problem and its application in pile foundation.

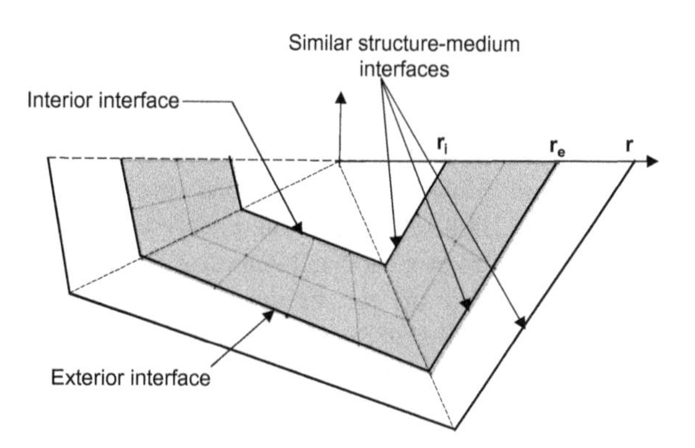

Fig. 17.1 Concept of CIFECM

Scaled Boundary Finite Element Method (SBFEM): This is an advanced version of CIFECM where spatial dimension at boundary is reduced by one, i.e., for a 3D problem only 2D boundary is required. This was used by Syed (2014) for nonlinear SSI as discussed later.

17.6.2 Nonlinearity of Soil

To analyze a nonlinear soil–pile interaction system in time domain, Matlock et al. (1978) have developed a unit load transfer curve also known as p-y curves. Various methods have been adopted to model the soil system in 1D such as Winkler models (Nogami and Konagai 1986, 1988; Nogami et al. 1992; Badoni and Makris 1996; El Naggar and Novak 1995, 1996), time-domain methods which are based on FEM formulations (Mylonakis and Gazetas 1999; Bentley and El Naggar 2000; Cai et al. 2000; Maheshwari et al. 2004a, b, 2005). However, in these methods, frequency dependence modeling of unbounded soil provided only approximate results, particularly in case of transient loading such as seismic loads. Nevertheless, the boundary conditions (local and transmitting) employed in these methods are not able to simulate the radiation boundary effects in case of oblique incidence of stress waves. However, these disadvantages of above methods can be improved using coupled finite element-boundary element (FE-BE) formulations.

Analytical analysis of linear and most equivalent linear systems using boundary integral methods (BIM) (Tajimi 1969; Kaynia and Kausel 1982; Mamoon et al. 1988; Fan et al. 1991; Miura et al. 1994; Maheshwari and Watanabe 2005) limited to steady-state response. To analyze a dynamic pile–soil interaction (DPSI), direct and transformed time-domain models have been proposed by Mamoon and Banerjee (1992). Cheung et al. (1995) have used boundary element method (BEM) in their model to evaluate the response of a single pile for horizontal excitation. Feng et al. (2003) employed time-domain BEM to evaluate the dynamic response of cylinder embedded in soil frictional slip at the interface. To capture the realistic nonlinear (3D inelastic) dynamic response of the frequency dependence of the numerical domain is to be accounted for, while considering the temporal and 3D spatial variations of numerical domain's response. To incorporate these requirements, coupled time–frequency domain and FE-BE coupling are employed. The authors have shown the advantages of FE-BE coupled model to evaluate the elastic (Emani and Maheshwari 2009) and cyclic nonlinear behaviors (Emani and Maheshwari 2008) of the 3D soil–pile systems.

Emani (2008) proposed and demonstrated a hybrid framework of analysis. This method is based on the hybrid frequency–time-domain (HFTD) formulation described by Wolf (1988). To satisfy the radiation conditions, the CIFECM (Wolf and Song 1996) is used and to account for material nonlinearity of soil, HISS model (Desai 2001) is employed (Maheshwari and Emani 2015).

Syed (2014) employed SBFEM to deal with boundary and used HiSS soil model for nonlinearity. Syed and Maheshwari (2017) and Maheshwari and Syed (2016) reported use of coupled FEM-SBFEM for nonlinear SSI for a soil–pile system. Syed and Maheshwari (2014, 2015) demonstrated the coupling and improvement in computational efficiency while using coupled FEM-SBFEM for nonlinear SSI analysis.

Firoj (2018) with nonlinear SSI for NPPs is reported by Firoj and Maheshwari (2018). The work is being extended for combined pile-raft foundation (CPRF) for NPPs as reported by Firoj and Maheshwari (2020).

17.6.3 SSI in Liquefiable Soil

Iida (1998) has used hypothetical Guerrero earthquake to analyze a 3D nonlinear soil–building interaction for various types of low to high-rise buildings and local sites effects are also incorporated. The results seem consistent with damage pattern observed in the Michoacan earthquake (1985). Wilson (1998) has evaluated the dynamic response of pile foundation in liquefying sand and soft clay for strong shaking in his doctoral thesis. Finn and Fujita (2002) reported the analysis and design result for pile foundations in liquefiable soils. Liyanapathirana and Poulos (2004) evaluated the effect of earthquake loading on liquefaction potential of soil deposits.

Liyanapathirana and Poulos (2005) have analyzed a piled system on Winkler foundation on liquefying soil with dynamically loaded beam. With the aid of Seed et al. (1976), Maheshwari et al. (2008) have reported response of pile foundation system for vertical loading with the effect of liquefaction phenomena. To evaluate the liquefaction potential of soil, Li et al. (2006) have performed shaking tests on the soil–pile–structure system. The experimental results compared with analytical solutions with equivalent linear soil model.

A five-story building that has tilted northeastward due to serious pile damage during the 1995 Kobe earthquake was studied by Uzuoka et al. (2007). Three-dimensional study was performed with elastoplastic soil medium. Soil–water coupled analysis was performed for soil–pile–building model. Sarkar (2009) investigated the three-dimensional soil–pile behavior under dynamic condition for the soil with liquefaction. The readers are referred to Maheshwari and Sarkar (2011, 2012), Sarkar and Maheshwari (2012a; b), Syed (2014), Syed and Maheshwari (2014, 2015), Maheshwari and Emani (2015), Maheshwari and Syed (2015).

17.7 Software Package for NPP Modeling

There is various software available for the finite element modeling of NPP structure on the soil.

17.7.1 SASSI

In many developed countries, System for Analysis of Soil-Structure Interaction (SASSI), a frequency-dependent program, is used to solve the 3D seismic SSI problem of NPPs. This enable the analyses of (a) wave propagation through the soil medium, (b) strain-dependent modulus reduction and damping property of soil and (c) input motion in free field using deconvolution procedure. Tabatabaie (2010) presents the recent development in the numerical modeling of NPP by considering the SSI effect using the SASSI (2010) program. The author presented the effect of foundation mesh refinement, effect of foundation embedment and effect of foundation flexibility.

17.7.2 LS-DYNA

LS-DYNA is a nonlinear dynamic analysis platform to solve the time-domain problem. LS-DYNA has a large number of martial library of soil (simulation of hysteretic behavior of soil), liquefaction soil model (pore pressure generation and loss of effective confining pressure) and geometrical nonlinearity (slip and separation). Various researcher (Wilford et al. 2010; Varma et al. 2015) used this software to deal with the SSI problem of NPP structures. Researcher concluded that the nonlinear soil–structure analysis has great effect on the seismic response of NPP structure.

17.7.3 ABAQUS

ABAQUS (2011) has the advantage over the LS-DYNA and SASSI in terms of efficient modeling of structural components (raft, piles and superstructure). Various researcher (Sextos et al. 2017; Firoj 2018) used this software for the modeling of NPP structure on CPRF and raft foundation. Researcher concluded that boundary condition, and geometrical and material nonlinearity has great effect on the seismic response of NPP structure.

17.8 Summary and Conclusions

A state of art on seismic soil–structure interaction for nuclear power plants is presented in this paper. It can be inferred that the soil–structure interaction is an important issue to be considered in the design of NPPs. Design may be unsafe if this effect is neglected. Some of the key conclusions are as follows.

1. The accurate modeling of boundary is very important for SSI during earthquakes. Both simplified and rigorous boundaries are available. Their application is case by case.
2. Interaction effects, e.g., soil–pile interaction (SPI) for a single pile and pile–soil–pile interaction (PSPI) for a pile group are must to be considered.
3. During seismic loading, the complex dynamic stiffness is frequency-dependent, and this characteristic needs to be modeled properly both in frequency and time domains.
4. During dynamic loading, the behavior of soil is nonlinear and this needs to be considered in the analysis.
5. For NPPs, recently combined pile-raft foundation (CPRF) is being used particularly for soft/alluvium soil conditions. The analysis of CPRF considering SSI is though complex but very important for safety of these structures.

Acknowledgements The contents of this manuscript are not original one rather collected from the published work of the authors which include Maheshwari (2014), Firoj and Maheshwari (2018) and few more as cited in this manuscript.

References

ABAQUS (2011) Abaqus analysis user's manual, Minneapolis, Minnesota, Dassault Systèmes Simulia Corp., USA
AERB (2005) Seismic qualification of structures, systems and components of Pressurised Heavy Water Reactors (Report No. AERB/SG/D-23). Atomic Energy Regulatory Board
ASCE (4-1998) Standard for seismic analysis of safety-related nuclear structures. ASCE 4–98 American Society of Civil Engineers
ASCE (2017) Minimum design loads and associated criteria for buildings and other structures. Reston, VA: 2. American Society of Civil Engineers, pp 1–889
Badoni D, Makris N (1996) Nonlinear response of single piles under lateral inertial and seismic loads. Soil Dyn Earthq Eng 15:29–43
Banerjee PK (1994) The boundary element methods in engineering. McGraw-Hill, London

Bentley KJ, El Naggar MH (2000) Numerical analysis of kinematic response of single piles. Can Geotech J 37:1368–1382
Berenger JP (1994) A perfectly matched layer of the absorption of electromagnetic waves. J Comput Phys 114(2):185–200
Beskos DE (1987) Boundary element methods in dynamic analysis. Appl Mech Rev 40:1–23
Bettess P (1977) Infinite elements. Int J Numer Methods Eng 11:233–250
Brebbia CA, Telles JCF, Wrobel LC (1984) Boundary element techniques. Springer, Berlin
Cai YX, Gould PL, Desai CS (2000) Nonlinear analysis of 3D seismic interaction of soil-pile-structure system and application. Eng Struct 22(2):191–199
Cheung YK, Tharn LG, Lie ZX (1995) Transient response of single pile under harmonic excitation. Earthq Eng Struct Dynam 24:1017–1038
Chore HS, Ingle RK, Sawant VA (2012) Parametric study of laterally loaded pile groups using simplified FE models. Coupled Syst Mech 1(1):1–18
Conte E, Troncone A, Vena M (2013) Nonlinear three-dimensional analysis of reinforced concrete piles subjected to horizontal loading. Comput Geotech 49:123–133
Dasgupta G (1982) A finite-element formulation for unbounded homogeneous continua. J Eng Mech ASCE 49:136–140
Desai CS (2001) Mechanics of materials and interfaces: the disturbed state concept. CRC Press LLC
Desai SS, Choudhury D (2015) Site-specific seismic ground response study for nuclear power plants and ports in Mumbai. Nat Hazard Rev 16(4):04015002–04015013
Dowrick D (1987) Earthquake resistance design. John Wiley & Sons
Emani PK (2008) Nonlinear dynamic soil-structure interaction analysis using hybrid methods. PhD Thesis, Dept. of Earthquake Eng., Indian Institute of Technology, Roorkee, India
Emani PK, Maheshwari BK (2008) Nonlinear analysis of pile groups using hybrid domain method. In: Procedings of 12th international conference of IACMAG, Goa, India, Paper No. 1590
Emani PK, Maheshwari BK (2009) Dynamic impedances of pile groups with embedded caps in homogeneous elastic soils using CIFECM. Soil Dyn Earthq Eng 29(6):963–973
Fan K, Gazetas G, Kaynia AM, Kausel E, Shahid A (1991) Kinematic Seismic response of single piles and pile groups. J Geotech Eng ASCE 117(12):1860–1879
Feng Y-D, Wang Y-S, Zhang Z-M (2003) Time domain BEM analysis of dynamic response of a cylinder embedded in soil with frictional slip at the interface. Soil Dyn Earthq Eng 23:303–311
Finn WDL, Fujita N (2002) Piles in liquefiable soils: seismic analysis and design issues. Soil Dyn Earthq Eng 22:731–742
Firoj M (2018) Nonlinear seismic soil structure interaction for nuclear power plants. PhD thesis in progress, Dept. of Earthquake Engineering, IIT Roorkee since Jan 2018
Firoj M, Maheshwari BK (2018) A review on soil structure interaction for nuclear power plants. In: Proceedings of 16th symposium on earthquake engineering, IIT Roorkee
Firoj M, Maheshwari BK (2020) Linear Spring Constants Of Soil For Pile Groups For The Nuclear Power Plants. Accepted for publication in the proceedings of 7th ICRAGEE, IISc, Bangalore
Gazetas G (1991) Formulas and charts for impedances of surface and embedded foundations. J Geotech Eng ASCE 117(9):1363–1381
Gazetas G, Mylonakis G (1998) Seismic soil structure interaction: new evidence and emerging issues. Geotechnical Special publication No. 75, pp 1119–1174
Ghosh S, Wilson EL (1969) Analysis of axi-symmetric structures under arbitrary loading (EERC Report No. 69–10). University of California, Berkeley, USA
Godbole PN, Viladkar MN, Noorzaei J (1990) Nonlinear soil-structure interaction analysis using coupled finite-infinite elements. Comput Struct 36(6):1089–1096
Iida M (1998) Three-dimensional non-linear soil building interaction analysis in the lakebed zone of Mexico city during the hypothetical Guerrero earthquake. Earthq Eng Struct Dyn 27:1483–1502
IS: 1893–2016—Part 1: criteria for earthquake resistant design of structures: general provisions and buildings. Bureau of Indian Standards, New Delhi (2016)

Kausel E (1994) Thin-layer method: formulation in the time domain. Int J Numer Methods Eng 37:927–941

Kausel E, Roesset JM (1975) Dynamic stiffness of circular foundations. J Eng Mech Div ASCE 101:771–785

Kausel E, Roesset JM, Waas G (1975) Dynamic analysis of footings in layered media. J Eng Mech Div ASCE 101:679–693

Kaynia AM, Kausel E (1982) Dynamic behavior of pile groups. In: Proceedings of 2nd international conference on numerical methods in offshore piling, Austin, Texas, pp 509–532

Kumar V (2013) Dynamic soil-structure interaction for nuclear power plant. M. Tech. dissertation, Dept. of Earthquake Engineering, Indian Institute of Technology, Roorkee, India

Kumar S, Raychowdhury P, Gundlapalli P (2015) Response analysis of a nuclear containment structure with nonlinear soil–structure interaction under bi-directional ground motion. Int J Adv Str Eng (IJASE) 7(2):211–221

Kwon OS, Einashai AS (2006) Fragility analysis of RC bridge pier considering soil-structure interaction. In: Structures congress 2006: structural engineering and public safety, pp 1–10

Li P, Lu X, Chen Y (2006) Study and analysis on shaking table tests of dynamic interaction of soil-structure considering soil liquefaction. In: Proceedings of the 4th international conference on earthquake engineering, Taipei, Taiwan

Liao ZP, Wong HL (1984) A transmitting boundary for the numerical simulation of elastic wave propagation. Soil Dyn Earthq Eng 2:174–183

Liao ZP, Liu JB (1992) Numerical instability of a local transmitting boundary. Earthq Eng Struct Dyn 21:65–77

Liyanapathirana DS, Poulos HG (2004) Assessment of soil liquefaction incorporating earthquake characteristics. Soil Dyn Earthq Eng 24:867–875

Liyanapathirana DS, Poulos HG (2005) Seismic lateral response of piles in liquefying soil. J Geotech Geoenviron Eng ASCE 131(12):1466–1478

Lysmer J, Kuhlemeyer RL (1969) Finite dynamic model for infinite media. J Eng Mech Div 95 (4):859–878

Lysmer J, Waas G (1972) Shear waves in plane infinite structures. J Eng Mech Div ASCE 98:85–105

Maheshwari BK (1997) Soil-structure-interaction on the structures with pile foundations-a three dimensional nonlinear dynamic analysis of pile foundations. PhD Dissertation, Dept. of Civil Engineering, Saitama University, Japan Sept

Maheshwari BK (2003) Three-dimensional finite element nonlinear dynamic analyses for soil-pile-structure interaction in the time domain. Research report submitted to Mid America Earthquake Center (NSF). Washington University, St. Louis, Missouri, Dept. of Civil Engineering

Maheshwari BK (2011) Advances in soil-structure interaction studies. In: Proceedings of Post-SMiRT-21 conference seminar on advances in seismic design of structures, systems and components of nuclear facilities, NPCI, BARC, Mumbai, India

Maheshwari BK (2014) Recent advances in seismic soil-structure interaction. In: Proceedings of the indian geotechnical conference held in Kakinada, Andhra Pradesh, pp 2463–2477

Maheshwari BK, Watanabe H (2005) Dynamic analysis of pile foundations: effects of material nonlinearity of soil. Electr J Geotech Eng 10(E), Paper No. 0585

Maheshwari BK, Watanabe H (2006) Nonlinear dynamic behavior of pile foundations: effects of separation at the soil-pile interface. Soils Found Jpn Geotech Soc, 46(4):437–448, Paper No 3234

Maheshwari BK, Watanabe H (2009) Seismic analysis of pile foundations using simplified approaches. Int J Geotech Eng 3(3):387–404

Maheshwari BK, Sarkar R (2011) Seismic behavior of soil-pile-structure interaction in liquefiable soils: parametric study. Int J Geomech ASCE 11(4):335–347

Maheshwari BK, Sarkar R (2012) Effect of soil nonlinearity and liquefaction on seismic response of pile groups. Int J Geotech Eng 6(4):497–506

Maheshwari BK, Emani PK (2015) Three dimensional nonlinear seismic analysis of pile groups using FE-CIFECM coupling in hybrid domain and HiSS plasticity model. Int J Geomech ASCE 15(3):04014055-1-12

Maheshwari BK, Syed NM (2016) Verification of implementation of HiSS soil model in the coupled FEM-SBFEM SSI analysis. Int J Geomech ASCE 16(1):04015034-1-8

Maheshwari BK, Truman KZ, El Naggar MH, Gould PL (2004a) Three-dimensional finite element nonlinear dynamic analysis of pile groups for lateral transient and seismic excitations. Can Geotech J 41:118–133

Maheshwari BK, Truman KZ, El Naggar MH, Gould PL (2004b) 3-D nonlinear analysis for seismic soil-pile-structure interaction. Soil Dyn Earthq Eng (Elsevier) 24(4):345–358

Maheshwari BK, Truman KZ, Gould PL, El Naggar MH (2005) Three-dimensional nonlinear seismic analysis of single piles using finite element method: effects of plasticity of soil. Int J Geomech 1(35):35–44

Maheshwari BK, Nath UK, Ramasamy G (2008) Influence of liquefaction on pile-soil interaction in vertical vibration. ISET J Earthq Technol 45(1):1–13

Mamoon SM, Banerjee PK (1992) Time domain analysis of dynamically loaded single piles. J Eng Mech Div ASCE 118:14–160

Mamoon SM, Banerjee PK, Ahmad S (1988) Seismic response of pile foundations. Technical Rep. No. NCEER-88–003, Dept. of Civil Engineering, State Univ. of Newyork, Buffalo, N.Y

Miura K, Kaynia AM, Masuda K, Kitamura E, Seto Y (1994) Dynamic behavior of pile foundations in homogenous and non-homogenous media. Earthq Eng Struct Dyn 23:183–192

Mylanokis G, Gazetas G (1999) Lateral vibrations and internal forces of grouped piles in layered soil. J Geotech Geoenviron Eng ASCE 125(1):16–25

Mylonakis G, Gazaetas G (2000) Seismic soil structure interaction: detrimental or beneficial. J Earthq Eng 4(3):277–301

El Naggar MH, Novak M (1995) Nonlinear lateral interaction in pile dynamics. Soil Dyn Earthq Eng 14:141–157

El Naggar MH, Novak M (1996) Nonlinear analysis for dynamic lateral pile response. Soil Dyn Earthq Eng 15:233–244

Nogami T, Konagai K (1986) Time domain axial response of dynamically loaded single piles. J Eng Mech ASCE 112(11):1241–1252

Nogami T, Konagai K (1988) Time domain flexural response of dynamically loaded single piles. J Eng Mech ASCE 114(9):1512–1525

Nogami T, Otani J, Konagai K, Chen HL (1992) Nonlinear soil-pile interaction model for dynamic lateral motion. J Geotech Eng ASCE 118(1):89–106

Noorzaei J (1991) Non-linear soil-structure interaction in framed structures. PhD Thesis, Civil Engineering Department, University of Roorkee, Roorkee, India

Sarkar R (2009) Three dimensional seismic behavior of soil-pile interaction with liquefaction. PhD Thesis, Dept. of Earthquake Eng., Indian Institute of Technology, Roorkee, India

Sarkar R, Maheshwari BK (2012a) Effects of separation on the behavior of soil-pile interaction in liquefiable soils. Int J Geomech ASCE 12(1):1–13

Sarkar R, Maheshwari BK (2012b) Effect of soil nonlinearity and liquefaction on dynamic stiffness of pile groups. Int J Geotech Eng 6(3):319–329

SASSI: System for Analysis of Soil-Structure Interaction, Version 8.3, MTR &Associates, Inc., Lafayette, California, March (2010)

Saxena N, Paul DK (2012) Effects of embedment including slip and separation on seismic SSI response of a nuclear reactor building. Nucl Eng Des 247:23–33

Seed HB, Martin PP, Lysmer J (1976) Pore-water pressure changes during soil liquefaction. J Geotech Eng Div ASCE 102(GT4):323–346

Seed RB, Dickenson SE, Riemer MF, Bray JD, Sitar N, Mitchell J et al (1990) Permeability report on the principal geotechnical aspects of the October 17, 1989 Loma Prieta Earthquake. Report No. UCB/EERC-90/05

Seed RB, Dickenson SE, Mok CM (1992) Recent lessons regarding seismic response analyses of soft and deep clay sites. In: Proceedings of seminar on seismic design and retrofit of bridges, Univ. of California, pp 18–39

Sextos AG, Manolis GD, Athanasiou A, Ioannidis N (2017) Seismically induced uplift effects on nuclear power plants. Part 1: Containment building rocking spectra. Nucl Eng Des 318:276–287

Spyrakos CC, Beskos DE (1986) Dynamic response of flexible strip-foundations by boundary and finite elements. Soil Dyn Earthq Eng 5(2):84–96

Syed NM (2014) Nonlinear seismic soil-structure interaction using scaled boundary finite element method. PhD Thesis, Dept. of Earthquake Eng., IIT Roorkee

Syed NM, Maheshwari BK (2014) Modeling using coupled FEM-SBFEM for three dimensional seismic SSI in time domain. Int J Geomech ASCE 14(1):118–129

Syed NM, Maheshwari BK (2015) Improvement in the computational efficiency of the coupled FEM-SBFEM approach for 3D seismic SSI analysis in the time domain. Comput Geotech 67:204–2012

Syed NM, Maheshwari BK (2017) Nonlinear SSI analysis in time domain using coupled FEM-SBFEM for a soil-pile system. Géotechnique 67(7):572–580

Tabatabaie M (2010) Recent advances in seismic soil-structure interaction analysis of nuclear power plants

Tajimi H (1969) Dynamic analysis of a structure embedded in an elastic stratum. In: Proceedings of 4th world conference on earthquake engineering, chile association on seismology and earthquake engineering, Santiago, Chile, vol 3, pp 53–69

Uzuoka R, Sento N, Kazama M, Zhang F, Yashima A, Oka F (2007) Three-dimensional numerical simulation of earthquake damage to group-piles in a liquefied ground. Soil Dyn Earthq Eng 27:395–413

Varma AH, Seo J, Coleman JL (2015) Application of nonlinear seismic soil-structure interaction analysis for identification of seismic margins at nuclear power plants (Report No. INL/EXT-15–37382). Idaho National Laboratory, Idaho Falls

Waas G (1972) Linear two-dimensional analysis of soil dynamics problems in semi-infinite layered media. PhD Dissertation, Univ. of California, Berkeley, CA

Wang X, Zhou Q, Zhu K, Shi L, Li X, Wang H (2017) Analysis of seismic soil-structure interaction for a nuclear power plant (HTR-10). Sci Technol Nucl Install 2017:1–13

Willford M, Sturt R, Huang Y, Almufti I, Duan X (2010) Recent advances in nonlinear soil-structure interaction analysis using LS-DYNA. In: Proceedings of the NEA-SSI workshop

Wolf JP (1985) Dynamic soil structure interaction. Prentice-Hall Inc., Englewood Cliffs, N.J., USA

Wolf JP (1988) Soil-structure-interaction in the time domain. Prentice-Hall Inc., Englewood Cliffs, New Jersey

Wolf JP (1991) Classification of analysis methods for dynamic soil-structure interaction. In: International Conferences on Recent Advances in Geotechnical Earthquake Engineering and Soil Dynamics, pp 1821–1832

Wolf JP (1994) Foundation vibration analysis using simple physical models. PTR Prentice-Hall, Englewood Cliffs, NJ, USA

Wolf JP, Weber B (1982) On a matrix Ricatti equation of stochastic control. SIAM J Appl Mech Soc Ind Appl Math 6:681–697

Wolf JP, Deeks A (2004) Foundation vibration analysis: a strength-of-materials approach. Elsevier, Oxford

Wolf JP, Song C (1996) Finite element modeling of unbounded domain. John Wiley & Sons Ltd., Chichester

Wolf JP, Song C (2000a) The scaled boundary finite element method—a premier: derivations. Comput Struct 78:191–210

Wolf JP, Song C (2000b) The scaled boundary finite element method—a premier: solution procedures. Comput Struct 78:211–225

Wolf JPV, Arx GAD, Barros FCP, Kakubo M (1981) Seismic analysis of the pile foundation of the reactor building of the NPP Angra 2. Nucl Eng Des 65(3):329–341
Wolf JP, Obernhuber P, Weber B (1983) Response of a nuclear power plant on aseismic bearings to horizontally propagating waves. Earthq Eng Struct Dyn 11(4):483–499
Xu J, Samaddar S (2009) Case study: effect of soil-structure interaction and ground motion incoherency on nuclear power plant structures. In: ASME pressure vessels and piping conference, pp 369–377

Chapter 18
Seismic Stability of Slopes Reinforced with Micropiles—A Numerical Study

Priyanka Ghosh, Surya Kumar Pandey, and S. Rajesh

18.1 Introduction

Slopes are either a naturally available soil profile which can be seen in most of the hilly regions or an engineered structure to serve various construction projects. Be it a natural or human-made structure, it needs to be analyzed carefully, which remains a challenging task in the field of geotechnical engineering. The failure of slopes under any condition may lead to tremendous loss to the society, which advocates for improving the soil to enhance the stability of slopes. Out of various ground improvement techniques, micropiles can be adopted to enhance the stability of such slopes. Micropiles are generally found to be versatile for serving various functions such as seismic retrofitting and underpinning (Elaziz and Naggar 2014; Elkasabgy and Naggar 2007; FHWA 2005; Kyung et al. 2017; Sun et al. 2013). Geotechnical engineers frequently use the stability charts proposed by Taylor (1937, 1948) to analyze a slope under the static condition. Various theoretical solutions were also recommended by different researchers (Bishop 1955; Chen 1975; Janbu 1954; Michalowski 1995, 2002; Spencer 1967) to determine the FOS of a slope under the static condition. However, these theories are mainly confined to static condition. To incorporate the effect of an earthquake, the theories mentioned above can be modified by including the seismic inertial forces. Therefore, an investigation on the static and the seismic stability of a slope reinforced with micropiles demands serious attention. Mononobe–Okabe theory (Mononobe and Matsuo 1929; Okabe 1926) marked the beginning of an evolution of the seismic analysis using the pseudo-static (PS) approach. After that, several researchers explored the seismic stability of a slope using the PS approach, which did not consider the effect of shear (V_s) and primary (V_p) wave velocities in the analysis and generated conservative results. In order to overcome the constraints posed by the PS approach, the original

P. Ghosh (✉) · S. K. Pandey · S. Rajesh
Department of Civil Engineering, IIT Kanpur, Kanpur 208016, India
e-mail: priyog@iitk.ac.in

T. G. Sitharam et al. (eds.), *Latest Developments in Geotechnical Earthquake Engineering and Soil Dynamics*, Springer Transactions in Civil and Environmental Engineering, https://doi.org/10.1007/978-981-16-1468-2_18

pseudo-dynamic (OPD) approach (Choudhury and Nimbalkar 2005; Ghosh 2008; Ghosh and Kolathayar 2011; Nimbalkar et al. 2006; Steedman and Zeng 1990) is considered in the present study. The seismic stability of a slope reinforced with micropiles is evaluated using the limit equilibrium method (LEM) considering c-ϕ soil. The study is performed by assuming a circular slip surface passing through the toe of the slope. The effect of different parameters such as horizontal (k_h) and vertical (k_v) seismic acceleration coefficients, slope angle (i), angle of internal friction of the soil (ϕ), amplification factor (f_a) and angle of inclination of micropile (θ_b) on the stability of a slope is explored in terms of FOS. Under the seismic condition, the stability of a slope with micropiles is found to be affected less compared to that of a slope without micropiles.

18.2 Problem Definition

The stability of a slope is expected to get improved with the use of micropiles. However, the study on the effect of seismicity on the stability of micropile-reinforced slope is limited. In this study, a finite slope of height (H) and inclination (i) reinforced with vertical as well as inclined micropiles is considered under the static and the seismic conditions (Fig. 18.1). The mechanical properties of the soil include the internal friction angle (ϕ), cohesion (c) and unit weight (γ). The limit equilibrium method, coupled with the OPD approach, is adopted in the analysis. The slip surface is reasonably assumed to be circular (Fellenius 1936), which passes through the toe of the slope. The main objective is to determine the factor of safety of the slope reinforced with micropiles under both static and seismic conditions.

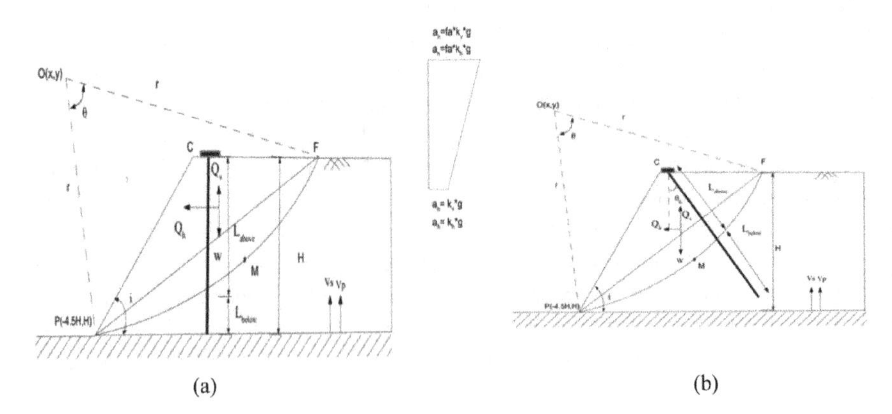

Fig. 18.1 Failure mechanism and associated forces with **a** vertical micropile and **b** inclined micropile

18.3 Assumptions

The following assumptions are made in the present study.

- The shear modulus of the soil is assumed to be constant throughout the height of the slope.
- The length of micropiles is considered to be uniform and always intersects the slip surface.
- The resistance of the pile cap is not considered in the analysis, and the micropile is assumed to be a fixed head pile.
- Location of the micropile is assumed at the top of the slope.
- Allowable displacement at the ground line is assumed 10% of the pile diameter (*d*) for the computation of the lateral capacity (Kyung and Lee 2018).
- The micropile is assumed to be a type-A-driven pile.

18.4 Methodology

18.4.1 Seismic Accelerations

In any earthquake event, the soil mass is subjected to seismic inertial forces developed due to the seismic accelerations. The easiest way to consider the seismic accelerations in the soil mass is the inclusion of uniform seismic acceleration coefficients throughout the soil body, as recommended by the PS approach. However, in reality, the seismic waves generated from any seismic event need not be in the same phase throughout the soil body. Moreover, these waves generally get amplified near the free surface. The phase change and amplifying nature of the seismic waves can be captured by the OPD approach, as proposed by Steedman and Zeng (1990). The OPD approach was also supported by a series of centrifuge experiments (Zeng and Steedman 1993). Considering these issues, the present investigation is performed using the OPD approach.

In the presence of a seismic excitation applied at the base of a slope, the soil mass at any depth (z) below the top surface and time (t) receives the horizontal (a_h) and the vertical (a_v) seismic accelerations, which can be expressed as

$$a_h(z,t) = \left[1 + \frac{H-z}{H}(f_a - 1)\right] k_h g \sin\left[\omega\left(t - \frac{H-z}{V_s}\right)\right] \tag{18.1}$$

$$a_v(z,t) = \left[1 + \frac{H-z}{H}(f_a - 1)\right] k_v g \sin\left[\omega\left(t - \frac{H-z}{V_p}\right)\right] \tag{18.2}$$

where H, f_a, V_s and V_p are the height of the slope, amplification factor, shear and primary wave velocity, respectively.

18.4.2 Stability Analysis with Vertical Micropiles

The stability of the slope is analyzed using the Fellenius method (Fellenius 1936), where the critical slip surface is obtained based on the minimum magnitude of the FOS. The mode of failure is considered as the toe failure, and hence, the circular slip surface always passes through the toe of the slope, as shown in Fig. 18.1a. The micropile of length L is placed vertically in such a way that it always intersects the slip surface. The horizontal (Q_h) and vertical (Q_v) seismic inertia forces are computed using the OPD approach as discussed earlier. The direction of Q_h and Q_v, as shown in Fig. 18.1a, is considered based on the recommendation in the literature. The forces acting on a micropile can be divided into two parts: axial and lateral forces. The axial force acting on a micropile is assumed to be equal to the axial capacity of the micropile at the limiting condition. The axial capacity of a micropile (P_{axi}) is generally governed by the geotechnical bond capacity and the structural capacity requirement. The allowable compressive load capacity of a micropile (P_G) based on the geotechnical bond requirement can be expressed as

$$P_G = \frac{\alpha_{\text{bond}} \pi d L_{\text{above}}}{\text{SF}} \tag{18.3}$$

where α_{bond} is the bond capacity between the pile and the soil, which depends on the type of pile; L_{above} is the length of the micropile above the slip surface, as shown in Fig. 18.1; and SF is the safety factor and generally taken as 2 as per FHWA (2005).

On the contrary, according to FHWA (2005), the allowable compressive load capacity of a type-A micropile (P_C) based on the structural requirement can be expressed as

$$P_C = 0.4 f_c A_{\text{grout}} + 0.47 f_y A_{\text{casing}} \tag{18.4}$$

where A_{grout} and A_{casing} are the cross-sectional area of the grout and the casing, respectively; and f_c and f_y are the compressive strength of the grout and the yield strength of the casing, respectively.

The axial capacity of a micropile (P_{axi}) is considered as the minimum of the capacity obtained from Eqs. 18.3 and 18.4. Similarly, the lateral capacity of a micropile (P_{lat}) can be determined based on the strength and the serviceability criteria (Murthy and Subba Rao 1995). Hence, by considering the equilibrium of forces, the FOS can be expressed as

$$FOS = \frac{(C_m + R \sin \phi) r + P_{\text{lat}} l_v + P_{\text{axi}} l_h}{Q_h \bar{y} + (W - Q_v) \bar{x}} \tag{18.5}$$

where C_m is the shear resistance mobilized along the slip surface, R is the reaction force exerted by the soil, r is the radius of the circular slip surface, $\bar{x}$ and $\bar{y}$ are the coordinates of the center of gravity of the failure wedge CFP with respect to the center of rotation O (Fig. 18.1), W is the self-weight of the failure wedge CFP, and l_v and l_h are the lever arms for the forces F_1 and F_2 respectively, where $F_1 = (P_{\text{lat}} - Q_h)$ and $F_2 = (P_{\text{axi}} + Q_v - W)$.

18.4.3 Stability Analysis with Inclined Micropiles

In case of an inclined micropile, the pile of length L is placed at a batter angle of θ_b with the vertical and passes through the circular slip surface as shown in Fig. 18.1b. Similar to a vertical micropile, the forces acting on an inclined micropile can be divided into two parts: axial and lateral forces. However, since the micropile is installed at a batter angle of θ_b, the axial force acts at an angle of θ_b with the vertical, whereas the lateral force is inclined at an angle, θ_b, with the horizontal. The axial capacity of an inclined micropile (P_{axi}) can be determined by following a similar procedure as mentioned for a vertical micropile. However, it is found to be challenging to predict the lateral capacity of an inclined micropile as the mobilization mechanism of the lateral resistance changes when the batter angle varies (Murthy and Subba Rao 1995; Reese and Welch 1975). Murthy and Subba Rao (Murthy and Subba Rao 1995) proposed a simplified approach to compute the lateral capacity of an inclined micropile (P_{lat}), where P_{lat} can be expressed based on the lateral capacity of a vertical micropile and the variation of the soil modulus. After determining the magnitude of P_{axi} and P_{lat} of an inclined micropile, the FOS for the slope can be determined from Eq. 18.5 just by replacing the respective parameters applicable to an inclined micropile.

18.5 Results and Discussion

Following the procedure, as discussed earlier, the numerical computations are performed by writing computer code in MATLAB. To obtain the minimum FOS, the value of t/T in the OPD approach and the location of the center of rotation (O) are varied, where T is the period of lateral shaking. The range of input parameters used in this study is given in Table 18.1.

The magnitudes of H/λ and H/η are chosen in such a way that $V_p/V_s = 1.87$, which is valid for most of the geological materials (Das 1993), where $\lambda = TV_s$ and $\eta = TV_p$. It is worth mentioning that H/λ and H/η represent the ratio of the time taken by the shear and the primary wave to travel the full height of the slope, respectively, to the period of lateral shaking (T).

The variation of FOS with k_h for a slope with vertical micropiles is presented in Fig. 18.2 for different values of ϕ. It can be seen that the FOS decreases

Table 18.1 Range of input parameters

Parameter	Range
ϕ	25–45°
f_a	1–1.6
k_h	0–0.2
k_v	0–k_h
i	30–45°
d	0.23–0.27 m
L/H	0.5–1
θ_b	−30–30°
c	5 kPa
γ	20 kN/m^3
f_c	27.6 MPa
f_y	552 MPa
α_{bond}	140 kPa

significantly with an increase in the magnitude of k_h and k_v. The recommended minimum static and seismic factors of safety for the micropiled structure as per FHWA (2005) are also presented in Fig. 18.2 just to show the limiting condition. The slope and the micropile parameters used in the analysis are given in Table 18.1.

The variation of FOS with L/H ratio for a slope with micropiles is shown in Fig. 18.3 for different values of ϕ and θ_b. It can be observed that the FOS increases with an increase in the magnitude of L/H ratio. This may be attributed to the fact that with an increase in the length of micropile, the length of micropile beyond the slip surface increases which offers higher pullout resistance due to the interaction between the grout and the soil. It can be also seen from Fig. 18.3b that the magnitude of FOS decreases with an increase in θ_b.

The variation of FOS with k_h for different values of H/λ and H/η is presented in Fig. 18.4. It can be observed that the FOS increases with an increase in the

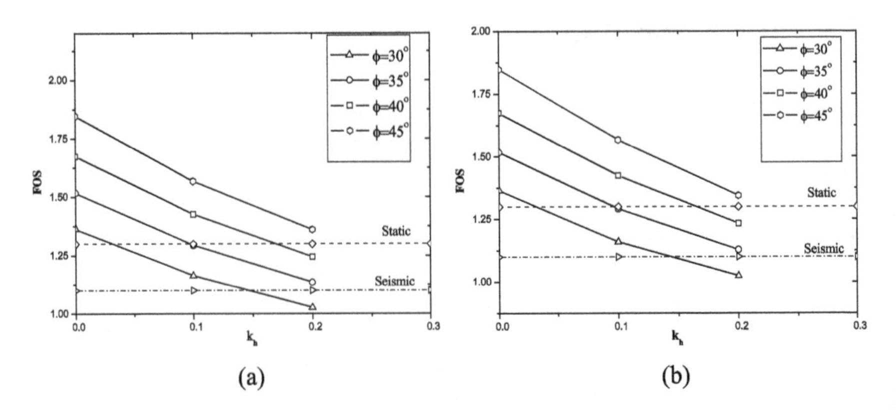

Fig. 18.2 Variation of FOS with k_h for different values of ϕ with i = 30°, d = 0.23 m, f_a = 1, H/λ = 0.3, H/η = 0.16 and L/H = 1. **a** k_v = 0.5k_h and **b** k_v = k_h

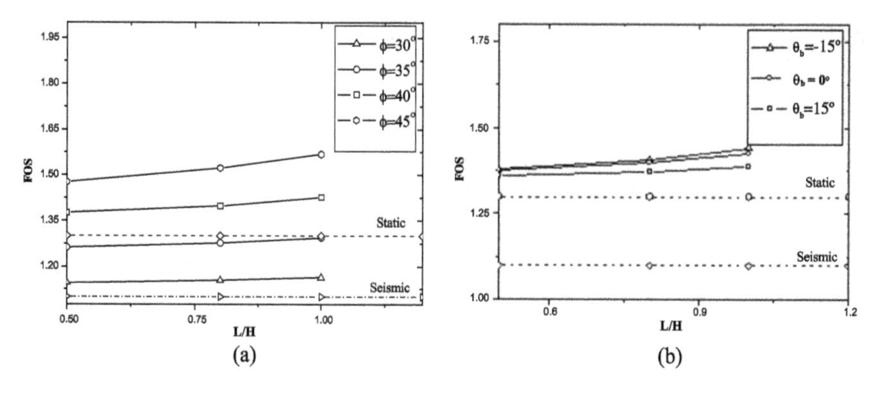

Fig. 18.3 Variation of FOS with L/H for **a** $\theta_b = 0°$ and **b** $\phi = 40°$ with $i = 30°$, $d = 0.23$ m, $f_a = 1$, $H/\lambda = 0.3$, $H/\eta = 0.16$, $k_h = 0.1$ and $k_v = 0.5k_h$

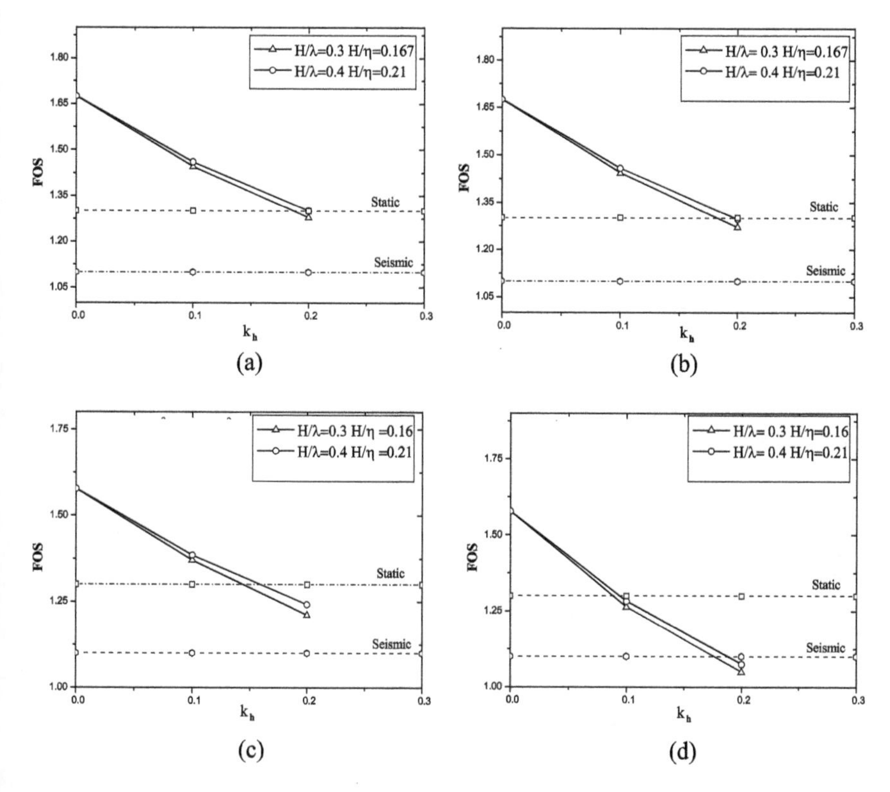

Fig. 18.4 Variation of FOS with k_h for different values of H/λ and H/η with $\phi = 40°$, $i = 30°$, $d = 0.23$ m, $f_a = 1$ and $L/H = 1$. **a** $\theta_b = 0$, $k_v = 0$, **b** $\theta_b = 0$, $k_v = 0.5k_h$, **c** $\theta_b = 15°$, $k_v = 0$, and **d** $\theta_b = 15°$, $k_v = 0.5k_h$

magnitude of H/λ and H/η. This may be attributed to the fact that with increase in H/λ and H/η, the velocity of shear and primary waves decreases, and thus, it reduces the effect of an earthquake.

The variation of FOS with k_h is shown in Fig. 18.5 for different values of i. It can be noted that the FOS decreases considerably with an increase in the magnitude of slope angle. This may be attributed to the fact that with an increase in i, the stability of a slope decreases which results in the reduction in FOS.

The variation of FOS with k_h is presented in Fig. 18.6 for different values of amplification factor (f_a). It can be seen that the FOS decreases with an increase in the magnitude of f_a. It may be attributed to the fact that with an increase in the amplification factor, the amplitude of acceleration increases, which in turn increases the seismic forces and, hence, the value of FOS decreases.

The variation of FOS with batter angle (θ_b) is presented in Fig. 18.7 for different values of ϕ under both static and seismic conditions. It is worth noting that positive

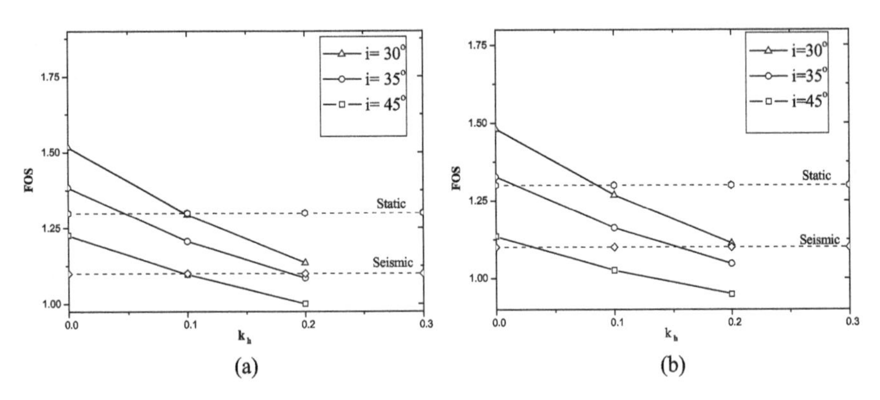

Fig. 18.5 Variation of FOS with k_h for different values of i with $\phi = 35°$, $d = 0.23$ m, $f_a = 1$, $H/\lambda = 0.3$, $H/\eta = 0.16$, $L/H = 1$ and $k_v = 0.5k_h$. **a** $\theta_b = 0$ and **b** $\theta_b = 15°$

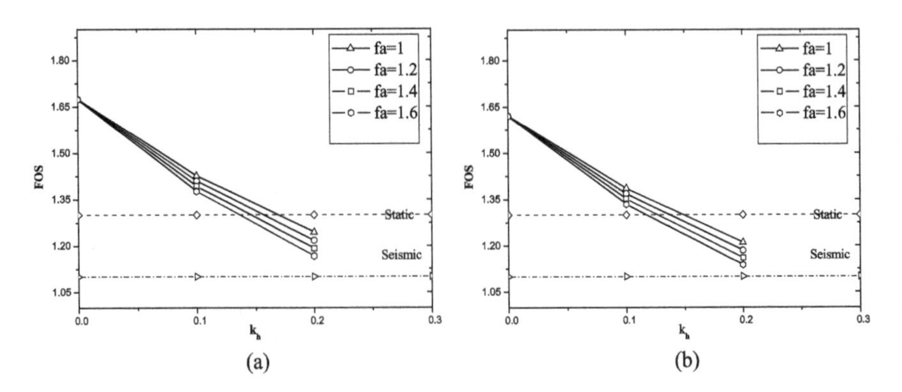

Fig. 18.6 Variation of FOS with k_h for different values of f_a with $\phi = 40°$, $i = 30°$, $d = 0.23$ m, $H/\lambda = 0.3$, $H/\eta = 0.16$, $L/H = 1$ and $k_v = 0.5k_h$. **a** $\theta_b = 0$ and **b** $\theta_b = 15°$

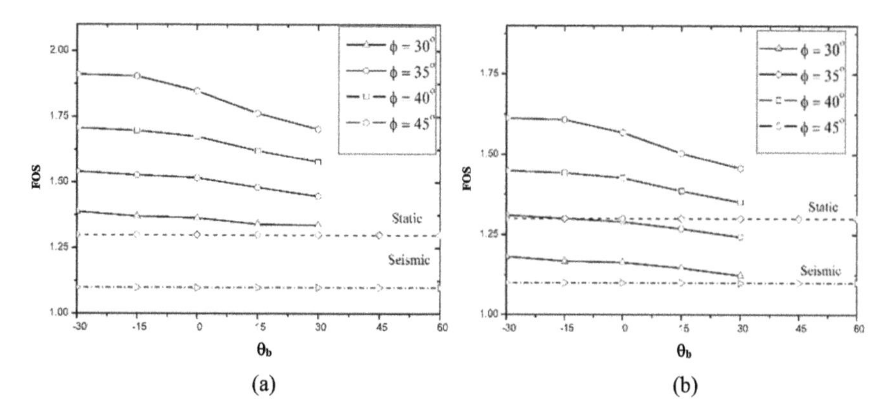

Fig. 18.7 Variation of FOS with θ_b for different values of ϕ with i = 30°, d = 0.23 m, f_a = 1, H/λ = 0.3, H/η = 0.16, L/H = 1 and $k_v = 0.5k_h$. **a** k_h = 0 and **b** k_h = 0.1

θ_b implies the angle between the axis of the micropile and the vertical direction in anticlockwise direction, whereas negative θ_b implies the angle in clockwise direction. It can be observed from Fig. 18.7 that the FOS decreases with an increase in the magnitude of θ_b. However, the reduction in the value of FOS is not found to be significant up to $\theta_b = -15°$. It may be attributed to the fact that the mobilized length (effective length) of the micropile above the slip surface decreases with an increase in the batter angle and, thus, there exists a reduction in the axial resistance.

In Fig. 18.8, the variation of FOS with k_h is presented for different values of micropile diameter (d). It can be observed from Fig. 18.8 that the FOS increases with an increase in the magnitude of d. This may be attributed to the fact that the axial and the lateral resistances of a micropile increase with an increase in the diameter of micropile.

Fig. 18.8 Variation of FOS with k_h for different values of d with i = 30°, ϕ = 35°, θ_b = 15°, f_a = 1, H/λ = 0.3, H/η = 0.16, L/H = 1 and $k_v = 0.5k_h$

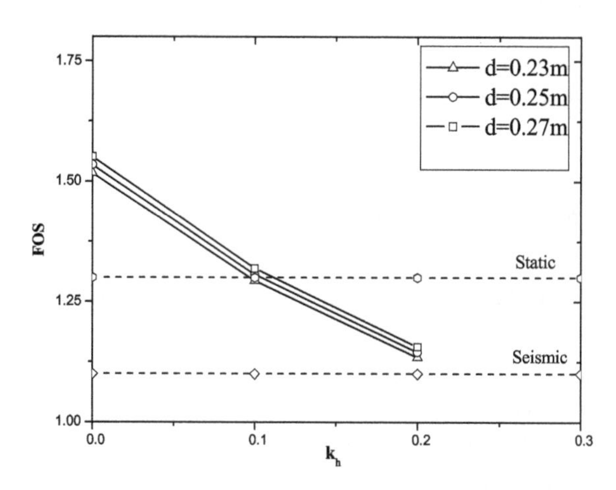

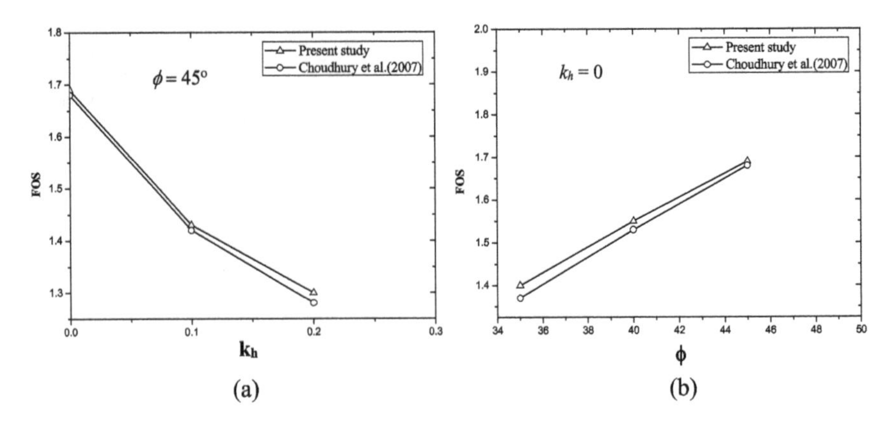

Fig. 18.9 Comparison of FOS for different values of **a** k_h and **b** ϕ with $i = 30°$, $d = 0.23$ m, $H = 10$ m, $c = 5$ kPa, $f_a = 1$, $H/\lambda = 0.3$, $H/\eta = 0.16$, $L/H = 1$ and $k_v = 0.5k_h$

18.6 Comparison

Studies on the seismic stability of a slope reinforced with micropiles are limited in the literature. Majority of the investigations available in the literature address the seismic slope stability analysis using the pseudo-static approach, which are unable to capture the time history and the phase effect of seismic accelerations. However, the present slope stability analysis was carried out by assuming a circular slip surface along with the original pseudo-dynamic approach in the presence of micropiles. Hence, an effort is made to obtain the seismic stability of a conventional slope without micropiles using the original pseudo-dynamic approach and compare the results with that obtained from the pseudo-static analysis available in the literature. In Fig. 18.9, the present results obtained for a slope without micropiles are compared with that reported by Choudhury et al. (2007) for different values of k_h and ϕ. It can be noticed that the present results compare reasonably well with that reported by Choudhury et al. (2007).

18.7 Conclusions

The following conclusions can be made from the results obtained from the present analysis.

- The FOS for a slope reinforced with micropiles decreases with an increase in the magnitude of k_h and k_v. Under both static and seismic conditions, the magnitude of FOS is found to increase by about 11% with an increase in ϕ from 30 to 45° at an interval of 5°.

- The FOS for a slope reinforced with micropiles increases with an increase in the length of micropiles. The magnitude of FOS is found to increase by 25% when the *L/H* ratio increases roughly by 25%. The enhancement in the FOS becomes more pronounced at a higher value of ϕ.
- The magnitude of FOS decreases with an increase in f_a but increases with an increase in H/λ and H/η.
- Under both static and seismic conditions, an increase in the diameter of micropiles from 0.23 to 0.27 m results around 3% higher FOS.
- Under the seismic condition, inclined micropiles are found to be more effective than vertical micropiles. The FOS generally decreases with an increase in the batter angle of micropiles.
- The FOS for a slope reinforced with micropiles is found to be conservative for the pseudo-static approach compared to the original pseudo-dynamic approach.

References

Bishop AW (1955) The use of the slip circle in the stability analysis of slopes. Geotechnique 5 (1):7–17

Chen WF (1975) Limit analysis and soil plasticity. Elsevier Science, Amsterdam, Netherlands

Choudhury D, Nimbalkar SS (2005) Seismic passive resistance by pseudo-dynamic method. Geotechnique 55(9):699–702

Choudhury D, Basu S, Bray JD (2007) Behaviour of slopes under static and seismic conditions by limit equilibrium method. In: Proceedings of Geo-Denver 2007: new peaks in geotechnics, Denver, Colorado, pp 1–10

Das BM (1993) Principles of soil dynamics. PWS-KENT, Boston, USA

Elaziz AYA, El Naggar MH (2014) Group behaviour of hollow-bar micropiles in cohesive soils. Can Geotech J 51(10):1139–1150

Elkasabgy MA, El Naggar MH (2007) Finite element analysis of the axial capacity of micropiles. 8th ISM Workshop, Toronto, ON, Canada

Fellenius W (1936) Calculation of stability of earth dams. In: Proceedings of 2nd Congress Large Dams, vol 4. Washington, USA, pp 445–462

FHWA (2005) Micropile design and construction. U.S. Department of Transportation, Federal Highway Administration, FHWA NHI-05-039

Ghosh P (2008) Seismic active earth pressure behind a non-vertical retaining wall using pseudo-dynamic analysis. Can Geotech J 45(1):117–123

Ghosh P, Kolathayar S (2011) Seismic passive earth pressure behind non vertical wall with composite failure mechanism: pseudo-dynamic approach. Geotech Geol Eng 29(3):363–373

Janbu N (1954) Application of composite slip surfaces for stability analysis. In: Proceedings of European conference on stability of earth slopes, Stockholm, Sweden, pp 43–49

Kyung D, Kim G, Kim D, Lee J (2017) Vertical load-carrying behavior and design models for micropiles considering foundation configuration conditions. Can Geotech J 54(2):234–247

Kyung D, Lee J (2018) Interpretative analysis of lateral load–carrying behavior and design model for inclined single and group micropiles. J Geotech Geoenviron Eng 144(1):04017105(1–11)

Michalowski RL (1995) Slope stability analysis: a kinematical approach. Geotechnique 45 (2):283–293

Michalowski RL (2002) Stability charts for uniform slopes. J Geotech Geoenviron Eng 128 (4):351–355

Mononobe N, Matsuo H (1929) On the determination of earth pressures during earthquake. In: Proceedings of world engineering conference, Tokyo, Japan, pp 177–185
Murthy VNS, Subba Rao KS (1995) Prediction of nonlinear behavior of laterally loaded long piles. In: Foundation Engineer, vol 1(2), New Delhi, India
Nimbalkar SS, Choudhury D, Mandal JN (2006) Seismic stability of reinforced-soil wall by pseudo-dynamic method. Geosynth Int 13(3):111–119
Okabe S (1926) General theory of earth pressures. J Jpn Soc Civil Eng 12(6):1277–1323
Reese LC, Welch RC (1975) Lateral loading of deep foundations in stiff clay. J Geotech Eng Div 101(7):633–649
Spencer E (1967) A method of analysis of the stability of embankments assuming parallel inter-slice forces. Geotechnique 17(1):11–26
Steedman RS, Zeng X (1990) The influence of phase on the calculation of pseudo-static earth pressure on a retaining wall. Geotechnique 40(1):103–112
Sun SW, Wang JC, Bian XL (2013) Design of micropiles to increase earth slopes stability. J Cent South Univ 20(5):1361–1367
Taylor DW (1937) Stability of earth slopes. J Boston Soc Civil Eng 24(3):197–246
Taylor DW (1948) Fundamentals of soil mechanics. Wiley, New York, USA
Zeng X, Steedman RS (1993) On the behaviour of quay walls in earthquakes. Geotechnique 43 (3):417–431

Chapter 19
Deformation Modulus Characteristics of Cyclically Loaded Granular Earth Bed for High-Speed Trains

Satyendra Mittal and Anoop Bhardwaj

19.1 Introduction

The development of a high-speed rail network has become a priority for every developed and developing country. As it is a fact that the transportation sector is the backbone of every economy, efforts are being made on an everyday basis to develop and find a sustainable approach for high-speed trains. India has the fourth-largest rail network in the world and since India is a developing economy, a new and efficient design methodology is a must need, to meet the growing needs of the country. The existing rail network of India still works at comparatively slower speeds and the design methodology has not been changed in a quite some time and if we compare the old (GE-1) and new design code (GE-14) (2007) we can find that the height of the embankment has been increased with the inclusion of the blanket layers. This is a common practice in most of the developing countries where the height of the embankment is increased to meet growing load and speed requirements. However, most of Europe's rail design methodologies are based on the deformation modulus. Indian Railways have also adopted a similar approach of UIC 719R (2008) in establishing the design as indicated in the GE-14 code (2007) of Research Design and Standard Organization (RDSO). As the situation stands there are no solid provisions to upgrade the existing design methodology or any evidence supporting the suitability of embankment for higher speeds and especially for the existing rail network. So, there is an interest in improving the railway embankment for higher speed as well as economical designs that can be implemented easily in the field on new as well as existing tracks. Many researchers have investigated the improvement of railway embankment by using various approaches such as chemical stabilizers, ground treatments with lime or cement and more recently as reinforcements. The use of geogrids is becoming more and more prominent in roads and now in railways too. The very first major contribution in the

S. Mittal (✉) · A. Bhardwaj
Indian Institute of Technology (IIT), Roorkee 247 667, India

T. G. Sitharam et al. (eds.), *Latest Developments in Geotechnical Earthquake Engineering and Soil Dynamics*, Springer Transactions in Civil and Environmental Engineering, https://doi.org/10.1007/978-981-16-1468-2_19

field of reinforcement for railways came from Jain and Keshav (1999), where series of empirical full-scale tests were conducted with the use of geogrids as reinforcements. The geogrids were installed at the bottom and within the embankment. The results were of great significance as the type of loading used for the experiments was of dynamic in nature (as a function of axle loads). As per the results reported, a single layer of geogrid reduces 40–20% of dynamic loads and two layers of geogrid reduce 60–30% at a depth of 0.90 m. Similarly, Shin et al. (2002) reported a total reduction of 47% in the settlements when one layer of geogrid and one layer of geotextile were used and also reported the critical number of cycles 'N_{cr}' after which no further settlements takes place in railway embankment. Indraratana et al. (2005) further investigated deformation characteristics of railway ballast and formation of soil. Innotrack (2006) guidelines for subgrade reinforcement with geosynthetics describe in detail, the numerical, laboratory and field testing methods to improve the performance of ballast embankment. Static tests and lightweight deflectometer were used to calculate the stiffness of the embankment. Reported results showed the improvement of 15% in deformation modulus in the presence of geogrids when placed under the ballast instead of sub-ballast. In this article, results of cyclic plate load test as per DIN 18,134 (2001) are reported and compared to European design standards for high-speed trains guidelines. The plate load test is selected because it is simple, widely used for other earth structures as a measure of quality and is covered with precise norms for different countries. Very few researchers have tried to investigate railway embankment using this test as the basis of performance, as European countries have already set up high-speed networks and standards up to a speed of 300kmph based on established parameters from this test (e.g., RIL 836) (AG, DN 2014; Alamaa 2016). Emersleben and Meyer (2008) used static plate load test and falling weight deflectometer to report the benefits of Geocells in road embankments by evaluating the embankment stiffness. Recently Minazek and Mulabdic (2013) tried to establish the benefits of plate load tests in determining the stiffness of reinforced embankment by using gravels on various types of geogrids. More recently Puri et al. (2017) studied the effect of blending sand with clay in field conditions to improve the bearing capacity and reduction of settlements under repeated loading conditions. Similarly, Mamatha and Dinesh (2018) have studied soil aggregate system but in context to the road pavement based proving the reliability of DIN 18,134 guidelines.

19.2 Methodology

The apparatus consists of a rigid square steel box made by using hard grade steel plates. This open box consists of smooth parallel walls and a back wall. The dimensions of the box are 1 m × 1 m × 1 m enabling to represent the embankment up to 1 m in height. As per DIN 18,134, the test can be conducted on the metal plate capable of sustaining max. load applied. The size in this experiment was limited to 200 mm instead of 300 mm (usually used) to eliminate the boundary effect for horizontal pressure bulb (5B). The tank was supported with a rigid loading frame capable of generating a load of 250 kN fitted with an electronic ram, controlled manually. To read the settlements of the entire plate, square plate was used which was mounted with four dial gauges at each corner. As per DIN 18,134, the max. load to be imparted on the loading plate should be capable of generating a load intensity of 500 kN/m^2. The load applied should be completed in six stages with equal load increments until the max. load intensity is reached. During the unloading phase, the load shall be removed in three stages to 50%, 25% and finally to approx. 2% of the max. load before starting the next cycle. The test is terminated when either the max. the design load is achieved or the designated settlement is reached. As per DIN 18,134 for road construction purposes 5 mm settlement is set (DIN 2001) while no such value is available for railway embankment, a similar 5 mm settlement can be used as these experiments involve the soil embankment. The test set up can be seen in Fig. 19.1.

UIC 719 R and RIL 836 (Alamaa 2016) provide the guidelines to achieve minimum deformation modulus for railway embankments to sustain high-speed lines or up to 300 kmph. UIC 719R contains guidelines and standards of earthworks and trackbed construction of railways lines for all members of the international union of railways and RIL 836 is a German guidelines for earthworks and geotechnical structural design and maintenance. The deformation modulus for first and second loading cycles is calculated using below equation as,

$$Ev = 1.5 \times r \times \frac{1}{a_1 + a_2.\sigma_{0\,\max}} \tag{19.1}$$

whereas E_v = deformation modulus, r = radius of plate, a_1, a_2 = constants from Eq. 19.2,

Fig. 19.1 Test setup used in the current experimental program

σ_{0max} = max. avg. normal stress below the loading plate in the respective cycle in MN/m^2.

The constants a_1 and a_2 can be calculated from the solution of second-degree polynomial equation for the settlement measurement as,

$$S = a_0 + a_1.\sigma_0 + a_2.\sigma_0^2 \quad (19.2)$$

whereas σ_0 = avg. normal stress below the plate in MN/m^2,
s = settlement of loading plate in mm,
a_0 = constant of second-degree polynomial in mm,
a_1 = constant of second-degree polynomial in mm/(MN/m^2),
a_2 = constant of second-degree polynomial in mm/(MN^2/m^4).

19.3 Materials and Preparation

The soil used for the blanket layer was cohesionless soil classified as silty sand (SM) whose grain size distribution is given in Fig. 19.2a. The ballast used in the test was provided by Indian Railways which was of the same gradation as used in actual railway formations having grain size from 20 to 65 mm. The grain size distribution of the same is given in Fig. 19.2b. The materials used to represent the sub-ballast that follows the gradation provided by the Indian Railways manual comprising of 4.75 mm grain size to 20 mm. Few further evaluated properties of the material used are shown in Table 19.1. Since the height of the tank is 1 m, the max. height for the subgrade layer can be 1 m and as per GE-14 the thickness of the soil subgrade is 500 mm for SQ2 and SQ3 categories for an axle load of 25 T. SQ1,

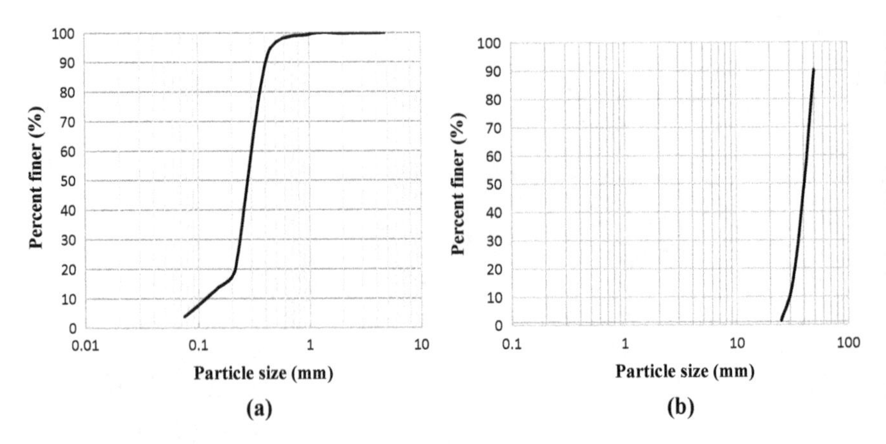

Fig. 19.2 **a** Particle size distribution curve for blanket layer **b** Particle size distribution curve for ballast

Table 19.1 Gradation and compaction strength parameters of material used

Material	Coeff. of uniformity (C_u)	Coeff. of curvature (C_c)	Classification	Density
Soil	2	1.38	SM	53% (I_d)
Ballast	1.5	0.9	Highly angular	16.7 (kN/m^3)

Table 19.2 Classification of soils and modulus as per GE-14, Indian Railways

Type	Classification	Blanket thickness (cm)	EV_2 for subgrade (MPa)	EV_2 for blanket layer (MPa)
SQ1	Fines >50%	100	45	100
SQ2	Fines 12–50%	75	45	100
SQ3	Fines <12%	60	45	100

SQ2 and SQ3 are the soil classification as per RDSO guidelines based on the percentage of fines as mentioned in Table 19.2, hence SQ3 represents a very good quality soil, SQ2 is average quality and SQ1 bad quality (usually avoided). Now as per RDSO guidelines, the thickness of the subgrade layers is decided based on the soil quality as mentioned in Table 19.2.

Apart from thickness GE-14 also mentions min. EV_2 (as per DIN 18,134) for subgrade layers as well as sub-ballast which can be seen in Table 19.2. From the objective of sustaining loads from the train, the height of the embankment comes to be in the order of 3.4–3.6 m from the ground level which ascertains lots of material (especially soil and aggregates). The study serves as a check to measure the quality of the subgrade for the mentioned design criteria as per GE-14. By following the stress isobar concept, the min. thickness for the soil subgrade in the tank was worked out to be 400 mm and which can be raised to 1000 mm. The plate load test was conducted and EV_1 and EV_2 were calculated to match the min. required values in GE-14 and to evaluate the height of embankment required based on deformation modulus values. Once the subgrade thickness is fixed, the inclusion of the blanket layer and ballast layer was also added to calculate the total height of the embankment.

19.4 Results and Discussion

The first few tests were conducted only on the soil as subgrade to be used in the embankment. The minimum possible thickness of 400 mm was achieved in the tank. Soil compaction is one of the most influential tasks while conducting plate load tests in the laboratory. The compaction effort was kept to ‘heavy compaction’ with the use of metal hammer weighing 8 kg. and was used to compact a single layer of 100 mm. Since the hammer used was a manual compactor, the number of blows can be increased to achieve heavy compaction. The relative density achieved for this type of soil was 53%. A preload was applied to the plate before the start of the test, generating the intensity of 10 kN/m^2 as per the recommendation of DIN 18,134. The dial gauges were reset to zero after the application of the preload before tests were performed. Table 19.3 shows the results that were obtained after the completion of the test program. The calculation of the deformation modulus was performed as per Eqs. 19.1 and 19.2. To solve for the constants a_1 and a_2, three additional equations were used which were based on the settlement and load for each step of the loading (total six steps) as observed in each test and are as follows:

$$a_0.n + a_1 \sum_{i=1}^{n} \sigma_{0i} + a_2 \sum_{i=1}^{n} \sigma_{0i}^2 = \sum_{i=1}^{n} s_i \quad (19.3)$$

$$a_0 \sum_{i=1}^{n} \sigma_{0i} + a_1 \sum_{i=1}^{n} \sigma_{0i}^2 + a_2 \sum_{i=1}^{n} \sigma_{0i}^3 = \sum_{i=1}^{n} s_i\sigma_{0i} \quad (19.4)$$

$$a_0 \sum_{i=1}^{n} \sigma_{0i}^2 + a_2 \sum_{i=1}^{n} \sigma_{0i}^3 + a_2 \sum_{i=1}^{n} \sigma_{0i}^4 = \sum_{i=1}^{n} s_i\sigma_{0i}^2 \quad (19.5)$$

After finding a_1 and a_2 from solving the above equations, EV_1 and EV_2 are calculated. Table 19.3 shows the results of tests conducted on various combinations of the embankment layers. The test was conducted at the minimum possible thickness of subgrade which was 400 mm and then increased to 500 mm to be on the safer side because of the stress isobar concept. Since the required EV_2 for the

Table 19.3 Deformation modulus values on the ballasted embankment in the laboratory

Thickness (mm)	EV_2 (MPa)	IR specification	Remarks
Subgrade (400)	61.47 > 45	Min. required EV_2 is 45 MPa	400 mm thickness is capable of required EV_2
Subgrade (500)	68.86 > 45	Min. required EV_2 is 45 MPa	500 mm thickness is capable of required EV_2
Subgrade (500) + blanket (200)	86.54 < 100*	Min. required EV_2 is 100 MPa	Required thickness of blanket should be increased
Subgrade (500) + ballast (300)	111.91 > 100*	Min. required EV_2 is 100 MPa	800 mm thickness capable of required EV_2
Subgrade (500) + blanket (200) + ballast (300)	138.94 > 100*	Min. required EV_2 is 100 MPa	1 m embankment, no EV_2 specifications for the inclusion of ballast from IR

* minimum EV_2 at top of blanket layer for 3.6 m high embankment as per Indian Railways

subgrade as mentioned in RDSO's GE-14 is 45 MPa and tests conducted show a value of more than 60 MPa, 500 mm thickness was selected as min. thickness of the subgrade. In the next step, the blanket material was added having a total thickness of 200 mm and then ballast material having a thickness of 300 mm instead of 350 mm (IR specifications) was added at the top making the total height of the embankment as 1 m. From Table 19.3, it is clear that the minimum required values set by IR are not directly linked to the height of the embankment and in most cases, the height of the embankment is much more than the required height based on deformation modulus values. To compare any existing recommendations from around the world, France's railway design requirements (Réseau ferré de France 2010) can be used where the design principle for ballasted track requirement for speeds up to 300 kmph is already working. The value on top of the embankment for the ballasted track is recommended to be 120 MPa, and the total height of the embankment is 900 mm(minimum) to 1200 mm (maximum) (Réseau ferré de France 2010). The availability of such types of standards and guidelines provides authenticity to these test trials and can help to achieve better and efficient designs. On comparison, it can be seen that deformation modulus of the order of 120 MPa can be achieved and embankment can be made fit for speeds up to 300 kmph. Also, it is evident that the height of the embankment is not a factor in achieving required embankment for high-speed trains.

Now, based on the current study, the required thickness based on the EV_2 values can be adopted as 800 mm and if blanket material is to be used thickness will be 1 m. Currently, as per GE-14, the minimum height of the embankment works out to be 3.45 m and the maximum is 3.6 m. On comparing with the height of the embankment which satisfies the min. required EV_2 values as per the current study, the percentage decrease can be calculated as,

$$\text{Original height of the embankment as per IR specification} = 3.45\,\text{m (min. height)}$$
$$\text{New height of the embankment as per current study} = 1\,\text{m (max. height)}$$
$$\text{Difference in height} = 2.45\,\text{m}$$

$$\text{Percentage decrease} = \frac{\text{difference in two number}}{\text{original number}} \times 100$$
$$= 71.01 \approx 71\%$$

19.5 Conclusions

The current study is an attempt to evaluate existing design methodology as adopted by Indian Railways using a cyclic plate bearing test as per DIN18134. The GE-14 code of RDSO establishes the height of embankment and different thicknesses based on the deformation modulus values based on this plate bearing test. The same test has been conducted in the laboratory on the various thickness of different layers (subgrade, blanket and ballast) of the embankment and the following conclusions can be made:

1. To meet the required deformation modulus, the height of the embankment adopted by IR can be reduced.
2. The total reduction in the height of the embankment based on the deformation modulus values is nearly 70%.
3. Due to a reduction in the embankment, construction time and mining material can be prevented from overusage.
4. Since existing design methodology is based on UIC 719R, a new methodology based on deformation modulus can be used as similar methodologies are adopted by European countries.
5. Based on the calculated values of deformation modulus in the laboratory, it can be assumed that based on the design values of deformation modulus (from RIL 836 or LGV), a track can be declared fit for high speed up to 300kmph.

After conducting this study, it is clear that the design methodology needs to be upgraded as per the growing requirements of the country. Merely increase in the height of the embankment will not meet the increasing load and speed demands of the second most populated country in the world. In the future, studies can be taken for use of reinforcement like geogrids under ballast and blanket will help further reduction in height of the embankment and will improve the quality of railways embankment for higher loads and speeds.

References

Alamaa A (2016) High-speed railway embankments: a comparison of different regulations

AG, DN (2014) RIL 836 Erdbauwerke und geotechnische Bauwerke planen, bauen und instand halten [Earthworks and geotechnical structures design, construction and maintainance]. Code. DB Netze AG, Frankfurt am Main, Germany, p 530 (in German)

DIN 18134:2001–09: Determining the deformation and strength characteristics of soil by plate loading test, Deutsche Institut für Normung (2001)

Emersleben A, Meyer N (2008) The use of geocells in road constructions over soft soil: vertical stress and falling weight deflectometer Measurements. In: Proceedings of 4th European geosynthetics conference, paper 132, Scottland

GE-14- RDSO- Guidelines for blanket layer provision on track formation. Design and Standard Organisation (RDSO), Ministry of Railways, India (2007)

Indraratna B, Sahin MA, Salim W (2005) Use of geosynthetics for stabilizing recycled ballast in railway track substructures. In: Proceedings of NAGS 2005/GRI 19 Cooperative Conference, 2005, pp 1–15. North American Geosynthetics Society, USA

Innotrack Integrated Project No. TIP5-CT-2006–031415: D2.2.6. GL, Guideline for subgrade reinforcement with geosynthetics (2006)

Jain VK, Keshav K (1999) Stress distribution in railway formation-A simulated study. In: Pre-failure deformation characteristics of geomaterials, pp 653–658

Mamatha KH, Dinesh SV (2018) Evaluation of strain modulus and deformation characteristics of geosynthetic-reinforced soil–aggregate system under repetitive loading. Int J Geotech Eng 12(6):546–555

Minažek K, Mulabdić M (2013) Pregled ispitivanja interakcije tla i armature u armiranom tlu pokusom izvlačenja. Građevinar 65(3):235–250

Puri P, Singh P, Garg P (2017) Effect of Sand on Strain Modulus (Ev2) Property of Clayey Soil

Réseau ferré de France (2010) Référentiel technique LGV dans le cadre de PPP ou de DSP tome 2—ouvrages en terre [Technical Reference LGV under PPP or DSP Volume 2—Earthworks]. Réseau ferré de France, Saint Denis, France, p 76 (in French)

Shin EC, Kim DH, Das BM (2002) Geogrid-reinforced railroad bed settlement due to cyclic load. Geotech Geol Eng 20(3):261–271

UIC (2008) UIC Code 719 R—Earthworks and track-bed for railway lines code. UIC, Paris, France, p 117

Chapter 20
Disturbance in Soil Structure Due to Post-cyclic Recompression

Ashish Juneja and A. K Mohammed Aslam

20.1 Background

When normally consolidated clays are subjected to low cyclic stress over prolonged time period, they tend to accumulate permanent strains because of pore pressure build-up. In extreme case, failure planes can develop within the soil mass when the above plastic strains exceed a threshold limits (Li and Selig 1996; Moses et al. 2003; Andersen 2009). Similar studies on sands (and silts to some extent) are few, except when these non-plastic deposits are investigated for their liquefaction potential that too under high cyclic load amplitude, e.g., (Wichtmann et al. 2005; Stamatopoulos 2010; Yang and Sze 2011). It is not uncommon for sands or silts to be subjected to tens of thousands of small amplitude cyclic loading when used as structural fill below machine foundations or wind power plants, or road and rail embankments. In such cases, the accumulated strains and stresses may not necessarily damage the foundation, possibly because of high safety factor and conservative soil parameters used in its design. However, the change in shear strength of these soils under such long-term repeated loading is important as subsequent development of infrastructure around the soil may become necessary. Notwithstanding the above, it is also important to assess any new load the embankment may bear in future.

Study of saturated sand to excess pore pressure generation and stiffness degradation and subsequent failure under cyclic load is well studied. The failure modes are dependent on the relative density of the sands and on whether there is stress reversal or not. Yang and Sze (2011) classified this into three patterns: (a) flow-type failure or strain softening characterized by sudden deformation in loose sand irre-

A. Juneja (✉)
Indian Institute of Technology Bombay, Powai, Mumbai 400 076, India
e-mail: ajuneja@iitb.ac.in

A. K. Mohammed Aslam
Yonsei University, Seoul 03722, South Korea

T. G. Sitharam et al. (eds.), *Latest Developments in Geotechnical Earthquake Engineering and Soil Dynamics*, Springer Transactions in Civil and Environmental Engineering, https://doi.org/10.1007/978-981-16-1468-2_20

spective of stress reversal, (b) cyclic mobility characterized by double-amplitude strain increasing to large values in medium dense and dense sand with stress reversal and (c) plastic strain accumulation characterized by plastic strain increasing in one direction to large values in medium dense and dense sand with no stress reversal. These modes of failure require that the stress levels be high. Such failures are not expected in the traffic loaded subsoils. Pore pressure and strain accumulation along with volumetric compression can be expected in such cases.

Soils subjected to such large number of repeated loading cycles, over time undergo complete excess pore pressure dissipation. This recompression is accompanied by volumetric compression of the soil, and the soil's shear behavior will be affected.

In the stress-controlled direct simple shear on plastic Drammen clay conducted by (Yasuhara and Andersen 1991), the excess pore pressures were drained and the stress path was compared with the recompression path during oedometer tests. The recompression index subsequent to the cyclic load was found to be different from the similar oedometer stress path recompression index with similar effective stresses. This was attributed to the fact that cyclic loading not only increases the pore water pressure but also disturbs the clay structure. They also noted that drainage makes soil more resistant to subsequent cyclic loading in the context that the pore pressures and cyclic shear strains were reducing in the subsequent cyclic loading. Similar strengthening due to drainage was also noted on slurry consolidated powdered Keuper Marl using a triaxial cyclic testing by O'Reilly et al. (1991). Similarly, by subjecting wet sedimented non-plastic silt to stress controller cyclic tests followed by recompression, Hyde et al. (2007) noted the improvement in cyclic strength of soil.

Similarly, in soils that to drainage periods in between or at the end of cyclic loadings, improvement in resistance to further loading and shear strength is seen, particularly in the case of normally consolidated soil (Yasuhara and Andersen 1991; Yasuhara 1994). Yasuhara (1994) also proposed that when the residual pore pressures from cyclic loading are dissipated the soil can be considered to be in an apparent overconsolidated state. This could be quantified using the excess pore pressure developed during cyclic loading.

Contrary to these, Wang et al. (2013) performed cyclic triaxial test on slurry consolidated low plasticity silt subjected to pore pressure ratios in the range of 0.35–1 followed by recompression and monotonic shearing. Even though the samples become denser during recompression, the shear strength was reduced for pore pressure ratio up to 0.7. The researchers attributed this to the damage to soil fabric by cyclic loading or recompression was not sufficient to increase the strength or inevitable variation in testing procedures. Also, liquefied samples reached critical state at much lower axial strain (13%) compared to the non-liquefied samples (around 25%).

The objective of this study is to formulate a methodology with which the post-cyclic recompression strength sands and silts. As a first approach, the strength was correlated using ordinary critical state soil parameters. Subsequently, some post-cyclic recompression paths were shown to be different compared to the paths of samples which were not subjected to prior cyclic loading. This leads to the introduction of new parameters to estimate the strength and stiffness.

20.2 Methodology

Sand (BS) consisted of rounded to sub-rounded particles of d_{50} equal to 0.3 mm, while silt (HS) consisted of sub-angular to angular particles of d_{50} equal to 0.05 mm. Table 20.1 shows the soil properties. Cyclic triaxial test was conducted on 100 mm diameter and 200 mm long samples. BS samples were prepared by pluviating air dry sand. In this procedure, the height of fall was fixed to obtain uniform samples of about 30% relative density. HS samples were prepared by tamping air dry soil in 10 equal layers using Ladd's (1978) under-compaction method. All the samples were saturated in the triaxial cell by first flushing them with carbon dioxide for about 30 min and then allowing de-aired distilled water to percolate from the bottom of the sample. Cell pressures and back pressures were then applied in increments until Skempton's (1954) B-factor of over 0.95 was achieved. The samples were tested under a mean effective stress of 100 kN/m^2. This mean effective stress, p'_0 and the above state of compaction resulted in samples which were of normally consolidated or slightly overconsolidated soil behavior (Georgiannou et al. 1990).

The samples were tested for their double-amplitude load-controlled cyclic triaxial strength by applying a uniform sinusoidal load at a frequency of 1 Hz. This frequency is widely accepted in cyclic triaxial testing since it allows adequate time for the generated excess pore pressures to distribute evenly within the sample (Kramer 1996). To study traffic loading, most authors have used the frequencies of

Table 20.1 Soil properties

Property	BS	HS
Specific gravity, G_s	2.65	2.65
Critical state friction angle, φ' degree	30	31
Dilatancy angle, ψ degree	3	–
Slope of normal compression line, λ	0.031	0.054
Slope of unloading reloading line, κ_0	0.006	0.008
NCL intercept at p'_0 = 1 kN/m^2, v_λ	1.873	1.924
Liquid limit, %	–	22
Plastic limit, %	–	19
Maximum dry unit weight, kN/m^3	–	18.8
Optimum water content, %	–	12
Minimum void ratio, e_{min}	0.479	–
Maximum void ratio, e_{max}	0.778	–
Classification (USCS)	SP	ML

0.1 Hz or 1 Hz, e.g., Paul et al. (2015) 0.1 Hz in clayey soil, Wichtmann et al. (2007) 0.1 Hz for drained triaxial in sand, Guo et al. (2013) 1 Hz in cyclic triaxial of clay, Chazallon et al. (2006) 1 Hz for modeling in granular material. Some have also stuck to higher frequencies like 5 Hz in triaxial test of ballast by Suiker et al. (2005) and 10–40 Hz in large triaxial test of ballast. Because of the high permeability, these tests can still yield good results irrespective of the high frequencies.

Cyclic shear stress, τ_{cyc}, varied from 2.5 to 20 kN/m^2 in different samples. These stress levels encompass the usual cyclic stress amplitude that is used to mimic the urban traffic loading (Chai and Miura 2002) and is significantly less than the cyclic shear stress applied in seismic studies (Yang and Sze 2011). The number of load cycles, N, varied from over a million to less than 10. In tests with large N, the cyclic shear stress was mostly too small to fail the samples.

20.3 Results and Discussion

20.3.1 Post-Cyclic Recompression

Excess pore pressures developed during cyclic loading were allowed to dissipate to permit the sample to reach the initial mean effective stress. During this stage, the volumetric compression was proportional to the dissipated pore pressure. The volumetric compression had an exponential relation with the pores pressure that is, volumetric compression significantly increased for excess pore pressure ratio r_u beyond 0.5. In all cases, volumetric compression however remained less than 1%. Sanín and Wijewickreme (2006) and Sanín (2010) observed that this value is insignificant to cause serious deformation. Figure 20.1a, b shows the soil behavior in compression space during the cyclic loading and recompression.

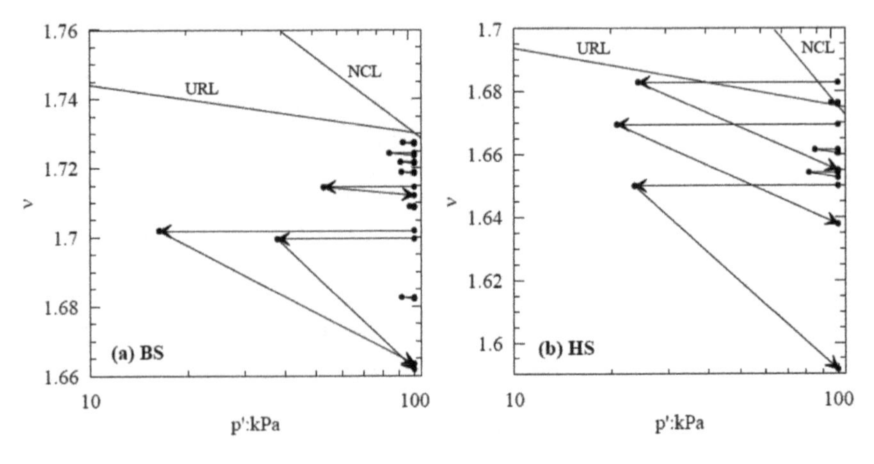

Fig. 20.1 Undrained loading followed by recompression during post-cyclic drainage

The recompression due to dissipation of excess pore pressure necessarily not flows the URL of the soil. Slopes of up to 3 times the slope of URL, κ, are observed. Soils that have high r_u values and the ones that liquefy show higher slopes. Similar observations have been noted by researches like Hyde et al. (2007) who observed 6–10 times the slope of URL in silt, Wang and Luna (2014) who observed 6 times the slope of URL in silt and Wang et al. (2015) who observed 5.5–6.7 times the slope of URL in silt.

20.3.2 Post-Cyclic Undrained Monotonic Strength

After the completion of recompression, the samples were tested in undrained monotonic loading up to about 25% axial strain. Figure 20.2a–d shows the stress–strain behavior during this for BS and HS. The stress–strain behavior is compared to that of soil not subjected to any cyclic loading. It was seen that there is an improvement in the strength of soil due to cyclic loading. This improvement is proportional to the cyclic stress ratio, *CSR* and total number of load cycles. For BS, the peak strength is seen in between 10 and 20% of axial strain and in between 20 and 30% for HS. The critical state is reached beyond 30% axial strain and the sample deformation and distortion are very prominent during this. So the peak strength was used as a key parameter in comparing the strength improvement. The peak strength was normalized to the initial mean effective stress in the study. Even though all tests were done under 100 kPa initial mean effective stress in the present study, they are important in relating the strength to the consolidation state of the soil. Figure 20.3a, b shows the peak shear strength improvement ratio defined as normalized peak strength of soils under repeated loading to the peak strength of soil not subjected to any repeated loading. The horizontal axis in the figure is the apparent overconsolidation ratio given by Yasuhara (1994) and Yasuhara et al. (2003) as

$$OCR_q = \frac{p'_0}{p'_0 - \Delta u} = \frac{1}{1 - r_u} \tag{20.1}$$

where p'_0 is the mean effective stress, Δu is the excess pore pressure and r_u is the excess pore pressure ratio.

For example, BS improved by about 59% when *CSR* was 0.05 for 50,000 cycles. Considerable improvement of up to 200% is seen in the case of HS soil. Beyond an OCR_q value of 1.5, there is a cap on the shear strength improvement ratio seen for both the soil. Strength improvement should be proportional to excess pore pressure developed during cyclic loading. Though a bit of scatter is seen in general, the strength improvement is proportional to the excess pore pressure ratio dissipated and to OCR_q.

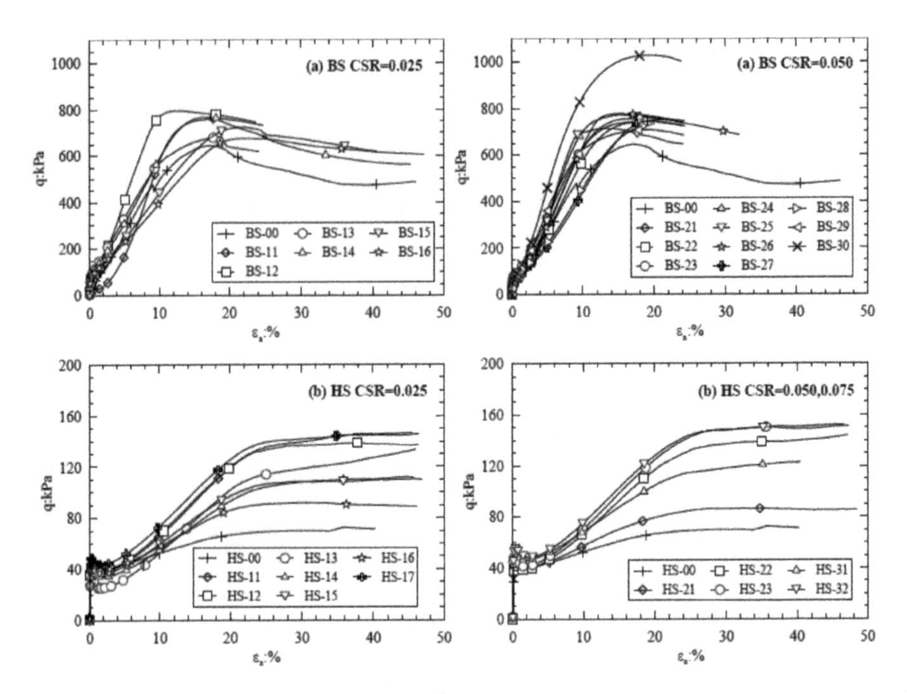

Fig. 20.2 Stress–strain behavior during post-cyclic shearing

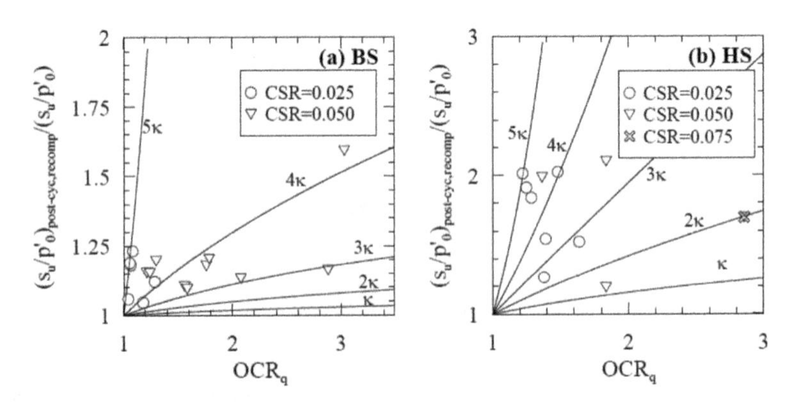

Fig. 20.3 Strength improvement ratio during post-cyclic shearing

Yasuhara (1994) and Yasuhara et al. (2003) gave the equation for the shear strength improvement ratio due to post-cyclic recompression as

$$\frac{\left(s_u/p_0'\right)_{\text{post-cyc,recomp.}}}{\left(s_u/p_0'\right)} = \left(\frac{1}{1-\Delta u/p_0'}\right)^{\frac{\Lambda_0 \kappa}{\lambda-\kappa}} = \left(\frac{1}{1-r_u}\right)^{\frac{\Lambda_0 \kappa}{\lambda-\kappa}} = \left(OCR_q\right)^{\frac{\Lambda_0 \kappa}{\lambda-\kappa}} \tag{20.2}$$

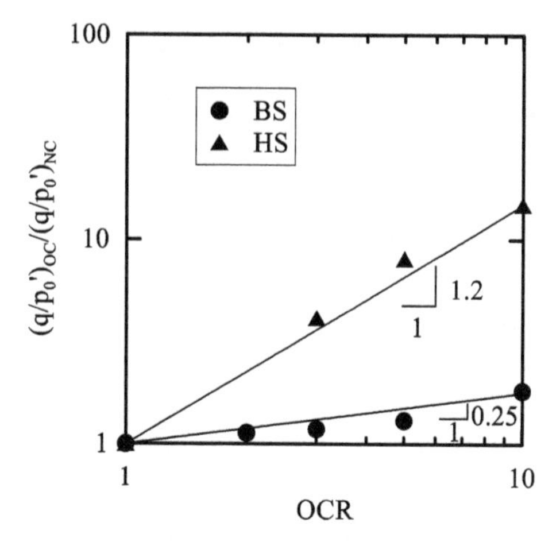

Fig. 20.4 Relationship between strength ratio and OCR using CU tests

where Λ_0 is strength ratio parameter from the Mitachi and Kitago's (1976) relation for strength of OC and NC soil as

$$\frac{\left(s_u/p_0'\right)_{OC}}{\left(s_u/p_0'\right)_{NC}} = OCR^{\Lambda_0} \tag{20.3}$$

Λ_0 was obtained from conventional CU triaxial tests on samples which did not have any history of cyclic loading. These are shown in Fig. 20.4. The figure is plotted in log–log scale to obtain the linear relationship between the strength ratio $\frac{\left(q/p_0'\right)_{OC}}{\left(q/p_0'\right)_{NC}}$ against OCR.

The strength improvement from Eq. (20.2) was plotted in Fig. 20.3a, b with URL slope of κ. As can be seen, the equation highly underpredicts the strength improvement. We can attribute this to the constant κ, assumed in the equation, which is not necessarily true. Additional curves using slopes of 2–5 times κ are also shown in the same figure. In BS and HS soils, strength improvement ratio lies in between 3 and 5 κ. The cyclic loading not only causes volumetric changes but also changes in the soil structure that causes it to have a much higher strength improvement.

20.4 Conclusion

The improvement in stress–strain behavior due to repeated loading for loose sand and normally consolidated silt was higher with higher cyclic stress amplitudes and with the total number of cycles of loading. For same number of cycles in a stage, there was improvement in stress–strain behavior with number of stages. The findings imply that post-cyclic recompression of loose sand and silt do not follow the original unloading reloading paths. The strength improvement can be modeled using simple relations relating it to the excess pore pressure developed during cyclic loading. However, using a higher slope for recompression line is recommended.

References

Andersen KH (2009) Bearing capacity under cyclic loading—offshore, along the coast, and on land. The 21st Bjerrum Lecture presented in Oslo, 23 November 2007. Can Geotech J 46(5): 513–535. https://doi.org/10.1139/T09-003

Chai JJ, Miura N (2002) Traffic-load-induced permanent deformation of road on soft subsoil. J Geotech Geoenviron Eng 128(11):907–916. https://doi.org/10.1061/(ASCE)1090-0241(2002)128:11(907)

Chazallon C, Hornych P, Mouhoubi S (2006) Elastoplastic model for the long-term behavior modeling of unbound granular materials in flexible pavements. Int J Geomech 6(4):279–289. https://doi.org/10.1061/(ASCE)1532-3641(2006)6:4(279)

Georgiannou VN, Burland JB, Hight DW (1990) The undrained behaviour of clayey sands in triaxial compression and extension. Geotechnique 40(3):431–449

Guo L, Wang J, Cai Y, Liu H, Gao Y, Sun H (2013) Undrained deformation behavior of saturated soft clay under long-term cyclic loading. Soil Dyn Earthq Eng 50:28–37. https://doi.org/10.1016/j.soildyn.2013.01.029

Hyde AFA, Higuchi T, Yasuhara K (2007) Postcyclic recompression, stiffness, and consolidated cyclic strength of silt. J Geotech Geoenviron Eng 133(4):416–423. https://doi.org/10.1061/(ASCE)1090-0241(2007)133:4(416)

Kramer SL (1996) Geotechnical earthquake engineering. In Prentice Hall, New Jersey

Ladd R (1978) Preparing test specimens using undercompaction. Geotech Test J 1(1):16. https://doi.org/10.1520/GTJ10364J

Li BD, Selig ET (1996) Cumulative plastic deformation for fine-grained subgrade soils. J Geotech Eng 122(12):1006–1013. https://doi.org/10.1061/(ASCE)0733-9410(1996)122:12(1006)

Mitachi T, Kitago S (1976) Change in undrained shear strength characteristics of saturated remolded clay due to swelling. Soils Found 16(1):45–58. https://doi.org/10.3208/sandf1972.16.45

Moses GGG, Rao SNS, Rao PPN (2003) Undrained strength behaviour of a cemented marine clay under monotonic and cyclic loading. Ocean Eng 30(14):1765–1789. https://doi.org/10.1016/S0029-8018(03)00018-0

O'Reilly MMP, Brown SSF, Overy RFR (1991) Cyclic loading of silty clay with drainage periods. J Geotech Eng 117(2):354–362. https://doi.org/10.1061/(ASCE)0733-9410(1991)117:2(354)

Paul M, Sahu RB, Banerjee G (2015) Undrained pore pressure prediction in clayey soil under cyclic loading. Int J Geomech 15(5):4014082. https://doi.org/10.1061/(ASCE)GM.1943-5622.0000431

Sanín MV (2010) Cyclic shear loading response of Fraser River delta silt. The University of British Columbia

Sanín MV, Wijewickreme D (2006) Cyclic shear response of channel-fill Fraser River Delta silt. Soil Dyn Earthq Eng 26(9):854–869. https://doi.org/10.1016/j.soildyn.2005.12.006

Skempton AW (1954) The pore-pressure coefficients A and B. Géotechnique 4(4):143–147. https://doi.org/10.1680/geot.1954.4.4.143

Stamatopoulos CAC (2010) An experimental study of the liquefaction strength of silty sands in terms of the state parameter. Soil Dyn Earthq Eng 30(8):662–678 (Elsevier). https://doi.org/10.1016/j.soildyn.2010.02.008

Suiker ASJ, Selig ET, Frenkel R (2005) Static and cyclic triaxial testing of ballast and subballast. J Geotech Geoenviron Eng 131(6):771–782. https://doi.org/10.1061/(ASCE)1090-0241(2005)131:6(771)

Wang S, Luna R (2014) Compressibility characteristics of low-plasticity silt before and after liquefaction. J Mater Civ Eng 26(6):4014014. https://doi.org/10.1061/(ASCE)MT.1943-5533.0000953

Wang S, Luna R, Yang J (2013) Postcyclic behavior of low-plasticity silt with limited excess pore pressures. Soil Dyn Earthq Eng 54:39–46 (Elsevier). https://doi.org/10.1016/j.soildyn.2013.07.016

Wang S, Luna R, Zhao H (2015) Cyclic and post-cyclic shear behavior of low-plasticity silt with varying clay content. Soil Dyn Earthq Eng 75:112–120. https://doi.org/10.1016/j.soildyn.2015.03.015

Wichtmann T, Niemunis A, Triantafyllidis T, Poblete M (2005) Correlation of cyclic preloading with the liquefaction resistance. Soil Dyn Earthq Eng 25(12):923–932. https://doi.org/10.1016/j.soildyn.2005.05.004

Wichtmann T, Niemunis A, Triantafyllidis T (2007) Strain accumulation in sand due to cyclic loading: drained cyclic tests with triaxial extension. Soil Dyn Earthq Eng 27(1):42–48. https://doi.org/10.1016/j.soildyn.2006.04.001

Yang J, Sze HHY (2011) Cyclic behaviour and resistance of saturated sand under non-symmetrical loading conditions. Géotechnique 61(1):59–73. https://doi.org/10.1680/geot.9.P.019

Yasuhara K (1994) Postcyclic undrained strength for Cohesive Soils. J Geotech Eng 120(11):1961–1979. https://doi.org/10.1061/(ASCE)0733-9410(1994)120:11(1961)

Yasuhara K, Andersen KH (1991) Recompression of normally consolidated clay after cyclic loading. Soils Found 31(1):83–94. https://doi.org/10.3208/sandf1972.31.83

Yasuhara K, Murakami S, Song BB-W, Yokokawa S, Hyde AFL (2003) Postcyclic degradation of strength and stiffness for low plasticity silt. J Geotech Geoenviron Eng 129(8):756–769. https://doi.org/10.1061/(ASCE)1090-0241(2003)129:8(756)

Chapter 21
Application of Soft Computing in Geotechnical Earthquake Engineering

Pijush Samui

21.1 Introduction

Uncertainty is quite common in geotechnical earthquake engineering. Hence, modeling of any phenomena is a challenging task. The use of various soft computing techniques has shown success for modeling various phenomena in geotechnical earthquake engineering such as liquefaction (Table 21.1), lateral spreading (Table 21.2), ground motion (Table 21.3), seismic slope stability (Table 21.4), pile foundation (Table 21.5) and retaining wall (Table 21.5). Zadeh (1992) introduced soft computing. Various types of soft computing are given below (Sharma and Chandra 2019):

- Fuzzy set
- Fuzzy logic
- Fuzzy theory
- Artificial neural network
- Swarm intelligence
- Evolutionary computing

The main purpose of this article is to review the application of soft computing techniques in several fields of geotechnical earthquake engineering. This article serves the application of various soft computing techniques {artificial neural network (ANN), adaptive neurofuzzy inference system (ANFIS), genetic algorithm (GA), support vector machine (SVM), relevance vector machine (RVM), least square support vector machine (LSSVM), particle swarm optimization (PSO), extreme learning machine (ELM), minimax probability machine (MPM), multivariate adaptive regression spline (MARS), genetic programming (GP), group method of data handling (GMDH)} for solving different problems in geotechnical

P. Samui (✉)
Department of Civil Engineering, NIT Patna, Patna, Bihar 800005, India
e-mail: pijush@nitp.ac.in

T. G. Sitharam et al. (eds.), *Latest Developments in Geotechnical Earthquake Engineering and Soil Dynamics*, Springer Transactions in Civil and Environmental Engineering, https://doi.org/10.1007/978-981-16-1468-2_21

Table 21.1 Modeling of seismic liquefaction by using different soft computing methods

Soft computing	In situ method	Number of dataset
Decision tree (DT) and logistic Regression (LR) (Gandomi et al. 2013)	SPT	620
ANN (Goh 1994)	SPT	85
ANN (Goh 1996)	CPT	109
ANN (Najjar and Ali 1998)	SPT	105
ANN (Goh 2002)	CPT and shear wave velocity	109 and 186
SVM (Pal 2006)	CPT and SPT	109 and 85
SVM (Zhao et al. 2007)	CPT and SPT	170 and 105
ANN (Hanna et al. 2007)	SPT	620
SVM (Goh and Goh 2007)	CPT	226
NN (Baykasoglu et al. 2009)	CPT	226
GP (Gandomi et al. 2011)	CPT	170
GP ANN (Kayadelen 2011)	SPT	569
GP (Gandomi and Alavi 2012)	CPT	170
GP (Muduli and Das 2013)	CPT	170
GP (Gandomi et al. 2013)	CPT	170
LSSVM (Hoang and Bui 2018)	CPT, SPT and shear wave velocity	185, 226 and 620
GP (Goharzay et al. 2017)	SPT	160
SVM, RVM and LSSVM (Samui 2014)	SPT	620
Fuzzy Logic (Rahman and Wang 2002)	SPT	205
SVM (Lee and Chern 2013)	CPT	466
Polynomial Model (Eslami et al. 2014)	CPT	182
ANN and SVM (Samui and Sitharam 2011)	SPT	288
MARS (Zhang and Goh 2016)	CPT	796
ANN (Rezaei and Choobbasti 2014)	SPT	30
LSSVM and RVM (Samui 2011)	SPT	85
ANN (Ramakrishnan et al. 2008)	SPT	23
ANN (Baykasoglu et al. 2009)	CPT	226
ANN (Kumar et al. 2014)	SPT	40
ANFIS (Yazdi et al. 2012)	CPT	182
ANN and ANFIS (Venkatesh et al. 2013)	SPT	159
ANN (Xue and Liu 2017)	CPT	166
SVM (Samui 2013)	CPT	134
ANN, ANFIS, SVM and PRIM (Kaveh et al. 2018)	CPT	444
GMDH (Kurnaz and Kaya 2019)	CPT	212
RVM (Samui and Karthikeyan 2014)	CPT	134
ELM (Samui et al. 2016)	CPT	109
RVM (Karthikeyan et al. 2013)	SPT	288

(continued)

Table 21.1 (continued)

Soft computing	In situ method	Number of dataset
ANN (Juang and Chen 1999)	CPT	170
SVM (Samui et al. 2011)	Shear wave velocity	186
ANN (Shahri 2016)	SPT and shear wave velocity	112
MPM (Samui and Hariharan 2014)	SPT	288
GP (Muduli et al. 2014)	CPT	226
RVM and LSSVM (Karthikeyan and Samui 2014)	Shear wave velocity	191
GMDH (Kurnaz and Kaya 2019)	SPT	451
ANN (Juang and Chen 1999)	CPT	963
Fuzzy template (Tesfamariam and Najjaran 2007)	Shear wave velocity	184

Table 21.2 Prediction of lateral spreading by using various soft computing techniques

Soft computing	Number of dataset
Neurofuzzy (Garcia et al. 2008)	448
M5 model tree algorithm (Avval and Derakhshani 2019)	484
GP (Javadi et al. 2006)	484
MARS (Goh and Zhang 2014)	467
ANN (Wang and Rahman 1999)	466
ANN (Baziar and Ghorbani 2005)	464
GMDH (Jirdehi et al. 2014)	250
LSSVM (Das et al. 2011)	228

earthquake engineering. For development of soft computing models, dataset is normalized in different ways such as 0 to 1 and −1 to +1. The various steps for development of soft computing have been shown in Fig. 21.1.

21.2 Liquefaction

Soft computing has taken the problem of prediction of seismic liquefaction potential of soil as a classification problem. Geotechnical engineers generally use different in situ techniques {standard penetration test (SPT), cone penetration test (CPT) and shear wave velocity technique} for determination of seismic liquefaction potential of soil. SPT, CPT, shear wave velocity, total vertical stress, effective vertical stress, magnitude, peck ground acceleration (PGA) and cyclic stress ratio (CSR) have been considered as inputs of soft computing techniques. Table 21.1 shows the application of various soft computation techniques for prediction of seismic liquefaction potential of soil. RVM, SVM, LSSVM and GP gave equation for prediction of

Table 21.3 Modeling of ground motion by using different soft computing techniques

Soft computing	Number of dataset
ANN (Gandomi et al. 2016)	179
Randomized ANFIS (Thomas et al. 2016)	3551
M5 and CART (Hamze-Ziabari and Bakhshpoori 2018)	2851
ANFIS (Kaveh et al. 2018)	2851
ANN (García et al. 2007)	1058
GP (Gnüllü 2012)	120
ANN and ANFIS (Mittal et al. 2011)	109
Deep learning (Derakhshani and Foruzan 2019)	12,556
M5 model tree (Kaveh et al. 2016)	2815
SVM (Thomas et al. 2017)	2815
GP-OLS (Gandomi et al. 2011)	2777
ANN (Gülkan and E. 2004)	112
ANN (Gülkan and Kalkan 2002)	47
ANN (Zaré and Bard 2002)	210
ANN (Ulusay et al. 2004)	221
GP (Cabalar and Cevik 2009)	47

Table 21.4 Prediction of seismic slope stability by using soft computing

Soft computing	Dataset
PSO-ANN (Gordan et al. 2016)	699
ANN and SVM (Xu et al. 2016)	2330
ICA-ANN, GA-ANN, ABC-ANN and PSO-ANN (Koopialipoor et al. 2019)	699
ANN (Tsompanakis et al. 2009)	66
ANN (Javdanian and Pradhan 2019)	103
ANN (Lin et al. 2009)	955
ANN, SVM (Turel 2011)	50
ANN (Peng et al. 2005)	–
ANN (Barkhordari and EntezariZarch 2015)	152

seismic liquefaction potential of soil. Most of the researchers adopted ANN for classifying liquefiable and non-liquefiable soil. ANN uses all training dataset for final prediction. However, RVM and SVM adopt only a portion of training dataset for final prediction. The developed soft computing techniques give charts for classifying liquefiable and non-liquefiable soil. RVM, SVM, MPM and LSSVM use kernel function for final prediction. ANN adopts activation function for prediction of output. GP employs any function for determination of output. Table 21.6 shows the tuning parameters of the different soft computing techniques.

Table 21.6 shows the tuning parameters of the various soft computing techniques.

Table 21.5 Application of soft computing to the other problems of geotechnical earthquake engineering

Soft computing	Dataset
ANN (Ahmad et al. 2007)	–
ANN (Zhiyong and Xi 1999)	–
BBO (Aydogdu 2017)	–
SVM (Samui et al. 2012)	26
IBSA (Nama et al. 2017)	–
ANN and ANFIS (Javdanian et al. 2015)	385
ANN (Panakkat and Adeli 2007)	–
ANN (Narayanakumar and Raja 2016)	461
ANN (Garcia and Romo 2004)	–
MARS (Samui and Kothari 2012)	21
ANN and ANFIS (Akbulut et al. 2004)	21

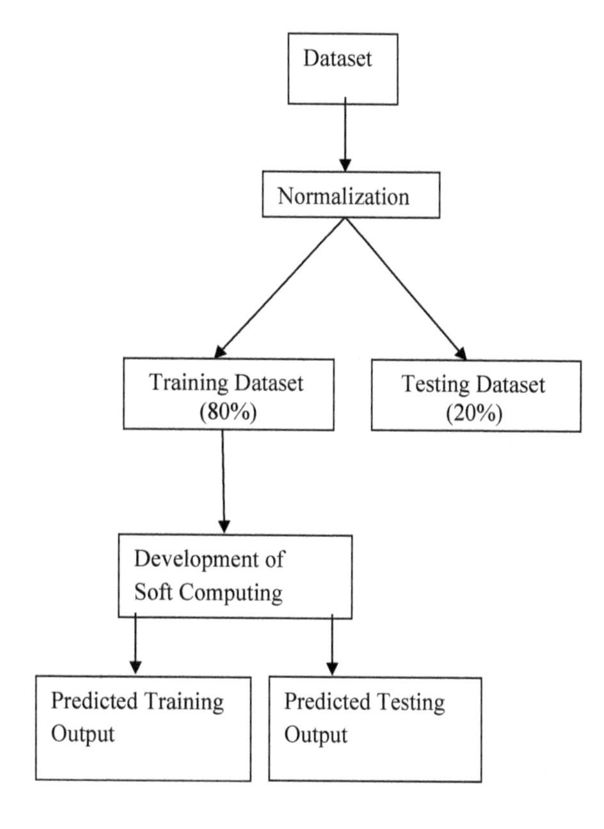

Fig. 21.1 Flowchart showing steps for developing soft computing

Table 21.6 Tuning parameters of various soft computing techniques

Model	Tuning parameters
ANN	Number of hidden layers, number of neurons in the hidden layer, activation function, number of epochs
SVM	Capacity factor, error insensitive zone, kernel parameter
RVM	Kernel parameter
MPMR	Error insensitive zone and kernel parameter
ANFIS	Type of membership function, number of epochs
GP	Number of population, number of generation, mutation frequency and crossover frequency
MARS	Number of basis function
GPR	Error terms and kernel parameter
LSSVM	Regularization parameter and kernel parameter
ELM	Activation function, number of neurons, number of initial training data used in the initial phase, range from which the size of data block randomly generated in each iteration
GMDH	Maximum number of neurons in a layer, maximum number of layers, selection pressure (in layers)

21.3 Lateral Spreading

The prediction of lateral spreading is a regression problem. Table 21.2 shows the application of various soft computing techniques for prediction of lateral spreading. The developed soft computing models gave equations for determination of lateral spreading of soil. Earthquake magnitude (M), nearest horizontal distance of the seismic energy source to the site (R), slope of ground surface (S_{gs}), cumulative thickness of saturated cohesionless soil layers where SPT number corrected for energy rating (N_{60}) is less than 15 (T_{15}), percentage of particles finer than 75 m for the granular materials contained within T_{15} (F_{15}) and average median grain size for granular materials within T_{15} ($D_{50\text{-}15}$) have been considered as inputs of the soft computing models.

21.4 Ground Motion

Prediction of ground motion is an important parameter for designing earthquake-resistant structure. Table 21.3 shows the application of different soft computing techniques for prediction of ground motion. Ground motion depends on earthquake magnitude (*M*), properties of soil, style of faulting and distance. Determination of ground motion has been treated as regression problem. The developed soft computing techniques give equations for prediction of ground motion.

21.5 Slope Stability

Prediction of status of slope during an earthquake is a complicated task. Table 21.4 illustrates the application of various soft computing techniques for determination of stability of slope under earthquake condition. Properties of soil, geometry of slope and parameters of earthquake have been taken as inputs of the soft computing models. The problem of prediction of status of slope has been taken as a classification technique. The prediction of factor of safety is a regression problem.

21.6 Other Fields of Geotechnical Earthquake Engineering

Florido et al. (2016) discussed various soft computing techniques for prediction of magnitude of earthquake. Samui et al. (2012) developed SVM for prediction of status of pile foundation during earthquake. Ahmad et al. (2007) developed ANN for determination of kinematic soil–pile interaction response parameters. Pseudo-dynamic active earth pressure was determined by a new optimization algorithm (Nama et al. 2017). Dynamic properties (shear modulus and damping ratio) of soil were also determined by soft computing (Javdanian et al. 2015). Samui and Kothari (2012) developed MARS models for prediction of damping ratio and shear modulus. Panakkat and Adeli (2007) developed ANN for prediction of earthquake magnitude.

21.7 Conclusion

The applications of various soft computing techniques for modeling different problems in geotechnical earthquake engineering have been described in this article. Dataset is required for developing soft computing models. Hence, the success of soft computing model depends on the quality and quantity of data. The developed soft computing models can reduce the cost of project by reducing the number of experiments. The different problems in geotechnical earthquake engineering can be easily solved by the developed soft computing techniques. It gives a handy and often highly reliable solution to various problems of geotechnical earthquake engineering. Probability of failure of geostructure can also be determined by using soft computing techniques. In summary, it can be concluded that various soft computing techniques can be used as alternative tools for modeling different problems in geotechnical earthquake engineering.

References

Ahmad I, El Naggar MH, Khan AN (2007) Artificial neural network application to estimate kinematic soil pile interaction response parameters. Soil Dyn Earthq Eng 27(9):892–905

Akbulut S, Hasiloglu AS, Pamukcu S (2004) Data generation for shear modulus and damping ratio in reinforced sands using adaptive neuro-fuzzy inference system. Soil Dyn Earthq Eng 24(11): 805–814

Avval YJ, Derakhshani A (2019) New formulas for predicting liquefaction-induced lateral spreading: model tree approach. Bull Eng Geol Env 78(5):3649–3661

Aydogdu I (2017) Cost optimization of reinforced concrete cantilever retaining walls under seismic loading using a biogeography-based optimization algorithm with Levy flights. Eng Optim 49(3):381–400

Barkhordari K, EntezariZarch H (2015) Prediction of permanent earthquake-induced deformation in earth dams and embankments using artificial neural networks. Civil Eng Infrastruct J 48(2): 271–283

Baykasoglu A, C¸evik A, Ozbakir L, Kulluk S (2009) Generating prediction rules for liquefaction through data mining. Expert Syst Appl 36(10):12491–12499

Baziar MH, Ghorbani A (2005) Evaluation of lateral spreading usingartificial neural networks. J Soil Dyn Earthquake Eng 25:1–9

Cabalar AF, Cevik A (2009) Genetic programming-based attenuation relationship: an application of recent earthquakes in turkey. Comput Geosci 35(9):1884–1896

Das SK, Samui P, Kim D, Sivakugan N, Biswal R (2011) Lateral displacement of liquefaction induced ground using least square support vector machine. Int J Geotech Earthq Eng (IJGEE) 2(2):29–39

Derakhshani A, Foruzan AH (2019) Predicting the principal strong ground motion parameters: a deep learning approach. Appl Soft Comput 80:192–201

Eslami A, Mola-Abasi H, TabatabaieShourijeh P (2014) A polynomial model for predicting liquefaction potential from cone penetration test data. Scientia Iranica 21(1):44–52

Florido E, Aznarte JL, Morales-Esteban A, Martínez-Álvarez F (2016) Earthquake magnitude prediction based on artificial neural networks: a survey. Croat Oper Res Rev 7(2):159–169

Gandomi H, Alavi AH (2012) A new multi-gene genetic programming approach to non-linear system modeling—part II: geotechnical and earthquake engineering problems. Neural Comput Appl 21(1):189–201

Gandomi AH, Alavi AH, Mousavi M, Tabatabaei SM (2011) A hybrid computational approach to derive new ground-motion prediction equations. Eng Appl Artif Intell 24(4):717–732

Gandomi AH, Fridline MM, Roke DA (2013) Decision tree approach for soil liquefaction assessment. Sci World J

Gandomi M, Soltanpour M, Zolfaghari MR, Gandomi AH (2016) Prediction of peak ground acceleration of Iran's tectonic regions using a hybrid soft computing technique. Geosci Front 7(1):75–82

Garcia SR, Romo MP (2004) Dynamic soil properties identification using earthquake records: a NN approximation. In: Proceedings of the 13th world conference on earthquake engineering. Vancouver, BC, Canada

García SR, Romo MP, Mayoral JM (2007) Estimation of peak ground accelerations for Mexican subduction zone earthquakes using neural networks. Geofísicainternacional 46(1):51–62

Garcia SR, Romo MP, Botero E (2008) Aneurofuzzy system to analyze liquefaction-induced lateral spread. Soil Dyn Earthq Eng 28(3):169–180

Gnüllü H (2012) Prediction of peak ground acceleration by genetic expression programming and regression: a comparison using likelihood-based measure. Eng Geol 141:92–113

Goh TC (1994) Seismic liquefaction potential assessed by neural networks. J Geotech Eng 120(9): 1467–1480

Goh TC (1996) Neural-network modeling of CPT seismic liquefaction data. J Geot Geoenviron Eng 122(1):70–73

Goh ATC (2002) Probabilistic neural network for evaluating seismic liquefaction potential. Can Geotech J 39(1):219–232

Goh TC, Goh SH (2007) Support vector machines: their use in geotechnical engineering as illustrated using seismic liquefaction data. Comput Geotech 34(5):410–421

Goh AT, Zhang WG (2014) An improvement to MLR model for predicting liquefaction-induced lateral spread using multivariate adaptive regression splines. Eng Geol 170:1–10

Goharzay M, Noorzad A, Ardakani AM, Jalal M (2017) A worldwide SPT-based soil liquefaction triggering analysis utilizing gene expression programming and Bayesian probabilistic method. J Rock Mech Geotech Eng 9(4):683–693

Gordan B, Armaghani DJ, Hajihassani M, Monjezi M (2016) Prediction of seismic slope stability through combination of particle swarm optimization and neural network. Eng Comput 32(1): 85–97

Gülkan P, Kalkan E (2002) Attenuation modeling of recent earthquakes in Turkey. J Seismol 6:397–409

Gülkan P, Kalkan E (2004) Site-dependent spectra derived from ground-motion records in Turkey. Earthq Spectra 4:1111–1138

Hamze-Ziabari SM, Bakhshpoori T (2018) Improving the prediction of ground motion parameters based on an efficient bagging ensemble model of M5′ and CART algorithms. Appl Soft Comput 68:147–161

Hanna M, Ural D, Saygili G (2007) Neural network model for liquefaction potential in soil deposits using Turkey and Taiwan earthquake data. Soil Dyn Earthq Eng 27(6):521–540

Hoang ND, Bui DT (2018) Predicting earthquake-induced soil liquefaction based on a hybridization of kernel Fisher discriminant analysis and a least squares support vector machine: a multi-dataset study. Bull Eng Geol Env 77(1):191–204

Javadi AA, Rezania M, Nezhad MM (2006) Evaluation of liquefaction induced lateral displacements using genetic programming. Comput Geotech 33(4–5):222–233

Javdanian H, Pradhan B (2019) Assessment of earthquake-induced slope deformation of earth dams using soft computing techniques. Landslides 16(1):91–103

Javdanian H, Jafarian Y, Haddad A (2015) Predicting damping ratio of fine-grained soils using soft computing methodology. Arab J Geosci 8(6):3959–3969

Jirdehi RA, Mamoudan HT, Sarkaleh HH (2014) Applying GMDH-type neural network and particle warm optimization for prediction of liquefaction induced lateral displacements. Appl Appl Math 9(2)

Juang CH, Chen CJ (1999) Cpt-based liquefaction evaluation using artificial neural networks. Comput Aided Civil Infrastruct Eng 14(3):221–229

Karthikeyan J, Samui P (2014) Application of statistical learning algorithms for prediction of liquefaction susceptibility of soil based on shear wave velocity. Geomat Nat Haz Risk 5(1): 7–25

Karthikeyan J, Kim D, Aiyer BG, Samui P (2013) SPT-based liquefaction potential assessment by relevance vector machine approach. Eur J Environ Civ Eng 17(4):248–262

Kaveh A, Bakhshpoori T, Hamze-Ziabari SM (2016) Derivation of new equations for prediction of principal ground-motion parameters using M5′ algorithm. J Earthq Eng 20(6):910–930

Kaveh A, Hamze-Ziabari SM, Bakhshpoori T (2018) Patient rule-induction method for liquefaction potential assessment based on CPT data. Bull Eng Geol Env 77(2):849–865

Kayadelen C (2011) Soil liquefaction modeling by genetic expression programming and neuro-fuzzy. Expert Syst Appl 38(4):4080–4087

Koopialipoor M, Armaghani DJ, Hedayat A, Marto A, Gordan B (2019) Applying various hybrid intelligent systems to evaluate and predict slope stability under static and dynamic conditions. Soft Comput 23(14):5913–5929

Kumar V, Venkatesh K, Tiwari RP (2014) Aneurofuzzy technique to predict seismic liquefaction potential of soils. Neural Netw World 24(3):249

Kurnaz TF, Kaya Y (2019) A novel ensemble model based on GMDH-type neural network for the prediction of CPT-based soil liquefaction. Environ Earth Sci 78(11):339

Lee CY, Chern SG (2013) Application of a support vector machine for liquefaction assessment. J Mar Sci Technol 21(3):318–324

Lin HM, Chang SK, Wu JH, Juang CH (2009) Neural network-based model for assessing failure potential of highway slopes in the Alishan, Taiwan Area: pre-and post-earthquake investigation. Eng Geol 104(3–4):280–289

Mittal A, Sharma S, Kanungo DP (2011) A comparison of ANFIS and ANN for the prediction of peak ground acceleration in Indian Himalayan region. In: Proceedings of the international conference on soft computing for problem solving (SocProS 2011) December 20–22. Springer, New Delhi, pp 485–495

Muduli PM, Das SK (2013) CPT-based seismic liquefaction potential evaluation using multi-gene genetic programming approach. Indian Geotech J

Muduli PK, Das SK, Bhattacharya S (2014) CPT-based probabilistic evaluation of seismic soil liquefaction potential using multi-gene genetic programming. Georisk Assess Manag Risk Eng Syst Geohazards 8(1):14–28

Najjar Y, Ali H (1998) On the use of BPNN in liquefaction potential assessment tasks. In: Attoh-Okine NO (ed) Proceedings of the international workshop on artificial intelligent and mathematical methods in pavement and geomechanical systems, pp 55–63

Nama S, Saha AK, Ghosh S (2017) Improved backtracking search algorithm for pseudo dynamic active earth pressure on retaining wall supporting c-Φ backfill. Appl Soft Comput 52:885–897

Narayanakumar S, Raja K (2016) A BP artificial neural network model for earthquake magnitude prediction in Himalayas, India. Circuits Syst 7(11):3456–3468

Pal M (2006) Support vector machines-based modelling of seismic liquefaction potential. Int J Numer Anal Meth Geomech 30(10):983–996

Panakkat A, Adeli H (2007) Neural network models for earthquake magnitude prediction using multiple seismicity indicators. Int J Neural Syst 17(01):13–33

Peng HS, Deng J, Gu DS (2005) Earth slope reliability analysis under seismic loadings using neural network. J Cent South Univ Technol 12(5):606–610

Rahman MS, Wang J (2002) Fuzzy neural network models for liquefaction prediction. Soil Dyn Earthq Eng 22(8):685–694

Ramakrishnan D, Singh TN, Purwar N, Barde KS, Gulati A, Gupta S (2008) Artificial neural network and liquefaction susceptibility assessment: a case study using the 2001 Bhuj earthquake data, Gujarat, India. Comput Geosci 12(4):491–501

Rezaei S, Choobbasti AJ (2014) Liquefaction assessment using microtremor measurement, conventional method and artificial neural network (Case study: Babol, Iran). Front Struct Civ Eng 8(3):292–307

Samui P (2011) Least square support vector machine and relevance vector machine for evaluating seismic liquefaction potential using SPT. Nat Hazards 59(2):811–822

Samui P (2013) Liquefaction prediction using support vector machine model based on cone penetration data. Front Struct Civ Eng 7(1):72–82

Samui P (2014) Vector machine techniques for modeling of seismic liquefaction data. Ain Shams Eng J 5(2):355–360

Samui P, Hariharan R (2014) Modeling of SPT seismic liquefaction data using minimax probability machine. Geotech Geol Eng 32(3):699–703

Samui P, Sitharam TG (2011) Machine learning modelling for predicting soil liquefaction susceptibility. Nat Hazards Earth Syst Sci 11(1)

Samui P, Kothari DP (2012) A multivariate adaptive regression spline approach for prediction of maximum shear modulus and minimum damping ratio. Eng J 16(5):69–78

Samui P, Karthikeyan J (2014) The use of a relevance vector machine in predicting liquefaction potential. Indian Geotech J 44(4):458–467

Samui P, Kim D, Sitharam TG (2011) Support vector machine for evaluating seismic-liquefaction potential using shear wave velocity. J Appl Geophys 73(1):8–15

Samui P, Bhattacharya S, Sitharam TG (2012) Support vector classifiers for prediction of pile foundation performance in liquefied ground during earthquakes. Int J Geotech Earthq Eng (IJGEE) 3(2):42–59

Samui P, Jagan J, Hariharan R (2016) An alternative method for determination of liquefaction susceptibility of soil. Geotech Geol Eng 34(2):735–738

Shahri AA (2016) Assessment and prediction of liquefaction potential using different artificial neural network models: a case study. Geotech Geol Eng 34(3):807–815

Sharma D, Chandra P (2019) A comparative analysis of soft computing techniques in software fault prediction model development. Int J Inform Technol 11(1):37–46

Tesfamariam S, Najjaran H (2007) Fuzzy template based modeling for assessing earthquake induced liquefaction. In: 2007 IEEE international conference on systems, man and cybernetics, pp 593–597

Thomas S, Pillai GN, Pal K, Jagtap P (2016) Prediction of ground motion parameters using randomized ANFIS (RANFIS). Appl Soft Comput 40:624–634

Thomas S, Pillai GN, Pal K (2017) Prediction of peak ground acceleration using ϵ-SVR, ν-SVR and Ls-SVR algorithm. Geomat Nat Haz Risk 8(2):177–193

Tsompanakis Y, Lagaros ND, Psarropoulos PN, Georgopoulos EC (2009) Simulating the seismic response of embankments via artificial neural networks. Adv Eng Softw 40(8):640–651

Turel M (2011) Soft computing based spatial analysis of earthquake triggered coherent landslides. Doctoral dissertation, Georgia Institute of Technology

Ulusay R, Tuncay E, Sonmez H, Gokceoglu C (2004) An attenuation relationship based on Turkish strong motion data and iso-acceleration map of Turkey. Eng Geol 74:265–291

Venkatesh K, Kumar V, Tiwari RP (2013) Appraisal of liquefaction potential using neural network and neuro fuzzy approach. Appl Artif Intell 27(8):700–720

Wang J, Rahman MS (1999) A neural network model for liquefaction induced horizontal ground displacement. J Soil Dyn Earthquake Eng 18:555–568

Xu C, Shen L, Wang G (2016) Soft computing in assessment of earthquake-triggered landslide susceptibility. Environ Earth Sci 75(9):767

Xue X, Liu E (2017) Seismic liquefaction potential assessed by neural networks. Environ Earth Sci 76(5):192

Yazdi JS, Kalantary F, Yazdi HS (2012) Prediction of liquefaction potential based on CPT up-sampling. Comput Geosci 44:10–23

Zadeh LA (1992) Foreword proceedings of the second international conference on fuzzy logic and neural networks. Iizouka, Japan, pp 13–14

Zaré M, Bard PY (2002) Strong motion data set of Turkey: data processing and site classification. Soil Dyn Earthq Eng 22:703–718

Zhang W, Goh ATC (2016) Evaluating seismic liquefaction potential using multivariate adaptive regression splines and logistic regression

Zhao HB, Ru ZL, Yin S (2007) Updated support vector machine for seismic liquefaction evaluation based on the penetration tests. Mar Georesour Geotechnol 25(3–4):209–220

Zhiyong Z, Xi Z (1999) A study of ANN based seismic response recognising system for pile foundation bridge pier. J North Jiaotong Univ (6):20

Chapter 22
Resilient Behavior of Stabilized Reclaimed Bases

Sireesh Saride and Maheshbabu Jallu

22.1 Introduction and Background

Recycling and reclaiming materials from existing distressed pavements, infrastructure such as bridges, retaining structures and buildings, and reusing them back into the civil engineering infrastructure is a practical scenario under the sustainability framework. These materials include reclaimed asphalt pavement (RAP), reclaimed concrete aggregate (RCA), construction and demolition (CD) waste and building waste (BW). Besides, several authors have enumerated different types of secondary and recycled materials which can be used in the pavement industry (Collins and Ciesilski 1994; Nehdi 2001; Saride et al. 2010). Using recycled and reclaimed materials in the pavement construction will encourage the conservation of natural resources and preservation of the environment (Taha et al. 1999). RAP is a removed and/or reprocessed pavement material and consists of high-quality aggregates coated with asphalt cement, which generally varies in the range of 4.5–6% (Sherwood 2001). Sherwood (2001) suggests that RAP is instrumental in increased stability and improved ride quality when used in the pavement base layers. However, various transportation agencies have restricted their bulk utilization as pavement base material is due to a lack of knowledge on their resilient behavior. McGarrah (2007) summarized that a majority of the State Department of Transportation (DoT) in the USA had allowed 20% RAP in the base course layer while a few DoTs have allowed even up to a maximum of 50%. In all these mixes, RAP was blended with virgin aggregate (VA).

Nevertheless, several researchers have pointed out that stabilizing RAP:VA mixes with conventional cementitious materials like cement and lime have yielded superior strength and stiffness properties (Taha et al. 2002; Puppala et al. 2009;

S. Saride (✉) · M. Jallu
Department of Civil Engineering, Indian Institute of Technology, Hyderabad,
TS 502285, India
e-mail: sireesh@ce.iith.ac.in

T. G. Sitharam et al. (eds.), *Latest Developments in Geotechnical Earthquake Engineering and Soil Dynamics*, Springer Transactions in Civil and Environmental Engineering, https://doi.org/10.1007/978-981-16-1468-2_22

Yuan et al. 2010). With the substantial availability of industrial by-products, stabilizing RAP:VA blends with these materials could not only reduce the cost of construction but also protect the environment. Especially, fly ash and ground-granulated blast furnace slag (GGBS) have been widely considered to stabilize the recycled bases (Saride et al. 2015; Arulrajah et al. 2017; Mohammadinia et al. 2017; Saride and Jallu 2020). To adopt cement-treated base (CTB) in pavements, several transportation agencies have stipulated guidelines in terms of target strength and stiffness of the materials after 7–28 days curing. State DoTs in the USA has suggested an unconfined compressive strength (UCS) of 2.07 MPa (300 psi) (Texas Department of Transportation 2014), while the Australian road agency has suggested a UCS of 1.5 MPa for the stabilized bases (Austroads 2006). According to the Indian Roads Congress (IRC), the minimum required UCS is 4.5 MPa/7 MPa for 7 and 28 days curing, respectively (IRC 2018). Besides, IRC also suggests a minimum resilient modulus (M_r) of 450 MPa. The resilient modulus is defined as a ratio of cyclic deviatoric stress to the recoverable (resilient) strain. However, no other transportation agency has recommended a minimum M_r value for the stabilized bases. Nevertheless, M_r value is the only mechanistic parameter used in any pavement design approach.

Janoo (1994) had conducted the falling weight deflectometer (FWD) test on various field test sections built with different RAP materials to back analyze the layer's M_r. Taha et al. (2002) have used the correlation between UCS and M_r to determine the M_r of the base layer. Bennert and Meher (2005) have reported that the M_r values also increased with an increase in the percentage of RAP, through a series of laboratory tests on RAP materials. Besides, the M_r tests on cement and cement-fiber-treated RAP mixes have shown the M_r values drastically improved against RAP mixes without stabilization. A 30% higher structural coefficient calculated based on the M_r of cement-treated RAP was noted (Puppala et al. 2009). Recently, Saride et al. (2016) have shown through UCS and M_r tests that these properties can be enhanced with an alkali activator used along with a low-calcium FA to stabilize the RAP:VA bases.

Though there are a decent number of studies on reclaimed bases, limited studies have characterized these materials and understood their resilient behavior, especially when stabilized with low-calcium fly ashes. Besides, limited studies are available on describing how to use this mechanistic behavior of reclaimed materials in the pavement design protocols. To note, several transportation agencies have adopted the M_r of pavement layers in the design.

22.1.1 Current Design Aspects

IRC Method of Design of Flexible Pavements. IRC proposes to design a flexible pavement as an elastic multilayer system (IRC 2018). The stresses and strains at critical locations are computed using the linear elastic solutions. The design thickness of the pavement is a function of elastic modulus of each pavement layer, traffic loading and the subgrade condition in terms of California bearing ratio (CBR). The pavement distresses that come from the repeated application of traffic are regarded as critical for the design, which is described in Fig. 22.1:

1. Vertical strains (compressive) on top of the subgrade (referred to as rutting strain),
2. Horizontal strains (tensile) at the bottom of the bituminous layer (referred to as fatigue strain),
3. Horizontal strain (tensile) at the bottom of the CTB layer.

Fatigue model. The asphalt layer experiences the flexural fatigue cracking if the tensile strain exceeds a limiting value. The relationship between the fatigue life of the pavement and the limiting tensile strain at the bottom of the bituminous layer (Fig. 22.1) for 90% reliability is given as:

$$N_f = 0.5161 \times C \times 10^{-04} \times [1/\varepsilon_t]^{3.89} \times [1/M_r]^{0.854} \tag{22.1}$$

where

N_f = fatigue life in the number of loading cycles,
$C = 10_M$; $M = 4.84\ [(V_{be}/V_a + V_{be}) - 0.69]$,

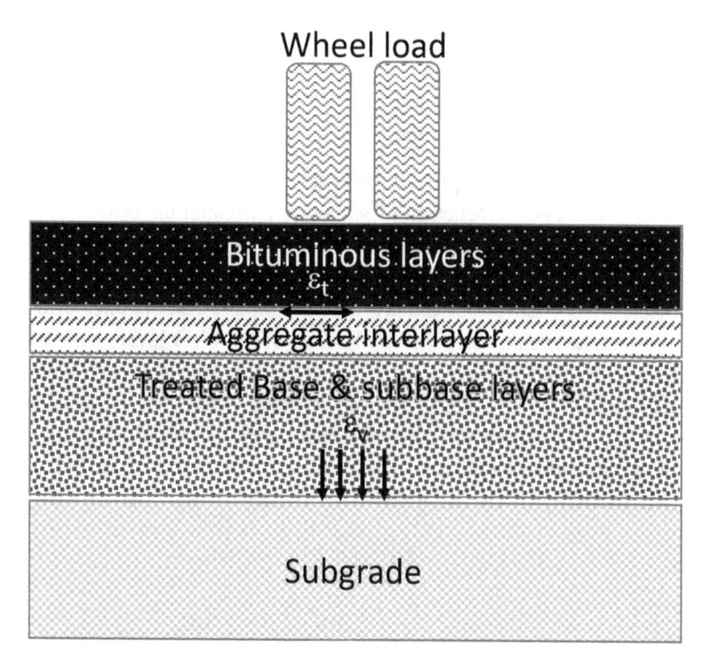

Fig. 22.1 Location of critical strains in a flexible pavement system

V_a = volume of air voids in the asphalt layer, V_{be} = volume of effective bitumen content in the asphalt layer,

ε_t = tensile strain at the bottom of the asphalt layer and

M_r = resilient modulus of the asphalt layer.

When a cement-treated base (CTB) is used, the fatigue life of the CTB shall also be verified at the bottom of the CTB layer, and cumulative fatigue damage has to be verified. The number of fatigue cycles can be obtained from Eq. 22.2.

$$N = \mathrm{RF}\left[\frac{\left(\frac{113000}{E^{0.804}} + 191\right)}{\varepsilon_t}\right]^{12} \tag{22.2}$$

where

N = number of loading cycles (each standard axle load cycle consists 0.8 MPa),

RF = reliability factor = 1 for roads with traffic >10 msa and 2 for other roads,

ε_t = tensile strain below the CTB layer (microstrain) and

E = resilient modulus of the CTB layer (MPa).

Rutting model. If the vertical strains in the subgrade exceed the limiting value, the pavement fails with rutting cracks on its surface. Hence, the limiting or allowable number of load repetitions to control the permanent deformation for 90% reliability is calculated using Eq. 22.3:

$$N = 1.41 \times 10^{-8} \times [1/\varepsilon_v]^{4.5337} \tag{22.3}$$

where

N = number of loading cycles and

ε_v = vertical strain in the subgrade.

Design procedure. The design procedure is a trial and error method in which the pavement layer thickness is a function of subgrade CBR and the design traffic. With a trial section, the strains at critical locations are obtained using the IITPAVE program. The volume of traffic, number of pavement layers, the thickness of each pavement layer and the elastic properties of each layer are the inputs. The obtained critical strains, along with horizontal tensile strains at the bottom of the CTB layer, will be compared with the limiting strains. If the computed strains are less than the limiting strains, then the assumed thicknesses are safe. The layer elastic modulus is calculated using available correlations between the layer thickness and the modulus of the supporting layer. The following relationships are being used to obtain the M_r of the subgrade.

The elastic modulus of the subgrade is calculated using Eqs. 22.4 and 22.5.

$$M_r(\mathrm{MPa}) = 10 \times \mathrm{CBR} \;\text{ for }\; \mathrm{CBR} < 5 \tag{22.4}$$

$$M_r(\mathrm{MPa}) = 17.6 \times (\mathrm{CBR})^{0.64} \text{for CBR} > 5 \tag{22.5}$$

The M_r of unbound granular subbase/base course layers is calculated as per Eq. 22.6.

$$M_{r\text{granular}} = 0.2(h)^{0.45} \times M_{r\text{support}} \tag{22.6}$$

where

h = thickness of the granular layer,

$M_{r\ \text{granular}}$ = resilient modulus of the granular layer (MPa),

$M_{r\ \text{support}}$ = resilient modulus of the supporting layer (MPa).

These expressions are valid for unbound base/subbase layers. However, when the granular layers are cement-treated, an M_r of 5000 MPa is suggested for the design. Besides, the modulus of rupture (flexural strength) of CTB shall be verified. IRC has specified the stabilization of natural aggregates with either cement or cementitious materials. Nevertheless, it is to be noted that the M_r of FA-based stabilizers may not yield an M_r of 5000 MPa, especially when the reclaimed bases or marginal bases are adopted, which needs to be understood.

AASHTO Design Methodology for Flexible Pavements. The American Association of State Highways and Transportation Officials (AASHTO) has been continuously evolving with the mechanistic pavement design procedures, in which M_r of every layer is taken into consideration. The following design procedure is adopted by the AASHTO pavement design guide (AASHTO 1993).

The empirical expression relating traffic, pavement structure and pavement performance for flexible pavements is:

$$\log_{10}(W_{18}) = (Z_R S_0) + 9.36\log_{10}(S_N + 1) - 0.2 + \frac{\log_{10}\left[\frac{\Delta PSI}{4.2-1.5}\right]}{0.4 + \frac{1094}{(S_N+1)^{5.19}}} + 2.32\log_{10}(M_r) - 8.0 \tag{22.7}$$

where

W_{18} = number of 18 kip equivalent single axle loads (ESALs),

Z_R = standard normal deviation (function of the design reliability level),

S_o = overall standard deviation (function of overall design uncertainty),

ΔPSI = allowable serviceability loss at the end of design life,

M_r = subgrade resilient modulus,

SN = structural number (a measure of required structural capacity).

The above parameters (first five) are the typical inputs to the design equation, and SN is the output. Equation (22.7) needs to be solved for the SN.

The structural number is defined as:

$$\text{SN} = a_1 D_1 + a_2 D_2 m_2 + a_3 D_3 m \tag{22.8}$$

where

a_1, a_2 and a_3 are the layer coefficients of the asphalt, base and subbase layers, respectively, which are the function of M_r of that particular layer.

m_1, m_2 and m_3 are the drainage coefficients of the asphalt, base and subbase layers, respectively.

D_1, D_2 and D_3 are the thicknesses of these layers.

According to the AASHTO (1993), a_2 and a_3 can be obtained from the following equations.

$$a_2 = 0.249 \times \log(M_r) - 0.977 \tag{22.9}$$

$$a_3 = 0.227 \times \log(M_r) - 0.839 \tag{22.10}$$

where

a_2 = layer coefficient of the base layer, a_3 = layer coefficient of subbase layer and M_r = resilient modulus of base and subbase layers in psi.

The asphalt layer coefficient (a_1) can be directly obtained from the design guide (AASHTO 1993).

In the case of CTBs, the base layer coefficient (a_2) shall be obtained from the nomogram provided, which relates the a_2 to strength (UCS) and stiffness (M_r) of the CTB.

The layer coefficients obtained from the nomogram are valid only for Portland cement-treated virgin aggregates. When the low-calcium-based stabilizers are adopted that too to treat secondary aggregate materials, the nomogram may not be useful. Very recently, Saride and Jallu (Saride and Jallu 2020) have demonstrated that the layer coefficients obtained from the nomogram yield an inferior pavement section, which would experience premature failures. Hence, a new set of expressions were proposed for determining the layer coefficients of FA geopolymer-stabilized RAP:VA bases.

Once the layer coefficients are calculated, the obtained SN (Eq. 22.8) shall be greater than the SN required (Eq. 22.7) for given traffic, subgrade M_r, serviceability and the reliability index of the pavement structure. Note that there may be many layer thickness combinations that can provide satisfactory SN values. However, in order to arrive at the optimal design, the economy must be considered.

Australian Design Methodology for Flexible Pavements. According to Austroads (2017), the elastic modulus is an important design parameter that induces the stress–strain response of the pavement layers against the traffic loading. The cement-treated aggregate layers are complex in nature, and their performance is a function of elastic modulus and the other mechanistic properties. Similar to other pavement design guidelines, Austroads (2017) also considers the rutting and fatigue performance criteria to predict when distress will occur.

Rutting criteria. The limiting subgrade rutting strain equation derived based on the mechanistic–empirical procedure for 90% reliability is given as

$$N = (9150/\mu\varepsilon)^7 \quad (22.11)$$

where
N = allowable number of standard axles,
$\mu\varepsilon$ = the vertical compressive strain on top of the subgrade.

Fatigue Criteria. To obtain the fatigue strains below the bituminous layers of conventional asphalt pavement is given in Eq. 22.12

$$N = \mathrm{SF}/\mathrm{RF}\left[\{6918(0.856\ V_b + 1.08)\}/E^{0.36}\mu\varepsilon\right]^5 \quad (22.12)$$

where
N = allowable number of load repetitions in terms of standard axles,
$\mu\varepsilon$ = tensile strain below the bituminous layer,
V_b = volume of bitumen in the asphalt (%),
E = elastic modulus of the bituminous layer (MPa),
SF = shift factor based on the laboratory and field fatigue life of the asphalt layer (approximate value = 6),
RF = reliability factor (3.9 for 90% reliability).

Cement-treated bases. For cement-treated bases, the fatigue criteria for CTBs alone need to be checked along with the fatigue criteria for bituminous layers. Equation 22.13 is used to obtain the number of load repetitions when CTBs are adopted.

$$N = (K/\mu\varepsilon)^{12} \quad (22.13)$$

where
N = number of allowable load repetitions,
$\mu\varepsilon$ = tensile strains below the CTB layer,
K = presumptive constant based on aggregate quality varies from 233 to 261 for cement-treated base quality aggregates to subbase quality natural gravels.

The design procedure involves first evaluating the material properties, estimating the traffic and climatic conditions, and then selecting a trial pavement section to analyze the allowable traffic. If the allowable traffic is less than the design traffic, the trial pavement section will be accepted. From the detailed literature search, it can be summarized that there are no definite guidelines to adopt recycled bases stabilized with low-calcium additives. This is due to the minimal understanding of the resilient behavior of stabilized recycled bases.

22.2 Materials and Methods

22.2.1 *Materials*

Reclaimed Asphalt Pavement (RAP). The RAP materials used in this study were sourced from a milling operation on a National Highway project near Rajahmundry (RAP-R) in India and Melbourne (RAP-M), Australia. The particle size distribution of RAP materials is shown in Fig. 22.2. The RAP materials can be classified under the A-1-a group (AASHTO 2009). The RAP was regraded to meet the gradation of wet mix macadam (WMM) proposed by the Ministry of Road Transport and Highways (MoRTH 2013). Modified Proctor compaction tests were performed to determine the maximum dry unit weight (MDU) and optimum moisture content (OMC) of the RAP material. The OMC and MDU of RAP-R and RAP-M were found to be 6.28% and 21.3 kN/m^3, and 5.63% and 21.4 kN/m^3, respectively.

Fly Ash (FA). Four distinct fly ashes (FAs) are obtained from Vijayawada (FA-V); Ramagundam (FA-R); and Neyveli (FA-N) thermal power plants in India and Queensland (FA-Q) from Australia. The grain size distribution of FAs is shown in Fig. 22.2. According to the Unified Soil Classification System (USCS), FA-V, FA-R and FA-N were classified as silty sand (SM), and FA-Q was categorized as

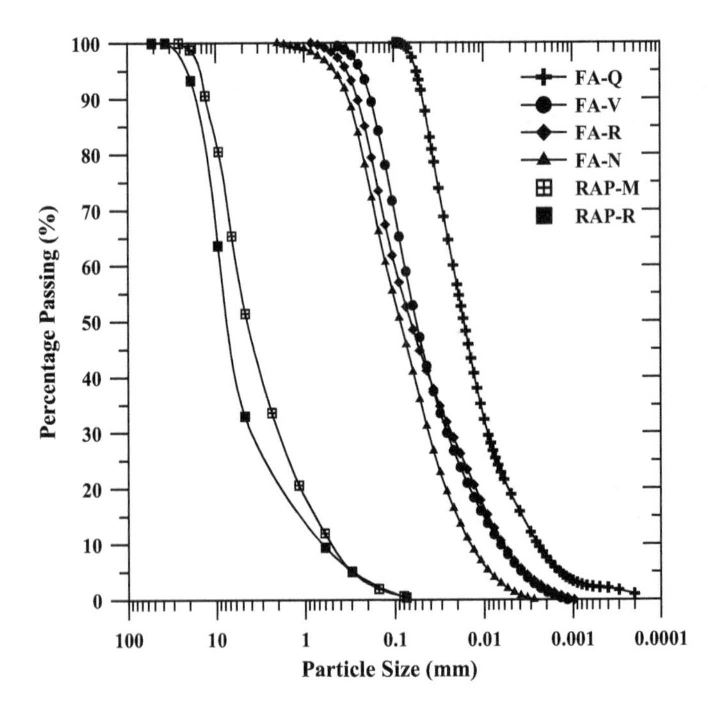

Fig. 22.2 Particle size distribution of RAP and FA

silt (M). Besides, as per ASTM C618 (2012a), FA-Q, FA-V and FA-R were classified as low-calcium '*class F*' (where $SiO_2 + Al_2O_3 + Fe_2O_3 > 70\%$ and $CaO < 10\%$), and FA-N was categorized under high-calcium '*class C*' (where $SiO_2 + Al_2O_3 + Fe_2O_3 > 70\%$ and $CaO > 10\%$) fly ashes.

Liquid Alkaline Activator (LAA). To activate the FAs to improve their reactive potential, liquid alkaline activator was prepared, which is a mixture of sodium silicate (Na_2SiO_3) and sodium hydroxide (NaOH) in a certain proportion. A 98% pure commercially available sodium hydroxide (NaOH) pellets were used. A laboratory-grade sodium silicate with a specific gravity of 1.5 was used for this purpose. A combination of Na_2SiO_3:NaOH is represented as liquid alkali activator (LAA) in this study. From the previous experience (Saride and Jallu 2020; Jallu et al. 2020), the NaOH molar concentration was also fixed at 0.5 M, 1 M, 3 M and 3 M for FA-R, FA-N, FA-Q and FA-V, respectively.

22.2.2 Test Methods

Before conducting the M_r and UCS tests on RAP:VA blends, modified Proctor compaction tests were conducted in accordance with the ASTM D1557 (2012b). A best-performing mix combination representing 60% RAP and 40% VA blended with 20% FA, denoted as 60R:40V + 20F, was selected based on the previous experience (Avirneni et al. 2016; Puppala et al. 2017). The maximum dry unit (MDU) weights and optimum moisture contents (OMCs) were found to be in the range of 21.5–22.5 kN/m^3 and 6–8%, respectively, for the FA-based RAP:VA mixes.

Resilient Modulus (M_r) Tests. M_r is a measure of material stiffness and a response to cyclic or repeated traffic loading. It is an important mechanical property used in the analysis and design of pavements. M_r is defined as a ratio of applied cyclic deviator stress to recoverable or resilient strain. M_r of material is tested using cyclic triaxial test equipment in which cyclic loads are applied on the specimen to simulate the traffic wheel loads experienced by the pavement layers.

In this study, the M_r tests were performed in accordance with AASHTO T 307–99 (2003) using a cyclic triaxial test setup as shown in Fig. 22.3. The cylindrical specimens of size 100-mm diameter and 200-mm height were prepared, and the specimens were cured for 28 days in a humidity chamber at 95% relative humidity and 27 °C. The samples were then subjected to different combinations of axial cyclic deviatoric stresses and confining stresses, which will simultaneously be applied for a prescribed number of load cycles. These dynamic stress conditions are suggested based upon the location of the element within the base/subbase layers in a pavement system, as specified by AASHTO (2003). The cyclic deviatoric stress was applied using a haversine load pulse represented in Fig. 22.4. The load pulse consists of a loading period of 0.1 s and a relaxation period of 0.9 s, maintaining a frequency of 1 Hz in each loading cycle.

Fig. 22.3 Cyclic triaxial test

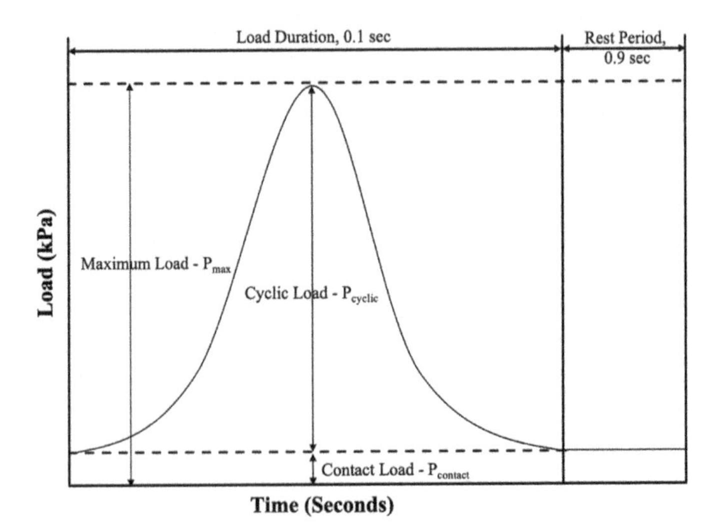

Fig. 22.4 A haversine load pulse applied in resilient modulus test

The axial deviator stress is composed of cyclic stress and contact stress, in which the cyclic stress constitutes 90% of the total axial deviatoric stress, and the remaining 10% stress constitutes the contact stress. The 10% contact stress is referred to as seating stress to maintain positive contact between the loading cap and the specimen. The test sequences consist of five confining pressures, and each confining pressure corresponds to different axial cyclic deviator stresses. Hence, every sample would be subjected to 15 different combinations of cyclic deviatoric stresses and confining pressures (Fig. 22.5). Under a combination, the axial cyclic deviatoric stress is repeated 100 times, and subsequently, the stress conditions are gradually increased on the specimen. During each test, the vertical deformations of the sample were recorded at the mid-height of the specimen through two linear variable deformation transducers (LVDTs).

The average resilient (elastic) vertical strains for the last five cycles (96–100) were recorded separately. The remaining plastic component of the vertical strain is referred to as rutting. The M_r was calculated from Eq. 22.14.

$$M_r = s_d / e_e \tag{22.14}$$

where

M_r = resilient modulus,

σ_d = cyclic deviatoric stress,

ε_e = axial elastic strain (average value taken for the last 5 cycles).

Generally, an average resilient modulus obtained from 15 different load combinations is reported. According to IRC:37 (IRC 2018), a minimum resilient modulus of 450 MPa is recommended at 28 days curing period for stabilized bases.

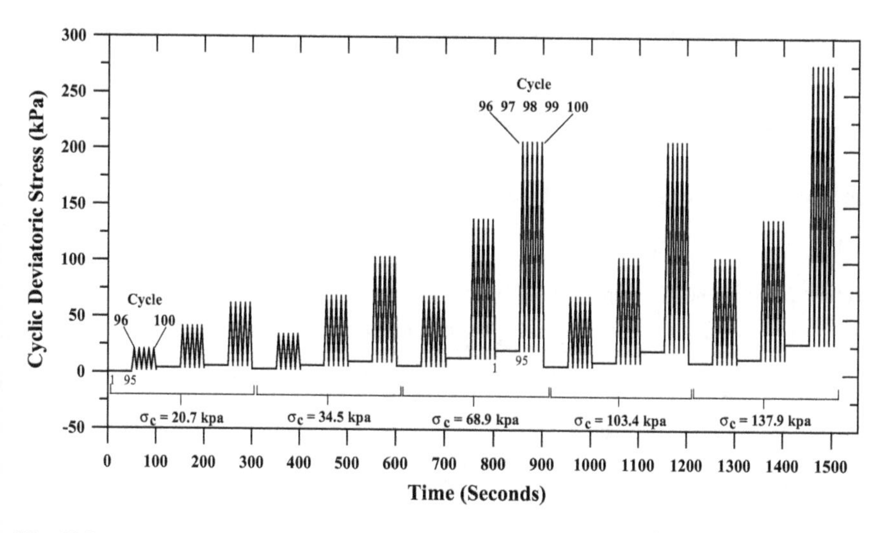

Fig. 22.5 Variation of axial cyclic deviatoric stress with time for different confining pressures in a resilient modulus test

UCS Studies. The UCS tests were conducted in accordance with the ASTM D1633 (2017) on 60R:40V + 20F specimens cured at 28 days. The UCS tests were performed to check the efficiency of the alkali-activated FA-stabilized RAP:VA mixes. The cylindrical specimens of 100 mm × 200 mm (diameter × length) size were prepared at MDU and OMC and cured at 27 °C temperature with 95% relative humidity. As per the IRC:37 (IRC 2018) recommendations, the 28-day cured specimen shall have a UCS $\geq$ 4.5 MPa for low-calcium-based stabilizers.

22.2.3 *Testing Program*

The following tests (Table 22.1) were conducted to understand the resilient behavior of reclaimed bases.

22.3 Results and Discussion

22.3.1 *M_r Studies*

The resilient modulus tests were initially performed on 100% RAP stabilized with raw FA to understand the resilient behavior of FA-stabilized RAP. These tests were conducted on specimens stabilized with all the FAs cured for 1, 7 and 28 days. The average M_r of the specimens stabilized with four different FAs is presented in Fig. 22.6. The M_r of 100% RAP has increased with an increase in the FA content by weight and curing time. The M_r of the 100% RAP is in the range of 250 to 450 MPa. Though the M_r reaches the target value of 450 MPa at 30% FA according to IRC:37 (IRC 2018), it is not feasible to add 30% FA by weight to the mix as it would decrease the aggregate content in a given volume of the specimen, which will eventually lead to low structural strength. Besides, relatively lower M_r values were observed for 100% RAP specimens perhaps due to insufficient bonding and pozzolanic reactions between the RAP and FA (Saride et al. 2016). Hence, it has

Table 22.1 Scheme of experiments performed in the study

RAP (%)	VA (%)	FA-Q, FA-V, FA-R, FA-N (%)	LAA (%)	M_r	UCS
100	0	10	0	O	⊗
100	0	20	0	O	⊗
100	0	30	0	O	⊗
60	40	20	0	⊗	O
60	40	20 (FA-Q, FA-V, FA-R)	50:50	O	O
60	40	20 (FA-N)	70:30	O	O

Note O—yes; ⊗ —no. The percent shown in the table is by weight

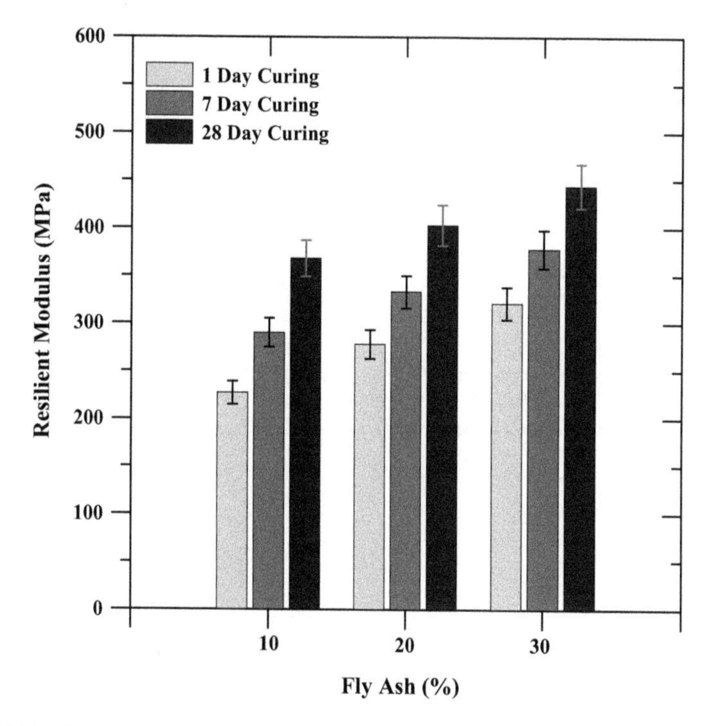

Fig. 22.6 Resilient modulus of 100% RAP with different FA dosages

necessitated adding VA to the matrix. To further enhance the mix stability, it was proposed to activate the FA with a liquid alkali activator (LAA). As discussed in Sect. 2.1, the LAA ratio (Na_2SiO_3:NaOH) was prepared in 50:50 and 70:30 proportions. It is to be noted that except for FA-N, the optimum LAA ratio was found to be 50:50 for remaining specimens with FA-V, FA-R and FA-Q; for FA-N, it was found to be 70:30. The variation in LAA was attributed to the chemical composition and reactive potential of the silica and alumina present in the FA.

The variation of M_r of FA geopolymer-stabilized 60R:40V + 20F specimens cured for 28 days is presented with respect to the cyclic deviatoric stress and different confining pressures (Fig. 22.7). The M_r of each mix has increased with an increase in cyclic deviatoric stress and effective confining pressure. Irrespective of the FA content, the improvement in M_r has reduced for axial cyclic deviatoric stress beyond 100 kPa. In other words, the increase in M_r is marginal for a higher stress combination. The LAA ratio has a significant influence on the M_r values. Except for FA-Q-stabilized RAP:VA mixes, the M_r of 60R:40 V + 20F mix stabilized with all other FAs has reached about 1600 MPa, which is about fourfold increase with respect to 100% RAP specimens. The 60R:40V + 20F-Q samples have obtained a M_r value of 1200 MPa, a threefold increase with respect to 100% RAP specimens.

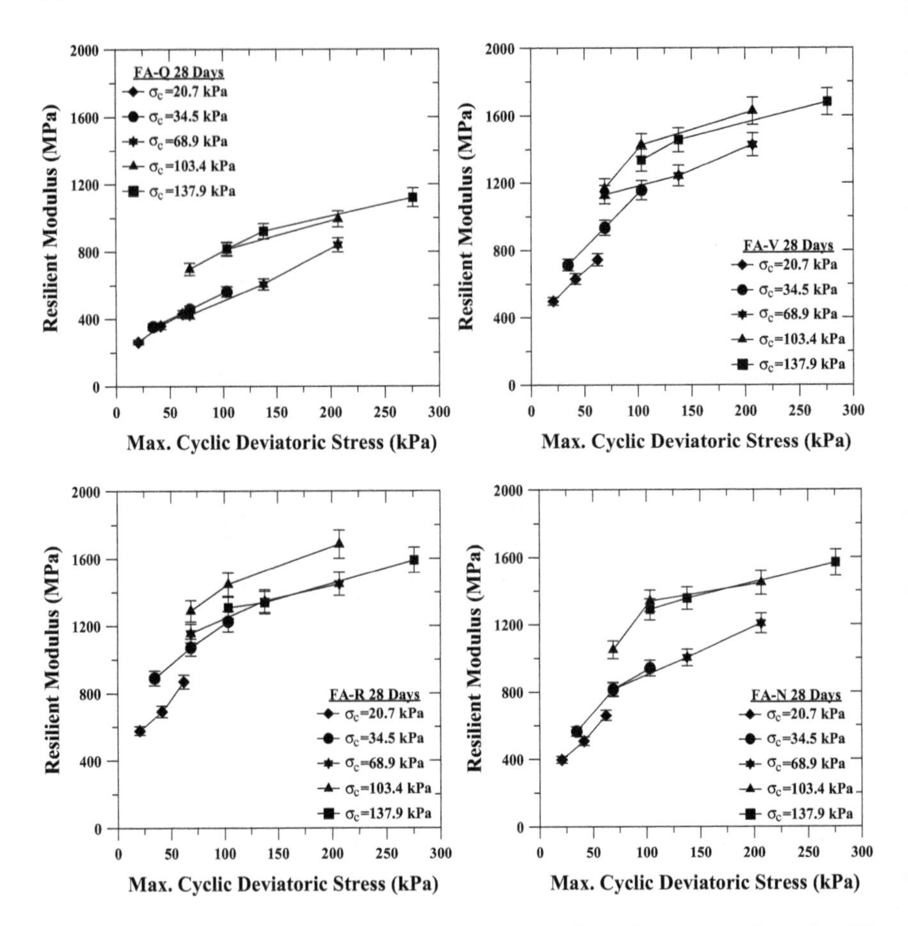

Fig. 22.7 Variation of resilient modulus with maximum cyclic deviatoric stress for various FA stabilized 60R:40V + 20F specimens

a. **UCS Studies**

Figure 22.8 presents the UCS of 28-day cured specimens with different FAs and LAA combinations. The UCS values of design mixes without activation are trivial ($\sim$0.5 MPa), indicating that the FAs have not produced enough pozzolanic compounds, which could enhance the strength of the mixes. With the LAA activation (LAA ratios between 50:50 and 70:30), the design mixes have shown significant strength improvement.

The improved performance of class F FA-treated specimens could be attributed to the dissolution of Si^{4+} and Al^{3+} ions in the alkali environment (Palomo et al. 1999; Rattanasak and Chindaprasirt 2009). A maximum UCS of 6.8 MPa was observed for an LAA ratio of 50:50 at 28-day curing period in FA-Q specimens and a minimum UCS of 5.5 MPa in FA-V specimens.

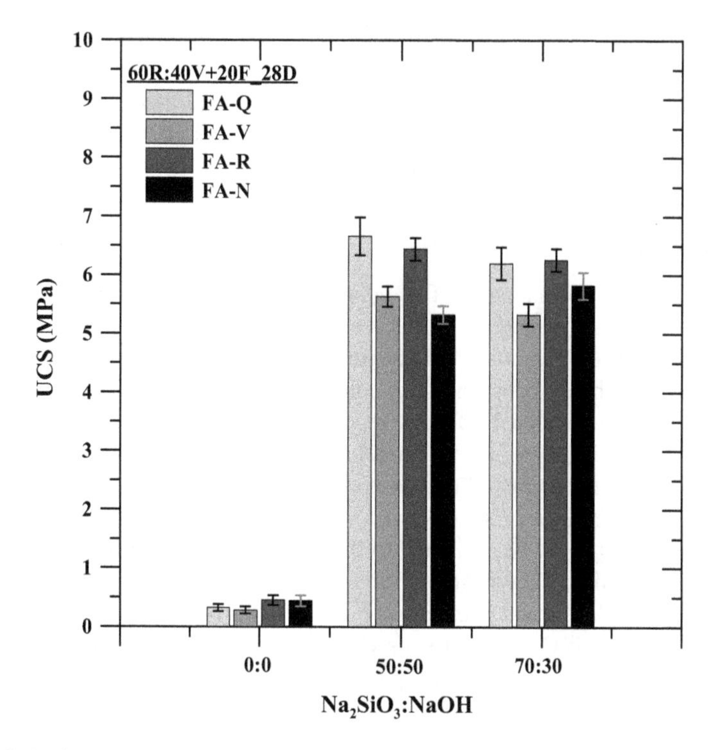

Fig. 22.8 Variation of UCS with LAA ratio for different FAs

These mixes have met the design requirement of CTBs of 4.5 MPa according to the IRC:37 (IRC 2018) and 1.5 MPa by Austroads (2006). However, these recommendations are for CTBs with conventional virgin aggregates. It is to be noted that for reclaimed materials, these standards seem high; nevertheless, to qualify as base materials, all the materials have to meet the criteria. Therefore, to enhance the reactive potential of FA, geopolymerization is inevitable, when the secondary aggregates are adopted in the pavement bases.

22.4 Design of Flexible Pavements with Stabilized Bases

The design procedures adopted by various transportation agencies can be seen in Sect. 1.1, which may be adopted to design the flexible pavements when stabilized bases are proposed. However, these design guidelines emphasize on stabilization of natural aggregates with calcium-rich additives like lime and cement as well as low-calcium stabilizers such as fly ash and steel slag. When secondary materials such as reclaimed pavement materials or weak/marginal aggregates are used, these guidelines may not provide complete direction for the design. In other words, these guidelines did not mention the use of marginal aggregates treated with low-calcium stabilizers. Hence, in this study, an attempt has been made to develop design charts based on the resilient modulus of FA-stabilized reclaimed bases. For this purpose, the general guidelines proposed in IRC:37 (IRC 2018) for conventional bases were followed, as the M_r values of these design mixes are not very high to meet the CTB specifications (M_r = 5000 MPa).

b. **Design Steps**

Step 1 To establish a range of M_r values for the reclaimed bases stabilized with Portland cement, FA, slag, cement kiln dust, etc., data were collected from the literature and are summarized in Table 22.2. The M_r ranges anywhere between 250 and 1500 MPa. However, the M_r of cement-treated RAP was reported in the range of 3380–3726 (Taha et al. 2002), which is much higher than the general range reported by other researchers (Puppala et al. 2009, 2012; Arulrajah et al. 2017). The higher range of M_r observed in the cement-stabilized RAP (Taha et al. 2002) could be attributed to the adopted correlation between UCS and modulus. They have not directly measured the M_r. The M_r of bituminous layers and granular subbase (GSB) is obtained as prescribed in the codes (IRC (Indian Roads Congress) 2018). The range of each pavement layer's elastic properties used in the analyses is presented in Table 22.3.

Table 22.2 Range of resilient modulus of stabilized reclaimed bases

S. no.	M_r range (MPa)	Type of base material	References
1	3381–3726	Cement-stabilized RAP	Taha et al. (2002)
2	420–525	Cement-fiber-treated RAP	Puppala et al. (2009)
3	243–510	FA-stabilized BRG, GAB and URM	Cetin et al. (2010)
4	400–460	Cement-treated RAP	Puppala et al. (2012)
5	300–550	FA-stabilized RAP:VA	Saride et al. (2015)
6	400–1220	CKD-stabilized RCA	Arulrajah et al. (2017)
7	300–750	CKD-stabilized RAP	Arulrajah et al. (2017)
8	325–600	FA-stabilized RAP	Mohammadinia et al. (2017)
9	350–800	CKD-stabilized CB	Mohammadinia et al. (2018)
10	300–1500	FA geopolymer-stabilized RAP: VA	Saride and Jallu (2020)

Note RAP: reclaimed asphalt pavement, VA: virgin aggregate, BRG: bank run gravel, GAB: graded aggregate base, URG: unpaved road gravel, RCA: recycled concrete aggregate, CB: crushed brick, FA: fly ash and CKD: cement kiln dust

Table 22.3 Input parameters used in the design

Layer	Modulus (MPa)	Poisson's ratio
Bituminous layer	3000	0.35
Aggregate interlayer	450	0.35
Stabilized base/subbase	300–1500	0.25
Subbase layer	250	0.35
Subgrade	30–72	0.35

Step 2 A trial thickness of each pavement layer is computed from the design catalogs given in the IRC:37 (IRC 2018) for relevant traffic and subgrade conditions or assumed accordingly for the materials which are not specified in the design catalogs.

Step 3 As discussed, the fatigue strain (ε_t) and rutting strain (ε_v) at critical locations, shown in Fig. 22.1, are computed using the IITPAVE program for the trial section considered in Step 2. These critical strains are compared with the allowable strains for a given traffic loading. Note that the fatigue strains mobilized under these stabilized recycled base layers (prescribed against CTBs in the code) have not verified separately. This is to avoid an uneconomical layer thickness of moderately treated base layers against CTBs whose M_r is suggested to be 5000 MPa. Therefore, in this present study, the design of stabilized reclaimed bases is proposed to design as conventional bases relatively with a higher M_r value. However, a drainage layer has to be considered above the stabilized base layers to provide sufficient drainage. This layer will also act as a stress absorbing interlayer, which can control the propagation of shrinkage cracks to the bituminous layers.

Step 4 For better pavement performance, the fatigue and rutting strains (obtained in Step 3) should be less than the limiting fatigue and rutting strain (obtained as per Sect. 1.1). If the mobilized strains were exceeding or very low compared to the limiting strains, the thickness of bituminous and stabilized base layers must be revised (increased/decreased) accordingly.

The design procedure explained is an iterative process, and it consumes a lot of man-hours. To overcome this limitation, design charts are proposed in this study for different traffic and subgrade conditions using a wide range of reclaimed materials with different moduli (Table 22.2), so that the designer can easily use them once the elastic modulus of each layer is established. Figure 22.9 presents the design charts for subgrade M_r of 30 MPa, 50 MPa, 60 MPa and 70 MPa, which corresponds to a CBR value of 3%, 5%, 7% and 9%, respectively, and for the asphalt layer M_r of 3000 MPa. For a given set of subgrade and other layer characteristics, and traffic conditions, an increase in the M_r of stabilized reclaimed base has reduced the required thickness of that layer. As expected, with an increase in the traffic volume, the required thickness of the base layer has increased. In addition, with an increase in subgrade stiffness, the base layer thickness has decreased. It is important to note

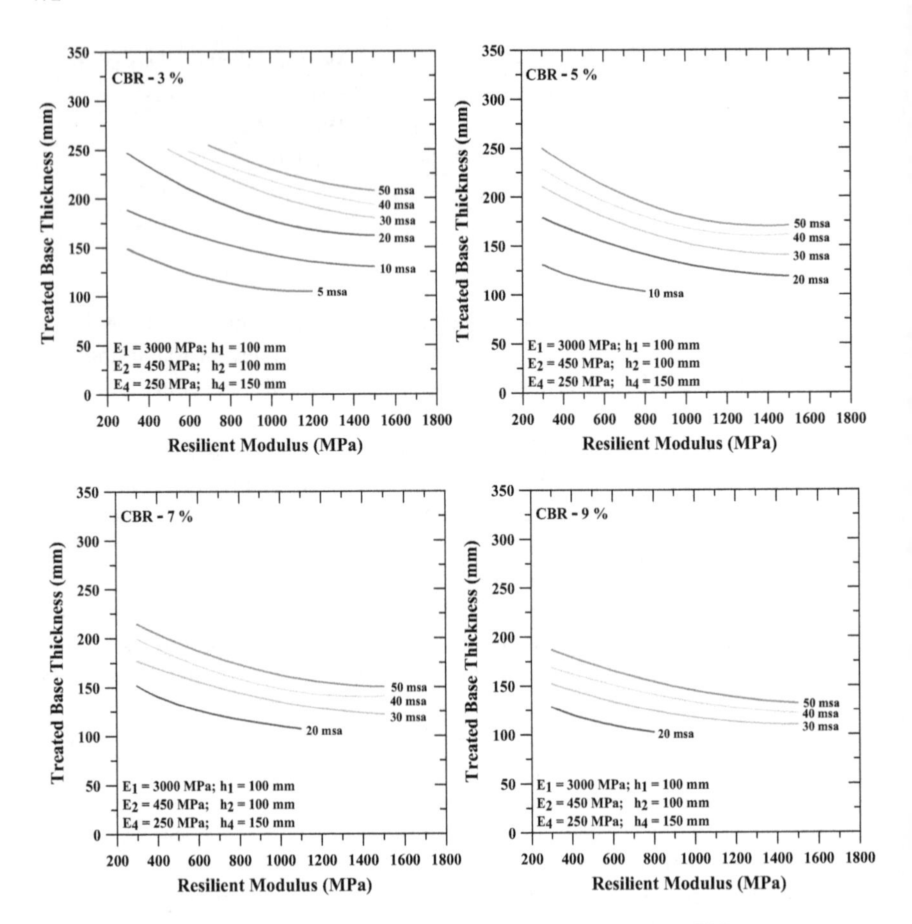

Fig. 22.9 Design charts for surface M_r of 3000 with a different subgrade CBR

that for a given low traffic volume, especially for 5 msa on the subgrade CBR values of 5, 7 and 9%, the design thickness of the base layer is observed to be lower than a minimum thickness suggested (IRC 2018). Hence, the prescribed minimum thickness of 100 mm is suggested for these cases. In addition, the thickness of the base layer beyond 250 mm would be uneconomical, and hence, it is considered as the upper limit for the treated base layers.

22.5 Conclusions

A series of resilient modulus (M_r) and unconfined compressive strength (UCS) tests were performed on RAP:VA bases stabilized with FA geopolymer. Four different FAs and two different RAP materials obtained from India and Australia were tested.

The resilient behavior of FA geopolymer-stabilized 60R:40V + 20F specimens cured for 28 days was determined, and the following conclusions were drawn.

The M_r of 100% RAP stabilized with fly ash (FA) was in the range of 250–450 MPa. An FA content of 20% by weight was considered as an optimum dosage for enhancing the resilient behavior of the base mixes.

The liquid alkali activator has a significant impact on the resilient modulus and strength of low-calcium FA-stabilized RAP bases. An LAA ratio of 50:50 is found to be optimum for most of the FAs to activate the silica and alumina ions. As high as a fourfold increase in M_r is noticed in the majority of design mixes tested against 100% RAP stabilized with 20% FA.

The resilient behavior of 60R:40V + 20F design mix has increased with an increase in the cyclic deviatoric stress and confining pressure. However, at higher confining and cyclic deviatoric stresses, the resilient modulus of all the combinations seems to become constant at around 1600 MPa.

The LAA ratio has a significant influence on the strength of the design mixes as well. A 12-fold increase in UCS is observed when an LAA ratio of 50:50 is adopted. Since the design mix 60R:40V + 20F meets both $\mathrm{M_r}$ and strength criteria, design charts were proposed for the FA geopolymer-stabilized reclaimed bases. A set of design charts based on $\mathrm{M_r}$ were proposed for moderately stabilized reclaimed bases.

References

AASHTO (1993) Guidelines for design of pavement structures. AASHTO, vol 1, Washington, DC

AASHTO (2003) Standard method of test for determining the resilient modulus of soils and aggregate materials. AASHTO T307–99, Washington, DC

AASHTO (2009) Classification of soils and soil-aggregate mixtures for highway construction purposes. AASHTO M145–91, Washington, DC

Arulrajah A, Mohammadinia A, D'Amico A, Horpibulsuk S (2017) Cement kiln dust and fly ash blends as an alternative binder for the stabilization of demolition aggregates. Constr Build Mater 145:218–225

ASTM (2012a) Standard specification for coal fly ash and raw or calcined natural pozzolan for use in concrete. ASTM C618, West Conshohocken, PA

ASTM (2012b) Standard test methods for laboratory compaction characteristics of soil using modified effort. ASTM D1557, West Conshohocken, PA

ASTM (2017) Standard Test Methods for Compressive Strength of Molded Soil-Cement Cylinders. ASTM D1633, West Conshohocken, PA

Austroads (2006) Guide to pavement technology. Part 4D: stabilised materials. AGPT04D/06, Sydney

Austroads (2017) Guide to pavement technology: part 2: pavement structural design. AGPT02–17

Avirneni D, Peddinti PR, Saride S (2016) Durability and long-term performance of geopolymer stabilized reclaimed asphalt pavement base courses. Const Build Mater 121:198–209

Bennert T, Meher A (2005) The development of performance specification for granular base and subbase material. Accessed online through http://www.state.nj.us/transportation/refdata/research/reports/FHWA-NJ-2005-003.pdf

Cetin B, Aydilek AH, Guney Y (2010) Stabilization of recycled base materials with high carbon fly ash. Res Conserv Recycl 54(11):878–892

Collins RJ, Ciesilski SK (1994) Recycling and use of waste materials and by-products in highway construction. Synthesis of Highway Practice 199, National Academy, Washington, D.C., pp 1–77
IRC (Indian Roads Congress) (2018) Guidelines for the design of flexible pavements. Indian code of practice. IRC:37–2018, New Delhi, India
Jallu M, Arulrajah A, Saride S, Evans R (2020) Flexural fatigue behavior of fly ash geopolymer stabilized-geogrid reinforced RAP bases. Const Build Mater 254:119–263
Janoo VC (1994) Layer coefficients for NHDOT pavement materials. Special Rep. 94–30, Prepared for NH Dept. of Transportation and U.S. Dept. of Transportation, Hanover, NH
McGarrah EJ (2007) Evaluation of current practices of reclaimed asphalt pavement/virgin aggregate as base course material. WA State Dept. of Transport. Rep. No. WA-RD 713.1, p 33
Mohammadinia A, Arulrajah A, Horpibulsuk S, Chinkulkijniwat A (2017) Effect of fly ash on properties of crushed brick and reclaimed asphalt in pavement base/subbase applications. J Hazard Mater 321:547–556
Mohammadinia A, Arulrajah A, D'Amico A, Horpibulsuk S (2018) Alkali-activation of fly ash and cement kiln dust mixtures for stabilization of demolition aggregates. Constr Build Mater 186:71–78
MoRTH (2013) Specification for road and bridge works, 5th Revision. Ministry of Road Transport and Highways (MoRTH), New Delhi, India
Nehdi M (2001) Ternary and quaternary cements for sustainable development. Concrete Int 23(4): 35–42
Palomo A, Grutzeck MW, Blanco MT (1999) Alkali-activated fly ashes: a cement for the future. Cem Concr Res 29(8):1323–1329
Puppala AJ, Saride S, Potturi A, Hoyos LR (2009) Resilient behavior of cement-fiber treated reclaimed asphalt pavement (RAP) aggregates as bases. Proceedings of International Foundation Congress and Equipment Expo, ASCE, GSP 187:433–440
Puppala AJ, Saride S, Williammee R (2012) Sustainable reuse of limestone quarry fines and RAP in pavement base/subbase layers. J Mater Civil Eng 24(4):418–429
Puppala AJ, Pedarla A, Chittoori B, Ganne VK, Nazarian S (2017) Long-term durability studies on chemically treated reclaimed asphalt pavement material as a base layer for pavements. Trans Res Rec 2657(1):1–9
Rattanasak U, Chindaprasirt P (2009) Influence of NaOH solution on the synthesis of fly ash geopolymer. Miner Eng 22(12):1073–1078
Saride S, Jallu M (2020) Effect of alkali activated fly ash on layer coefficients of reclaimed asphalt pavement bases. J Transp Eng Part B Pavements. https://doi.org/10.1061/JPEODX.0000169
Saride S, Puppala AJ, Williammee R (2010) Assessing recycled/secondary materials as pavement bases. Gr Improv 163(1):3–12
Saride S, Avirneni D, Javvadi SCP (2015) Utilization of reclaimed asphalt pavements in Indian low-volume roads. J Mater Civil Eng 28(2):04015107
Saride S, Avirneni D, Challapalli S (2016) Micro-mechanical interaction of activated fly ash mortar and reclaimed asphalt pavement materials. Constr Build Mater 123:424–435
Sherwood PT (2001) Alternate materials in road construction: a guide to the use of recycled and secondary aggregates, 2nd edn. Thomas Telford, London
Taha R, Ali G, Basma A, Al-Turk O (1999) Evaluation of reclaimed asphalt pavement aggregate in road bases and subbases. Transportation Research Record 1652, Washington, DC, pp 264–269
Taha R, Ali A, Khalid A, Muamer A (2002) Cement stabilization of reclaimed asphalt pavement aggregates for road bases and subbases. J Mater Civil Eng 14(3):239–245
Texas Department of Transportation (2014) Standard specifications for construction and maintenance of highways, streets, and bridges (Item 276). TxDOT, Austin, TX
Yuan D, Nazarian S, Hoyos LR, Puppala AJ (2010) Cement treated RAP mixes for roadway bases. Report. FHWA/TX-10/0–6084–1, p 124

Chapter 23
Computing Seismic Displacements of Cantilever Retaining Wall Using Double Wedge Model

Prajakta R. Jadhav and Amit Prashant

23.1 Introduction

Several incidences of cantilever retaining wall failure under earthquake loading have been documented (Ortiz 1982; Ross et al. 1969; Lai 1998; Elnashai et al. 2010), which mainly occurred due to excessive translational and rotational displacements. These failures have occasionally caused distress in adjacent structures as well as the infrastructure constructed on the backfill retained by these walls. Presently, the codal provisions (1994; EN 1998; Anderson et al. 2008) recommend these walls to be analyzed and designed like gravity retaining walls by assuming the soil above the heel as part of the wall. Such an ad hoc arrangement was adopted due to lack of sufficient understanding of the seismic response of these walls and unavailability of a simple but reasonably accurate model for prediction of seismic displacements. This paper presents a displacement-based design methodology to predict sliding and rotational displacements in these walls by considering a realistic failure mechanism in the backfill.

During Alaska earthquake in 1964, several bridge abutments of cantilever retaining wall type underwent sliding as well as tilting failure (Ortiz 1982; Ross et al. 1969). In 1998, Lai (1998) reported that a cantilever retaining wall got uplifted and tilted away from the backfill during the Hyogoken Nanbu earthquake in 1995. In 2010, Elnashai et al. (2010) reported settlement of about 70 cm of backfill retained by a wall of height 7 m, thus developing cracks in approach road. Conventionally, these walls are idealized as gravity retaining walls for stability against seismic loads and designed using force-based approach, mostly using Mononobe–Okabe (M-O) method (Okabe 1926; Mononobe and Matsuo 1929). This M-O method requires an assumed value of seismic coefficient to represent the dynamic loading induced by an earthquake, which involves a lot of uncertainties.

P. R. Jadhav · A. Prashant (✉)
Civil Engineering, Indian Institute of Technology Gandhinagar, Ahmedabad, India
e-mail: ap@iitgn.ac.in

T. G. Sitharam et al. (eds.), *Latest Developments in Geotechnical Earthquake Engineering and Soil Dynamics*, Springer Transactions in Civil and Environmental Engineering, https://doi.org/10.1007/978-981-16-1468-2_23

This seismic coefficient is often estimated based on the peak acceleration of the earthquake, thus ignoring the other characteristics of the earthquake motion, such as time period and dominant frequency. Hence, the M-O method can significantly underpredict or overpredict the forces, as the case may be, for a given earthquake motion. For the design of the wall components, such an analysis is required to get a fair estimate of bending moments and shear forces in the wall. However, from the perspective of the overall stability of the wall, a force-based analysis may not be sufficient, and a displacement-based analysis may provide a better sense of the expected damage against a seismic event. Hence, it is imperative to understand the expected deformation modes in cantilever retaining walls subjected to seismic loading and capture the realistic mechanism of backfill deformations.

In current practice, the seismic deformations of cantilever retaining wall are determined by analyzing them as gravity retaining walls. It assumes a vertical plane passing through the heel and considers the soil mass between the vertical plane and the wall-stem as part of the wall itself. The seismic displacements are mostly computed by following the Newmark sliding block analysis (Newmark 1965) for the wall with locked-in soil above the heel and an active earth pressure acting on the vertical plane at the heel. However, at this vertical plane, the earth pressure force may be maximum, but the yield acceleration need not be minimum. Richards and Elms (1979) extended the Newmark's method for slopes (Newmark 1965) and applied to gravity retaining walls. They assumed the wall and the adjacent soil wedge traverse with the same yield acceleration during earthquake loading. Zarrabi-Kashani (1979) proposed a method of estimating permanent displacements in gravity retaining walls by giving due consideration to the possibility of relative movement between wall and adjacent soil wedge. However, this methodology did not ensure to have only tangentially downward movement of soil wedge causing outward displacements of the wall, which was accounted for in the Modified-Zarrabi's model presented by Jadhav and Prashant (2017). The question is whether it is reasonable to adopt these design philosophies of gravity walls for cantilever retaining walls, which has a relatively more complex geometry. This geometry may influence the mechanism responsible for displacements in cantilever retaining walls. Experimentally, it has been observed that there is a formation of v-shaped rupture planes in the backfill of cantilever retaining walls (Watanbe et al. 2003; Penna et al. 2014). In the present study, a double wedge model (Jadhav and Prashant 2019) has been formulated to predict the displacement of cantilever retaining wall with due consideration to the v-shaped mechanism in the backfill. In lack of sufficient research on walls with shear key, they are currently designed similar to the walls without shear key. There is also some ambiguity about the position of shear key and the failure plane to be considered at the wall-base for stability analysis. Under seismic loading, the effect of shear key is reflected in the failure modes exhibited by the wall and needs due consideration. With due consideration to the responsible mechanism, a displacement-based design methodology for walls with shear key is also presented here.

23.2 Double Wedge Model for Sliding Displacements

Figure 23.1 schematically presents the double wedge model with the formation of v-shaped rupture planes in the backfill for a cantilever retaining wall. The v-shaped wedge is expected to evolve from top of its heel end when the wall is subjected to ground acceleration at its base. The rupture plane extending toward the backfill is referred to as inner rupture plane, and the one extending toward the wall-stem is called outer rupture plane. The soil mass enclosed between the outer rupture plane and the wall-stem is considered to be a part of the wall itself, and it is termed as 'wall with locked soil mass.' In a way, the proposed model primarily deals with the seismically induced motion of these two distinct wedges, viz. soil wedge and wall with locked soil mass, and so it is termed as 'double wedge model.'

In Fig. 23.1, k_h and k_v represent horizontal and vertical ground acceleration coefficients, respectively. k_{hy}' and k_{vy}' represent horizontal and vertical absolute acceleration coefficients, respectively, of soil wedge with respect to fixed datum. Similarly, k_{hy}'' and k_{vy}'' represent horizontal and vertical absolute acceleration coefficients, respectively, of wall with locked soil mass with respect to fixed datum. The model considers both the wedges as rigid plastic and hence the acceleration is uniform throughout the wall-height. Also, these wedges are assumed to traverse on a non-deformable base. This model computes displacements in the wall not only due to earthquake loading but also due to the thrust exerted by the tangentially downward movement of soil wedge on the wall with locked soil mass. The soil wedge is not allowed to move normal to rupture planes but only in the tangentially downward direction. This condition of tangential movement is ensured by equating the normal components of yield acceleration along the rupture planes. In a way, this model considers two distinct yield accelerations for soil wedge and wall with locked soil mass unlike Newmark sliding block model, as shown in Fig. 23.1. Depending upon the position of outer rupture plane, the model considers two different cases, viz. case 1: when outer rupture plane is not intersecting the backface of wall, and case 2: when outer rupture plane is intersecting the backface as shown in Fig. 23.1, and yield acceleration for wall with locked soil mass is computed

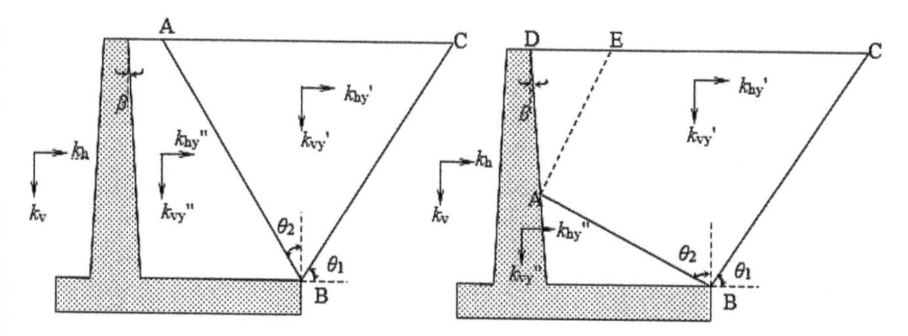

Fig. 23.1 Schematic representation of double wedge model for cantilever retaining wall **a** outer rupture plane is not intersecting the wall-stem, **b** outer rupture plane is intersecting the wall-stem

differently for each case. Equations (23.1) and (23.2) are used for computing yield acceleration of wall with locked soil mass for case 1 and case 2, respectively. When outer rupture plane intersects the backface of the wall, the earth pressure force on AD part of the wall-stem is computed using Mononobe–Okabe theory represented by P_1 in Eq. (23.2). The small wedge along AD is considered to be a part of the soil wedge, and hence it is assumed to have the same yield acceleration as that of soil wedge. The horizontal and vertical yield accelerations in soil wedge are computed using Eqs. (23.3) and (23.4), respectively.

$$k_{hy}\prime\prime = \frac{\left[\begin{array}{r}\frac{W\cos\theta_1 sin(\phi+\theta_2-\theta_1)\tan\phi - \cos(\phi+\theta_2-\theta_1)}{W_s(\sin(\phi+\theta_2)\tan\phi_b - \cos(\phi+\theta_2))}\tan\phi_b(1-k_v) - \frac{k_h\tan\theta_1\tan\theta_2}{1+(\tan\theta_1\tan\theta_2)}\\ +(\sin\theta_1-\cos\theta_1\tan\phi)(k_v\cos\theta_1+k_h\sin\theta_1-\cos\theta_1)\end{array}\right]}{\frac{W\cos\theta_1 sin(\phi+\theta_2-\theta_1)\tan\phi-\cos(\phi+\theta_2-\theta_1)}{W_s(\sin(\phi+\theta_2)\tan\phi_b-\cos(\phi+\theta_2))} + \frac{1}{1+(\tan\theta_1\tan\theta_2)}} \tag{23.1}$$

$$k_{hy}\prime\prime = \frac{\left[\begin{array}{r}\frac{W cos\theta_1 T_2 tan\phi_b(1-k_v)}{W_s(\sin(\phi+\theta_2)\tan\phi_b-\cos(\phi+\theta_2))} - \frac{k_h\tan\theta_1\tan\theta_2}{1+(\tan\theta_1\tan\theta_2)} - \frac{P_1T_1\cos\theta_1}{W_s} + \frac{P_1T_2T_4\cos\theta_1}{W_sT_3}\\ +(sin\theta_1 - cos\theta_1 tan\phi)(k_h sin\theta_1 + k_v\cos\theta_1 - \cos\theta_1)\end{array}\right]}{\frac{1}{1+(\tan\theta_1\tan\theta_2)} + \frac{W cos\theta_1 T_2}{W_s(\sin(\phi+\theta_2)\tan\phi_b-\cos(\phi+\theta_2))}} \tag{23.2}$$

Here,

$$T_1 = \sin(\delta+\beta-\theta_1)\tan\phi - \cos(\delta+\beta-\theta_1)$$

$$T_2 = \sin(\phi+\theta_2-\theta_1)\tan\phi - \cos(\phi+\theta_2-\theta_1)$$

$$T_3 = \sin(\phi+\theta_2)\tan\phi_b - \cos(\phi+\theta_2)$$

$$T_4 = \sin(\delta+\beta)\tan\phi_b - \cos(\delta+\beta)$$

$$k_{hy}\prime\prime = k_{hy}\prime + (k_{hy}\prime - k_h)\tan\theta_1\tan\theta_2 \tag{23.3}$$

$$k_{vy}\prime = k_v + \tan\theta_1\left(k_h - k_{hy}\prime\right) \tag{23.4}$$

It is assumed that $k_{vy}'' = k_v$ since the weight of the wall is significant enough to allow no movement normal to wall-base. θ_1 and θ_2 are the inclinations of the inner and outer rupture planes with the horizontal and vertical planes, respectively. W_s and W refer to the weight of soil wedge and wall with locked soil mass, respectively. ϕ_b represents interface friction angle wall with foundation soil and L and H represent length of heel and height of wall, respectively.

The implementation of this model is similar to that of Newmark's sliding block analysis for slopes. The steps for computing the displacement are given below:

1. During ground motion, at a particular time, θ_2 values are varied from 0 to $\tan^{-1}(L/h)$ for case 1 and beyond $\tan^{-1}(L/h)$ for case 2.
2. For a particular value of θ_2, the initial value of a parameter $\alpha = \tan^{-1}(k_{hy}'/1 - k_{vy}')$ is computed by assuming $k_{vy}' = 0$ and $k_{hy}' = k_h$. Here, k_h is earthquake acceleration at that time instant.
3. The values of θ_1 are varied from value of ϕ up to angle <90°. By solving the force triangle, the earth pressure force P is computed for case 1 and case 2, respectively. The values of P are obtained for a particular value of θ_2 and a range of θ_1 values. The value of θ_1 at which earth pressure force P is maximum is recorded.
4. For the obtained θ_2 and θ_1, the weights of soil wedge and wall with locked soil mass are computed. Accordingly, the value of k_{hy}'' is calculated using Eqs. (23.1) and (23.2) for case (1) and case (2), respectively.
5. Using these values of θ_1, θ_2 and k_{hy}'', the values for k_{hy}' and k_{vy}' are computed using Eqs. (23.3) and (23.4), respectively, which provides the value of α. Compare the considered value of α with its computed value, and check if the difference is less than a tolerance value of 0.01. If this criterion is not fulfilled, repeat the steps 3–5 using the then computed values of α for that value of θ_2.
6. Fulfilling this criterion indicates that, the obtained values of k_{hy}'', k_{hy}' and k_{vy}' are the yield acceleration values at that time instant. Compute displacements if the ground acceleration is greater than the yield acceleration of wall with locked soil mass using the methodology similar to that of Newmark sliding block analysis.
7. Repeat the above steps for the complete duration of earthquake motion.

Also, it is to be ensured that the wall movement takes place in the outward direction only, as it is done in case of Newmark's analysis (Newmark 1965). This condition is ensured along the wall-base by equating velocities of wall and ground. It ensures either no relative movement of the wall along the ground or relative movement of the wall to occur only in the outward direction. In double wedge model, as soil wedge and wall with locked soil mass traverse with different accelerations, this condition is ensured along both the rupture planes such that velocity of the wall is always greater than or equal to that for the soil wedge (Jadhav and Prashant 2017).

Double wedge model takes less than a minute for analysis on a general use computer. This model has been validated (Jadhav and Prashant 2019) with several case studies and has been observed to predict sliding displacements closer to actual values in comparison with the other conventional models.

23.3 Finite Element Model for Computing Rotational Displacements

Two-dimensional plane strain analysis of cantilever retaining wall system has been performed to study the combined translational cum rotational behavior of wall as the double wedge model can compute only sliding displacements. The idea is to

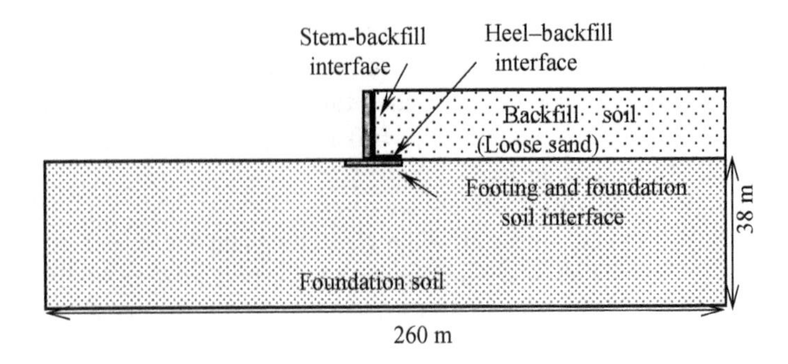

Fig. 23.2 FE model description for cantilever retaining wall system

develop a simple methodology to estimate rotational displacements at the wall-top based on the sliding displacements computed from double wedge model.

The FE model of cantilever retaining wall system is shown in Fig. 23.2, which has been developed in GiD and analyzed in OpenSees. The retained soil has loose sand, and the foundation soil has layers of medium-dense sand, dense sand and very dense sand. The wall is modeled using 1D elastic beam-column elements and soil by using quad elements of pressure-dependent multiyield model. The three interfaces have been modeled using zero-length interface elements. The earthquake signal is applied at the base of the model domain in the form of equivalent shear forces (Zhang et al. 2008). Radiation dampers are provided along the boundaries to prevent spurious reflections along with Rayleigh damping of 1 and 5% for soil (nonlinear material model) and wall elements, respectively, and a small amount of numerical damping. The model is run for the duration of earthquake loading and different components of displacements have been recorded. This FE model has been validated with shake table tests from Kloukinas et al. (2015).

For a wall of height 12 m and heel length 4.5 m, the analyses were performed on different classes of foundation soil as identified by AASHTO LRFD manual (2012). The wall underwent combined translational cum rotational failure with significant displacements when placed on medium-dense sand which is usually encountered in practice. Hence, the next part of the study primarily considered medium-dense sand only. Two different wall geometries each with three different toe-widths were subjected to four earthquake motions of different characteristics; each scaled to six different PGAs ranging from 0.12 to 0.72 g. Total 144 cases were analyzed, and both peak displacements and residual displacements were computed.

Double wedge model analyses for all these cases were also performed by subjecting the wall to free-field motions, and the obtained residual sliding displacements relative to free field from FE model were compared with those from double wedge model. The sliding displacements consistently matched at higher PGAs., but under-predicted the displacements at lower PGAs. This deviation at low PGA can be attributed to the fact that, at lower PGAs, the strength of soil beneath wall gets mobilized locally in FE model and is not captured in double wedge model with the assumed non-deformable base. Accounting for this scenario, an equivalent base

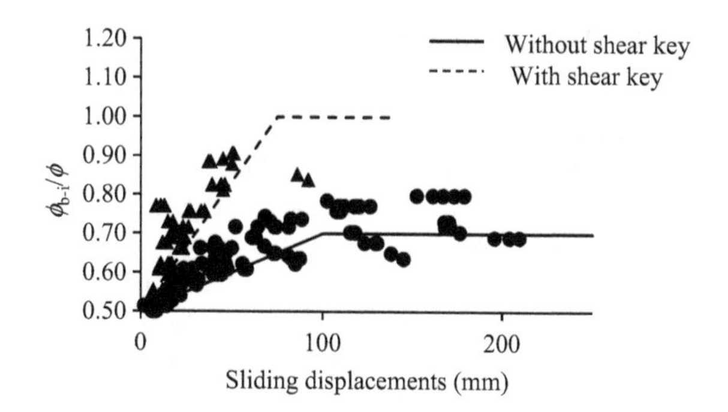

Fig. 23.3 Variation of ratio of equivalent base interface friction angle to sliding displacements in mm. Data points marked in circle represent wall without shear key and those with triangle represent wall with shear key

interface friction angle ($\phi_{b\text{-}i}$) is proposed to be used. The interface friction angles for all the 144 cases were back-calculated by matching displacements with FE model and the values of the ratio $\phi_{b\text{-}i}/\phi$ have been plotted with residual sliding displacements, as shown in Fig. 23.3. Looking at the data in Fig. 23.3, a bilinear curve has been proposed to perform further analysis. The values of this ratio vary from 0.5 for zero displacements to 0.67 for 100 mm displacements and remain constant at 0.67 after that. The proposed curve has been chosen not on the basis of trend line for data points but such that it would result in a slightly conservative estimate of displacements with less probability of exceedance from the actual values. In order to capture this interface deformability, the double wedge model analysis is somewhat modified, wherein it is proposed to use this curve for $\phi_{b\text{-}i}$ values instead of directly using a single ϕ_b value as $(2/3)\phi$. Firstly, the steps for double wedge model are implemented using ϕ_b value as 2/3 ϕ to get displacement X. If X is less than 100 mm, then the steps are repeated for values of ϕ_b corresponding to X in the proposed curve. The iterations are repeated until the difference between consecutive displacements is less than a tolerance value of ± 10 mm. If the value of X in the first step itself is greater than 100 mm, then the same value is considered as the sliding displacements. After incorporating this correction for wall placed on medium-dense sand, the displacements computed using double wedge model have been observed to show a closer match with those from FE model at all PGAs.

The ratios of both residual as well as peak rotational displacements to sliding displacements relative to free field for all the 144 cases were calculated. It was observed that values show uniform distribution at all the PGAs, varying from 2 to 4 for peak displacements and 1 to 2 for residual displacements. The idea is to propose a suitable value of this ratio such that, this ratio once multiplied with the sliding displacements from the double wedge model would yield rotational displacements. Accordingly, corresponding to a 5% probability of exceedance of this ratio for peak and rotational displacements, this ratio has been suggested as 3 and 2, respectively. Also, the ratios of peak to residual sliding displacements relative to free field were observed to be either 1 or less than 1. Finally, it is recommended to take the peak sliding displacements the same as the residual sliding displacements computed from the double wedge model.

23.4 Design of Cantilever Retaining Walls with Shear Key

Cantilever retaining walls with shear keys are presently being designed like those without shear key. A shear key is placed at the base of cantilever retaining walls for harnessing the sliding displacements. The failure modes in these walls can change in the presence of shear key, which poses the need to understand the involved mechanism due to shear key. There is a lot of ambiguity regarding the position of shear key and about the failure planes to be considered in the stability analysis (Jadhav and Prashant 2020). In view of this, FE model of cantilever retaining wall with shear key was developed similar to that of walls without shear key. Interface elements were provided along the shear key to study the relative movement between wall and soil in the vicinity. The developed model was validated (Jadhav and Prashant 2020) with centrifuge tests from Ortiz (1982). Following the same procedure, three different cantilever retaining wall systems were developed by varying the position of shear key at heel, toe and stem. It was observed that wall with shear key below the wall-heel performed better than the other configurations. It was so probably due to the part 1 shown in Fig. 23.4 helping in resisting the overall sliding as well as rotational displacements in the wall. Accordingly, the horizontal plane at the depth of shear key has been proposed as the failure plane to be used in the stability analysis.

In order to propose a simple displacement-based design methodology, the double wedge model was modified to incorporate the effect of shear key. In this model, the rupture planes are assumed to evolve at the depth of shear key, and the outer rupture plane touches the uppermost edge of the wall-heel, as shown in Fig. 23.4. Along with part 2, shown in Fig. 23.4, part 1 and part 3 are also considered locked with the wall. The passive pressure P_p in the presence of shear key is considered from the wall-toe base to the shear key depth. The corresponding passive wedge is assumed to have the same yield acceleration like that of the wall with locked soil mass. By using all other steps similar to that of the wall without shear key, Eqs. (23.5) and (23.6) have been derived to compute yield acceleration of wall

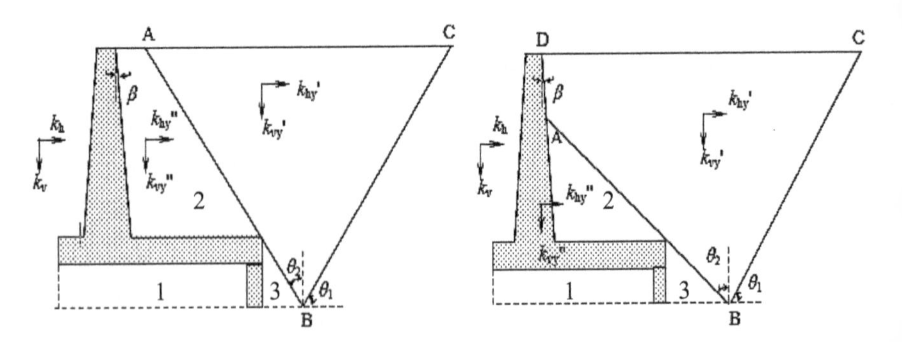

Fig. 23.4 Double wedge model for wall with shear key when outer rupture plane is not intersecting and when it is intersecting the backface of wall

with locked soil mass for case (1) and (2), respectively. Equations (23.3) and (23.4) are for computing yield acceleration in soil wedge (Jadhav and Prashant 2020).

$$k_{hy}// = \frac{\left[\begin{array}{l} C + \dfrac{DW\cos\theta_1}{W_s(\sin(\phi+\theta_2)\tan\phi - \cos(\phi+\theta_2))}\tan\phi_b(1-k_v) \\ + \dfrac{D\cos\theta_1(\cos\phi_b - \sin\phi_b\tan\phi_b)}{W_s(\sin(\phi+\theta_2)\tan\phi_b - \cos(\phi+\theta_2))}P_p - \dfrac{k_h\tan\theta_1\tan\theta_2}{1+\tan\theta_1\tan\theta_2} \end{array}\right]}{\frac{1}{1+\tan\theta_1\tan\theta_2} + \frac{DWcos\theta_1}{W_s(\sin(\phi+\theta_2)\tan\phi - \cos(\phi+\theta_2))}} \tag{23.5}$$

$$k_{hy}// = \frac{\left[\begin{array}{l} C + \dfrac{DW\cos\theta_1}{W_s(\cos(\phi+\theta_2) - \sin(\phi+\theta_2)\tan\phi_b)}\tan\phi_b(1-k_v) \\ + \dfrac{DP_p\cos\theta_1(\cos\phi_b - \sin\phi_b\tan\phi_b)}{W_s(\cos(\phi+\theta_2) - \sin(\phi+\theta_2)\tan\phi_b)} + \dfrac{DP_1(\sin(\delta+\beta)\tan\phi_b - \cos(\delta+\beta))\cos\theta_1}{W_s(\cos(\phi+\theta_2) - \sin(\phi+\theta_2)\tan\phi_b)} \\ - \dfrac{P_1\cos\theta_1 D(\sin(\delta+\beta-\theta_1)\tan\phi - \cos(\delta+\beta-\theta_1))}{W_s} - \dfrac{k_h\tan\theta_1\tan\theta_2}{1+\tan\theta_1\tan\theta_2} \end{array}\right]}{\frac{1}{1+\tan\theta_1\tan\theta_2} + \frac{DW\cos\theta_1}{W_s(\cos(\phi+\theta_2) - \sin(\phi+\theta_2)\tan\phi_b)}} \tag{23.6}$$

Here,

$$C = (\sin\theta_1 - \cos\theta_1\tan\phi)(k_v\cos\theta_1 + k_h\sin\theta_1 - \cos\theta_1)$$

$$D = \sin(\phi+\theta_2-\theta_1)\tan\phi - \cos(\phi+\theta_2-\theta_1)$$

In order to incorporate the interface deformability, the bilinear curve shown in Fig. 23.3 for walls with shear key is proposed to be used, which has $\phi_{b\text{-}i}/\phi$ ratio ranging linearly from 0.5 at 0 mm displacements to 1 at 75 mm and then remains constant at 1 for displacements beyond 75 mm. The upper-limit value of this ratio has to be considered as 1 since the failure plane for wall with shear key does not pass through the interface of wall and soil but at the depth of shear key with soil–soil interface. Also, the two data points which show displacements of about 85 mm correspond to same earthquake motion which has significant displacement causing peaks at the initial time instants only as compared to other earthquakes. However, there is rare probability of occurrence of such an earthquake motion and hence 75 mm is considered as the critical value beyond which the value of ratio is to be taken as 1. The model implementation for walls with and without shear key remains the same.

Combined translational cum rotational behavior of these walls with shear key was studied by running 64 cases of FE analysis by considering two different heights of walls and two of toe-widths and subjecting them to four different earthquake motions scaled to different PGAs. The peak displacement factor and residual displacement factor to be multiplied with the sliding displacements computed from the double wedge model have been proposed as 3.5 and 2.5, respectively. These factors are higher than those for walls without shear key since the presence of shear key reduces the sliding displacements and induces more rotations in these walls. The peak sliding displacements are recommended to be taken as same as residual sliding displacements.

The simplified methodology for computing both sliding as well as rotation displacements in the wall with and without shear key is as follows:

1. For earthquake motion corresponding to design PGA of the site under consideration, compute sliding displacements using the double wedge model.
2. Compute residual and peak rotational displacements by multiplying displacements from double wedge model with 2 and 3, respectively, for walls without shear key, and with 2.5 and 3.5, respectively, for walls with shear key. The estimated displacements are a bit conservative, in general.

The displacements have been computed using proposed method for both the walls with and without shear key and compared with those from FE model. Some of these cases are shown in Table 23.1. At lower PGA, both the displacements are convergent with those from FE model. At higher PGAs, the sliding displacements using double wedge model vary within an acceptable tolerance of about 15–20 mm. The estimated rotational displacements vary by about 70 mm as these are computed using factors corresponding to $\sim$5% probability of exceedance. Currently, as there is no simplified methodology to compute in-plane rotational displacements in these walls and as the proposed methodology gives reasonably conservative estimate, this method can be rendered acceptable.

Table 23.1 Comparison of predicted displacements with FE displacements in mm

Wall type	Without shear key		With shear key	
Bed-rock PGA	0.12 g	0.36 g	0.12 g	0.36 g
FE sliding disps	15	68	9	24
Double wedge model	13	85	7	34
FE in-plane rotational disps	29	123	12	43
Proposed method	26	170	18	85

23.5 Case Study: Shake Table Tests at the University of Bristol

Kloukinas et al. (2015) performed shake table tests on a model of aluminum cantilever retaining wall and subjected to earthquake signal shown in Fig. 23.5 scaled to PGA of 0.3 and 0.5 g. The wall is retaining Leighton-Buzzard silica sand of height 0.57 m and is resting on the same sand layer of thickness 0.5 m. The properties required for analysis are given in Fig. 23.5.

The displacements under both the PGAs have been computed using the proposed simplified methodology and conventional Richards and Elms model (Richards and Elms 1979) for gravity retaining wall, and the predictions have been compared with the actual displacements, as listed in Table 23.2. At both the PGAs, the displacements computed using the proposed methodology show a better prediction than conventional model with measured displacements. The difference could be due to approximate digitization of the data points of the earthquake signal. However, as the proposed method has predicted reasonably conservative displacements than conventional model, it can be rendered reliable in the estimation of sliding and rotational displacements in cantilever retaining walls.

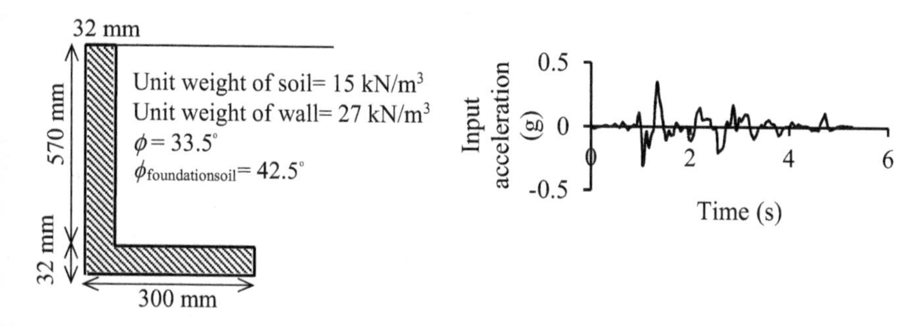

Fig. 23.5 Geometry of wall and earthquake signal from shake table tests used as case study

Table 23.2 Comparison of predicted displacements with actual displacements in mm

PGA (g)	Actual		Conventional model		Proposed model	
	Sliding	Rotational	Sliding	Rotational	Sliding	Rotational
0.3	3	6	0.3	–	4	8
0.5	10	25	5	–	20	40

23.6 Conclusion

An analytical double wedge model which simulates the experimentally observed v-shaped rupture planes in the backfill of cantilever retaining wall has been presented for computing sliding displacements in the walls with and without shear key. Residual displacement factor and peak displacement factor of 2 and 3, respectively, are recommended to be multiplied with the sliding displacements from double wedge model for estimating residual and peak rotational displacements, respectively, in case of the walls without shear key. These factors are recommended to be 2.5 and 3.5, respectively, for walls with shear key. The suitability of this methodology for estimating the seismic displacements has been verified by obtaining comparable displacements from the proposed method than the conventional method with the actual measurements in a case study. In a way, the proposed methodology which is based on a double wedge model that simulates the mechanism responsible for displacements is deemed to be reliable for seismic design of cantilever retaining wall with and without shear key.

References

AASHTO-LRFD (2012) Bridge design: american association of state highway and transportation officials, Specifications. Washington, DC

Anderson D, Martin G, Lam I, Wang J (2008) Seismic analysis and design of retaining walls, buried structures, slopes, and embankments, NCHRP

Earth retaining structures, British Standard (1994)

Elnashai AS, Gencturk B, Kwon OS, Al-Qadi IL, Hashash Y, Roesler JR, Kim SJ, Jeong SH, Dukes J, Valdivia A (2010) The Maule (Chile) earthquake of February 27, 2010: consequence assessment and case studies, Mid-America Earthquake (MAE) Center, Research Report 10-04, Department of Civil and Environmental Engineering, University of Illinois at Urbana-Champaign

EN 1998-5:2004—Eurocode 8: Design of structures for earthquake resistance—part 5: foundations, retaining structures and geotechnical aspects, Eurocode 8 (2004)

GiD. CIMNE software, https://www.gidhome.com/ accessed in Oct 2018

Jadhav P, Prashant A (2017) Computation of permanent sliding displacements of retaining wall during seismic loading. In: 3rd international conference on performance-based design in earthquake geotechnical engineering, Vancouver, Canada, 16–19 July 2017

Jadhav PR, Prashant A (2019) Double wedge model for computing seismic sliding displacements of cantilever retaining walls. Soil Dyn Earthquake Eng 116:570–579

Jadhav P, Prashant A (2020a) Seismic translational and rotational displacements of cantilever retaining wall resting on medium-dense sand. ASCE (Under review)

Jadhav P, Prashant A (2020b) Computation of seismic translational and rotational displacements of cantilever retaining wall with shear key. Soil Dyn Earthquake Eng 130:

Kloukinas P, di Santolo AS, Penna A, Dietz M, Evangelista A, Simonelli AL, Taylor C, Mylonakis G (2015) Investigation of seismic response of cantilever retaining walls: limit analysis vs shaking table testing. Soil Dyn Earthquake Eng 77:432–445

Lai S (1998) Rigid and flexible retaining walls during Kobe earthquake. In: Proceedings, 4th international conference on case histories in geotechnical engineering 10, 8–15 March, 1998, St. Louis, Missouri

Mononobe N, Matsuo H (1929) On the deformation of earth pressure during earthquakes proceedings, world engineering conference, pp 9–177

Newmark NM (1965) Effects of earthquakes on dams and embankments. Geotechnique 15 (2):139–160

Okabe S (1926) General theory of earth pressure. J Japan Soc Civil Eng 12(1)

Ortiz AL (1982) Dynamic centrifuge testing of a cantilever retaining wall. Ph.D. thesis, California Institute of Technology, Pasadena

Penna A, Scotto A, Kloukinas P, Taylor C, Mylonakis G, Evangelista A, Simonelli AL (2014) Advanced measurements on cantilever retaining wall models during earthquake simulations. 20th IMEKO TC4 international symposium and 18th international workshop on ADC modelling and testing research on electric and electronic measurement for the economic upturn Benevento, pp 127–132, Italy, 15–17 Sept 2014

Richards R Jr, Elms DG (1979) Seismic behavior of gravity retaining walls. J Geotech Geoenviron Eng 105:14496

Ross GA, Seed HB, Migliaccio RR (1969) Bridge foundation behavior in Alaska earthquake. J Soil Mech Found Div 95(4):1007–1036

The Open System for Earthquake Engineering Simulation (OpenSees). PEER, software. http://opensees.berkeley.edu/OpenSees/home/about.php, accessed in Oct 2018

Watanbe K, Munaf Y, Koseki J, Tateyama M, Kojima K (2003) Behaviors of several types of model retaining walls subjected to irregular excitation. Soils Found 43(5):13–27

Zarrabi-Kashani K (1979) Sliding of gravity retaining wall during earthquakes considering vertical acceleration and changing inclination of failure surface. Diss. Massachusetts Institute of Technology

Zhang Y, Conte JP, Yang Z, Elgamal A, Bielak J, Acero G (2008) Two-dimensional nonlinear earthquake response analysis of a bridge-foundation-ground system. Earthquake Spectra 24 (2):343–386

Chapter 24
Importance of Site-Specific Observations at Various Stages of Seismic Microzonation Practices

Abhishek Kumar

24.1 Introduction

Seismic microzonation is an approach of assessing the combined effect of EQ generated ground shaking and all possible induced effects for a region/site of interest. Thus, seismic microzonation not only provides probable seismic hazard scenario at bedrock level but also possible induced effects taking local soil characteristics into account. Typical components of a seismic microzonation study (as shown in Fig. 24.1) include assessment of various geological, geotechnical and geophysical characteristics of region/site in order to arrive at seismic hazard and quantification of induced effects toward estimating seismic design parameters and identification of zones in terms of any possible impact from seismic activity. Depending upon the site/region to be studied for seismic microzonation, specific components/parameters can be chosen. Further, depending upon the availability of above-mentioned parameters, various thematic layers are prepared. A seismic microzonation is thus characterizing the study area in terms of relative effect of EQ as well as induced effects, within the study area. This helps in identification of regions belonging to low to moderate to high EQ shaking and corresponding induced effects. Further, once microzonation study is performed, which gives combined effect of EQ till ground surface, taking building classification and building use's details into account, one can perform seismic vulnerability and risk assessment of the study area. The beauty of seismic microzonation study is that it gives a complete picture of probable scenario from bedrock to ground surface during to possible future EQ scenario. With Government of India's initiative about the development of infrastructure in different parts of India, seismic microzonation studies if performed can help in city planning and identification of locations within city, which are most appropriate for hospitals, schools, gas refineries, residential

A. Kumar (✉)
Indian Institute of Technology, Guwahati, India
e-mail: abhiak@iitg.ac.in

T. G. Sitharam et al. (eds.), *Latest Developments in Geotechnical Earthquake Engineering and Soil Dynamics*, Springer Transactions in Civil and Environmental Engineering, https://doi.org/10.1007/978-981-16-1468-2_24

areas, alignment of metros, alignment of water supply and sewage collection framework, etc. Seismic microzonation maps once developed are also useful for decision making in case of a natural disaster, in order to supply essential items to the affected area. Most feasible road networks, which can be used for supplying the medicine and other essential items, will be available at ease.

Though the steps to-be-followed for seismic microzonation are well-established, there are newer observations/inputs in various steps/thematic layers, which if considered in ongoing as well as future studies can enhance their accuracy and can minimize the effect of accessing remote locations in terms of both source and site information. In the present paper, detailed discussion on various site-specific observations and how these observations can be incorporated in existing seismic microzonation studies are discussed as can be observed from further sections.

24.2 Identification of Seismic Sources

Identification of seismic sources is important as this is the first step whether attempting deterministic seismic hazard analysis (DSHA) or probabilistic seismic hazard analysis (PSHA). This is primarily important in case the analysis is based on linear seismic sources. In general, known seismic sources within the desired seismotectonic region are collected from the literature. It must be mentioned here that

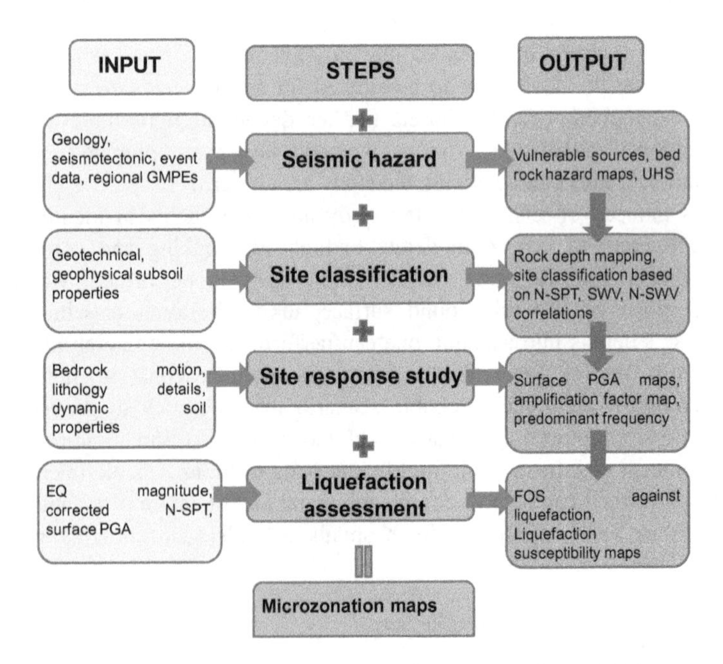

Fig. 24.1 Flow chart for seismic microzonation for plains (as per Kumar 2013)

the identification of seismic sources is done based on in situ geological, geophysical, geomorphological and seismological investigations and is beyond the scope of seismic microzonation study. Such investigations are usually done in specific locations/area as a part of seismic source identification project and under seismic hazard assessment of important structures such as nuclear power plant or dams. Hence, it can be said that the identification of seismic sources is not a regular practice and is done may be once in 20 years or more. Two issues which arise here are: (1) possibility of the presence of seismic sources where no such investigations are done; and (2) newly developed seismic sources after last time investigations on identification of seismic sources were done. In both the cases, determination of worst scenario-related seismic source in DSHA or controlling EQs scenario based on PSHA will be based on incomplete/not-updated source information. Understanding the characteristics of ground, particularly for inaccessible locations, is possible and accurate by means of remote sensing technique. This way, depending upon the spectral information received for the surface by the satellite-mounted sensors, ground characterization can be done. Another advantage of using remote sensing data is its ability to capture temporal variation/ modifications, which make it possible to understand the changes happening in a region over a period. As far as identification of seismic sources are concerned, Landsat data which can be used for assessing land surface temperature through thermal remote sensing plays a very important role (Bektas and Filiz Emine 2016; Ghulam 2010; Jeevalakshmi et al. 2017; Meng et al. 2019). Ongoing focal mechanism along the faults lead to increase in thermal characteristics from nearby seismic sources, which can give a possible idea that a seismic source might exist. Such observation, however, can also be verified based on river profiles and stream gradients using the hypsometric curve, stream length ratio, basin asymmetry factor and other factors, obtained from remote sensing observations. Thus, it is recommended in this work that in addition to revising existing literature for known seismic sources and EQ catalogue, one must refer to remote sensing based outcomes and update the source information as well. Similar to the need for reassessment of seismic hazard of a region after numerous major to great EQ occurrence, available information on known seismic sources also needs regular updates.

24.3 Seismic Source Characterization

It is observed that majority of high to moderate seismically active regions are surrounded by regions of different tectonic settings and geodynamics. However, while characterizing such regions into seismic source zones, areas are selected arbitrarily. In general, information on past EQ events as well as event densities in an area is used for seismic source zonation and subsequently for seismic activity.

It must be mentioned here that seismic activity is directly dependent on the EQ catalogue which is directly a function of predefined and arbitrarily chosen seismic source zones. Hence, calling such source zonation to be based on seismic activity is inappropriate. Baro and Kumar (2017), while trying to understand tectonic setting of the seismotectonic region around the Shillong Plateau (SP) concluded that the entire region has variation in terms of (1) rupture characteristics, (2) slip rate, (3) tectonic setting and (4) dominating focal mechanism (see Fig. 24.2). As a result, the entire seismotectonic region was divided into four seismic source zones. Further, based on past EQ information in each source zone, the seismic activity of each source zone was determined. Hence, present work recommends that the demarcation of seismic source zones should be done prior the determination of seismic activity parameters, based on governing factors such as those mentioned above. In addition, source characterization based on seismic activity itself may be inappropriate since it is highly dependent on the source zone demarcation.

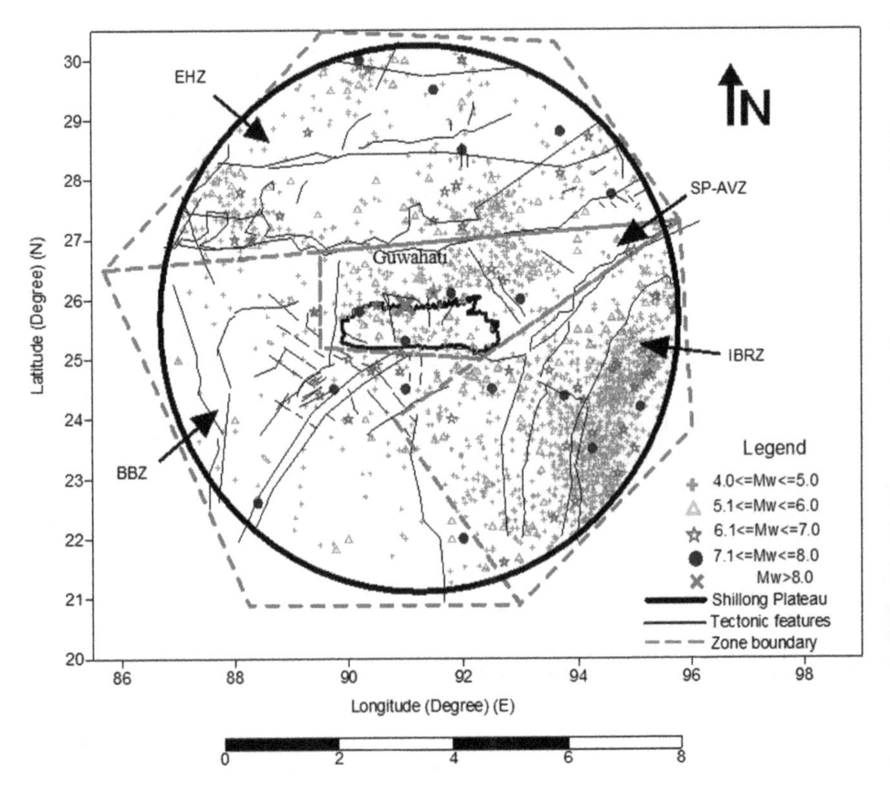

Fig. 24.2 Source characterization of the seismotectonic region of Shillong Plateau into four seismic source zones (modified after Baro and Kumar 2017)

24.4 Declustering of EQ Catalogue

Once the past EQ events within the seismotectonic region are collected, identification of dependent and independent events is done. This is known as declustering (van Stiphout et al. 2016) and is done so that main events and their dependent events can be separated. Further, using main events alone, correct estimation of duration of completeness of the catalogue is done. In addition, depending upon the declustered EQ catalogue, its seismic activity is determined. Thus, it can be said that depending upon the method used for declustering the EQ catalogue, seismic activity, which has been used for source zone identification, can vary. In other words, the choice of declustering method controls the outcome of seismic activity. In most of the seismic hazard studies existing in the literature, method for declustering is selected randomly. Borah and Kumar (2018) did comparative study about the effect of randomly selected declustering process on the estimated seismic activity parameters of same seismotectonic area having same initial EQ catalogue. As per Borah and Kumar (2018), past EQ events within 500 km radial distance from Guwahati were collected from the literature. EQ catalogue consisted of 6202 EQ events starting from 825 AD to April 2018. Three methods of declustering used were (1) Gardner and Knopoff (1974) method, (2) a modified version of this method by Uhrhammer (1976) and (3) Reasenberg (1985). While the first two methods were window-based methods in which the dependent events are identified in a space–time window, the largest event is considered as independent event while others are considered as dependent of above independent event and are removed. While Gardner and Knopoff (1974) method proposed original correlations for space and time windows, the method was modified by Uhrhammer (1976) proposing new space and time windows. Reasenberg (1985) on the other hand gave linked window method in which two events are considered in a cluster. When a new event is associated with an event, which was previously associated in a cluster, then the new event becomes the member of the cluster. If two events from two different clusters fall within the interaction zone, then the two clusters are merged to form a cluster. After declustering, a number of independent events obtained based on method by Gardner and Knopoff (1974) method, Uhrhammer (1976) and Reasenberg (1985), from the same catalogue were 1314, 2664 and 3067, respectively. Based on these three different catalogues, Borah and Kumar (2018) found that the duration of completeness using Stepp (1972) was different even for same magnitude class. Further, b parameter obtained from above methods was 0.91, 0.93 and 0.96, respectively. It must be mentioned here that b parameters are also used to estimate maximum potential EQ possible on a seismic source in addition to seismic hazard assessment. Thus, slight change in b parameter can change maximum potential EQ magnitude and subsequently the seismic hazard values at a site significantly. In addition, depending upon the duration of completeness, 'a' parameter of the catalogue will change. It must be mentioned here that in addition to above three methods, numerous other methods for declustering exist and thus one must be careful while selecting a particular method.

24.5 Effect of Input Motion Characteristics While Assessing Local Site Effect

The presence of local soil changes the ground motion characteristics between bedrock and surface significantly. This phenomenon is called as local site effect and is responsible for significant alteration in ground motion characteristics at the surface with respect to bedrock. Importance of local soil in amplifying the bedrock motion can be understood from the fact that even at larger distances from the epicenter, significant amplitude of ground motion as well as damages had been witnessed globally during various equations (1985 Michoacan EQ, 1989 Loma Prieta EQ 1999, Chamoli EQ, 2001 Bhuj, 2011 Sikkim EQ and 2015 Nepal EQ). While the seismic hazard values at the bedrock level are determined based on detailed seismic hazard assessment, determination of local site effect requires regional ground motion records as well as dynamic soil properties on in situ soil. Dynamic soil properties of soil are the shear modulus (G) and the damping ratio (β). Being dynamic properties, these change with shear strain (γ) developed. Dynamic soil property curves represent variation of G and β with γ. While a very low γ offers very high G (also known as G_{max}), at higher values of γ (>0.2%), the response of soil is governed by higher β value. Thus, it can be said that though the soil remains same, depending upon the γ generated in the soil during a particular EQ, its role in amplifying the bedrock motion changes. Kumar et al. (2016) while trying to understand the response of local soil at a typical location in Delhi, India, found that the soil caused amplification as high as 6–7 when input motion has peak horizontal acceleration (PHA) of 0.008 g. Similarly, the effect of local soil in amplifying bedrock motion is minimal for PHA $\geq$ 0.53 g. In another work, Kumar et al. (2015) concluded that low amplification for higher PHA and higher amplification of low PHA values, as observed for Delhi (which was later verified by Kumar et al. 2017 for Nepal) should be given due importance while assigning weights and ranks to PHA and soil amplification in seismic microzonation practice. At present, both PHA and amplification factors are considered independently, and in general, higher values of each of the two parameters are assigned higher ranks while estimating hazard index. However, works by Kumar et al. (2015, 2016 and 2017) confirmed that both PHA and amplification are interdependent and it is not possible for both parameters to have higher values simultaneously. To arrive to such decision, above works also recommended a set of 30 globally recorded ground motions to be used collectively in the absence of regional ground motion records, in order to account the effect of all possible variation on ground motion characteristics on local site effects.

In another work, Mondal and Kumar (2017) highlighted that most of ground response analyses are done considering default range of frequency content if input motion up to 15 Hz. However, in case the input ground motion has significantly larger amplitude even after 15 Hz, use of input motions should not be restricted to 15 Hz but even higher frequency contents should be considered in ground response analyses.

24.6 Assessment of Liquefaction Potential

Due to EQ shaking, primarily in locations with high groundwater table, there is development of excess pore pressure, which during seismic excitation reduces the effective stress to a minimal value. As a result, the soil loses its shear strength and flows almost like liquid. This phenomenon is known as liquefaction. In seismic microzonation practice, factor of safety against liquefaction is considered a very important thematic layer with higher ranks. Identification of liquefiable zones is seismic microzonation practices are based on grid-wise in situ information. In actual practice, it may happen that many important locations cannot be covered in in-situ investigations due to the presence of built-up area or permission restrictions. Similarly, remote locations many a times are not covered in in-situ investigations. Subsurface properties, being very complex and change drastically, observations related to liquefaction occurrence made at one places cannot be applied to a larger area especially where in situ investigations could not be done as explained earlier. In such cases, identification of liquefiable zones based on observing change in surface moisture before and after an EQ as radiance obtained from Landsat remote sensing data can be found useful and should be done in addition to in situ investigations.

While seismic microzonation studies attempt to determine the factor of safety against liquefaction, guidelines for improvement of these sites are always required by field engineers. As a result, though seismic microzonation identifies zones of possible liquefaction, what to do with these zones in order to improve their in situ shear strength is a requirement further. Though this work does not come under the objectives of seismic microzonation, it plays a vital role in terms of infrastructural growth. If such maps are available, these can help the client as well as designer to take necessary measures during construction itself so that losses due to liquefaction occurrence can be reduced to a remarkable extent. In this direction, Kumar and Srinivas (2017) proposed three empirical correlations (depending upon Fine content values). These correlations can be used both ways, for the estimation of factor of safety against liquefaction of in situ soil as well as to determine improved standard penetration test (SPT) N value to be achieved from ground improvement. This improved SPT N value at the site will ensure that the site will not undergo liquefaction during most likely occurring seismic event in the future and thus other associated damages can be minimized significantly.

24.7 Conclusion

With Government of India's focus on 'Collective efforts inclusive growth,' numerous projects targeting for overall infrastructure development have been launched in the recent years. These projects also target for the development of smart cities across India. In every one to two decade, significant finance in India is mobilized toward rehabilitation works after a natural disaster hits. India, being prone to EQs too, while one side the money in being invested in infrastructural development, other side lots of money goes in rehabilitation and retrofitting works of existing infrastructure after an EQ. To minimize EQ induced damages, seismic microzonation of many important cities is under progress. Present paper discusses five important stages of seismic microzonation study, where careful observations need to be made in decision making, based on remote sensing-based information and other important observations. These will help in significant improvement in current seismic microzonation practices as well as in accurate assessment of hazard indices for the study area.

References

Baro O, Kumar A (2017) Seismic source characterization for the Shillong plateau in northeast India. J Seismol. https://doi.org/10.1007/s10950-017-9664-2

Bektas B, Filiz Emine ME (2016) Land surface temperature retrieval from landsat 8 TIRS-a case study of Istanbul. In: EGU general assembly conference abstracts 18, p 15631

Borah N, Kumar A (2018) Studying and comparing the declustered EQ catalogue obtained from different methods for Guwahati region NE India, Paper no. 56, TH-6. In: Proceedings of Indian geotechnical conference 2018, IISc Bangalore

Gardner JK, Knopoff L (1974) Is the sequence of earthquakes in Southern California, with aftershocks removed, Poissonian? Bull Seismol Soc America 64(5):1363–1367. https://doi.org/10.1785/0120160029

Ghulam A (2010) Calculating surface temperature using landsat thermal imagery. Calculating Surf Temp Using Landsat Thermal Imagery 1(1):1–9

Jeevalakshmi D, Narayana Reddy NS, Manikiam B (2017) Land surface temperature retrieval from LANDSAT data using emissivity estimation. Int J Appl Eng Res 12(20):79–87

Kumar A (2013) Seismic microzonation of Lucknow based on region specific GMPEs and Geotechnical field studies. PhD Thesis, Indian Institute of Science, Bangalore

Kumar A, Srinivas BV (2017) Easy to use empirical correlations for liquefaction and no liquefaction conditions. Geotech Geol Eng 35(4):1383–1407

Kumar A, Harinarayan NH, Baro O (2015) High amplification factor for low amplitude ground motion: assessment for Delhi. Disaster Adv 8(12):1–11

Kumar A, Baro O, Harinarayan NH (2016) Obtaining the surface PGA from site response analyses based on globally recorded ground motions and matching with the codal values. Nat Hazards 81(1):543–572

Kumar A, Harinarayan NH, Baro O (2017) Nonlinear soil response to ground motions during different earthquakes in Nepal, to arrive at surface response spectra. Nat Hazards 87(1):13–33

Meng X, Cheng J, Zhao S, Liu S, Yao Y (2019) Estimating land surface temperature from landsat-8 data using the NOAA JPSS enterprise algorithm. Remote Sens 11(2):1–18

Mondal JK, Kumar A (2017) Impact of higher frequency content of input motion upon equivalent linear site response analysis for the study area of Delhi. Geotech Geol Eng 35(3):959–981

Oommen T, Baise LG, Gens R, Prakash A, Gupta RP (2013) Documenting earthquake-induced liquefaction using satellite remote sensing image transformations. Environ Eng Geosci 19 (4):303–318

Reasenberg P (1985) Second-order moment of central California seismicity, 1969-1982. J Geophys Res 90:5479–5495

Stepp J (1972) Analysis of completeness of the earthquake sample in the Puget sound area and its effect on statistical estimates of earthquake hazard. In: Proceedings of international conference on microzonation, Seattle, USA, pp 897–910
Uhrhammer RA (1976) Characteristics of northern and central California Seismicity. Earthquake Notes
van Stiphout T, Zhuang J, Marsan D (2016) Seismicity declustering, community online resource for statistical seismicity analysis, (February 2012). https://doi.org/10.5078/corssa-52382934

Chapter 25
Influence of Bio- and Nano-materials on Dynamic Characterization of Soils

K. Rangaswamy, Geethu Thomas, and S. Smitha

25.1 Introduction

Soils have a unique nature which needs to take care according to site conditions. Geotechnical engineers face challenges when they deal with constructions in soils prone to various conditions like differential settlement, swelling and shrinkage, liquefaction, slope instability, foundation failure, heave, etc. The conditions need to be considered by detailed analysis and the suitable remedy should be suggested which minimize the risk of failure. There are many methods for soil improvement such as soil reinforcement, mechanical improvement and chemical stabilization. Among them, chemical stabilization is the conventional method of ground improvement. Previously use of traditional additives (cement, lime, flyash, etc.) was common and beneficial which can produce negative impacts on the environment when used in higher dosages. Caustic behavior of lime and cement can pollute the groundwater (Eujine et al. 2017; Eujine et al. 2016), and the carbon emission during the pozzolanic reaction could significantly contaminate the environment (Thomas and Rangaswamy 2020). The introduction of non-traditional additives invaded into the field of stabilization since they could overcome the ill effects of traditional additives.

International Union for Pure and Applied Chemistry (IUPAC) defines biopolymers as the substance composed of one type of biomacromolecules. They are produced by living organisms and consist of repeating units of monomers. They have been used in construction field from early times itself. Sticky rice mortar was used in the ancient Chinese civilization as a binding material in mortar (FuWei et al. 2009). The application of biopolymers in the field of geotechnical engineering has become an emerging research topic. In the past decade, several studies were conducted on biopolymer-treated soils whose results prove them to be an efficient and

K. Rangaswamy (✉) · G. Thomas · S. Smitha
Department of Civil Engineering, NIT Calicut, Kozhikode, Kerala 673601, India
e-mail: ranga@nitc.ac.in

T. G. Sitharam et al. (eds.), *Latest Developments in Geotechnical Earthquake Engineering and Soil Dynamics*, Springer Transactions in Civil and Environmental Engineering, https://doi.org/10.1007/978-981-16-1468-2_25

sustainable solution for soil stabilization. Biopolymers like chitosan, agar, xanthan, guar, etc., have been used for improving the shear strength (Lee et al. 2019; Smitha and Sachan 2016), erosion control (Yeong et al. 2020; Lee et al. 2020) and hydraulic conductivity control (Chang et al. 2019) in soil. They have been proved to be effective in both sandy and clayey soil.

If one could reduce the size of a particle to nano-size, a modified behavior of that object could be identified which gives a reformed and in-depth analysis from the microsized level particulate studies that are almost similar to the bulk materials itself. The idea of the nano-studies was first introduced during a radical lecture by Feynman (1960). The lecture became the milestone for a revolution in the field of nanotechnology, which was new to the world. Givi et al. (2013) in their investigation observed that the reduction in the size of particle in nanoscale could improve the chemical and physical characteristics of soil considerably even in the presence of a small quantity of nanoparticle (Givi et al. 2013). Nanoparticles have various applications in the civil engineering field mainly in concrete and in soil. In construction using concrete, the nanoparticle can increase the performance of concrete (Sobolev et al. 2006), abrasion resistance of pavement (Li et al. 2006) and mechanical properties of hardened cement paste (Qing 2007). In soil stabilization technique, the inclusion of nano-material can improve the durability of expansive soils (Azzam 2014), index properties of weak soils (Hanson et al. 2016; Thomas and Rangaswamy 2020; Majeed et al. 2014), reduce swelling characteristics of sensitive clay (Ali and Jahangiri 2015; Pham and Nguyen 2014). Nano-silica (Biricik and Nihal 2014; Thomas and Rangaswamy 2020; Hanson et al. 2016), carbon nanotubes (Correia and Rasteiro 2016), nano-titanium dioxide (Li et al. 2006), nano-cement, nano-clay (Subramani 2016), nano-copper, nano-magnesium dioxide (Raihan et al. 2012; Taha et al. 2015) are some among the commonly used nano-additives in the field of soil improvement.

Weak soils have to be improved for both static loadings as well as dynamic loading. The previously mentioned literature is dealing with static loading conditions but there are a few studies which handle the efficacy of additives during dynamic loading conditions. Among them, a few studies are using traditional additives like cement (Subramaniam and Banerjee 2014; Fatahi et al. 2013) and lime (Fahoum and Aggour 1996), and the soil improvement using nanoparticle under dynamic loading is fewer (Choobbasti et al. 2015). Dynamic soil properties are mainly such as shear modulus, damping ratio and Poisson's ratio. Dynamic soil properties can be measured by in situ wave measurement with and without boreholes, laboratory tests for small-strain properties (wave transmission tests, small-scale cyclic loading tests), laboratory tests for medium to large strain (simple shear test, torsional simple shear test, cyclic triaxial test). Soft soils are prone to excessive lateral deformations and possess high peak ground acceleration under dynamic loading. The dynamic soil behavior of soft soils needs to be considered more precisely and site specifically in order to minimize the impact of dynamic loading.

In the current study, the effect of biopolymer and nano-material on the dynamic properties of silty sand and soft clay has been explored. This has been carried out by performing few strain-controlled cyclic triaxial tests on treated and untreated soil samples. The dynamic properties considered are pore pressure response, secant shear modulus and damping ratio for both the soils. The dynamic response of two different soils under large strain range and the influence of two non-traditional additives on the dynamic soil properties are presented in this paper.

25.2 Materials and Methods

25.2.1 Materials

The materials used to carry out the experiments in the present work are silty sand, soft clay, agar biopolymer and nano-silica.

Sandy Soil The soil was collected from Wayanad district in Kerala. Silty sand was chosen since it is prone for liquefaction failure when exposed to dynamic loads under saturated, undrained condition. The soil was classified as silty sand (SM) as per Indian Standard soil classification system. Further details of the soil can be found in Smitha et al. (2019).

Clayey soil The location chosen for the soil collection was a piling site at Kaloor, Kochi, Kerala. The location is known to have weaker soils in saturated condition due to its marine neighborhood. These soils are not favorable for construction even under static loading which requires relatively deeper foundations. The soil belongs to the category of clay with high compressibility (CH) as per Indian Standard soil classification system and is having a low compressive strength (28 kPa) corresponding to soft soils (Thomas and Rangaswamy 2019).

Biopolymer For the present study, agar biopolymer was chosen due to its higher mechanical strength and stability and due to the perceptible difference in its gelling and melting temperature. Agar is obtained from certain species of algae like Gelidiella which could be found in marine waters of tropical and subtropical regions. It is composed of agarose and agaropectin where agarose is the gelling agent (McHugh 2003). It is known to be stable over a wide range of temperature and is resistant against microbial action (Khatami and O'Kelly 2013; Wu 1990) and therefore may be durable to some extent. For the current study, agar was obtained from the manufacture Urban Platter. Its specifications are provided in Smitha and Rangaswamy (2020).

Nanoparticle Nano-silica was selected for the present study since their efficacy has already been proven in altering static and dynamic behavior of soils. The nano-sized silica particles have high surface energy which accelerates the pozzolanic reaction and increase the soil stiffness and other index and engineering properties of soil. The smaller the nanoparticle size, higher be the reactivity of additive (Givi et al. 2013). 17 nm nano-silica was chosen for the present investigation, which has a specific gravity around 2.2–2.4 and purchased from Astrra chemicals, Chennai (Thomas and Rangaswamy 2020). 1% of portland pozzolana cement is also used along with the nano-silica to avoid delay in the initial strength gain that was manufactured by Ramco Cements Ltd, India (specific gravity of 3.05, initial and final setting time of 38 min and 7.5 h, respectively).

25.2.2 Experimental Program

Sample preparation Triaxial specimens of treated and untreated silty sand and clayey soil of 50 mm diameter and 100 mm height were prepared.

Preparation of silty soil specimen Untreated silty sand specimens of relative density 30% were prepared by air pluviation method by applying vacuum during sample preparation process. Agar-treated soil samples were prepared by adding hot agar solution in oven-dry soil at the required percentage and then placing in mold after which they were extracted and kept for curing for the required curing period. The detailed procedure for agar-treated soil sample preparation is illustrated in the form of schematic diagram (Fig. 25.1). Silty sand was treated with 2% agar biopolymer by weight, and it was cured for 7 days. It was saturated by following the conventional method of saturating sandy soil followed in other research studies (Arab and Belkhatir 2012; Boominathan et al. 2010), that includes water circulation, CO_2 saturation and forced (or pressure) saturation.

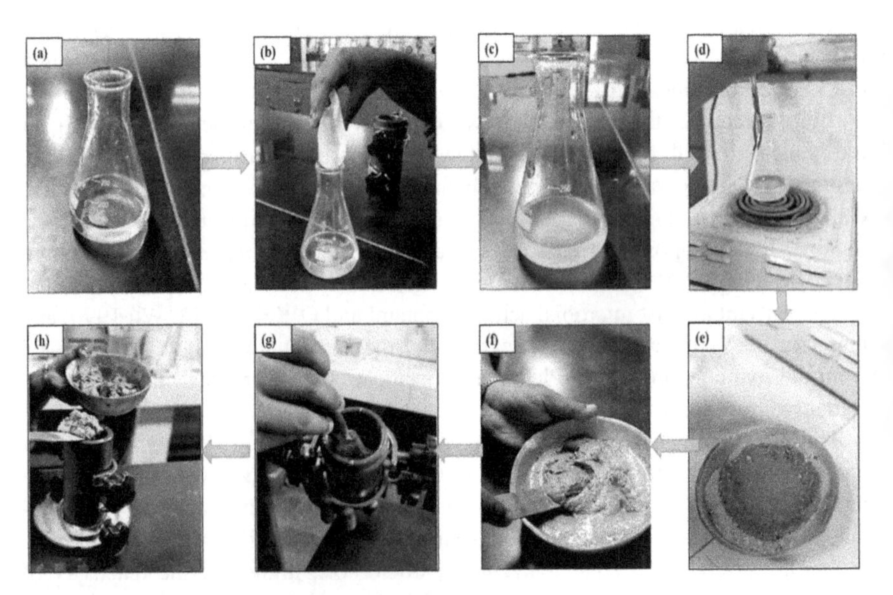

Fig. 25.1 Sample preparation **a** Water measured in a conical flask. **b** Transferring agar powder into water. **c** Agar–water mix. **d** Heating the agar–water mix to make it to a solution. **e** Hot agar solution poured into the oven-dry soil. **f** Mixing of agar solution with soil. **g** Applying grease to the inner walls of the mold. **h** Placing the treated soil in split mold

Preparation of clayey soil specimens Both treated and untreated clay specimens were extruded from a cell prepared by slurry consolidation method (Liu et al. 2017; Wang et al. 2011). The oven-dried, powdered and sieved through 425 micron IS sieve clay soil is mixed with water around 1.2 times the liquid limit (LL = 91%) and mixed thoroughly to get a uniform slurry. The slurry is poured into a container of size 23 cm diameter and 30 cm height with perforations on sides and bottom. The sides and bottom are covered with filter paper to avoid leakage of slurry in the initial stages. Preloading is applied in increments over a loading pad with perforations until desired density and curing days are achieved. The settlement of the slurry is monitored using dial gauges which can measure a maximum of 50 mm settlement. After 30 days, the samples were extruded using samplers and kept undisturbed in desiccator. 1% cement and 1% nano-silica are added to the dry soil and mixed with hands to get uniform blend and mixed with water to get the treated slurry. All other steps are similar to that of untreated soil. Sample is saturated by back-pressure and confining pressure increments until Skepton's pore pressure parameter (B) reaches around 0.98. Throughout the process of this forced saturation, the difference between cell pressure and back pressure was maintained as 20 kPa. Theoretically, the parameter B reaches the value of 1, and the soil is considered as completely saturated. Since for a saturated soil in an isotropic stress condition, any change in the total mean principle stress will affect the excess pore pressure of the soil consequently without altering the effective stress value (Skempton 1954; Cai et al. 2013; Pandya and Sachan 2019). After saturation process, specimen has to go through consolidation process until the deformation versus time curve become a plateau. Saturation and consolidation processes together took around 2 days to complete then the consolidated specimen tested for shearing.

Testing Program Two series of consolidated undrained strain-controlled cyclic triaxial tests were performed after saturating the soil specimens. Consolidated undrained cyclic triaxial tests were performed on untreated silty sand (0% agar) and 2% agar-treated soil. Similar tests were performed on untreated clayey soil and treated clay soil mixed with 1% nano-silica and 1% cement (cured for 28 days). All the tests were performed as per ASTM: D5911-92. The samples were under an effective confining pressure of 100 kPa during loading. The cyclic loading was applied in the form of sinusoidal waves at a frequency of 1 Hz and cyclic strain amplitude of 1 mm.

The effects of biopolymer and nano-silica treatment on dynamic properties as well as pore pressure response were studied. The major dynamic properties of soil include secant shear modulus (G) and damping ratio (D). G and D are essential parameters that are fundamental for all geotechnical engineering problems. They were found for 1st, 2nd, 3rd, 4th, 5th, 10th, 15th, 20th, 30th, 40th and 50th cycles in both sandy and clayey soil for both treated and untreated conditions. Shear modulus was estimated by first finding secant Young's modulus (E_{sec}) as given in Eqs. 25.1 and 25.2 where σ_d is the deviatoric stress and $\in$ is the cyclic axial strain. Poisson's ratio, ν, can be taken as 0.5 for saturated soil (Lambe and Whitman 1969).

$$E_{sec} = \frac{\sigma_d}{\in} \tag{25.1}$$

$$G = \frac{E_{sec}}{2(1+v)} \tag{25.2}$$

Damping represents energy dissipation. It was found out from hysteresis loop, i.e., the plot between deviatoric stress and axial strain using Eq. 25.3 where it is expressed as damping ratio in percentage.

$$D = \left(\frac{\text{Area of hysterisis loop}}{2\pi(\text{Area of Triangle})}\right) \times 100 \tag{25.3}$$

25.3 Results and Discussions

Shear strength of soil describes the magnitude of the shear stress that a soil can sustain against overlying structures. The shear resistance of soil is a result of friction and interlocking of particles and possibly cementation or bonding at particle contacts. In this study, undrained response and shear strength of biopolymer-treated silty sand were evaluated at optimum dosage of 2% biopolymer and with 7 days of curing. CU triaxial tests were carried out to get the strength characteristics and thereby to draw conclusions on improvement of silty sand soil in terms of strength and liquefaction resistance. The soft clay before and after treatment with nanoparticles were also evaluated to get better knowledge about the stabilization effect on the dynamic soil properties such as shear modulus, damping ratio and pore pressure response.

25.3.1 *Effect of Biopolymer Treatment on Silty Sand*

Effect on pore-pressure Silty sand has very negligible cohesion. When it is fully saturated, during cyclic loading pore pressure tends to rise up as the number of cycles progresses. Pore pressure goes on rising till it reaches a value equal to the confining pressure and at that condition liquefaction failure in soil occurs. In the study, the pore pressure ratio, i.e., the ratio of pore pressure to effective confining pressure was found and plotted against number of cycles in Fig. 25.2. The testing was continued till 200 cycles, but the variation in pore pressure and other parameters was very negligible after 50th cycle. Hence, all the results are plotted for 50 cycles of loading only. In similar studies like that of Hazirbaba and Omarow (2018) where sandy soil is stabilized with geofibre, the number of cycles considered is up to 50 cycles only. From figure, it can be seen that the pore pressure ratio changes cyclically as the cyclic loading progresses in both untreated and treated silty soil. It

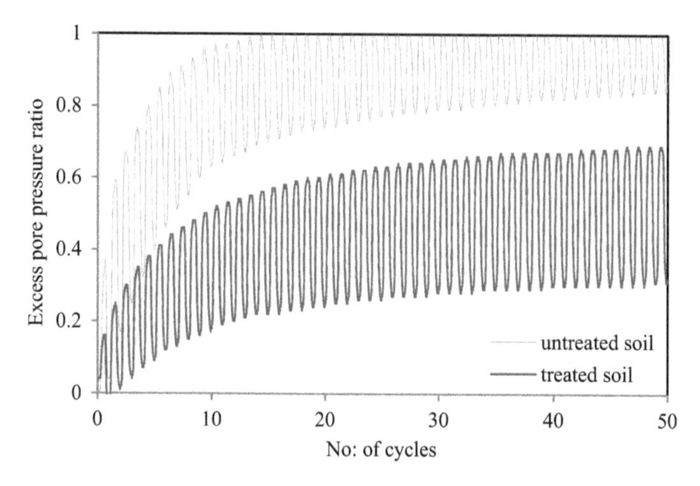

Fig. 25.2 Excess pore pressure ratio versus no: of cycles of untreated and agar biopolymer-treated silty sand

is clear that the pore pressure ratio reached unity at the 14th cyclic which is considered as the start of initial liquefaction. But for agar-treated soil at the same cyclic amplitude, the pore pressure ratio never reached unity even after 50 cycles of loading or pore water pressure will never be equal to confining pressure in treated soil. Furthermore, the maximum excess pore pressure built up is just 0.67, which is 33% less than untreated silty sands. This implies that agar treatment has the potential to control excess pore pressure buildup in silty sand and prevent liquefaction failure.

The potential of agar hydrogel inside treated soil to control pore water pressure buildup during cyclic loading may be due to the fact that agar gel occupied the pore spaces of soil. On curing the agar would loss its water of hydration and become a firm plastic-like structure. But being hydrophilic in nature, during saturation of soil the dried, solid agar would absorb water and swell to some extend again filling the pore spaces of soil. Thus, when the soil is subjected to cyclic compression and extension instead of water that take up pressure, it is now the hydrogel that is taking pressure. Therefore, the pore pressure will be lower.

Effect on shear modulus and damping ratio Secant shear modulus and damping ratio expressed as percentage at particular number of cycles is tabulated in Table 25.1. Considering shear modulus it can be perceived that as the number of cycles progresses G decreases in both treated and untreated soil. But the amount of decrease is much less in treated soil when compared to untreated silty soil. It decreased by about 64% from first cycle to 50th cycle in untreated silty sand; whereas the decrease was just 12% in treated soil up to 50th cycle of loading. This indicated that agar treatment enhanced the strength and stiffness properties of soil.

Damping ratio also decreased with increase in number of cycles as is the general trend during cyclic loading in soil. The damping ratio of treated soil was found to be

Table 25.1 Dynamic properties of untreated and 2%, 7-day cured agar-treated silty sands at different number of cycles

No: of cycles (*N*)	Untreated soil		Treated soil	
	G (kg/cm^2)	*D* (%)	*G* (kg/cm^2)	*D* (%)
1	9.4	56.8	16.1	64.2
2	8.8	54.7	15.8	64.0
3	8.2	55.6	15.6	64.0
4	7.8	53.3	15.5	62.1
5	7.6	50.3	15.3	62.5
10	6.4	46.7	15.0	62.0
15	5.6	45.7	14.8	58.7
20	5.0	42.3	14.7	58.5
30	4.3	41.3	14.4	54.9
40	3.7	39.2	14.4	53.0
50	3.4	38.1	14.2	51.7

higher as seen from Table 25.1. This was because there would be a phase difference between the rigid sand particles and soft or ductile agar hydrogel, which would result in higher energy dissipation. Similar result has been demonstrated in previous work of Im et al. (2017) where resonant column test of biopolymer-treated soil had been carried out. Higher D will be advantageous in reducing the impact of earthquake waves in agar-treated silty soil.

25.3.2 *Effect of Nano-material on Soft Clay*

Effect on pore-pressure In saturated clays, the pore pressure buildup shows an increase in trend as the number of cycle increases. Increase in excess pore pressure may rise up to its maximum and gradually reaches a steady value as the number of cycle increases. Unlike cohesionless soils, clay might not be reaching a value of pore pressure equal to the confining pressure under moderate number of cycles of loading. Excess pore pressure is considered to be related to the cyclic degradation in the case of normally consolidated clays (Matasovic and Vucetic 1995). The testing was continued up to 200 cycles. Refer Fig. 25.3 for comparing the pore pressure response obtained for soft clay before and after treatment. Peak value of pore pressure ratio for untreated soil was around 0.5 which is reduced to 0.3 due to the inclusion of additives and a curing period of 28 days. The reduction in pore pressure ratio due to the stabilization is around 40% and the growth of pore pressure becomes reduced after 40 cycles but not getting a constant value until 200 cycles. This indicates that the inclusion of nanoparticles has a role in reducing the pore pressure buildup of soft clay.

The possible mechanism behind the changes in soil properties due to the stabilization is explained in the following section. During the process of curing, the

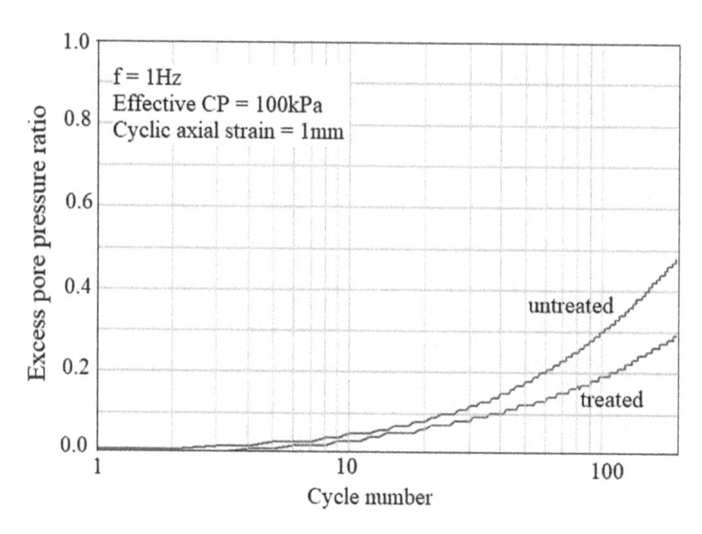

Fig. 25.3 Excess pore pressure ratio versus no: of cycles of untreated and soil–cement with nanoparticles

cement reacts with water to produce calcium silicate hydrate (C-S-H) gel and calcium hydroxide in the soil matrix. The C-S-H gel is responsible for the strength gain and increase in bonding and filling of available air voids whereas calcium hydroxide has no role in the strength gain. The nano-silica reacts with undesirable calcium hydroxide to form C-S-H gel which causes secondary form of gel formation adding up to the strength gain. Thus, the quantity of calcium hydroxide is consumed by the nanoparticles to produce more gel. The nano-silica acts as a nucleus of the gel and sticks around the clay flakes (Thomas and Rangaswamy 2020). During curing, the matrix becomes deficient of moisture and the introduction of more water during the saturation process allows the gel to take up the water and retain around the soil flakes. Due to this, the cyclic loading takes by the gel around the soil particles which are considered to be elastic that reduces the water pressure in the remaining voids.

Effect on shear modulus and damping ratio The shear modulus of the soil goes on reducing with the number of cycle progresses and reaches a steady value (Table 25.2). The degradation of the shear modulus of cohesive soils is dependent on many factors such as overconsolidation ratio, plasticity index, effective confining pressure, cyclic strain, frequency of loading and number of loading cycles. In the present study, the treatment reduced the plasticity index (PI) of soil from 58 to 13.7 in 28 days of curing. The PI plays a major role in the dynamic characteristics of soil. The shear modulus of soft clay increased around 53% with the influence of additives at cycle 1. The soil's plasticity reduced due to the stabilization process which consequently resulted in more strength and stiffness of the clay.

Damping ratio denotes the energy dissipation capacity of soil during dynamic loading. The increase in number cycles does not show significant change in the

Table 25.2 Dynamic properties of untreated and soil–cement with nanoparticles cured for 28 days at different number of cycles

No. of cycles (*N*)	Untreated soil		Treated soil	
	G (kg/cm^2)	*D* (%)	*G* (kg/cm^2)	*D* (%)
1	18.95		28.97	
2	18.82		28.55	
3	17.46		27.98	
4	17.01		27.44	
5	16.72		26.95	
10	15.21	12.01	24.62	21.28
15	13.85		22.71	
20	13.28		21.40	
30	12.50	(Not significant change in damping ratio for moderate cyclic shear strain and *N*)	20.19	(Not significant change in damping ratio for moderate cyclic shear strain and *N*)
40	11.59		19.99	
50	11.44		19.50	

damping ratio of the clay soil but the reduction in PI improved the damping ratio considerably. This increase in damping ratio of the soil after treatment clearly explains that the stabilization using nano-silica can dissipate more energy during dynamic loading. The reduction in voids, better particle-to-particle bonding in treated soil also, and the particle rearrangement without hindering the particle orientation due to permanent plastic deformations (Seethalakshmi and Sachan 2019; Pandya and Sachan 2019) within the soil matrix after treatment might be the reasons for increasing the strain energy dissipation capacity.

25.4 Conclusions

The following are the major findings from the study:

- There was a significant decrease in excess pore water pressure buildup during cyclic loading in silty sand treated with biopolymer. About 33% reduction in pore pressure ratio was obtained at 1% cyclic axial strain, whereas in soft clays treated with cement and nanoparticle, pore pressure buildup could be reduced up to 40% at cyclic axial strain of 1%.
- Both the soils showed an enhancement in secant shear modulus and damping ratio upon treatment.
- The increase in shear modulus and damping ratio and decrease in pore pressure ratio when compared with untreated soil irrespective of the soil type show that the method of using biopolymer and nano-material may be considered as a reliable ground improvement technique when the site is prone to dynamic loading conditions.

References

Ali PM, Jahangiri S (2015) Effect of nano silica on swelling, compaction and strength properties of clayey soil stabilized with lime. J Appl Environ Biol Sci 5 (zhang 2007):538–548

Arab A, Belkhatir M (2012) Fines content and cyclic preloading effect on liquefaction potential of silty sand: a laboratory study. Acta Polytech Hung J 9(4):47–64. https://doi.org/10.1007/s13369-013-0700-4

Azzam WR (2014) Durability of expansive soil using advanced nanocomposite stabilization. Int J Geomater 7(1):927–937

Biricik H, Nihal S (2014) Comparative study of the characteristics of nano silica-, silica fume- and fly ash-incorporated cement mortars. Mater Res 17(3):570–582. https://doi.org/10.1590/S1516-14392014005000054

Boominathan A, Rangaswamy K, Rajagopal K (2010) Effect of non-plastic fines on liquefaction resistance of poorly graded Gujarat sand. Int J Geotech Eng 4:241–253

Cai Y, Gu C, Wang J, Juang CH, Xu C, Hu X (2013) One-way cyclic triaxial behavior of saturated clay: comparison between constant and variable confining pressure. J Geotech Geoenviron Eng 139:797–809

Chang I, Tran A, Cho GC (2019) Introduction of biopolymer-based materials for ground hydraulic conductivity control. World Tunnel Congress (WTC) 2019 At: Naples, Italy

Choobbasti AJ, Ali V, Saman SK (2015) Mechanical properties of sandy soil improved with cement and nanosilica. De Gruyter Open Open Eng 5:111–116. https://doi.org/10.1515/eng-2015-0011

Correia AAS, Rasteiro MG (2016) Nanotechnology applied to chemical soil stabilization. Procedia Engineering 143 (Advances in Transportation Geotechnics 3. In: The 3rd international conference on transportation geotechnics (ICTG 2016)), pp 1252–1259. https://doi.org/10.1016/j.proeng.2016.06.113

Eujine GN, Chandrakaran S, Sankar N (2016) The engineering behaviour of enzymatic lime stabilised soils. Proc Inst Civil Eng Ground Improv 170(1):1–11. https://doi.org/10.1680/jgrim.16.00014

Eujine GN, Chandrakaran S, Sankar N (2017) Accelerated subgrade stabilization using enzymatic lime technique. J Mater Civ Eng ASCE 29(9):1–7. https://doi.org/10.1061/(ASCE)MT.1943-5533.0001923

Fahoum K, Aggour MS (1996) Dynamic properties of cohesive soils treated with lime. J Geotech Eng 122(5):382–389

Fatahi B, Fatahi B, Le TM, Khabbaz H (2013) Small-strain properties of soft clay treated with fibre and cement. Geosynthetics Int 20(4):286–300

Feynman R (1960) There's plenty of room at the bottom, reprint from speech given at annual meeting of the american physical society. Eng Sci 23:22–36

FuWei Y, BingJian Z, ChangChu P, YuYao Z (2009) Traditional mortar represented by sticky rice lime mortar—one of the great inventions in ancient China. Sci China Ser E: Technol Sci 52:1641–1647

Givi AN, Rashid SA, Aziz FNA (2013) Influence of 15 and 80 nano-SiO_2 particles addition on mechanical and physical properties of ternary blended concrete incorporating rice husk ash. J Exp Nanosci 8(1):1–18. https://doi.org/10.1080/17458080.2010.548834

Hanson JL, Nazli Y, Amro EB, Ryne M, Jared SS (2016) Determination of the index properties of clay soils in the presence of nanoparticles. Geo-Chicago 2016 GSP 26, ASCE, pp 441–450

Hazirbaba K, Omarow M (2018) Excess pore pressure generation and post-cyclic loading settlement of geofiber-reinforced sand. Gradevinar 70(1): 11–18. https://doi.org/10.14256/JCE.1683.2016

Im J, Tran ATP, Chang I, Cho GC (2017) Dynamic properties of gel-type biopolymer-treated sands evaluated by resonant column (RC) tests. Geomech Eng 12(5):815–830. https://doi.org/10.12989/gae.2017.12.5.815

Khatami HR, O'Kelly BC (2013) Improving mechanical properties of sand using biopolymers. J Geotech Geoenviron Eng 139:1402–1406. https://doi.org/10.1061/(ASCE)GT.1943-5606.0000861
Lambe TW, Whitman RV (1969) Soil mechanics. Wiley, New York, NY
Lee S, Im J, Cho GC, Chang I (2019) Laboratory triaxial test behavior of Xanthan gum biopolymer—treated sands. Geomech Eng 17(5):445–452. https://doi.org/10.12989/gae.2019.17.5.445
Lee S, Kwon YM, Cho GC, Chang I (2020) Investigation of biopolymer treatment feasibility to mitigate surface erosion using a hydraulic flume apparatus. Geo-Congress 2020: Minneapolis, Minnesota. 10.1061/9780784482834.006
Li H, Zhang MH, ping Ou J (2006) Abrasion resistance of concrete containing nano-particles for pavement. Wear 260(11–12):1262–1266. https://doi.org/10.1016/j.wear.2005.08.006
Liu W, Xiaowei T, Yang Q (2017) A slurry consolidation method for reconstitution of triaxial specimens. Geotech Eng 21:150–159. https://doi.org/10.1007/s12205-016-0199-9
Majeed ZH, Taha MR, Jawad IT (2014) Stabilization of soft soil using nanomaterials. Res J Appl Sci Eng Technol 8(4):503–509
Matasovic N, Vucetic M (1995) Generalized cyclic-degradation-pore-pressure generation model for clays. J Geotech Eng 121(1):33–42
McHugh DJ (2003) A guide to the seaweed industry. Food and Agriculture Organization of The United Nations, Rome
Pandya S, Sachan A (2019) Effect of frequency and amplitude on dynamic behaviour, stiffness degradation and energy dissipation of saturated cohesive soil. Geomech Geoeng 1–15. 10.1080/17486025.2019.1680885
Pham H, Nguyen QP (2014) Effect of silica nanoparticles on clay swelling and aqueous stability of nanoparticle dispersions. J Nanopart Res 16(2137):1–11. https://doi.org/10.1007/s11051-013-2137-9
Qing Y (2007) Influence of nano-SiO 2 addition on properties of hardened cement paste as compared with silica fume. Constr Build Mater 21:539–545. https://doi.org/10.1016/j.conbuildmat.2005.09.001
Raihan M, Omer T, Eldeen M (2012) Influence of nano-material on the expansive and shrinkage soil behavior. J Nanopart Res 14(1190):1–13. https://doi.org/10.1007/s11051-012-1190-0
Seethalakshmi P, Sachan A (2019) Dynamic behaviour of micaceous sand with varying mica content and its association with compactability, compressibility and monotonic shear response. Int J Geotech Eng 1–17. 10.1080/19386362.2019.1589159
Skempton AW (1954) The pore pressure coefficients A and B. Geo-technique 4(4):143–147
Smitha S, Rangaswamy K (2020) Effect of biopolymer treatment on pore pressure response and dynamic properties of silty sand. J Mater Civil Eng 32(8). https://doi.org/10.1061/(asce)mt.1943-5533.0003285
Smitha S, Sachan A (2016) Use of Agar biopolymer to improve the shear strength behavior of Sabarmati sand. Int J Geotech Eng 10(4):387–400
Smitha S, Rangaswamy K, Keerthi DS (2019) Triaxial test behaviour of silty sands treated with Agar biopolymer. Int J Geotech Eng 1–12. https://doi.org/10.1080/19386362.2019.1679441
Sobolev K, Flores I, Hermosillo R, Torres-martínez LM (2006) Nanomaterials and nanotechnology for high-performance cement composites. Nanotechnology of concrete: recent developments and future perspectives (ACI Session). Denver, USA, pp 91–118
Subramani V (2016) Soil stabilization using nano materials. Int J Res Appl Sci Eng Technol 4:641–645
Subramaniam P, Banerjee S (2014) Factors affecting shear modulus degradation of cement treated clay. Soil Dyn Earthquake Eng 65:181–188. https://doi.org/10.1016/j.soildyn.2014.06.013
Taha MR, Jawad IT, Majeed ZH (2015) Treatment of soft soil with nano-magnesium oxide. Nanotechnol Constr 1:1–9
Thomas G, Rangaswamy K (2019) Strength behavior of enzymatic cement treated clay. Int J Geotech Eng 00(00):1–14. https://doi.org/10.1080/19386362.2019.1622854

Thomas G, Rangaswamy K (2020) Strengthening of cement blended soft clay with nano-silica particles. Geomech Eng 20(6):505–516. https://doi.org/10.12989/gae.2020.20.6.505

Wang SY, Luna R, Stephenson RW (2011) A slurry consolidation approach to reconstitute low-plasticity silt specimens for laboratory triaxial testing. Geotech Test J 34(4):1–79. https://doi.org/10.1520/GTJ103529

Wu C (1990) Training manual on gracilaria culture and seaweed processing in China. FAO Fishery Technical Paper, China

Yeong M, Kwon SH, Kwon T-H, Cho GC, Chang I (2020) Surface-erosion behaviour of biopolymer-treated soils assessed by EFA. Géotech Lett 10(2):1–7

Chapter 26
Dynamic Characterization of Lunar Soil Simulant (LSS-ISAC-1) for Moonquake Analysis

Kasinathan Muthukkumaran, T. Prabu, and I. Venugopal

26.1 Introduction

In recent years, various space missions are interested in exploring the answers for the longstanding questions about the origin of the Moon, environmental conditions, formations, and geotechnical properties of the lunar regolith. At the same time, the lunar missions like Luna, Surveyors, Apollo, etc., brought back the lunar soil samples to Earth for determining its characterization, mineral composition, and geotechnical properties (Carrier et al. 1973). A small quantity of brought back lunar soil was not enough for the extensive research work about the lunar regolith, hence triggered the researchers to find an alternative material, which is called lunar soil simulant (LSS). During the past decades, countries like the USA, Russia, China, Korea, and Japan, etc., have developed such lunar soil simulants indigenously using different materials like rocks (Anorthosite and basaltic), volcanic ash and some other minerals along with artificial materials (Florez et al. 2015; He et al. 2013; Jiang et al. 2012; Li et al. 2009; Ryu et al. 2018; Zeng et al. 2010a, b). The simulants were developed to emulate either the mineral composition or the geotechnical properties of the lunar regolith. The simulants were generally used to assess the geotechnical properties such as particle size distribution, plasticity index, compaction characteristics, shear strength, and compressibility behavior of the lunar soil (Carrier et al. 1991; Mitchell et al. 1973). The rover wheel and lunar soil interaction behavior were also experimented using these simulants for the design of rover wheels and other in situ exploration machines (Oravec 2009; Tao et al. 2006;

K. Muthukkumaran (✉) · T. Prabu
Department of Civil Engineering, National Institute of Technology, Tiruchirapalli, Tamilnadu, India
e-mail: kmk@nitt.edu

I. Venugopal
C&MG LEOS, U R Rao Satellite Centre, Indian Space Research Organization, Bangalore, India

T. G. Sitharam et al. (eds.), *Latest Developments in Geotechnical Earthquake Engineering and Soil Dynamics*, Springer Transactions in Civil and Environmental Engineering, https://doi.org/10.1007/978-981-16-1468-2_26

Yu et al. 2012). The shear strength parameters like angle of internal friction and cohesion of the lunar soil etc. play an important role in the design of lunar vehicles apart from the in situ resource utilization (ISRU) programs (Oravec 2009; Tao et al. 2006; Yu et al. 2012). These parameters were determined by conducting direct shear tests and triaxial tests using the simulant.

Various space research organizations of advanced countries have already initiated R &D works towards Moon colonization. In general, the shear strength parameters and deformability of the lunar soil are taken as a major concern for the design of such exploration equipment. While constructing lunar structures on the Moon, the dynamic behavior of the lunar soil is an essential input rather than the static properties alone for evaluating design criteria for the foundation system of any lunar structure. The behavior of lunar soil under dynamic loading is entirely different from static loading. Therefore, it is important to ascertain the dynamic behavior of the lunar soil under dynamic loading, especially under different moonquake conditions. Moonquakes (Taylor 2007; Nakamura 1847; Weber 2014) are different from the terrestrial earthquakes because of the spatial variability of the lunar surface. However, shear modulus and damping ratios are the prime parameters for the dynamic assessment. This paper explains the determination of shear modulus (G), damping ratio (ξ), and Poisson's ratio (υ) of Lunar Soil Simulant under different moonquake conditions using the cyclic triaxial tests. The simulant LSS-ISAC-1, which has been jointly developed by the Indian Space Research Organization (ISRO) Bangalore, National Institute of Technology (NIT) Tiruchirappalli, and Periyar University, Salem was used for this study. The simulant LSS-ISAC-1 was indigenously developed for the Chandrayaan missions and towards futuristic R&D works.

26.2 Moonquakes

The lunar regolith has deep dry strata for a few meters at the top due to the absence of water and a very fractured rock crust beneath thereafter. Therefore, the lunar seismic signals have a large degree of wave scattering and very few attenuations. Because of this, the Moon "rings like a bell," whenever the moonquakes occur (Taylor 2007). In general, moonquakes happen as a result of releasing the stored energy and tidal energy, not due to the release of tectonic energy. The moonquakes are classified into three different types, such as deep moonquakes, shallow moonquakes, and impacts (Nakamura 1847). Deep moonquakes are the most frequent seismic event which occurs most of the time than the other two. These moonquakes occur at a deeper depth, approximately about 700–1200 km from Moon and but not strong enough to cause any ground shaking. Nakamura (1847) described that the deep moonquakes take place to release the stored energy without a significant release of tectonic energy. Deep moonquakes happen with respect to the tidal periodicity once in a month or seven and a half months (Nakamura 1847). The shallow moonquakes are the most energetic seismic event compared to the deep moonquakes and release more energy. The release of seismic energy from the Moon is commonly assumed to be small, only ~2 × 10^{10} J/yr compared to Earth's 1017–1018 J/yr, and the actual average lunar seismic energy could be as high as 10^{14} J/yr (Carrier et al. 1991). Shallow moonquakes occur at a shallow depth of about 100 km, and this appears to be similar to terrestrial earthquakes. There is no such periodicity observed for the shallow moonquakes. The third type of moonquakes is "impacts," which occur regularly due to the impact of meteoroids. The Apollo seismometers were deployed to measure the moonquakes over a period of the past 8–10 years. The seismometers have observed and recorded 13,000 moonquakes comprising deep moonquakes, shallow moonquakes, and impacts during that period. The magnitude of the moonquakes was measured based on the Richter scale; the measured magnitude of the deep moonquakes was less than 3, mostly lesser than one. For example, a deep moonquake was recorded on 23rd May 1970, at 13.09 h, and the magnitude was measured as 2. The deep moonquake occurred at a focal depth of 800 km, and the moonquake continued for an hour (from 13.09 h to 14.00 h) (Carrier et al. 1991). The average observed shallow moonquakes have a magnitude of 2.4–4.1 (Weber 2014). The largest shallow moonquake occurred on 3rd January 1975 at the epicentral location 29°N and 98°W with a location error of 1°. The recorded magnitude was 4.1, and the earthquake focal depth was around 1–5 km (Carrier et al. 1991). Also, the maximum magnitude recorded during the shallow moonquakes was 5 and 5.5, which is strong enough to shake the buildings and other objects on the surface of the Moon or Earth. However, the moonquakes can sustain for ten minutes owing to its dry, cold, and rigid nature of the regolith. In general, the earthquake vibrations die down within half to one minute, and the maximum observed period is not more than two minutes. Therefore, the dynamic response on the Moon and Earth would not be similar to each other. Hence it needs to be dealt with separately in further research works.

26.3 Geotechnical Properties of Lunar Soil Simulant

The ISRO has jointly developed a lunar soil simulant for carrying out Soil Simulation studies for Rover & Lander. The new lunar simulant was preferred to emulate the properties of Apollo-16 lunar soil samples in all aspects of mineralogy, chemistry, and a few geotechnical properties. The Periyar University, Salem, Tamilnadu has done the geological study and found an appropriate Anorthosite rock belt at Sittampundi, Salem, Tamilnadu, which tallies with actual lunar soil samples. The EDAX and XRD test was performed to find the minerals and chemical composition of the Sittampundi Anorthosite rocks. The study revealed that the minerals and chemical composition of the respective rock samples was matching with the Apollo-16 lunar samples. The identified rock samples were pulverized into different sizes and mixed with different proportions to obtain the optimum mix-matching with the gradation of the Apollo-16 lunar soil. The

Table 26.1 Geotechnical properties of the LSS-ISAC-1 (Venugopal et al. 2020a, b)

ASTM Standard	Geotechnical properties	Values
D421	Specific gravity, G_s	2.70
D6913	Fines, %	40–48
ASTM D2487	Soil classification (USCS)	Silty sand
D4914	Bulk density, ρ (g/cm^3)	1.50
D4253-e1 and D4254	Relative density, %	63
	Maximum density, ρ_{max} (g/cm^3)	1.75
	Minimum density, ρ_{min} (g/cm^3)	1.18
D698-e2 and D1557-e1	Maximum dry density, ρ_{dmax} (g/cm^3)	1.45
	Optimum moisture content, %	12.77
D3080/IS2720&D2850	Cohesion stress, c (kPa)	0.456
	Angle of internal friction, ϕ (°)	38°
D2435	Compression index, C_c	0.036
	Swelling index, C_s	0.001

identified optimum mix was subjected to various laboratory tests to determine the geotechnical properties of the simulant. The experiments were performed as per the ASTM standards, and the results are given in Table 26.1. The obtained geotechnical properties of the optimum mix were matching with the properties of the Apollo-16.

26.4 Dynamic Properties

The dynamic properties like shear modulus, damping ratio, and Poisson's ratio of the LSS were determined by conducting the cyclic triaxial test. The bender element tests were also carried out to obtain the maximum shear modulus (G_{max}) of the LSS.

26.4.1 *Cyclic Triaxial Test*

In order to simulate the moonquake intensity at the laboratory level, the cyclic triaxial test was preferred. The dynamic properties like shear modulus (G), damping ratio (ξ), and Poison's ratio (υ) of the LSS-ISAC-1 were assessed by conducting the cyclic triaxial test at varying frequencies and different confining pressures. The samples were prepared at the measured bulk density and tested as per the method given in the ASTM D5311-M13 standards (2013). The prepared samples were tested at frequencies of 0.5, 1, 1.5, and 2 Hz and at different confining pressures ranging from 0.5 to 2 kPa. The sample preparation for the cyclic triaxial test, its confinement chamber with applied confining pressure and the loading assembly are shown in Fig. 26.1a–c respectively.

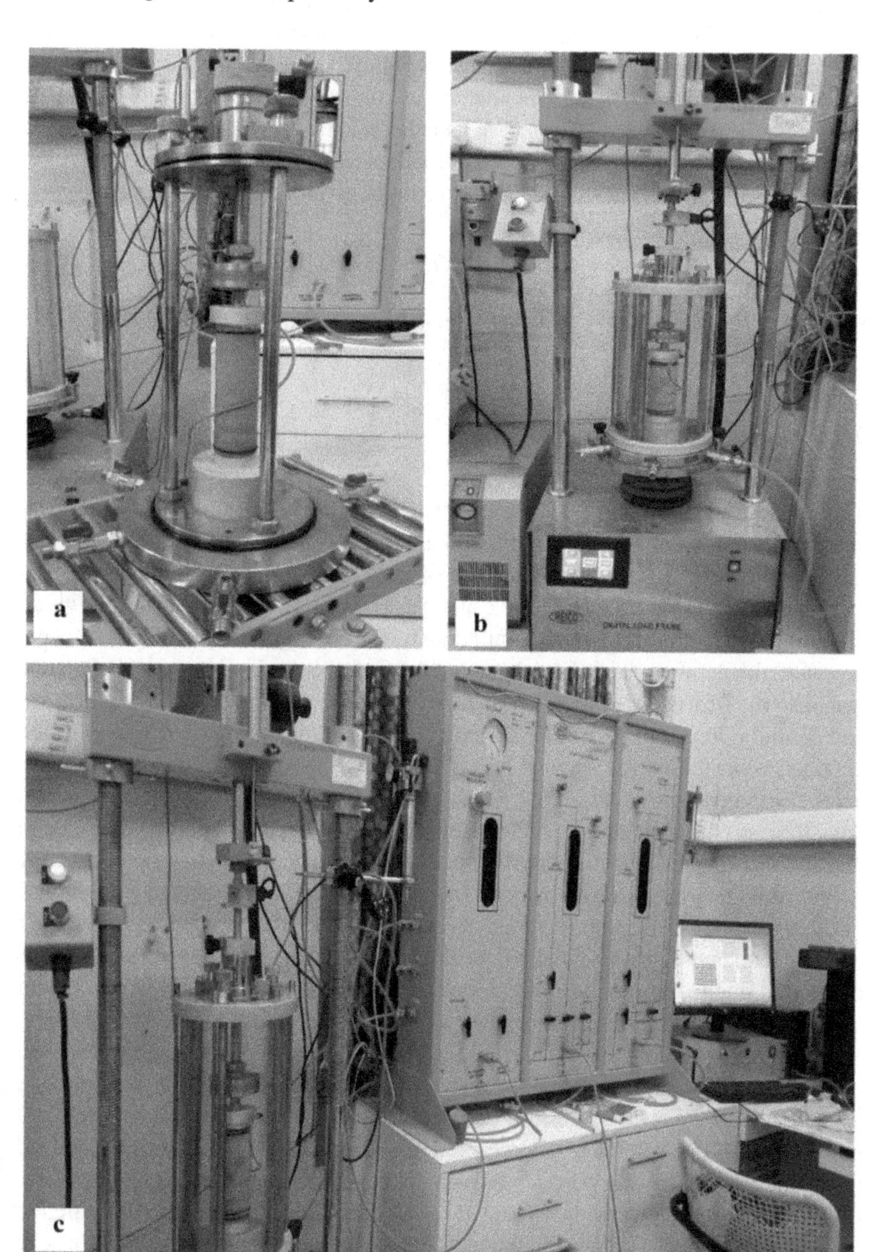

Fig. 26.1 Cyclic triaxial test of LSS-ISAC-1 **a** Sample preparation. **b** Confining chamber with sample. **c** Loading assembly of cyclic triaxial test

26.4.2 *Bender Element Test*

The bender element test was performed to measure the shear wave velocity of the simulant to find the maximum shear modulus. The maximum shear modulus is the small strain material property of the soil, which is necessary for the dynamic analysis, especially when the foundation is subjected to dynamic loading. The bender element test was performed as per the standards mentioned in the ASTM WK23118 (2311). The maximum shear modulus ($G_{\max}$) was calculated with the measured shear wave velocity using the formula given below,

$$G = \rho V_s^2 \tag{26.1}$$

where ρ is the bulk density of the soil, and V_s is the shear wave velocity.

26.5 Results and Discussions

The cyclic triaxial test was performed on the LSS-ISAC-1 sample with the measured bulk density of 1.5 g/cm^3. Initially, the test was performed for a frequency of 0.5 Hz with a varied confining pressure of 0.5–2 kPa. Subsequently, the samples were subjected to the frequency of 1 and 1.5 Hz. The test was performed with an amplitude of 2 mm and loaded continuously for 30 no of cycles. The tests were repeated three times to ensure the reliability of the results. A typical dynamic response (hysteresis loop) of the LSS-ISAC-1 with an applied frequency of 1.5 Hz and confining pressure of 100 kPa is shown in Figs. 26.2 and 26.3. These plots are obtained from the digital acquisition system (DAS) associated with the cyclic triaxial equipment. From Fig. 26.2, it is observed that there is no change in the deviatoric stress with an increase in loading cycles. The reason behind this is due to the absence of pore water pressure. Generally, the deviatoric stress and effective stress would decrease with an increase in the loading cycle in saturated soil because of the development of pore pressure. Since the LSS-ISAC-1 was tested in its bulk density, which is almost equal to its dry density. The pore pressure would be developed due to the presence of a very minimum amount of natural water content (available in the terrestrial atmosphere as moisture) present in the LSS-ISAC-1. However, this pore pressure values might be negligible, and even the samples can be oven-dried before performing the test to avoid the development of pore pressure because the atmospheric moisture content and natural moisture content was not present in the lunar surface. The shear modulus (G) values are directly obtained from the DAS, and the values observed to be in the range of 3000–8000 kPa. Likewise, the results were also obtained for all other frequency levels with different confining pressures. The results obtained for the applied frequency of 1.5 Hz with different confining pressure are presented in Table 26.2.

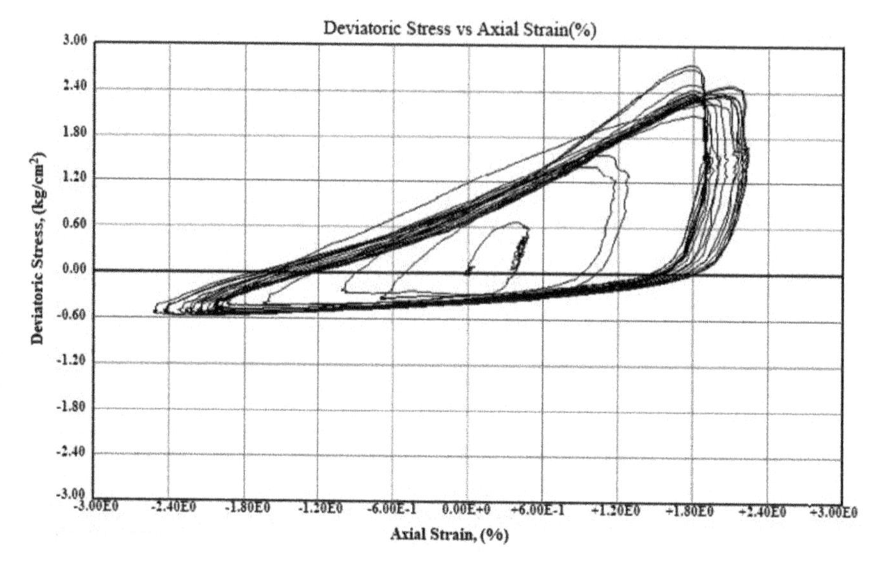

Fig. 26.2 Hysteresis loop of the cyclic triaxial test for LSS-ISAC-1 (Freq = 1.5 Hz, CP = 100 kPa)

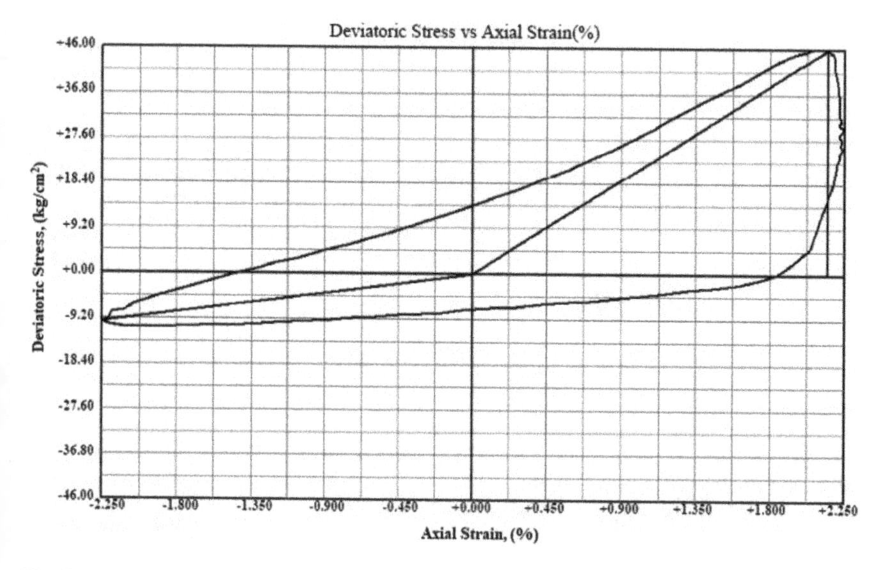

Fig. 26.3 Hysteresis loop for the single cycle No.15 (Freq = 1.5 Hz, CP = 100 kPa)

The dynamic response of the simulant for each cycle is obtained and analyzed using the digital acquisition system. The dynamic properties such as shear modulus and damping ratio of the simulant were directly obtained from the DAS. The damping ratio value can be calculated using the formula given below,

Table 26.2 Cyclic triaxial test results

Frequency (Hz)	Amplitude (mm)	Confining pressure (kPa)	Poisson's ratio	Damping ratio (%)	Shear modulus (*G*) (kPa)	Young's modulus (*E*) (kPa)
1.5	2	50	0.32	18–25	2100–3200	4500–7800
1.5	2	100	0.32	17–23	2200–3300	5500–8900
1.5	2	150	0.32	11–14	5200–6900	14,300–18,200
1.5	2	200	0.32	14–16	6000–8300	18,600–21,200

$$\text{Dampling ratio}(\xi) = \frac{1}{4\Pi}\frac{A_L}{A_\Delta} \tag{26.2}$$

where AL = area enclosed by the hysteresis loop. $A\Delta$ = area of the triangle.

The A_L denotes the energy dissipated by the soil during its deformation, and it is a measure of internal damping within the soil mass (Nakamura 1847). The AΔ represents the maximum strain energy stored. The calculated damping ratio of the LSS-ISAC-1 for the applied frequencies is in the range of 15–25%. The calculated maximum shear modulus (Table 26.3) values lie in the range of 2600–14,000 kPa. The typical shear wave velocity obtained from the bender element test is shown in Fig. 26.4. It is noted that the shear wave velocity will increase if the material density is more. The density of the simulant was increased with an increase in confining pressure.

From the test results, it is observed that the dynamic properties of shear modulus and damping ratio are influenced by factors like the type of soil, loading cycles, confining pressure, and applied frequency. The Poisson's ratio of the LSS-ISAC-1 was calculated using the standard correlation (Eq. 26.3) between the determined shear modulus and obtained Young's modulus.

$$E = 2G(1 + \upsilon) \tag{26.3}$$

Table 26.3 Bender element test results

Confining pressure (kPa)	Time	Shear wave velocity (m/s)	Shear modulus (G_{max}) (kPa)
0	1.73 ms	41.61	2682
50	1.01 ms	71.28	7872
100	910 μs	79.12	9699
150	860 μs	83.72	10,860
200	780 μs	92.30	13,200
250	750 μs	96	14,279

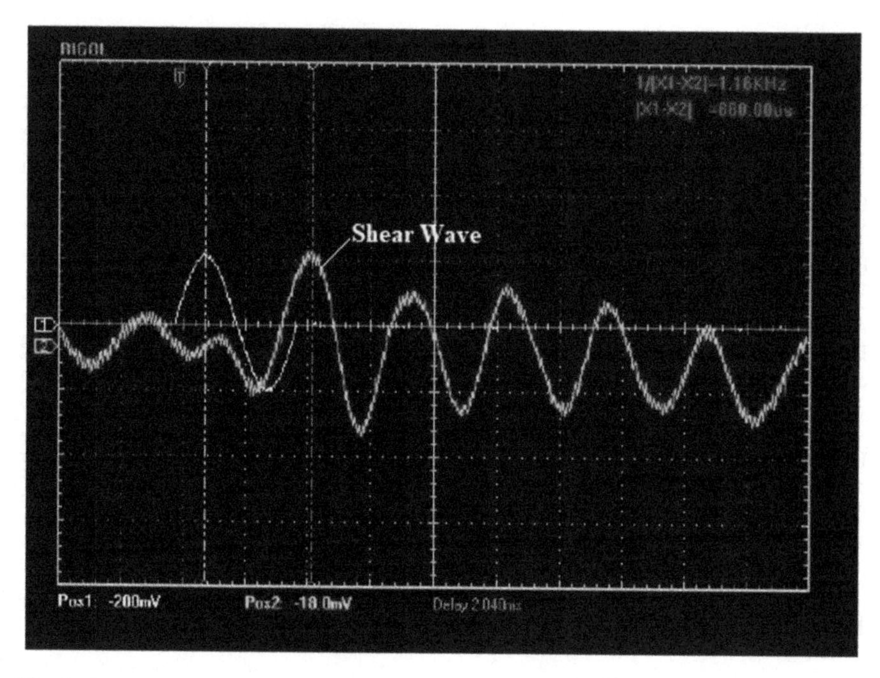

Fig. 26.4 A typical shear wave obtained from the bender element test for the LSS-ISAC-1

where E is Young's modulus or modulus of elasticity, G is the shear modulus, and υ is the Poisson's ratio. The calculated Poisson's ratio of the LSS-ISAC-1 was 0.32–0.34.

26.6 Conclusions

Understanding the physical and chemical properties of the lunar regolith is very much important for the futuristic moon colonization, lunar explorations, and other lunar missions. However, the available lunar returned samples are very much less and far too small to be used for macroscale research programs. Therefore, a new lunar soil simulant (LSS-ISAC-1) was developed by the Indian Space Research Organization, India, in large quantities for extensive research. The geotechnical properties of LSS-ISAC-1 were verified to represent the real characteristics of Apollo-16 lunar soil. The index/physical properties of the LSS-ISAC-1 were examined by conducting a series of laboratory tests and summarized in Table 26.1. The geomechanical properties of the LSS-ISAC-1 would be considered for the rover wheel design and interaction studies.

The dynamic characterization of the lunar soil is very much imperative when the lunar structures would be built on the lunar surface. The dynamic response of the lunar structures would be assessed based on the dynamic properties of the lunar soil. Therefore, the dynamic properties such as shear modulus (G) and damping ratio (ξ) of the lunar soil were estimated by using the lunar simulant LSS-ISAC-1. The cyclic triaxial test was performed effectively to simulate the moonquake conditions at the laboratory level. The dynamic properties were assessed based on the moonquake conditions in terms of different frequencies and confining pressures. These dynamic properties of the LSS-ISAC-1 would be a promising result for ISRO on research towards the futuristic moon colonization.

Acknowledgements The work has been supported and funded by the U R Rao Satellite Centre of Indian Space Research Organization under the ISRO-RESPOND Project No: 426. The authors are thankful to Dr.P. Kunhikrishnan, Director and Dr. M. Annadurai, Former Director, URSC, Indian Space Research Organization for providing lunar soils, and anorthosite samples and extended support for the success of the Research work. The authors also thank Dr. S. Anbazhagan, Professor, Periyar University, Salem for his extensive work and support for identifying the anorthosite rock beds and for elaborate efforts in pulverizing rock samples into required gradations from 30 microns to 1000 microns.

References

ASTM (2013) Standard test methods for cyclic triaxial test of soils. ASTM D5311- M13, West Conshohocken, PA

ASTM: New Test Method for Determination of Shear Wave Velocity by Bender Element. ASTM WK23118—West Conshohocken, PA

Carrier III WD, Olhoeft GR, Mendell W (1991) Physical properties of the lunar soil. In: Heiken G, Vaniman D, French B (eds) Lunar sourcebook: Auser's guide to the Moon. Cambridge, University Press, New York, pp 475–594

Carrier WD, Mitchell JK, Mahmood A (1973) The nature of lunar soil. Journal of the Soil Mechanics and Foundation Division 99:813–832

Florez, E., S, Roslyakov, S., Iskander, M., Baamer, M.: Geotechnical properties of BP-1 lunar regolith simulant. Journal of Aerospace Engineering, 28(5): 04014124 (2015)

He C, Zeng X, Wilkinson A (2013) Geotechnical properties of GRC-3 lunar simulant. Journal of Aerospace Engineering 26(3):528–534

Jiang MJ, Li LQ, Liu F, Sun YG (2012) Properties of TJ-1 lunar soil simulant. Journal of Aerospace Engineering 25(3):463–469

Li YQ, Liu JZ, Yue ZY (2009) NAO-1: Lunar highland soil simulant developed in China. Journal of Aerospace Engineering 22(1):53–57

Mitchell JK, Houston WN, Scott RF, Costes NC, Carrier WD, Bromwell LG (1973) Mechanical properties of lunar soil-Density, porosity, cohesion, and angle of internal friction. In: Lunar Science conf., MIT Press, Houston, pp 3235–3253

Nakamura Y (1980) Shallow moonquakes: how they compared with earthquakes. In: Proc. 11th Lunar planet. sci. conference, USA, pp 1847–1853

Oravec HA (2009) Understanding the mechanical behavior of lunar soils for the study of vehicle mobility. Ph.D. thesis, Case Western Reserve University, Cleveland, OH

Ryu BH, Wang CC, Chang I (2018) Development and Geotechnical engineering properties of KLS-1 lunar simulant. J Aerosp Eng 31(1):04017083

Tao J, Wang L, Wu F (2006) Mechanical analysis of wheel-soil interaction of lunar rover. Mach Des Manuf 12:56–57

Taylor SR (2007) Encyclopedia of the solar system, 2nd edn

Venugopal I, Muthukkumaran K, Annadurai M, Prabu T, Anbazhagan S (2020a) Study on geomechanical properties of lunar soil simulant (LSS-ISAC-1) for chandrayaan mission. Advance Space Res 66:2711–2721. https://doi.org/10.1016/j.asr.2020.08.021
Venugopal I, Prabu T, Muthukkumaran K, Annadurai M (2020b) Development of a novel lunar highland soil simulant (LSS-ISAC-1) and its geotechnical properties for chandrayaan missions. Planet Space Sci 194:105–116. https://doi.org/10.1016/j.pss.2020.105116
Weber RC (2014) Encyclopedia of the solar system, 3rd edn
Yu X, Fang L, Liu J (2012) Interaction mechanical analysis between the lunar Rover wheel-leg foot and lunar soil. International workshop on information and electronics engineering (IWIEE). Procedia Engineering, vol 29, pp 58–63
Zeng X, He C, Oravec H, Wilkinson A, Agui J, Asnani V (2010a) Geotechnical properties of JSC-1A lunar soil stimulant. J Aerosp Eng 23(2):111–116
Zeng X, He C, Wilkinson A (2010b) Geotechnical properties of NT-LHT-2 M lunar highland simulant. J Aerosp Eng 213–218. 10.1061/(ASCE)AS.1943-5525.0000026

Chapter 27
Dynamic Response of Monopile Supported Offshore Wind Turbine in Liquefied Soil

Sumanta Haldar and Sangeet Kumar Patra

27.1 Introduction

Wind energy has shown unprecedented growth in the production of renewable energy in India. In response to growing energy demand from a low-carbon energy source, the Government of India has recently proposed to double its renewable energy capacity (http://www.renewableenergyworld.com). This would include exploiting India's 7500 km-long coastlines for the production of low-carbon energy from offshore wind power. A reliable quantity of renewable energy can be produced by offshore wind turbines (OWT) due to stable wind conditions at offshore sites. Monopile is a common choice as a foundation for offshore wind turbines (OWTs), which is a slender member with 3–6 m outer diameter, 30–40 m long, and it is economical at shallow water depth (Cui and Bhattacharya 2014). Various environmental loads due to wind and wave, dynamic loads due to the rotor, out of balance mass, and the blade shielding effect act on the tower and foundation of OWT (Adhikari and Bhattacharya 2012). Several countries, such as the USA, China, India and South East Asia and are investing more in offshore wind energy, which is located at potentially active seismic zones (Risi et al. 2018; Wang et al. 2018). OWT structures are often installed in seismic areas which are subjected to seismic loading during its operational period. Silty sand liquefies due to the seismic loading, which leads to change the responses of an OWT system. Therefore, a comprehensive study on the dynamic response of OWT in the liquefiable deposit is essential to provide a rational design guideline.

The design of offshore wind turbines depends on complex dynamic wind and wave loading, water depth, site condition, soil properties, installation cost and time, and climate variation. The offshore wind turbine is designed considering serviceability limit state, fatigue limit state, ultimate limit state, accidental limit state and

S. Haldar (✉) · S. K. Patra
Indian Institute of Technology Bhubaneswar, Jatni, Bhubaneswar 752050, India
e-mail: sumanta@iitbbs.ac.in

T. G. Sitharam et al. (eds.), *Latest Developments in Geotechnical Earthquake Engineering and Soil Dynamics*, Springer Transactions in Civil and Environmental Engineering, https://doi.org/10.1007/978-981-16-1468-2_27

buckling. The serviceability limit state (SLS) criteria are the permissible limit of rotation at the tower top and monopile head at sea bed level. The SLS criterion impacts the design of foundation and material consumption (Cui and Bhattacharya 2016). The offshore wind turbine components are generally designed to sustain up to its service life of 20 years if the design fatigue life is not specified (DNVGL-ST-0126 2016).

Failure of various offshore structures due to seismic liquefaction during strong seismic events has been reported in the literature (Groot et al. 2006; Sumer et al. 2007; Ye and Wang 2015). The effect of liquefaction of seabed on the design implication on offshore structures is limited. Various guidelines for the design of OWT structures, namely Risø (2001), Lloyd (2010), IEC 614000-1 (2005) and DNVGL-ST-0126 (2016), recommended seismic load combined with operational loads. However, there is a scarcity of comprehensive design guidelines for the OWT structure subjected to seismic loading (Kaynia 2018).

This study examines the behavior of monopile supported OWT in liquefiable soil. The depth of soil liquefaction is influenced by the various ground motion parameters, such as duration, intensity, peak ground acceleration of earthquake and local site condition (Unjoh et al. 2012). The dynamic response of the OWT is examined due to the effect of such ground motion parameters. Maximum mudline rotation and a maximum bending moment of monopile in fore-aft and side-to-side directions are investigated at operational and parked conditions subjected to the various seismic strong motions. The analysis is carried out in a two-layered sand deposit having different relative densities using a finite element model in *OpenSees* (Mazzoni et al. 2006).

27.2 Methodology

27.2.1 Numerical Model of OWT

The numerical analysis is carried out in a two-dimensional (2D) finite element (FE) model in *OpenSees* (Mazzoni et al. 2006). A schematic diagram of the numerical model is presented in Fig. 27.1. The free-field motion of soil is simulated using a shear column. Monopile and tower are modeled as the Euler–Bernoulli beam element. The interaction between soil and pile is modeled using liquefiable *p*-*y* springs (*PyLiq1* in *OpenSees*). A 5 MW reference wind turbine and its propertied are assumed from the National Renewable Energy Laboratory (NREL) (Jonkman et al. 2009). The OWT is assumed to be installed in 20 m water depth from mean sea level (MSL).

The soil is modeled as a plane strain element, and the entire domain is discretized using 9-node quadrilateral element with solid–fluid coupling in *OpenSees* (Mazzoni et al. 2006). The four corner nodes have three degrees of freedom, namely two translational degrees of freedom and one fluid pressure. The other five nodes

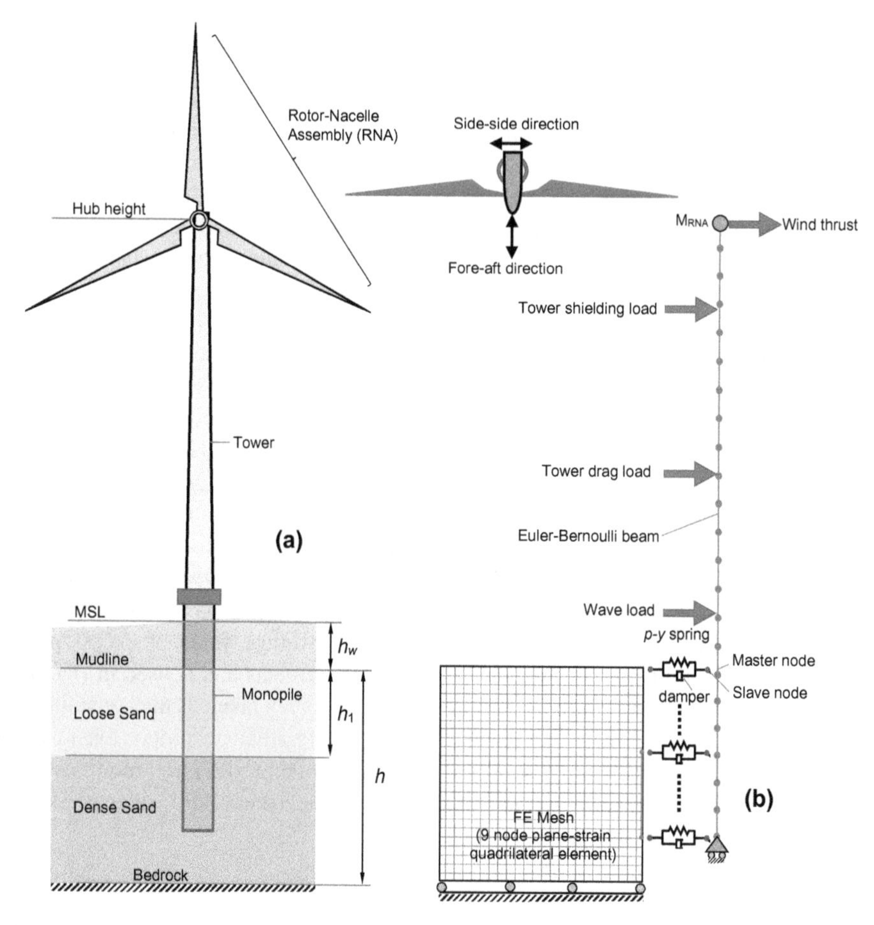

Fig. 27.1 **a** Schematic of the offshore wind turbine installed in potentially liquefiable deposit, **b** numerical model of soil–monopile–tower interaction analysis

have two translational degrees of freedom. The *PressureDependMultiYield2* material model in *OpenSees* is used as the constitutive model of soil (Yang et al. 2003, 2008). The model is based on the multi-yield-surface plasticity theory (Prevost 1985). The soil element can simulate the dynamic response of solid–fluid, fully coupled material, based on Biot's theory of porous medium. The elastic–plastic material used to simulate the response of pressure-sensitive soil materials under general loading conditions. The characteristics include dilatancy and non-flow liquefaction or cyclic mobility due to cyclic loading conditions. The Drucker–Prager failure criterion is used to represent the yield surface.

The monopile and tower are modeled by using the linear Euler–Bernoulli beam element with the structural properties of NREL 5 MW reference OWT (Jonkman et al. 2009). The rotor diameter of the reference turbine = 126 m, hub height above MSL varies from 80 to 90 m, range of rotor frequency 0.12–0.22 Hz, cut-in wind speed is 3 m/s, rated wind speed is 11.4 m/s, rotor–nacelle assembly (RNA) mass is 350 t, density of the tower is 8500 kg/m^3, Young's modulus of tower is 210×10^9 Pa, yield strength of steel is 355×10^6 Pa. The cross section for the monopile is considered to be uniform and is modeled as a series of interconnecting beam elements, as shown in Fig. 27.1. The tapered configuration of the tower is modeled, and each element in a tower considered to have uniform cross section (Cui et al. 2012). Lumped mass and rotary inertia are also defined in each monopile and tower node. The nodes in the tower and monopile have two translational and one rotational degree of freedom. The total mass of the rotor–nacelle assembly (RNA) is considered as a lumped mass (M_{RNA}) at the tower top. The length of each beam element is considered to be 0.5 m based on a convergence study.

The interaction between pile and soil column is modeled using *p-y* springs. The *p-y* spring enables the relative movement between soil and pile (Boulanger et al. 1999). The soil deformation due to seismic motion acts as displacement to the *p-y* spring at the common node of soil and spring. Similarly, the force developed on the monopile due to operational loads on tower and monopile is transferred to the soil column through connected nodes of *p-y* spring with soil and pile. *PyLiq1* uniaxial *zero-length* material object (Boulanger et al. 1999) in *OpenSees* is used to represent the force–displacement behavior of *p-y* spring. The uniaxial material uses the API-RP-2GEO (2011)-based *p-y* behavior. The average value of excess pore water pressure ratio (R_u) is obtained from solid soil elements and is used in *PyLiq1* material model. The constitutive response of *PyLiq1* is then taken as the constitutive response of API-RP-2GEO (2011)-based *p-y* curve modified in proportion to the mean effective stress within the solid soil elements. A factor of ($1-R_u$) modifies the ultimate soil resistance (p_{ult}) and tangent modulus. The excess pore water pressure ratio (R_u) is estimated as,

$$R_u = 1 - \frac{p'}{p'_c} \tag{27.1}$$

where p'_c mean effective consolidation stress prior to undrained loading and $p\prime$ is the average mean effective stress. The residual peak resistance (p_{res}) is the minimum peak resistance when the adjacent soil element liquefies, and it is considered as 10% of p_{ult}. The damping of pile and soil is modeled by using the viscous dashpots between monopile and soil column, which are parallel to each soil spring. Difference sources of damping of the OWT system, namely aerodynamic damping, hydrodynamic damping, material damping of steel and radiation damping of soil, are considered. A total 9% of damping is considered as reported in the literature (Bisoi and Haldar 2015). About 2% of Rayleigh damping is assigned to the soil domain to improve numerical stability (Yang and Elgamal 2002).

27.2.2 *Method of Analysis*

The soil column consists of a two-layered soil deposit. The top layer, having thickness h_1, consists of loose sand; the bottom layer consists of relatively dense sand. The overall depth of soil (h) is 50 m, where the bedrock is assumed to be present. To conduct the dynamic analysis, a time integration using the average acceleration method with the Newmark integrator is used in *OpenSees* solver (Mazzoni et al. 2006). The vertical displacements are restricted for the nodes at the base of the soil column. Every pair of remaining nodes in soil column which share the same vertical position is imposed equal translational degrees of freedom to achieve a simple shear deformation. A Lysmer and Kuhlemeyer (1969) dashpot is used to account for the finite rigidity of the underlying elastic medium. A viscous damper is used at the base of the soil column in the horizontal direction. The damping coefficient of the damper is estimated as the product mass density and shear wave velocity of the bedrock and the base area of the soil column. The velocity–time history is estimated from the acceleration time history of the seismic motion and applied at the base of the soil column in horizontal direction. This includes a time history of horizontal force at the base of the soil column, which is proportional to the velocity–time history of the ground motion (Joyner and Chen 1975). The density of the bedrock is assumed to be 2500 kg/m^{3,} and the shear wave velocity of bedrock is considered to be 940 m/s. The size of the FE mesh of the soil column is considered to be 0.5 m × 0.5 m.

27.3 Loads on OWT

27.3.1 *Wind and Wave Load*

Wind and wave load, operational loads due to rotor and mass imbalance, and seismic load act on the OWT structure. Two distinct components of wind load are (i) wind thrust acting on the hub (F_{hub}) and (ii) the wind thrust acting on the tower (F_{tower}). Total wind thrust on the hub is estimated based on radius of the rotor and thrust coefficient. The thrust load at hub height acts in the fore-aft direction of an OWT. In the fore-aft direction, thrust on the tower is estimated, integrating the velocity profile along with the tower over the height from MSL to the tip of the blade. In the side-to-side direction, the thrust on the tower is estimated overall height of the tower is considered. 3P load, $F_{\text{tower,3P}}$ acting on fore-aft direction, is calculated by integrating velocity profile over the height of the tower obstructed by the blades. 1P load ($F_{1\text{P}}$) due to mass and aerodynamic imbalances of the rotor acts on the side-to-side direction of the OWT structure. The wave load on the OWT structure is calculated based on Morison's equation. A description of wind and wave loads is summarized in Table 27.1.

Table 27.1 Summary of wind and wave load on the wind turbine

Load type	Equation	Remarks
Wind load at hub (F_{hub}) in fore-aft direction	$F_{hub} = 0.5\rho_a \pi R_T^2 U^2 C_T^2(\lambda_s), \lambda_s = V_r R_T / U$ $U = \bar{U} + u, u = 2\sigma_{u,\sigma_u = I_u \bar{U}}$ (frequency of this load is 0.01 Hz)	$\sigma_u = I_u \bar{U}$ is the thrust coefficient and depends on tip speed ratio, λ_s which is adopted from Jara (2006)
Wind thrust acting on tower in fore-aft ($F_{fa,tower}$) and side-to-side direction ($F_{ss,tower}$)	$F_{fa,tower} = \int^{H-R_T} 0.5\rho_a C_D D_T(z) U^2(z) dz$ (Van Binh et al. 2008) $F_{ss,tower} = \int_0^H 0.5\rho_a C_D D_T(z) U^2(z) dz$	$U(z) = U_{ref}(z/z_{ref})^{0.12}$ (DNVGL-ST-0126 2016). Tower drag is considered as static load
3P load in fore-aft direction	$F_{3P} = R_A \int_{H-R_T}^{H} 0.5\rho_a C_D D_T(z) U^2(z) dz$ $R_A = A_B / A_T$	Frequency of 3P load is three times of rotor frequency (1P frequency)
1P load in side-to-side direction	$F_{1P} = I_m \Omega^2$ $I_m = mr$, where m = imbalance mass in kg assumed to be placed at a distance r from the center of the hub and Ω is the angular frequency of the rotor blade	It is due to mass and aerodynamic imbalances of the rotor (Arany et al. 2015). 1P load is dynamic in nature, and the frequency is the rotor frequency
Wave load, F_{wave} in side-to-side and fore-aft direction	$F_{wave} = C_M \rho \frac{\pi}{4} D^2 \int_{h_w}^{\eta(t)} \ddot{u} dz + C_D \rho \frac{D}{2} \int_{h_w}^{\eta(t)} \dot{u}\|\dot{u}\| dz$ $\dot{u} = \frac{h_w \pi}{T_w} \frac{\cos h(k(z_1 + d_w))}{\sin h(kh_w)} \cos(kx - \omega t)$ $\ddot{u} = \frac{2h_w \pi}{T_w} \frac{\cos h(k(z_1 + d_w))}{\sin h(kh_w)} \sin(kx - \omega t)$	$\eta(t) = 0.5h_w \cos(kx - \omega t)$ $\eta(t)$=surface wave profile, d_w = water depth, C_M = mass coefficient (=2), C_D = drag coefficient (=0.7), D = monopile diameter, ρ = mass density of water

R_T = rotor radius; ρ_a = density of air at 15 °C and 1 atm pressure; V_r = speed of rotor in rad/s; U = wind speed at hub level in m/s; $\bar{U}$ = mean wind speed; u = fluctuating component of mean wind speed; I_u = turbulence intensity; σ_u = standard deviation of wind speed; U_{ref}= reference wind speed estimated from 10 min mean wind speed at height z_{ref} = 10 m; D_T (z) = tower diameter at height z from MSL; C_D = drag coefficient of the tower (= 0.7); A_B = blade area; A_T = area of tower top; H = hub height from MSL; $\dot{u}, \ddot{u}$ are the velocity and acceleration of wave induced water; T_w = wave period; ω = wave frequency in rad/s; z_1 = depth below MSL; k = wave number; x = direction of wave propagation of sea wave

27.3.2 *Seismic Load*

Strong motion acceleration time history data are obtained from the center for engineering strong motion data database (https://strongmotioncenter.org) and strong motion virtual data center (https://strongmotioncenter.org/vdc/scripts/search.plx). Five accelerograms are downloaded and scaled to 0.1, 0.2, 0.3 and 0.4 g peak acceleration values. The ground motion parameters of the seismic records, moment

Table 27.2 Ground motion parameters for scaled accelerograms

Accelerogram	M_w	T (s)	SMA (g)				Arias Intensity (m/s)			
			0.1 g	0.2 g	0.3 g	0.4 g	0.1 g	0.2 g	0.3 g	0.4 g
Chalfant Valley (1986)	5.6	40	0.024	0.048	0.072	0.096	0.043	0.17	0.40	0.72
Town of Big Bear Lake (2008)	5.1	46	0.035	0.070	0.105	0.140	0.045	0.18	0.38	0.68
Parkfield (2004)	6.0	21	0.037	0.074	0.111	0.148	0.044	0.18	0.40	0.71
Niigata (1964)	7.5	87	0.033	0.066	0.099	0.132	0.103	0.41	0.93	1.65
Kobe (1995)	7.2	41	0.077	0.154	0.231	0.308	0.140	0.57	1.28	2.27

magnitude (M_w), duration (T) and sustained maximum acceleration (SMA) are given in Table 27.2. The selected strong motions are applied at the base of the soil column.

27.4 Parameters

Two-layered soil deposits having a topsoil layer with 40% relative density (RD) and bottom layer with 60% relative density are considered. Two different values of thickness of the topsoil layer are 5 m, and 10 m (i.e., $h_1/h = 0.1$ and 0.2) are considered. The overall depth of the soil layer (i.e., the depth of bedrock) is 50 m. Embedded length of monopile (L_p), namely 30 and 40 m, is considered. The outer diameter (D_o) and wall thickness (t_p) of monopile are 6.0 m and 0.09 m, respectively. The tower is considered a tapered section with 70 m height (H) above MSL. The diameter and thickness of the tower at the base are chosen to be 6.0 and 0.027 m, respectively. The diameter and thickness of the top of the tower are considered to be 4.0 m and 0.019 m, respectively. The summary of all the parameters is presented in Table 27.3. In this study, different load combinations for fore-aft and side-to-side directions are considered for the operational and parked condition of OWT. In operational conditions, OWT produces power and rotates at rated wind speed, whereas the power production halts in parked conditions. In the case of operational condition, the wind thrust at the hub, 3P load, tower drag and wave load act in the fore-aft direction, while 1P load, tower drag and wave load act in the side-to-side direction. In parked condition, tower drag and wave load act on OWT both in fore-aft and side-to-side directions. Combined operational and seismic load and parked condition combined with seismic load are also considered.

Table 27.3 Summary of model parameters

Category	Parameters	Value
Loading parameters	Mean wind speed, $\bar{U}$	10 m/s
	Turbulence intensity, I_u	12% (IEC 614000-1 2005)
	Rotational speed of OWT, Ω	0.75 rad/s
	Wave height, d_w	2 m
	Water depth, h_w	20 m
	Wave period, T_w	10 s
Soil parameters	Mass density (kg/m^3)	1800 (RD = 40%) and 2000 (RD = 60%)
	Low-strain shear modulus G_{max} (Pa)	9×10^7 (RD = 40%) and 11×10^7 (RD = 60%)
	Friction angle, ϕ (°)	32 and 35
	Phase transformation angle, ϕ_{PT} (°)	26

27.5 Results and Discussion

27.5.1 Depth of Liquefaction

Depth of liquefaction for all strong motion records is scaled to a peak acceleration values 0.1–0.4 g, which are represented as peak acceleration and are applied at the base of the soil column. The zone below the mudline, which produces excess pore pressure ratio greater than 0.8 (Beaty and Byrne 2011), is considered to be the depth of liquefaction. The calculated depth of liquefaction (h_L) is normalized with an overall depth of soil (h) for all five seismic records. The normalized depth of liquefaction (h_L/h) is plotted for peak acceleration 0.1–0.4 g and all the strong motion records and presented in Fig. 27.2. A marginal increase in depth of liquefaction is observed due to the increase in the thickness of the loose sand layer for a

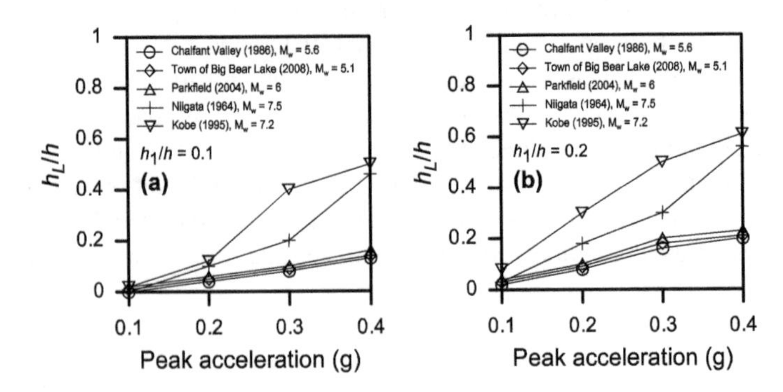

Fig. 27.2 Effect on normalized depth of liquefaction for: **a** $h_1/h = 0.1$, **b** $h_1/h = 0.2$

particular value of peak acceleration or arias intensity. For example, $h_L/h = 0.2$ and 0.28 for $h_1/h = 0.1$ and 0.2, respectively, when peak acceleration is 0.3 g for Niigata (1964) earthquake. This marginal increase in depth of liquefaction is attributed to a reduction of stiffness of the upper soil layer. The depth of liquefaction increases with an increase in peak acceleration (or Arias intensity) as expected. It is interesting to note, for Chalfant Valley (1986), Town of Big Bear Lake (2008) and Parkfield (2004) records, depth of liquefaction (h_L) is not comparable although arias intensities are similar. It is observed that the sustained maximum acceleration (SMA) of these records play a key role in increasing the depth of liquefaction. SMA is defined as third highest absolute value of acceleration, and it provides a rational representation of damage during a seismic event rather than peak acceleration (Kramer 1996). SMA of Parkfield (2004) motion is found to be more than the other two records. Hence, the depth of liquefaction is found to more in this case. The arias intensity and SMA of Kobe (1995) motion are more than Niigata (1964); hence, the depth of liquefaction is observed to greatest for Kobe (1995) motion.

The semiempirical method of evaluating the depth of liquefaction is a conventional approach proposed by Idriss and Boulanger (2006). The normalized depth of liquefaction by FE analysis is compared with the semiempirical method by Idriss and Boulanger (2006) for three strong motion records and presented in Fig. 27.3. It can be observed from Fig. 27.3 that the estimated depth of liquefaction based on the semiempirical method is reasonably good agreement with the FE analysis results for low to moderate range of peak acceleration (i.e., 0.1–0.2 g). However, the semiempirical method overpredicts the depth of liquefaction at high peak acceleration values (e.g., 0.3–0.4 g). The present study lends attention to assess the depth of liquefaction for OWT analysis and design using a semiempirical approach at moderate to high-intensity earthquake, as this method lacks incorporation of ground motion parameters, such as arias intensity and sustained maximum acceleration (SMA).

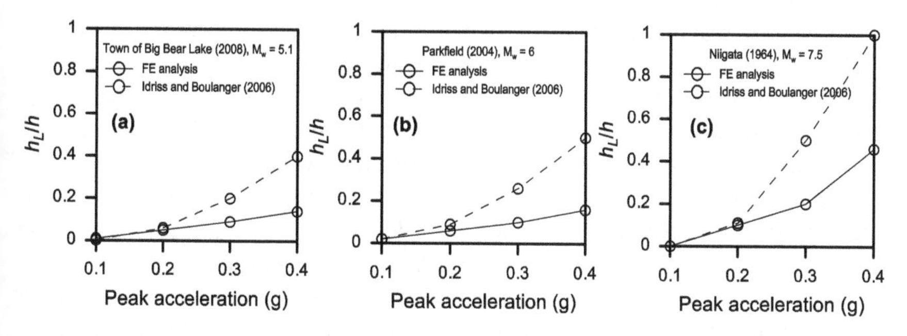

Fig. 27.3 Comparison of the depth of liquefaction below the mudline by 3D-FE model and semiempirical method for the strong motion: **a** Town of Big Bear Lake (2008), **b** Parkfield (2004) and Niigata (1964)

27.5.2 Responses of OWT

The serviceability limit state criterion is important for the design of monopile supported OWT. This criterion requires the estimation of the rotation of the monopile at the mudline level over the lifetime of the OWT structure (Arany et al. 2017). DNVGL-ST-0126 (2016) specifies that the limiting value for the permanent rotation is 0.25°. In this study, the maximum rotation of monopile at mudline is obtained at operational condition, parked condition, and combined with seismic loading and in fore-aft and side-to-side directions. In the fore-aft direction, maximum mudline rotation of monopile due to operational and seismic load ($\theta_{m,fore-aft,op-eq}$) is normalized with the maximum mudline rotation of monopile at operation load ($\theta_{m,fore-aft,op}$) which is presented in Fig. 27.4a, b for all the seismic motions, $h_1/h = 0.1$ for $L_D = 30$ and 40 m. Similarly, maximum rotation of monopile at mudline at the parked condition and seismic load ($\theta_{m,fore-aft,pr-eq}$) is normalized with a maximum mudline rotation of monopile at parked condition ($\theta_{m,fore-aft,pr}$) which is presented in Fig. 27.4c, d. It is observed that maximum rotation of monopile at mudline increases, if PBA increases. As PBA increases, the depth of liquefaction below mudline also increases. Hence, more depth of monopile remains unsupported. It causes a decrease in stiffness of soil surrounding the monopile and an increase in the rotation of monopile at mudline level.

Figure 27.4 shows that the $\theta_{m,fore-aft,pr-eq}/\theta_{m,fore-aft,pr}$ is more than that of $\theta_{m,fore-aft,op-eq}/\theta_{m,fore-aft,op}$ for all higher peak acceleration values (i.e., 0.3–0.4 g). It is because $\theta_{m,fore-aft,op}$ is significantly higher than that of $\theta_{m,fore-aft,pr}$ as the magnitude of loading on OWT in operational condition is considerably higher than that of parked condition. This effect is found to be marginal at low to moderate peak acceleration (i.e., 0.1–0.2 g). It is also observed from Fig. 27.4a–d that maximum rotation of monopile at mudline decreases significantly due to increasing embedded depth of monopile at high peak acceleration (i.e., 0.3–0.4 g); however, marginal decrease in the maximum rotation is observed in case of low to moderate peak acceleration (i.e., 0.1–0.2 g).

Normalized maximum rotation of monopile at mudline is observed to be more in side-to-side direction than that of the fore-aft direction as can be seen from Fig. 27.5a, b. This is because $\theta_{m,side-side,op-eq}$ is found to be less than $\theta_{m,side-side,op}$, and hence, amplification of response due to seismic load is found to be highest in side-to-side conditions. Increase in the normalized maximum rotation of monopile at mudline level at parked condition is found to be identical in fore-aft and side-to-side directions (i.e., $\theta_{m,fore-aft,pr-eq}/\theta_{m,fore-aft,pr}$ and $\theta_{m,side-side,pr-eq}/\theta_{m,side-side,pr}$) (see Fig. 27.5b, d). No significant difference amplification of the maximum rotation is observed in case of operational and parked conditions (i.e., $\theta_{m,side-side,op-eq}/\theta_{m,side-side,op}$ and $\theta_{m,side-side,pr-eq}/\theta_{m,side-side,pr}$) in side-to-side direction (see Fig. 27.5c, d) due to marginal contribution of 1P load than that of the seismic load.

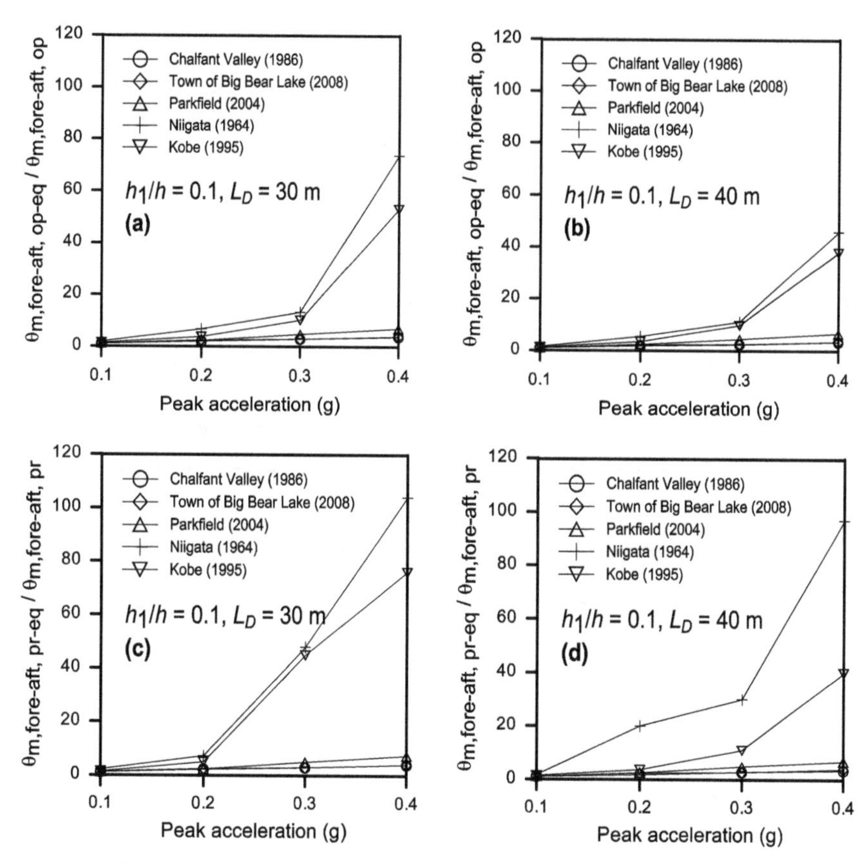

Fig. 27.4 Normalized maximum mudline rotation of monopile at difference peak acceleration for the selected strong motion records for fore-aft direction due to operational and seismic loading when **a** L_D = 30 m, **b** L_D = 40 m and due to parked and seismic loading when **c** L_D = 30 m, **d** L_D = 40 m

27.6 Concluding Remarks

The impact of earthquake-induced liquefaction on the response of the OWT structure is studied, and possible design implications are investigated. The analysis is carried out considering two different embedded length of monopile installed in two-layered soil sand having loose sand at top layer underlain by medium dense sand. The response of a 5 MW OWT considering the load cases in operational and parked conditions in conjunction with a seismic event is examined. The following conclusion is drawn.

Ground motion parameters, such as arias intensity and sustained maximum acceleration (SMA) of the seismic accelerogram, affect the depth of liquefaction of soil. The depth of liquefaction increases if arias intensity and sustained maximum

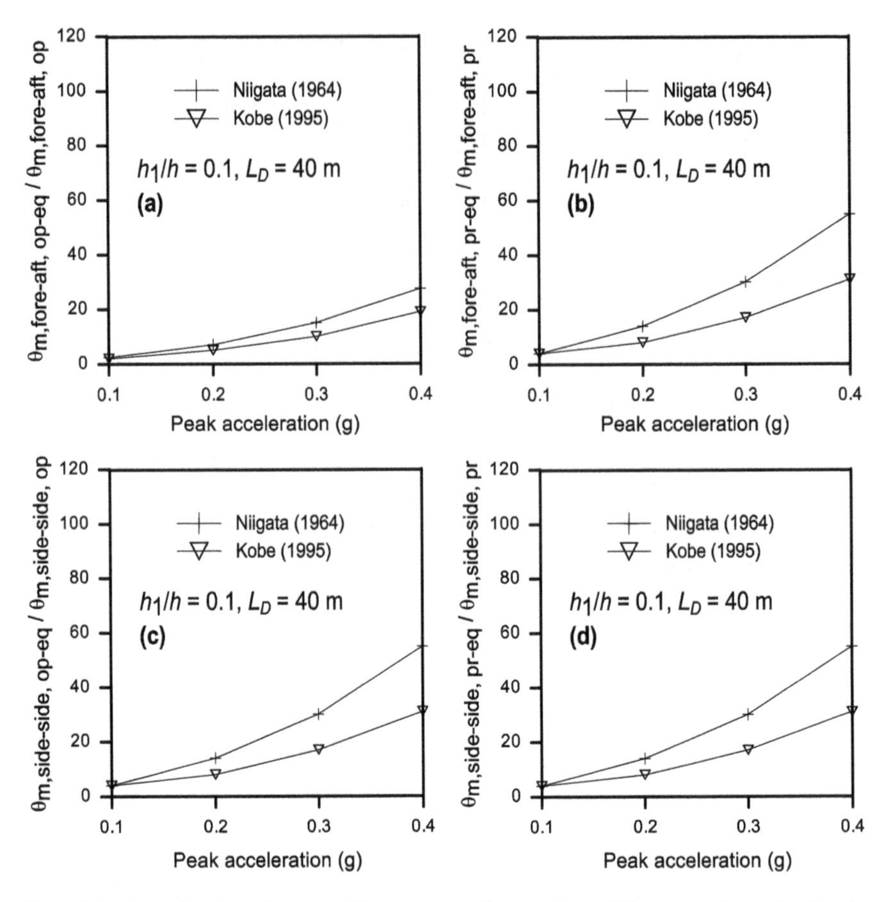

Fig. 27.5 Normalized maximum mudline rotation of monopile at difference peak acceleration for the selected strong motion records for fore-aft direction due to **a** operational and seismic loading and **b** parked and seismic loading; and side-to-side direction due to **c** operational and seismic loading and **d** parked and seismic loading when $L_D = 40$ m

acceleration of seismic motion increases. For a constant arias intensity, the depth of liquefaction increases due to an increase in SMA. Semiempirical method of prediction of depth of liquefaction found to be conservative in case of medium to a high-intensity seismic event. It is essential for the site-specific design of OWT.

Amplification of maximum mudline rotation of monopile in side-to-side direction due to the seismic event is found to be more in parked condition than that of operational condition combined with a seismic event. The dynamic response of OWT increases significantly due to combined action of seismic, wind and wave load during parked conditions than that occurs during the combined action of seismic, wind and wave load during the operational condition. However, the design of OWT is found to be critical in the fore-aft direction considering seismic load, as the monopile rotation is highest in this case.

At both operational and parked condition of OWT, the wind and wave load govern over seismic load at peak acceleration = 0.1–0.2 g, whereas seismic load governs over wind and wave loads when peak acceleration = 0.3–0.4 g. Hence, a critical combination of wind, wave and seismic load shall be considered for the site-specific design of OWT.

An increase in the embedded length of monopile is beneficial for the design of monopile supported OWT in liquefied soil deposit when subjected to seismic loads as the maximum rotation decreases due to an increase in length.

References

Adhikari S, Bhattacharya S (2012) Dynamic analysis of wind turbine towers on flexible foundations. Shock Vibr 19(1):37–56

API-RP-2GEO (2011) Geotechnical and foundation design considerations, American Petroleum Institute, Washington, DC, USA

Arany L, Bhattacharya S, Macdonald J, Hogan SJ (2017) Design of monopiles for offshore wind turbines in 10 steps. Soil Dyn Earthquake Eng 92:126–152

Beaty MH, Byrne PM (2011) UBCSAND constitutive model. Version 904aR, Itasca UDM Web Site, 69

Bisoi S, Haldar S (2015) Design of monopile supported offshore wind turbine in clay considering dynamic soil–structure-interaction. Soil Dyn Earthquake Eng 73:103–117

Boulanger RW, Curras CJ, Kutter BL, Wilson DW, Abghari A (1999) Seismic soil-pile-structure interaction experiments and analyses. J Geotech Geoenviron Eng ASCE 125(9):750–759

Cui L, Bhattacharya S (2016) Soil–monopile interactions for offshore wind turbines. In: Proceedings of the institution of civil engineers-engineering and computational mechanics, vol 169(4), pp 171–182

Cui C, Jiang H, Li YH (2012) Semi-analytical method for calculating vibration characteristics of variable cross-section beam. J Vib Shock 31(14):85–88

DNVGL-ST-0126 (2016) Design of offshore wind turbine structures. DET NORSKE VERITAS

Elgamal A, Yang Z, Parra E, Ragheb A (2003) Modelling of cyclic mobility in saturated cohesionless soils. Int J Plast 19(6):883–905

Idriss IM, Boulanger RW (2006) Semi-empirical procedures for evaluating liquefaction potential during earthquakes. Soil Dyn Earthquake Eng 26:115–130

IEC 61400-1 Ed. 3 (2005) wind turbines—part 1: design requirements. Geneva, International Electrotechnical Commission

Jara FAV (2006) Model testing of foundations for offshore wind turbines. Ph.D. thesis, University of Oxford, UK

Jonkman J, Butterfield S, Musial W, Scott G (2009) Definition of a 5-MW reference wind turbine for offshore system development. No. NREL/TP-500-38060, National Renewable Energy Lab. (NREL), Golden, CO, United States

Joyner WB, Chen AT (1975) Calculation of nonlinear ground response in earthquakes. Bull Seismol Soc Am 65(5):1315–1336

Kaynia AM (2018) Seismic considerations in design of offshore wind turbines. Soil Dyn Earthquake Eng 124:399–407

Kramer SL (1996) Geotechnical earthquake engineering. Prentice Hall, NJ

Lloyd G (2010) Guideline for the certification of wind turbines. Germanischer Lloyd, Hamburg

Lysmer J, Kuhlemeyer RL (1969) Finite dynamic model for infinite media. J Eng Mech Div ASCE 95(4):859–878

Mazzoni S, McKenna F, Fenves GL (2006) Open system for earthquake engineering simulation (OpenSees) user manual. Pacific Earthquake Engineering Research Center, University of California, Berkeley. http://opensees.berkeley.edu/

Prevost JH (1985) A simple plasticity theory for frictional cohesionless soils. Int J Soil Dyn Earthquake Eng 4(1):9–17

Risi DR, Bhattacharya S, Goda K (2018) Seismic performance assessment of monopile-supported offshore wind turbines using unscaled natural earthquake records. Soil Dyn Earthquake Eng 109:154–172

Risø (2001) Guidelines for design of wind turbines. Copenhagen: Wind Energy Department of Risø National Laboratory and Det Norske Veritas

Unjoh S, Kaneko M, Kataoka S, Nagaya K, Matsuoka K (2012) Effect of earthquake ground motions on soil liquefaction. Soils Found 52(5):830–841

Van Binh L, Ishihara T, Van Phuc P, Fujino Y (2008) A peak factor for non-Gaussian response analysis of wind turbine tower. J Wind Eng Ind Aerodyn 96(10–11):2217–2227

Wang P, Zhao M, Du X, Liu J, Xu C (2018) Wind, wave and earthquake responses of offshore wind turbine on monopile foundation in clay. Soil Dyn Earthquake Eng 113:47–57

Yang Z, Elgamal A (2002) Influence of permeability on liquefaction-induced shear deformation. J Eng Mech ASCE 128(7):720–729

Yang Z, Elgamal A, Parra E (2003) Computational model for cyclic mobility and associated shear deformation. J Geotech Geoenviron Eng ASCE 129(12):1119–1127

Yang Z, Lu J, Elgamal A (2008) OpenSees soil models and solid-fluid fully coupled elements user manual. San Diego, University of California

Chapter 28
Nonlinear Ground Response Analysis: A Case Study of Amingaon, North Guwahati, Assam

Arindam Dey, Shiv Shankar Kumar, and A. Murali Krishna

28.1 Introduction

The response of any structure depends on its regional seismicity, source mechanism, geology and local soil conditions. In order to assess the response spectrum, one-dimensional ground response analysis (GRA) can be conducted employing various methods (linear, equivalent linear and nonlinear GRA) in both time and frequency domains. Several studies have been reported in the literature on seismic ground response analyses and estimation of spectral parameters for various Indian cities such as Dehradun (Ranjan 2005), Kolkata (Govindaraju and Bhattacharya 2011), Mumbai (Phanikanth et al. 2011), Guwahati (Kumar and Murali Krishna 2013; Raghukanth et al. 2008). This paper reports the outcome of the 1D nonlinear ground response analysis (GRA) conducted for the site of Amingaon, Guwahati, based on seven different borehole stratigraphy available at the study site. The GRA for Amingaon locality was carried out using scaled strong motions pertaining to peak ground acceleration (PGA) 0.18 and 0.36 g, scaled from the seismic ground motion recorded during the 2011 Sikkim Earthquake at the IIT Guwahati campus.

A. Dey (✉)
Indian Institute of Technology Guwahati, Guwahati, Assam 781039, India

S. S. Kumar
National Institute of Patna, Patna, Bihar 800005, India

A. Murali Krishna
Indian Institute of Technology Tirupati, Tirupati, Andhra Pradesh 517506, India

T. G. Sitharam et al. (eds.), *Latest Developments in Geotechnical Earthquake Engineering and Soil Dynamics*, Springer Transactions in Civil and Environmental Engineering, https://doi.org/10.1007/978-981-16-1468-2_28

28.2 Methodology of GRA

Seismic ground response analysis is carried out to estimate the seismic response of stratified soil (in terms of acceleration, peak ground acceleration or PGA profiles, stress and strain histories, and response spectra), when subjected to an input bedrock motion. The commonly applied methodologies include linear, equivalent linear and nonlinear GRA (Kramer 1996). Linear analysis methodology assumes a constant shear modulus for each soil layer. Equivalent linear method approximates the nonlinear behavior of the soil (i.e., shear modulus and damping is strain dependent) in terms of equivalent and strain compatible secant shear modulus and damping ratio, corresponding to the chosen effective shear strain (Kramer 1996). Though this method is computationally convenient and provides reasonable results, it is incapable to represent the change in soil stiffness that occurs realistically during an earthquake (Park 2003). Both the linear and equivalent linear methods operate on the frequency domain making use of transfer functions. In nonlinear method, the intricate but realistic stress–strain behavior of soil is modeled for accurate measurement of soil behavior using direct numerical integration of the equation of motion in the time domain. The mathematical details of each of the above models are well-documented in Kramer (1996). In the present study, DEEPSOIL, a commercially available software (Hashash et al. 2016), is used to conduct nonlinear ground response analyses. The software utilizes a pressure-dependent hyperbolic model (modified Matasovic–Kondner–Zelasko, MKZ, model) as backbone curve that defines the interrelationship between the developed stresses and strains in the soil medium during a cyclic loading–unloading phenomenon, which in turn are governed by Masing and Extended Masing rules (Park 2003). Further, details about the adopted model are available in (Matasovic 1992). A set of material curves are defined in DEEPSOIL for defining the strain-dependent modulus reduction and damping ratio for different soils, following the standard propositions of Seed and Idriss (1970) and Vucetic and Dobry (1988). These standard propositions are directly used in this study for the corresponding sandy and clayey soils, respectively.

28.3 Study Area and Site Characterization

Guwahati City is the largest metropolitan city of Assam, at an elevation of 50–60 m above mean sea level (MSL), and situated between the bank of the Brahmaputra River and foothills of the Shillong Plateau. Topographically, the city comprises several hills and isolated hillocks made of Precambrian granitic rocks. Guwahati is surrounded by six seismic blocks, namely the Shillong Plateau, Assam Valley, eastern Himalaya, Mishmi thrust, Bengal basin and Indo-Burmese range (Raghukanth, S.T.G., Sreelatha, S., Dash, S.K.: Ground motion estimation at Guwahati city for an Mw 8.1 earthquake in the Shillong Plateau. Tectonophysics

448, 98–114 2008). The city has experienced several devastating earthquakes of magnitudes ranging M_w 5-8.7 (Nath et al. 2008). Thus, guided by the prevalent seismicity level, two scaled earthquake motions, (PGA 0.18 g and 0.36 g, representing design basis earthquake, DBE, and maximum considerable earthquake, MCE, respectively) are chosen to analyse and conduct 1D ground response analysis on seven typical sites of varying local stratigraphy, namely BH1-7 at Amingaon (AMGN).

Extensive data are collected for the Amingaon region in North Guwahati. These include borehole data and laboratory test data providing information such as thickness of subsoil strata, standard penetration test (SPT) values, index properties of soil deposits. Table 28.1 shows two typical soil profiles observed at different locations of Amingaon. The SPT-*N* values at 1.5 m interval are available up to bottom of the borehole. At AMGN BH1 site, it is observed that the soil is predominantly clay mixed with silt and sand up to depth of about 6.5 m, which is underlain by a sandy stratum. The softer layers, characterized by less SPT-N values, are observed up to a depth of 4.5 m, beyond which stiff soil is recorded. At BH2-BH3 sites, the soil profile is dominated by adulterated clay up to a depth of

Table 28.1 Typical soil profile at Amingaon, North Guwahati **a** BH1. **b** BH7

(a)

Layer No.	Thickness (m)	Soil Type	Depth (m)	Field Density (kN/m^3)	Average Field Density (kN/m^3)	SPT N-value	Shear Wave Velocity (m/s)	Average Shear Wave Velocity (m/s)
1	0.5		0.5		19.1			80.39
2	0.5	Clay	1					
3	0.5		1.5			5	160.78	
4	0.5		2	19.1				169.74
5	1		3		19.3	7	178.70	
6	0.5		3.5	19.5				178.70
7	1.0		4.5		19.7	7	178.70	
8	1.5		6			11	205.95	192.33
9	0.5		6.5	19.9				223.17
10	1.0	Sand	7.5		19.65	18	240.39	
11	0.5		8	19.4				251.76
12	1		9		19.55	24	263.12	
13	1.5		10.5			27	273.04	268.08
14	0.5		11	19.7				273.04

(b)

Layer No.	Thickness (m)	Soil Type	Depth (m)	Field Density (kN/m^3)	Average Field Density (kN/m^3)	SPT N-value	Shear Wave Velocity (m/s)	Average Shear Wave Velocity (m/s)
1	0.5		0.5		17.0			60.0
2	0.5		1					
3	0.5		1.5			2	120	
4	0.5	Clay	2	17.0				134.95
5	1.0		3		18.0	4	149.9	
6	1.5		4.5			5	160.786	155.343
7	0.5		5	19.0				173.573
8	1.0		6		19.35	8	186.36	
9	1.5		7.5			13	217	201.68
10	0.5		8	19.7				226.56
11	1.0		9		20.0	17	236.12	
12	1.5		10.5			21	252.32	244.22
13	0.5		11					265.775
14	1.0		12	20.3	20.3	29	279.23	
15	0.5		12.5					279.775

11 m. The stiffness of soil is observed to be gradually increasing along the depth of the borehole. At BH4-BH6 sites, the soil is predominantly clay up to 3.5 m, 5 m and 9.5 m, respectively, which is underlain by sandy strata. However, BH7 exhibited only clay soil present up to 12.5 m depth. The groundwater table present at BH1–BH3 site is 0.2 m below ground level (GL), whereas it is 0.5 m below GL at BH4–BH5 and 1.5 m below GL at BH6–BH7. The shear wave velocity is estimated from the N-value obtained from standard penetration test (SPT), as per the proposition of Imai and Tonouchi (1982). In practice, the borelogs were terminated wherever the hard strata were encountered. The local geology of the site mostly comprises of medium stiff soils underlain by rocky granitic strata. Hence, the elastic bedrock having very high shear wave velocity ($\approx 5 \times 10^3$ m/s) is considered at the bottom of borehole to generate rigid bedrock condition.

28.4 Strong Motion

In order to study the influence of PGA on the nonlinear response of the soil profile, the strong motion records of 2011 Sikkim earthquake (PGA 0.02 g), recorded at IIT Guwahati (IITG) station, have been scaled-up to PGA 0.18 and 0.36 g. Figure 28.1 highlights the input motions, as well as their Fourier amplitude spectrum, which quantitatively represents the energy distribution over the frequency bandwidth. Table 28.2 summarizes the strong motion characteristics, as obtained using SEISMOSIGNAL (2020).

28.5 Results and Discussions

Ground response analyses using nonlinear (NL) analysis are carried out for all the boreholes at Amingaon site, and the results are obtained in terms of acceleration time history, strain time history and stress time history. It is observed that the PGA of the input motion is amplified at ground surface, the amplification being lesser for the higher PGA bedrock motion. The 0.18 g bedrock motion is amplified by 1.83–2.72 times, while the 0.36 g input motion is amplified by 1.47–2.22 times. This feature is attributed to the role played by higher damping ratio (i.e., more dissipation of energy) induced by higher strains developed in the soil strata during the cyclic loading of higher intensity. The variation of PGA along the depth is presented in Fig. 28.2. The notable deviations in the PGA and surface amplification exhibit the effect of local site geology, chosen input motion on the acceleration response recorded at surface. Relating the PGA distribution with the soil profiles, nonlinear analyses maintain a tangent shear modulus, which changes at smaller time intervals, thereby governing the dynamic stress–strain characteristics of the soil substrata.

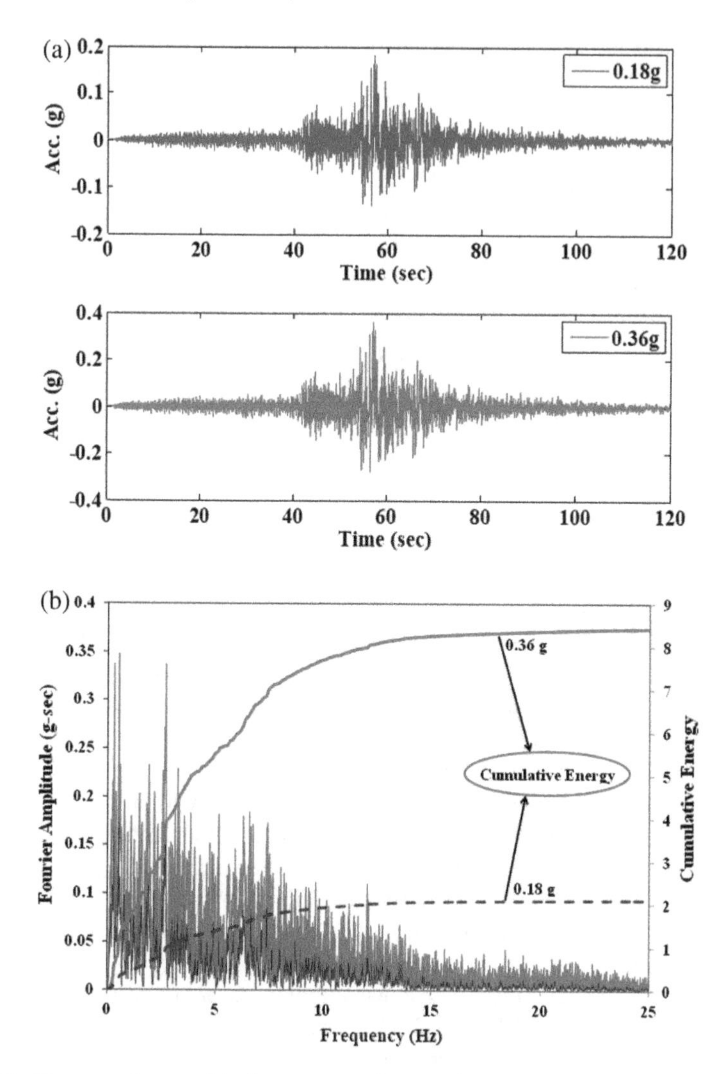

Fig. 28.1 Strong motions used in the present study **a** acceleration-time history. **b** Fourier spectra

Table 28.3 summarizes the results of ground response analysis obtained for AMGN site. Figure 28.3 represents the ground level response spectrum obtained at Amingaon site considering 5% damping for two different input motion PGA. Considering different boreholes BH1–BH7, it is observed that the spectral acceleration obtained from the 0.18 g PGA motion is lesser than that obtained for the 0.36 g PGA input motion. Higher PGA motions usually induce higher strains

Table 28.2 Strong motion parameters for scaled-up Sikkim strong motion used for AMGN site

Strong motion parameters	Scaled PGA = 0.18 g	Scaled PGA = 0.36 g
Magnitude	6.8	
Site class	B	
Distance from source (km)	400	
Predominant period (s)	0.38	
Mean period (s)	0.64	
Bracketed duration (s)	116.32	
Significant duration (s)	34.89	
Arias intensity (m/s)	0.793	3.174
Specific energy density (cm^2/s)	7217.55	28,870.22
Cumulative absolute velocity (cm/s)	1375.73	2751.45
v_{max}/a_{max} (s)	0.193	

within the deformable soil media and thus generate higher damping in the propagating medium. As a result, higher PGA motions are usually associated with generating lesser amplification in the soil system (Basu et al. 2019). The comparatively higher value of spectral acceleration is observed with the 0.36 g motion owing to the higher energy distribution within the cut-off frequency range. Such outcomes can be successfully applied for earthquake-resistant design of geotechnical structure resting on a stratified deposit. This indicates that for 0.36 g motion, the structures constructed in the region should be given due attention for seismic retrofitting or make them earthquake resistant, as they would be prone to damage for higher magnitude motions. Figure 28.4 highlights the maximum stress ratio profile, obtained at Amingaon sire based on the NL analysis. In general, a comparative with equivalent linear analysis is generally provided. However, for the sake of brevity, the same is not highlighted in this manuscript. Similar exercise has been reported in Kumar et al. (2014) and Basu and Dey (2016).

Based on NL analysis, Fig. 28.5 shows a typical shear strain history at Layer 3 of the boreholes BH6 and BH7 at AMGN site. It can be observed that the residual strain level at the end of the seismic motion obtained for 0.36 g PGA motion is higher than that obtained for the 0.18 g PGA motion, attributed to the higher energy content of the former. The residual strain observed at Layer 3 for both the boreholes is noted to be higher than the other layers, owing to the presence of predominant soft soil stratum at 1.5 m from surface level that exhibits cyclic mobility. Though stiffer soil stratum also shows some residual strain, the same is quite less than that

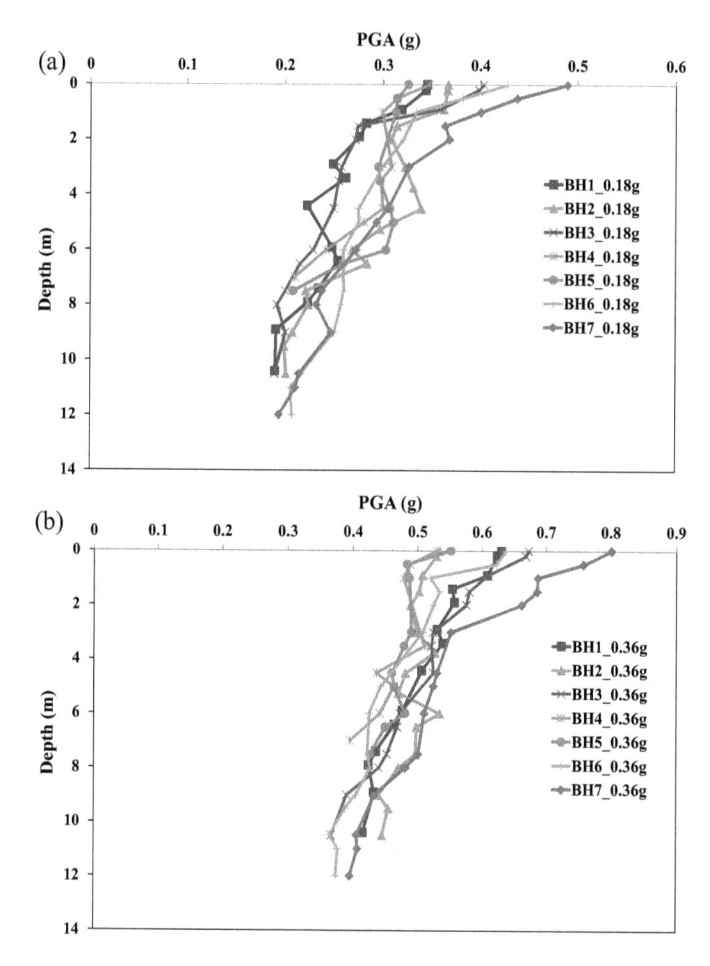

Fig. 28.2 Profile of PGA at AMGN site for different input motions **a** 0.18 g. **b** 0.36 g

Table 28.3 Summary of ground response at Amingaon site

Parameter		BH1	BH2	BH3	BH4	BH5	BH6	BH7
PGA	0.18 g	0.35	0.37	0.40	0.35	0.32	0.43	0.5
	0.36 g	0.63	0.53	0.67	0.53	0.55	0.64	0.8
PSA	0.18 g	1.83	2.0	1.94	1.62	1.58	2.29	2.1
	0.36 g	2.96	2.60	3.58	2.17	2.13	3.29	4.4

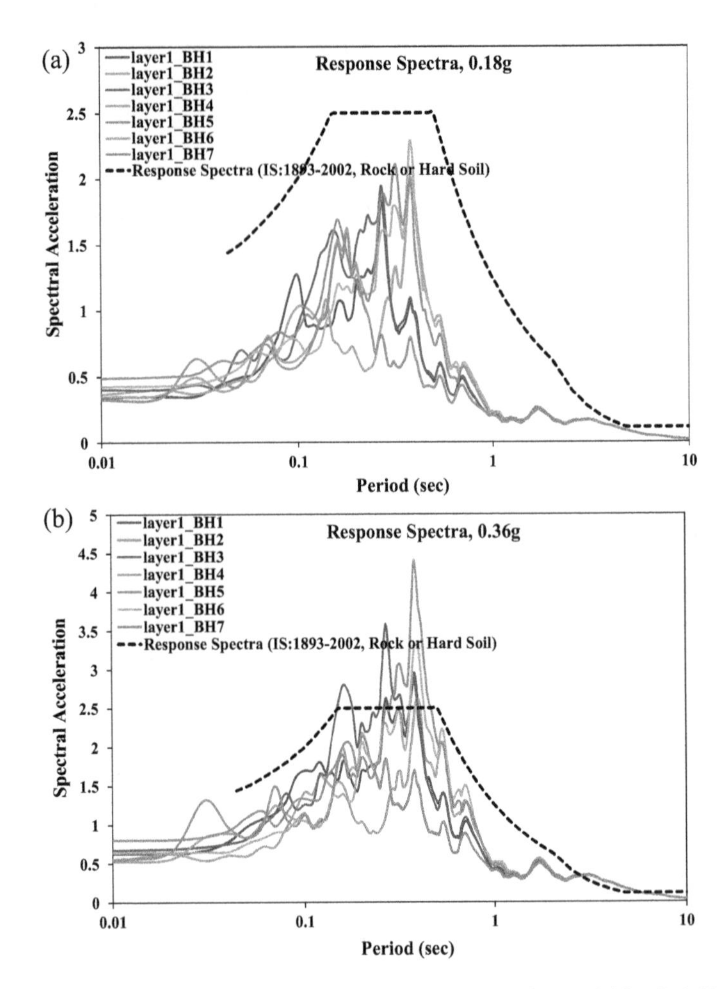

Fig. 28.3 Response spectra at AMGN site for different input motions **a** 0.18 g. **b** 0.36 g

observed in soft soil. A trial exercise using a scaled-up 0.88 g input motion exhibited severe residual strain in Layer 7 of BH2, as shown in Fig. 28.6a. As the layer is mostly clayey, the onset of cyclic mobility is exhibited, as shown by the hysteresis plot in Fig. 28.6b.

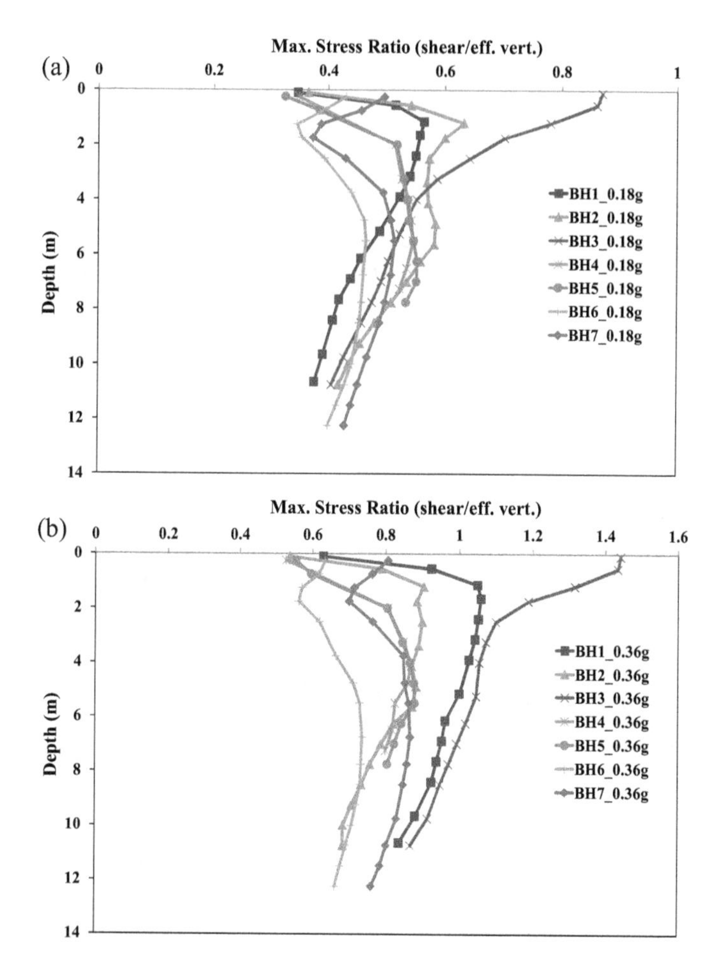

Fig. 28.4 Maximum shear stress ratio profile at AMGN site for input motions **a** 0.18 g. **b** 0.36 g

28.6 Conclusions

One-dimensional ground response analyses is conducted at Amingaon, North Guwahati, using two widely scaled-up strong motions. It had been observed that the response is strongly influenced by the local site geology, as well as the strong motion characteristics. The study clearly indicated that higher PGA bedrock motions would amplify to lesser extent as the motion reaches that ground surface,

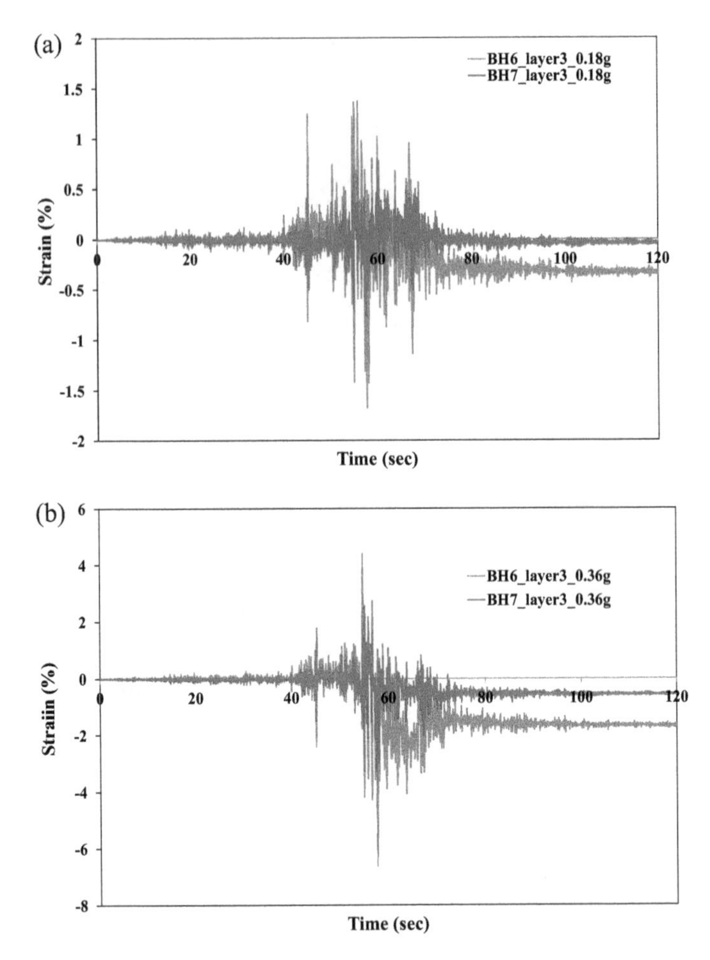

Fig. 28.5 Strain history at Layer 3 of BH6 and BH7 for input motions **a** 0.18 g. **b** 0.36 g

which is attributed to the higher damping, initiated by the development of higher shear strains in the soil substrata. The influence of the input motion is reflected on the response spectra and pseudo-spectral acceleration. Similarly, higher residual strains are obtained for higher intensity motion, to the extent of exhibiting localized cyclic mobility in the perched soft layer present in the substrata.

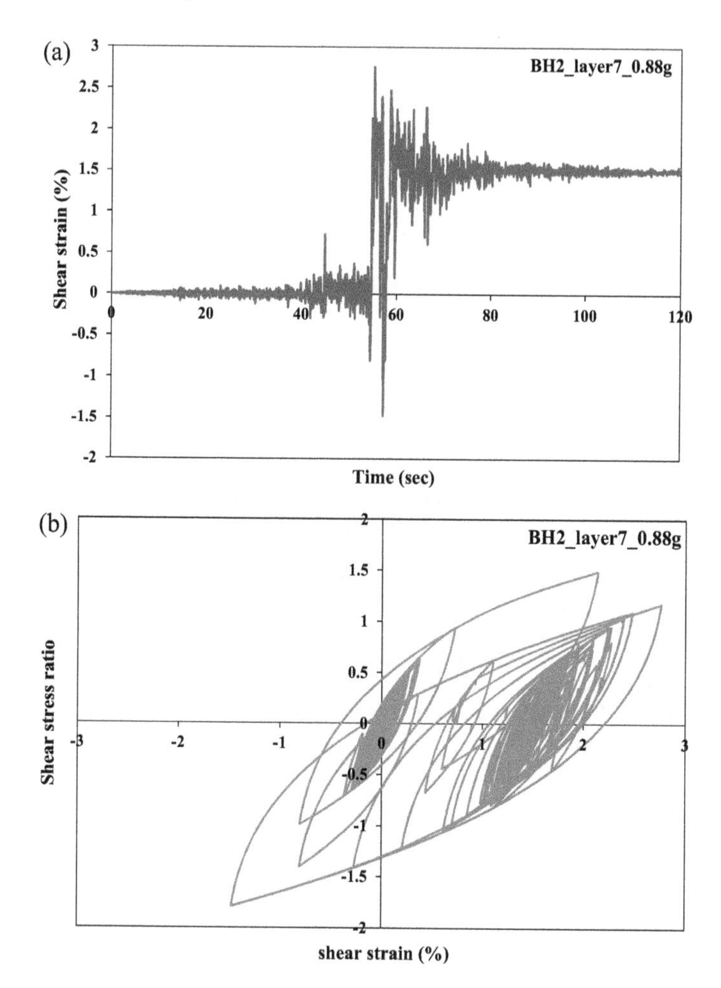

Fig. 28.6 Response of Layer 7 of BH2 subjected to scaled-up 0.88 g input motion **a** Strain history. **b** Hysteresis plot

References

Basu D, Dey A (2016) Comparative 1D ground response analysis of homogeneous sandy stratum using Linear, Equivalent Linear and Nonlinear Masing approaches. National seminar on geotechnics for infrastructure development, Kolkata, India, pp 1–7

Basu D, Madhulatha B, Dey A (2019) A time-domain nonlinear effective-stress non-Masing approach of ground response analysis of Guwahati city, India. Earthquake Eng Eng Vibr 18 (1):61–75

https://www.seismosoft.com, last accessed 2020/02/13

Govindaraju L, Bhattacharya S (2011) Site-specific earthquake response study for hazard assessment in Kolkata city, India. Nat Hazards 61:943–965

Hashash YMA, Groholski DR, Phillips CA, Park D, Musgrove M (2016) DEEPSOIL Version 6.0, Tutorial and user manual

Imai T, Tonouchi K (1982) Correlation of N-value with S-wave velocity and shear modulus. In: Proceedings of the 2nd european symposium of penetration testing, Amsterdam, pp 67–72

Kramer SL (1996) Geotechnical earthquake engineering, 1st edn. Prentice Hall, New Jersey (NJ), USA

Kumar SS, Murali Krishna A (2013) Seismic ground response analysis of some typical sites of Guwahati City. Int J Geotech Earthquake Eng 4(1):83–101

Kumar SS, Dey A, Murali Krishna A (2014) Equivalent linear and nonlinear ground response analysis of two typical sites at Guwahati city. In: Indian geotechnical conference, Kakinada, India, pp 603–612

Matasovic N (1992) Seismic response of composite horizontally layered deposits. Ph.D. thesis, University of California, Los Angeles

Nath SK, Thingbaijam KKS, Raj A (2008) Earthquake hazard in Northeast India—a seismic microzonation approach with typical case studies from Sikkim Himalaya and Guwahati city. J Earth Syst Sci 117:809–831

Park D (2003) Estimation of non-linear seismic site effects for deep deposits of the Mississippi embayment. Ph.D. thesis, University of Illinois at Urbana-Champaign

Phanikanth VS, Choudhury D, Rami Reddy G (2011) Equivalent-linear seismic ground response analysis of some typical sites in Mumbai. Geotech Geol Eng 29(6):1109–1126

Raghukanth STG, Sreelatha S, Dash SK (2008) Ground motion estimation at Guwahati city for an M_w 8.1 earthquake in the Shillong plateau. Tectonophysics 448:98–114

Ranjan R (2005) Seismic response analysis of Dehradun city, India. M.Sc thesis, Indian Institute of Remote Sensing, NRSA, Dehradun, India

Seed HB, Idriss IM (1970) Soil moduli and damping factors for dynamic response analysis. Technical report EERC-70-10, Earthquake Engineering Research Centre, University of California, Berkeley

Vucetic M, Dobry R (1988) Degradation of marine clays under cyclic loading. J Geotech Eng ASCE 114(2):133–149

Lightning Source UK Ltd.
Milton Keynes UK
UKHW020613070722
405511UK00001B/13